Silicon Compilation

Silicon Compilation

Edited by
DANIEL D. GAJSKI
UNIVERSITY OF CALIFORNIA AT IRVINE

Addison-Wesley Publishing Company

READING, MASSACHUSETTS ▪ MENLO PARK, CALIFORNIA ▪ NEW YORK ▪ DON MILLS, ONTARIO
WOKINGHAM, ENGLAND ▪ AMSTERDAM ▪ BONN ▪ SYDNEY ▪ SINGAPORE ▪ TOKYO ▪ MADRID
BOGOTÁ ▪ SANTIAGO ▪ SAN JUAN

Library of Congress Cataloging-in-Publication Data

Silicon compilation.

Bibliography: p.
Includes index.
1. Integrated circuits—Very large scale integration—Design and construction—Data processing. 2. silicon compilers. I. Gajski, Daniel D.
TK7874.S52 1988 621.395 87-1365
ISBN 0-201-09915-2

The SDL2000™ core microprocessor shown on the cover is used with the permission of Silicon Compiler Systems, Inc., New Jersey. This microprocessor was created entirely with Silicon Compiler Systems tools, and was functional on first silicon.

ABCDEFGHIJ-HA-8987

Contributors

Misha Burich
Silicon Compilers, Inc.

Robert K. Bryton
University of California, Berkeley

Raoul Composano
IBM T.J. Watson Research Center

Francky Catthoor
Interuniversity MicroElectronics Center

Edward Cheng
Silicon Compilers, Inc.

Vince Corbin
Independent Consultant

Hugo De Man
Catholic University of Louvain
Interuniversity MicroElectronics Center

Giovanni DeMicheli
Stanford University

J. van Eijndhoven
Eindhoven University of Technology

Daniel Gajski
University of California, Irvine

Gert Goossens
Interuniversity MicroElectronics Center

Jin H. Kim
Trimeter Technologies Corporation

Thaddeus J. Kowalski
AT&T Bell Laboratories

Mindy Lam
SDA Systems

Hung-fai Stephen Law
SDA Systems

Stanley Mazor
Silicon Compilers, Inc.

R.H.J.M. Otten
Delft University of Technology

Jan Rabaey
University of California, Berkeley

Jay R. Southard
Algorithmic Systems Corporation

Warren Snapp
Boeing Electronics Company

Donald E. Thomas
Carnegie-Mellon University

Jan Vanhoof
Interuniversity MicroElectronics Center

Graham Wood
SDA Systems

Preface

The first workshop on silicon compilation was organized in 1984 at the Design Automation Conference in Albuquerque. Although one hundred people were expected to attend, one thousand showed up. The audience was very interested but skeptical. They did not believe that production-quality chips could be generated by silicon compilers. At that time, there were no test cases to contradict their disbelief.

Today, the commercial silicon compiler market is estimated at around 20 million dollars. It is expected to grow to 500 million dollars by 1990. There are hundreds of chips designed completely or in part by silicon compilers. Contrary to the situation three years ago, the design community today believes in silicon compilation as a viable concept that will help us in solving design crises induced by the design complexities offered by VLSI technology. However, there is little agreement concerning what silicon compilation really is, what form a silicon compiler will take, and how they will impact present design methodologies.

A *silicon compiler* may be viewed as a translator from one level of design to another, or from one representation to another, on the same level. A *level* can be as abstract as product specification or as concrete as mask geometries. Representations can be behavioral, functional, structural, physical, or anything in between. Silicon compilation concepts can be applied equally to gate-array, standard-cell, or full-custom technologies.

In a much broader sense, silicon compilation can be thought of as methodology that brings computer engineering and computer science together. To handle design complexities available with VLSI technology, we want to capture designs and design knowledge and reuse them as frequently as possible. We need IC and design description languages for capturing designs, data bases for storing designs, algorithms and expert systems to perform synthesis, graphics to allow human control over the design process, and design environments to allow designers to write silicon compilers. Thus, silicon compilers can be classified according to choices made in each of these areas.

Because of the broad spectra of design levels, technologies, and environments covered by silicon compilation and the youth of silicon compilation as a discipline, we have not used the standard textbook approach, in which each chapter gives a survey of standard algorithms and techniques for each phase or level of design. In the present approach, each chapter represents a wide horizontal slice through the design process. Presently, several books exist and many more will be written, I believe, in which each chapter is devoted to traditional CAD problems such as partitioning, placement, routing, layout verification, layout compaction, logic design, simulation, circuit modeling, timing analysis, test generation, fault modeling, and design databases.

Since silicon compilers are tools that must deal with most of the aforementioned problems in one way or another and may be seen as representing a very narrow vertical slice through the design process, this book is constructed orthogonally to standard CAD books. After a short introduction on essential issues of silicon compilation, each chapter is devoted to a different silicon compiler. This selection, although not complete in may ways, includes several commercial and research compilers, each of them used to generate many designs. The compilers were selected to cover different levels of design and different strategies toward silicon compilation.

This book is intended for professionals in the field, users and tool builders alike. The authors assume basic knowledge of standard CAD techniques (such as placement, routing, and compaction), CAD tools (such as simulators, time analyzers, and design rule checkers), and design languages. Also, this book can be used as a textbook for a second undergraduate course in design automation or in a graduate-level silicon compilation course after students have become familiar with traditional design automation subjects.

Acknowledgments

Finally, I would like to thank all the contributors for taking time to write and patiently revise the manuscripts many times over to fit into the context of this book. Special thanks are due our reviewers: Jeffrey Fox, Silc Technologies, Inc., and Jon Solworth, Cornell University. I am also grateful to Peter Gordon and his staff for constant encouragement in bringing this book to publication.

Daniel D. Gajski
Irvine, California

Contents

CHAPTER 1

Introduction to Silicon Compilation

Daniel D. Gajski
Donald E. Thomas

1.1 Solving the Design Crisis

The design crisis of the 1980s was induced by advances in VLSI technology that allowed chip complexities never encountered before. The time it took to design chips became as long as chip lifetimes, and this situation led to a bottleneck in the product development cycle.

There are basically two approaches to solving the design crisis. One school of thought holds that the design process is difficult to automate and that the human designer is the main source of design knowledge. Thus, these proponents believe that the solution lies in improving the productivity of the human designer by providing sophisticated tools for design capture, verification, analysis, and optimization. Examples of such tools are logic simulators, time analyzers, design rule checkers, and compactors. These tools are usually integrated with CAD environments for schematic or layout capture.

The other school of thought holds that human design knowledge can be captured and that silicon compilers can be written to generate VLSI designs automatically. *Silicon compilers* are programs that generate layout data from some higher-level description. This higher-level description can be a symbolic layout, a circuit schematic, a set of Boolean expressions, a logic schematic, a behavioral description of a microarchitecture, an instruction set, or just an algorithm for signal processing. Silicon compilers contain knowledge of how to synthesize higher-level constructs from basic components and how to arrange components and interconnections on silicon. This knowledge is stored in the form of design rules that conform to constraints imposed by the technology used.

The advantages of silicon compilation are apparent in three areas: usability, quality, and profitability.

Silicon compilers expand the scope of designers in the construction of application-specific integrated circuits (ASICs). They allow system designers to work on higher levels of abstraction without detailed knowledge of IC design and processing technology.

Silicon compilation improves design quality through "correct by construction" capability. Every design component instantiated by a silicon compiler is free of errors and satisfies all technology dependent design rules. The design errors introduced by the designer on higher, more abstract levels are easier to detect and correct. In addition to generating a correct design, silicon compilation enforces good design practices, comparable to those used by the best designers in the field.

Silicon compilation increases design productivity by decreasing the number of components the designer must deal with by several orders of magnitude. Instead of dealing with millions of polygons or transistors, the designer must deal with only two dozen microarchitectural blocks. This increased productivity improves competitiveness in the marketplace by permitting earlier entry or quick redesign to satisfy different niche markets.

The weakness of silicon compilation at the present time is in the areas of silicon efficiency, process independence, and integrability.

Silicon compilers are intended to cover as wide a spectrum of applications as possible. Because of this flexibility, the designs are not as compact as human generated designs, which are tuned to one application or to one process. However, this problem will disappear as soon as languages and environments for writing compilers become widely available and designers start writing specialized compilers for niche markets.

Different compilers must be written for different processing technologies, such as CMOS, bipolar, or GaAs. Furthermore, process independence within a single technology has not been completely achieved. Some compilers must be fine-tuned when process design rules change.

The main problem with silicon compilers today lies in their integration into standard, well-established CAD environments and methodologies. A smooth, evolutionary transition from the schematic/layout capture environment to a silicon compilation environment would speed the wide acceptance of silicon compilation.

Silicon compilation technology is revolutionary in two aspects. First, it introduces a true hierarchy into the design process. Although present systems allow a hierarchical design capture, they do not preserve it through the rest of the design process since all tools are on a much lower level of design. For example, a design must be flattened to the gate level before it is sent to the logic simulator. On the other hand, the designer and tools in silicon compilation–based systems operate on a higher, abstract level without ever going to leaf cells, transistors, or mask geometries. This means that the designer is able to evaluate the design, make tradeoffs, and estimate the consequences of those tradeoffs without looking into details of the design. This requires tools such as estimators, constraint propagators, and tradeoff analyzers to span several levels of design. These tools are becoming available through silicon compilation.

Second, silicon compilation is the driving force behind the integration of several areas of computer science, such as design description languages, design algorithms, design data bases, graphics, CAD environments, and knowledge-based technology, into a new discipline called "science of design."

1.2 Domains of Description

The term *silicon compilation* was introduced by Dave Johannsen at Caltech [Joha79], where he used it to describe the concept of assembling parameterized pieces of layout. The terms has gained popularity recently throughout the IC CAD community, where it has been used in a variety of different contexts.

In a narrow sense, it is an extension of the standard-cell approach in which standard cells are replaced by parameterized cell compilers that allow users to customize the cell functionality, electrical, and layout parameters. In the case of simple cells such as NAND and NOR gates, the user can specify the number of inputs, choose among several drive buffers, and select the position of some I/O ports on the boundary of the cell. More recently, compilers for microarchitectural components such as ROMs, RAMs, PLAs, and ALUs were added to the basic SSI set. Since the number of options has increased with the compiler complexity, special forms or menus have been provided for specification of compiler parameters.

In a much broader sense, silicon compilation can be defined as a translation process from a higher-level description into layout. Here a higher-level description means a description in which some level of detail is obscured from the user and that is not just a textual equivalent of the layout. This translation process can be broken down into several steps, and each step can be considered a compiler for the lower level. This way we can define a logic compiler that translates a description into a set of logic gates and flip-flops or a microarchitecture compiler that translates an instruction set description into a set of registers, buses, and ALUs.

To represent different approaches to silicon compilation, we will use the *Y-chart*, a tripartite representation of design shown in Figure 1.1 [Gaku83] [Wath85]. The axes in the Y-chart represent three different domains of description: behavioral, structural, and physical. Along each axis are different levels of the domain of description. The farther from the center of the Y, the more abstract the level of the description. Thus, at the center would be the fully specified system. Representing the design tools as arcs along a domain's axis or between the axes illustrates what information is

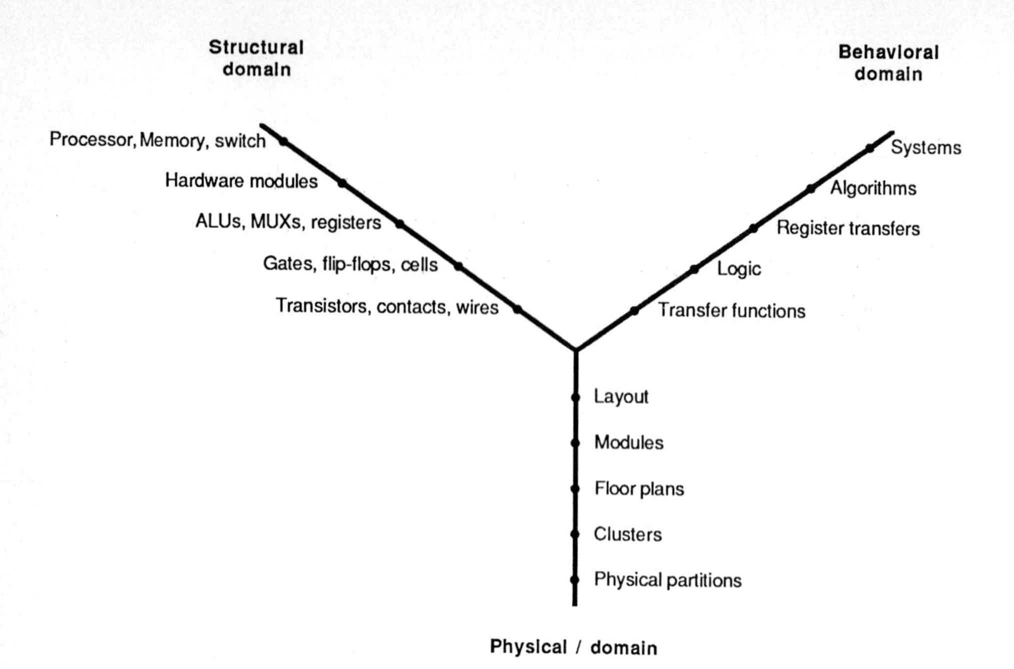

Figure 1.1 Levels of design in the tripartite representation.

used by each tool and what information is generated by use of the design tool. Verification tools such as simulators are represented by loops starting and ending at the same point on one of the axes.

Consider extending this chart by drawing circles concentric with the Y. Each circle intersects the Y axis at a particular level of representation within a domain. The circle could be viewed as representing all the information known about the design at some point in time. The outer circle is the system level, the next is the algorithmic level, followed by the microarchitectural, logic, and circuit levels. Table 1.1 lists these levels. Although these names are not used commonly by all CAD developers and users, we will use them in the following sections to illustrate the levels at which design is being carried out.

In the behavioral domain we are interested in what a design does, not in how it is built. The design is treated as a black box with a specified set of inputs, a set of outputs, and a set of functions describing the behavior of each output in terms of inputs and time. In addition to stating functionality, a behavioral description includes an interface description and constraints imposed on the design. The interface description specifies the I/O ports and timing relationships or protocols among signals at those ports. Constraints specify technological relationships that must hold for the design to be acceptable.

It is interesting to note the different time units used on each level of behavioral domain. Overall performance or transfer rates are used at the system level. Only the order of execution is specified on the algorithmic

Table 1.1 Design Objects on Different Domain Levels

	Behavioral Domain	*Structural Domain*	*Physical Domain*
System level	Performance specs.	CPUs Memories Switches Controllers Buses	Physical partitions
Algorithmic level	Algorithms (manipulation of data structures)	Hardware modules Data structures	Clusters
Microarchitectural level	Operations Register transfers State sequencing	ALUs MUXs Registers Microsequencer Microstore	Floor plans
Logic level	Boolean equations FSM	Gates Flip-flops Cells	Cell, module plans
Circuit level	Transfer functions, timing	Transistors Wires Contacts	Layout

level, while time is specified in terms of states or clocks on the microarchitectural level. The logic level uses propagation delays, setup, and hold times, while the circuit level uses signal behavior in continuous time.

To describe behavior, we use transfer and timing diagrams on the circuit level and Boolean expressions on the logic level. On the microarchitectural level, we use register transfer (or finite-state machine) description, which specifies for each control state the condition to be tested, all register transfers to be executed, and the next control state to be entered. An algorithmic description defines all the data structures and a sequence of transformations to manipulate them. At an algorithmic level, variables or data structures are not bound to registers or memories, and operations are not bound to any functional units or control state. The system, or architectural, description defines broad operational characteristics using performance specifications and data transfer rates and is not concerned with how data are manipulated or which algorithm is used.

A structural representation is the bridge between the functional and physical representations. It is one-to-many mapping of a functional representation onto a set of components and connections under constraints such as cost, area, and time. Sometimes a structural representation such as a logic or circuit schematic may serve as a functional description. The most commonly used levels of structural representation can be identified in terms of the basic structural elements used. On the circuit level the basic elements are transistors, resistors, wires, and contacts, while gates and flip-flops are the basic elements on the logic level. ALUs, registers, RAMs, and ROMs can be used to represent register-transfer as well as algorithmic structures. However, at the algorithmic level we generally group microarchitectural components into data paths, control units, and data storage; we are more concerned with synchronization and communication among the components than with their implementation. Processors, memories, and switches are used on the system level.

The physical representation ignores, as much as possible, what the design is supposed to do and binds its structure in space (physical design) or to silicon (geometric design). The structure-to-geometry mapping can be defined as a two-step process. In the first step, usually called symbolic or topological layout, relative or approximate positions for all structural elements are determined. The absolute positions are determined in the second step after substitution of layouts for symbols and compaction. Although symbolic layout can be viewed as an independent design representation, we have included it in the geometric representation to simplify our representation space. The most commonly used levels in physical representation are mask geometries, cells and floor plans. Note again that floor planning may cover several distinct levels in two other representational domains. On the system level we may have physical partitioning and placement on the board and cabinet levels, while on the geometrical level we may have placement and floor planning of blocks of different sizes.

Figure 1.2 shows examples of behavioral, structural, and physical domain descriptions. The behavioral description shown, which uses the ISPS language [Bar81], is a conditional subroutine call instruction in a computer. In this description, depending on the value of the variable c.bit, either the statement labeled "0" or the group of statements labeled "1" is executed. In the former case, the program counter is incremented, and in the latter case, a subroutine address (M[pc]) is moved into the program counter (pc) after the return address (pc + 1) is pushed onto the stack (M[sp]). A more complete description would include all of the machine's instruction and addressing modes.

It should be noted that behavioral operators such as + and − in Figure 1.2(a) do not necessarily correspond to functional blocks in Figure 1.2(b) in a one-to-one fashion. Certainly the behavior specified in the statement must be implemented, but it may be implemented in the microarchitectural representation as a shared operator used to implement several other behavioral operators also. This emphasizes the fact that the behavioral description contains very little structural information. At this level of behavioral

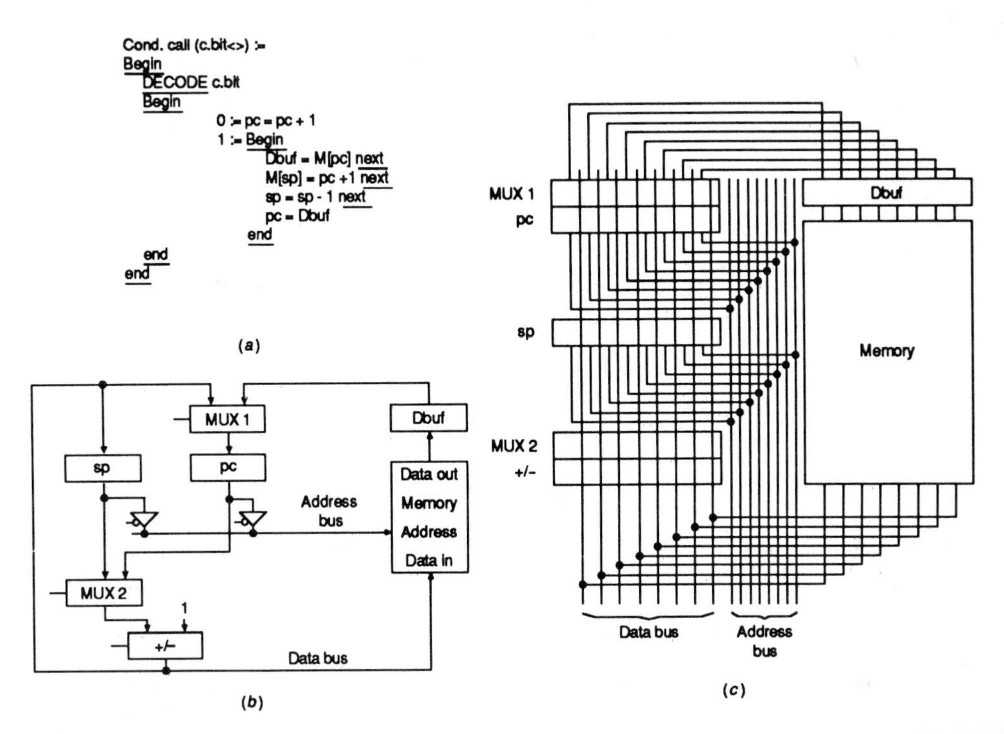

Figure 1.2 An example of domain descriptions. (a) Behavioral domain. (b) Structural domain. (c) Geometric domain.

description, there may be different microarchitectural representations that can implement a specific behavior. Similarly, a structural description contains very little information about physical representation. For each structural description there may be many floor plans in the physical domain. One floor plan is shown in Figure 1.2c.

The Y chart can be used to represent different levels of silicon compilation. Generally speaking, silicon compilers are programs that translate a structural or behavioral design description into the layout area. This translation process is usually divided into two steps. The translation from behavior to structure is usually called *synthesis* since the design description is synthesized from a predefined set of components. The translation from the structure into geometry is called floor planning, placement and routing, or cell layout, depending on the level of design.

An ideal silicon compiler is represented in Figure 1.3. It consists of four compilers. The *system compiler* decomposes a program or an algorithm into a set of communicating processes. A *processor compiler* decomposes each process into a set of microarchitectural components or modules. Each *module compiler* generates a regular or irregular array of cells. A *cell compiler* decomposes the cell into gates, transistors, and, eventually, mask polygons for different mask layers.

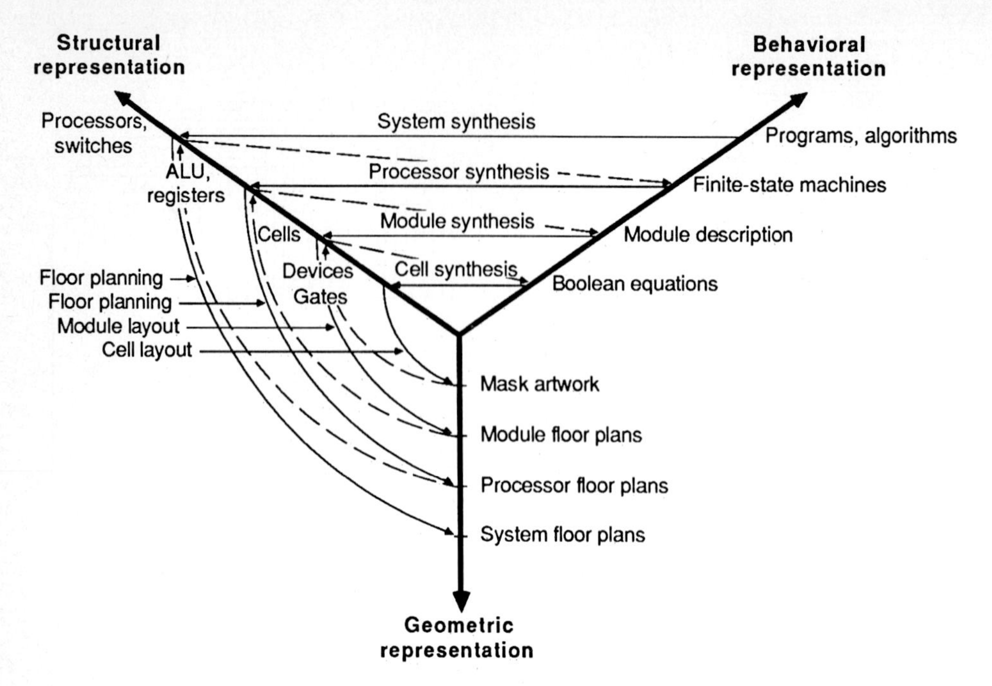

Figure 1.3 Ideal silicon compilers.

Each synthesis step is followed by a placement and routing step. The footprint of each component is obtained by instantiating the component through a component compiler. Thus, the final layout is obtained hierarchically through recursive calls to component compilers, which return the layout in terms of their respective components.

Both steps in translation process can be further divided into two phases to simplify the design process and to delay binding to a particular technology. In the case of design synthesis, a design description can be synthesized from generic or abstract components. This structural description is later resynthesized or optimized using real components. One example of such a system on the logic level is SOCRATES [Greg86]. Similarly, the translation to geometry can be achieved by first going through the symbolic layout phase, in which only the relative positions of components are determined. In the second phase the components are compacted according to technology-dependent design rules. One example of a layout system is MULGA [West81].

Silicon compilation is not necessarily restricted to custom designs. It can be applied equally to standard-cell and gate-array semicustom methodologies. In the case of standard cells, the compiler would decompose the initial description into cells through system, processor, and module synthesis and then place and route those cells; the layout of each individual cell would be obtained from a standard-cell library. The same strategy can be used for gate arrays, where the entire design is decomposed into gates before placement and routing.

1.3 Cell Compilation

A *cell compiler* translates a cell behavioral description into mask layout. A cell is usually viewed as a single-bit logic or storage function of a microarchitectural component or a circuit of SSI or MSI complexity. However, many cell compilers generate layouts from a structural description, that is, a circuit or a logic schematic. Since layout generation, even from a schematic, is difficult, the compilation process is usually simplified by constraining the layout architecture, layout dimensionality, cell input description, output format, or the method for device sizing.

A *layout architecture* or a *layout style* is defined by a set of global layout rules and by the compiler's ability to interface with the external environment, that is, to satisfy constraints on cell size, shape, and I/O port positions. Global rules specify topological relationships between layout objects (transistors, contacts, and wires), orientation of objects, and layer assignment. One such global constraint would be the restriction of all vertical wires to the polysilicon layer and all horizontal wires to the metal layer, or the restriction of all p devices to one row and all n devices to another, parallel row.

Generally, the layout architecture is either fixed or flexible. In a *fixed architecture* the layout pattern is defined at the time of compiler writing. It can be viewed as a template of place holders for layout objects whose presence or absence determines the functionality and the interface of the cell. Thus, the main problem is to find an assignment of layout objects to place holders that minimizes some aspect of design, such as area or time delay. In a fixed architecture the compiler user has very little freedom in controlling the layout shape and pin positions. Consequently, the finished layout does not fit very well into the global environment, since additional routing channels are needed to interface the designed cell with its neighbors.

In a *flexible* cell *architecture* the designer is allowed to specify the layout shape and pin positions freely. This allows for a top-down design strategy in which the design is partitioned into cells and interfaces are defined before a particular cell layout is generated. This in turn allows cell connection through abutment and routing through or over cells. This top-down strategy may generate denser layouts because of the global control over the layout space. In contrast, the bottom-up strategy optimizes cells before they are used. The bottom-up strategy may result in floor plans with some wasted areas, caused by shape incompatibility, and large routing areas, caused by pin position incompatibility.

The layout can be restricted to one dimension only. In a *one-dimensional layout, architectural* components such as gates and transistors are placed in a linear array with connections among components indicated on a set of parallel tracts over or next to the array. As the circuit complexity increases, the size of the array and the average wire length increases linearly with the number of components. In a *two-dimensional architecture*, however, they increase proportionally to the square root of the number of components.

Furthermore, the one-dimensional style produces, for large circuits, long, narrow cell shapes that are difficult to use during floor planning.

An input description to a cell compiler consists of two parts. The first part defines the *environmental constraints* that must be satisfied in order to interface with the neighboring cells. The environmental constraints include cell size and I/O pin specification. The size can be fixed either in both dimensions or in one dimension only, with the other left open. Usually, the size is approximated by an aspect ratio. The pin specification includes the location, width, and layer assignment for each pin. The pin location can be defined in terms of (a) a side of the cell only, (b) a relative order on a side of the cell, (c) a position interval on a side of the cell, or (d) an absolute position.

The second part of the input description specifies the *cell functionality* to be implemented. A functional description usually consists of a set of Boolean equations or finite-state machine expressions. It uses abstract operators that have to be mapped onto real components by the synthesizer. A structural description consists of a list of components and their interconnections. The components may range from transistors, gates, or flip-flops to some other, high-level constructs. To simplify the layout architecture and the corresponding compilation process, the number of component types can be restricted to a small number, for example, two transistor types in the CMOS technology or one gate type in some gate array technologies. However, grouping components in terms of well-known layout patterns makes layout more efficient and the compilation process more complex.

The output from a cell compiler can be in either geometric or symbolic form. A *geometric layout description* is a ready-for-fabrication artwork that satisfies technology-dependent design rules. A *symbolic layout description* uses symbols representing transistors, contacts, and wires as well as their relative positions. However, neither the absolute positions nor exact component sizes are defined. A postcompilation step is needed to expand symbols into their equivalent geometries and to compact the geometric representation according to the spacing rules particular to the technology. A symbolic representation relieves a cell compiler from design rule considerations and makes the compilation process simpler. Process simplicity, in turn, allows use of more complex algorithms to improve silicon efficiency and execution times for large circuits. Furthermore, a cell compiler can easily be retargeted to different fabrication processes if binding to process-dependent spacing rules is delayed to the compaction stage of the compilation process.

Since circuit parasitics may retard the circuit speed, the logically correct layout may not always satisfy the performance requirements. To cope with this problem, the size of the transistors, power lines, and connections should be adjusted according to their loads. The sizing information may be specified by the user in the input specification or determined by the compiler from the performance requirements and the parasitic parameters extracted from the layout. In the former case, a separate circuit extractor and a size optimizer are needed; in the latter case, they are part of the compiler.

1.3.1 Fixed Layout Architectures

Generally speaking, cell compilers with fixed layout architecture attempt to regularize placement of randomly connected logic or circuit schematics, which are usually found in control and "glue" logic. Thus, these cell compilers are not very efficient for well-structured components such as registers and memories.

The fixed layout methods can be roughly categorized according to the number of dimensions in which the cell grows with the number of components. Linear arrays such as the Weinberger array, complex cells, and the gate matrix have provided a fertile ground for many research studies. Methods pertaining to quadratic layout growth such as standard cells, PLAs, and SLAs have been most popular in practice.

The Weinberger array [Wein67] was an initial attempt to stylize the layout implementation of multilevel combinatorial logic by limiting the number of components to only one: a NOR (or NAND) gate. Regularization is achieved by laying the gates in a one-dimensional array, one gate in each column. Each gate column consists of two vertical metal wires. One is connected to the pullup transistor and serves as the gate output port. The other is connected to the ground power line. In a real application, two neighboring gates share a common ground wire. All nets in the schematic are laid horizontally across all gates, in polysilicon. A transistor is formed by intersection of a diffusion segment connecting output and ground lines and a vertical polysilicon extension from a horizontal poly line. A NOR gate schematic and its equivalent Weinberger layout are shown in Figure 1.4.

The size of a Weinberger array layout is proportional to the product of the number of gates and the number of nets. In the layout shown, each gate occupies a single column and each net occupies a single row. This approach, although very simple, generates a sparse array. In practice, the number of columns and rows can be reduced by having two gates share one column and by having two or more nets share one row. For example, the array in Figure 1.4b can be reduced if columns are shared by gates A and C, gates B and E, and gates D and F, and rows are shared by n_6 and n_{10} and nets n_7 and n_9 (see Figure 1.5).

Thus, the compilation process for the Weinberger array architecture consists of input translation, folding, column assignment, and row assignment. The input translation task converts the input Boolean expressions or schematic into NOR expressions or schematic and also minimizes the number of NOR gates. Folding uses column sharing to reduce the number of columns. A pair of gates can be folded if they do not connect to the same net and if their input nets are not interleaved in the final layout. Since interleaving is not allowed, the folding of one pair of gates may prevent other pairs from being folded. The column assignment, or placement, orders the folded or simple gates so that the number of rows, that is, the width of the routing channel, is minimal. The width of the routing channel is approximated by the maximum number of nets that cross any column.

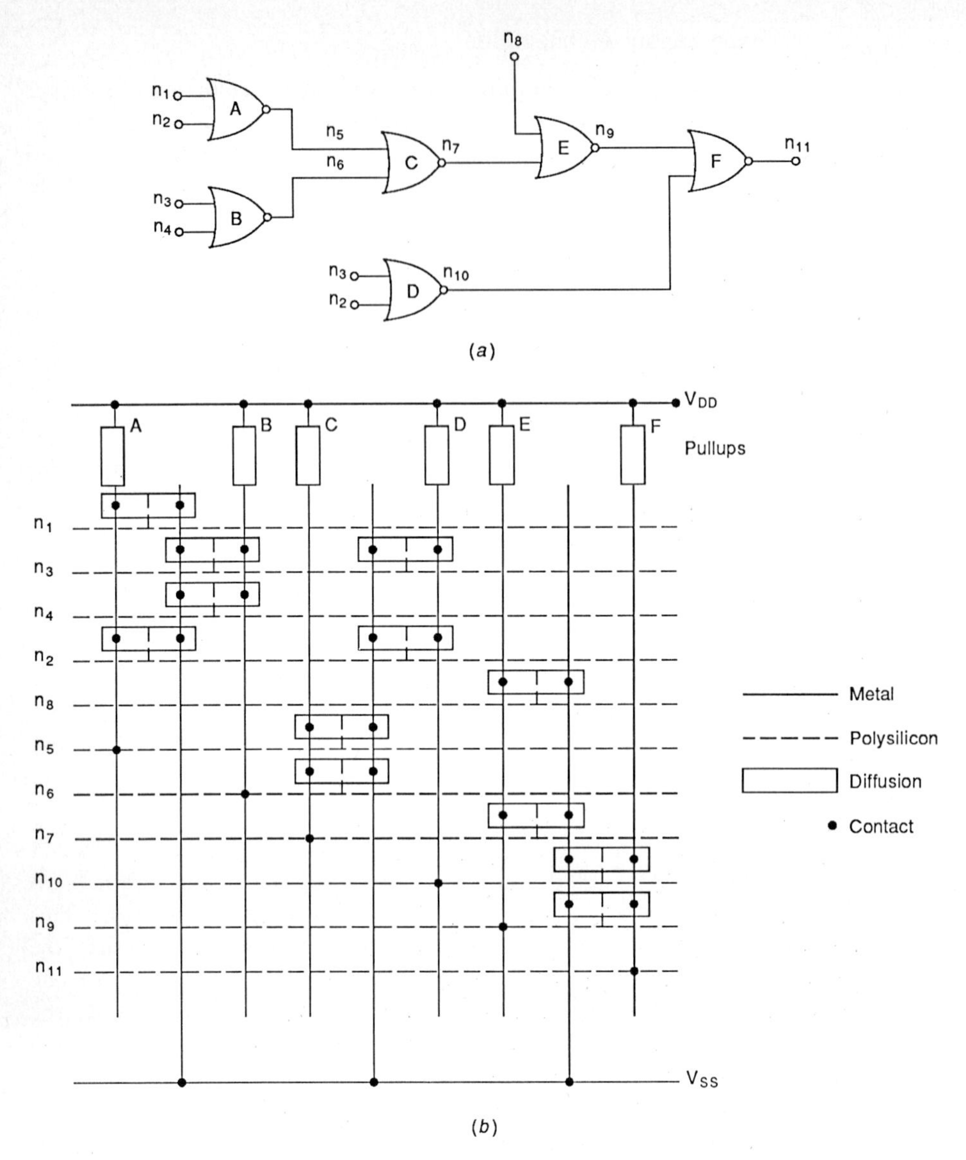

Figure 1.4 Weinberger array architecture. (a) Schematic. (b) Layout.

The object of row assignment is to assign nets to row using as few rows as possible.

Another linear architecture restricted to CMOS transistor schematics is the complex-gate architecture. All CMOS complex gates can be laid out using a single row of n transistors and a single row of p transistors aligned at the common gate connections. Simpler complex gates may be laid out using an unbroken row of transistors called a "line of diffusion" in which source-drain connections of two transistors are made by abutment in the diffusion layer.

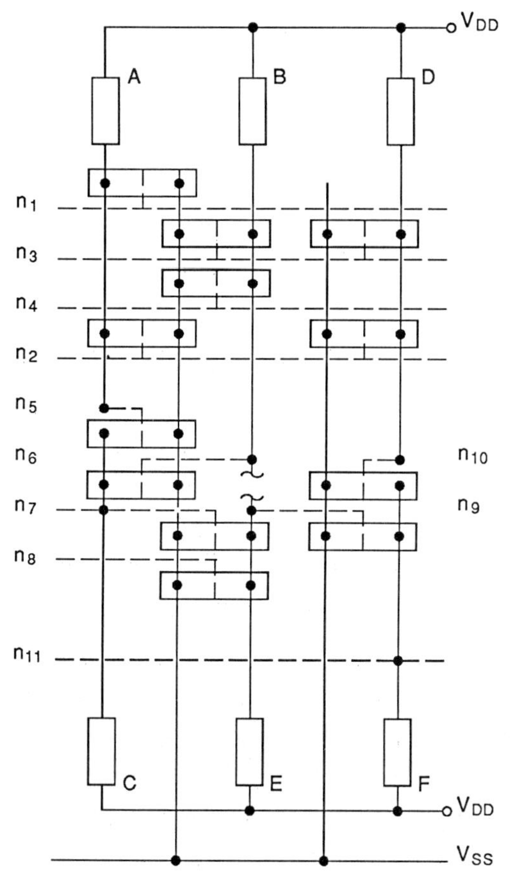

Figure 1.5 Folded Weinberger array of Figure 1.4.

The technique of finding a line of diffusion for a complex gate is based on finding a Euler path in the graph obtained from the circuit schematic (Ueha79]. The vertices in such a graph correspond to all the nets connecting transistors, and the edges represent the transistors. As an example, a complex-gate schematic and its corresponding graph are shown in Figure 1.6a and b. The connection of edges in the graph mirrors the series-parallel connections of transistors in the schematic. Each edge is labeled with the signal connected to the gate of the corresponding transistor. Note that the p and n parts of the graph are dual of each other as are the p and n parts of the schematic. The basis of this layout architecture is the fact that two adjacent edges in the graph have a common source-drain connection that can be made through abutment. Therefore, a Euler path, that is, a path that traverses all the edges but none of the edges twice, in the p and n parts of the graph, will guarantee a layout with two uninterrupted lines of diffusion. If p and n transistors are aligned, the gates can be connected only through a vertical polysilicon connection without any horizontal routing. This requires the order of the edges in the p and n Euler paths to be the same. In our example, both Euler paths have the same labeling: <A,B,C,D>. Since two-terminal nets are connected through abutment, the multiterminal nets must be routed in the channel between the p and n lines of diffusion. For the sample layout shown in Figure 1.6c, nets n_2 and n_3 are joined through abutment, while nets n_1 and n_{out} are routed in the channel.

The gate matrix architecture [Lope80] extends the style of the complex cell layout to more than one complex cell. It uses the structure of aligned transistors along vertical polysilicon lines and of one row of transistors for each pullup or pulldown of a complex gate. Since the ordering of polysilicon lines for one complex gate may not be suitable for the other, some pullups and pulldowns may occupy more than one row. A transistor schematic and its corresponding gate matrix layout are shown in Figure 1.7a and b. Each gate net, such as A, B, C, or D, and each output net, such as net Z in Figure 1.7a, is assigned one vertical polysilicon line that defines one column of the layout. All transistors using the same gate net are placed in a single column. Connections among transistors are made by abutment (line of diffusion) or by horizontal metal lines. If a pullup or a pulldown is assigned to more than one row, the corresponding rows are connected with a short vertical diffusion line placed between poly lines. For example, pulldown N_2 is placed in rows d, e, and f. Two diffusion runs are used to connect these three rows, one between columns B and C and the other to the left of column A. A layout is not realizable if there is an overlap of two diffusion lines or of a transistor and a diffusion line in the same column. An unrealizable layout can be made realizable by row permutation or, if that is not successful, by stretching the layout horizontally and inseting another vertical poly or diffusion line. Connections to V_{DD} and V_{SS} are made in the second metal layer. They are indicated by arrows in Figure 1.7b.

The compilation process consists of synthesis and topological optimization. Synthesis consists of partitioning Boolean equations into a set of complex gates or, if the input description already contains one equation per complex gate, of factoring equations to minimize the number of transistors. The factorization is limited by such technological constraints as the allowed number of transistors in series and in parallel in each pullup or pulldown. Topological optimization consists of column assignment, net assignment, and row assignment. The first two seek to minimize the number of rows since the number of columns is fixed after the synthesis phase. Column assignment attempts to order columns so that the maximum number of pullups or pulldowns can be laid in one row or less. For example, pulldown N_2 in Figure 1.7b can be laid in one row if column D is to the right of column B and columns C and Z are to the left of column A. A permutation satisfying those requirements uses only two rows instead of four, as shown in Figure 1.7c. Net assignment reduces the number of rows by sharing them among two or more pullups or pulldowns. Row assignment improves the realizability of the layout, as explained earlier.

The *standard-cell architecture*, presently the most popular for ASIC circuits, divides the layout area into a set of parallel rows separated by routing

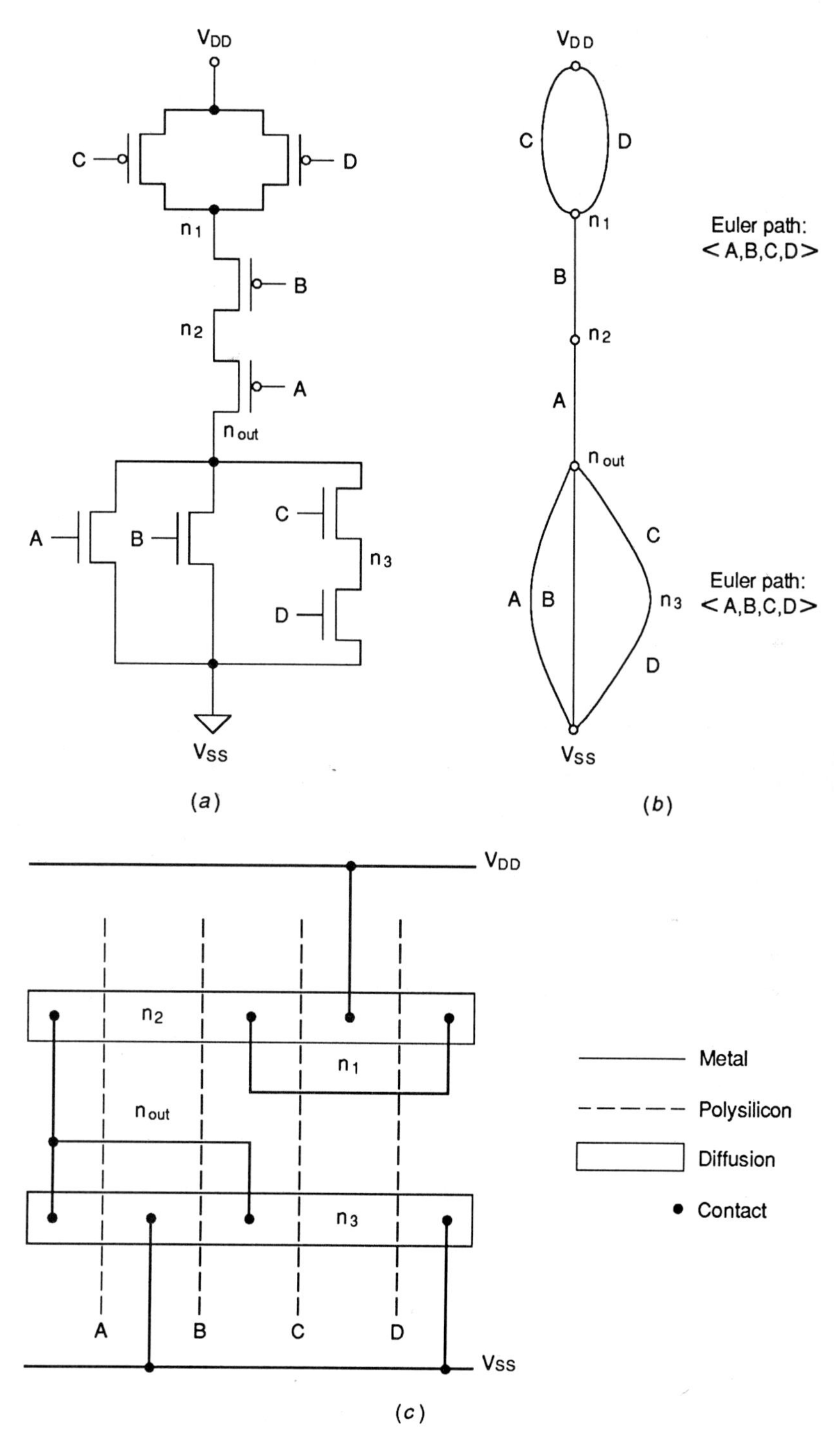

Figure 1.6 Complex-gate layout architecture. (a) Schematic. (b) Circuit graph. (c) Layout.

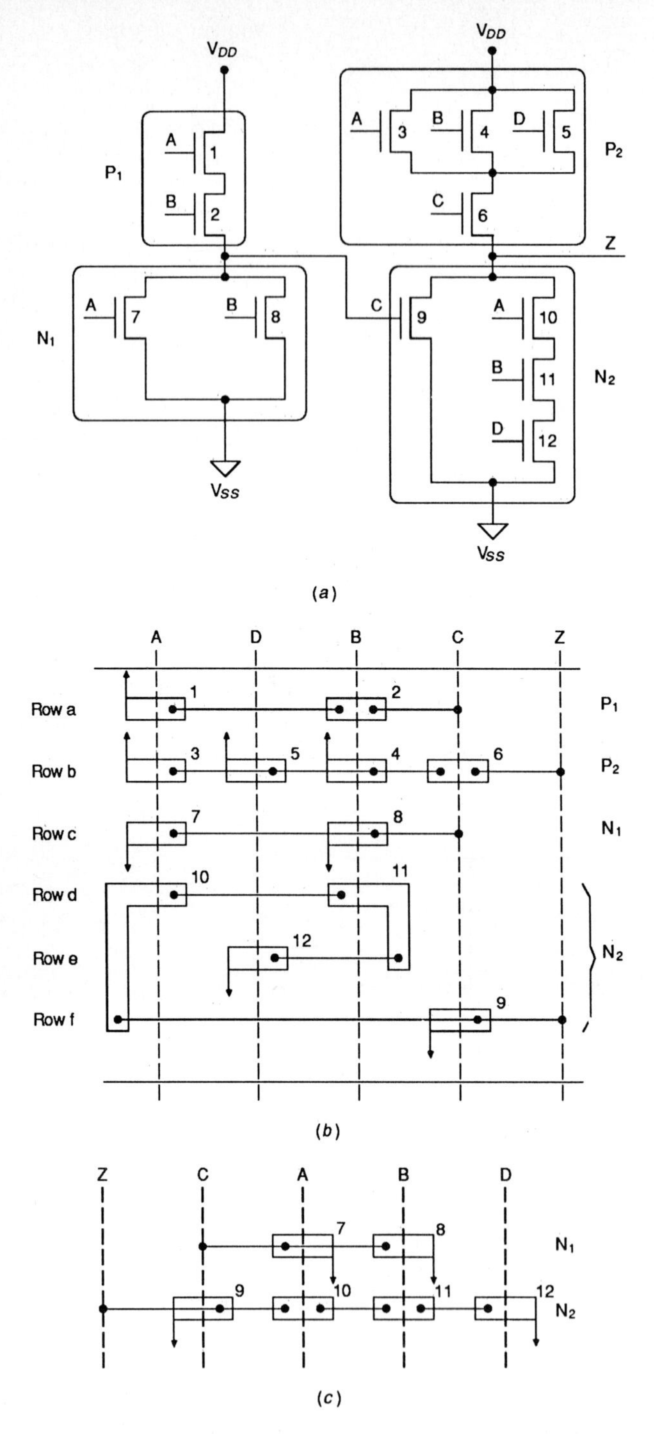

Figure 1.7 Gate matrix layout architecture. (a) Schematic. (b) Layout. (c) Optimized layout (n part only).

channels, as shown in Figure 1.8. Each row consists of a number of predefined leaf cells from the standard-cell library, which contains at least one standard cell for each component in the circuit schematic. Leaf cells all have the same, fixed height but may have different widths. The I/O pins are on the top and bottom boundaries of each cell. Power pins are on the left and right boundaries so that cells can be abutted.

The compilation process consists of placement and routing. Placement is the assignment of standard cells to a position in a row, and routing is the assignment of an interconnection to a position in the routing channel between two rows. The objective of placement and routing is to minimize the average wire length. Since a net length is not known during placement, an estimate is used. A good approximation of net length is the sum of the height and width of the net's enclosing rectangle. The enclosing rectangle of a net is the smallest rectangle that covers all cells connected to the net.

Although the standard-cell concept is simple, it has been proved that both placement and routing problems are NP-complete. Many heuristic algorithms for placement and routing have been developed in the past 20 years. The most popular placement algorithms are based on a pairwise exchange of cells that improves some objective function such as total wire length, average wire length, or number of wires crossing the cut in the middle of the layout. This hill climbing strategy can be replaced by simulated anealing, which allows an exchange of cells without improvement in the objective function with a certain probability that decreases with time. The most popular routing algorithms attempt either to maximize the placement of horizontal segments in each track, as with the left-edge algorithm

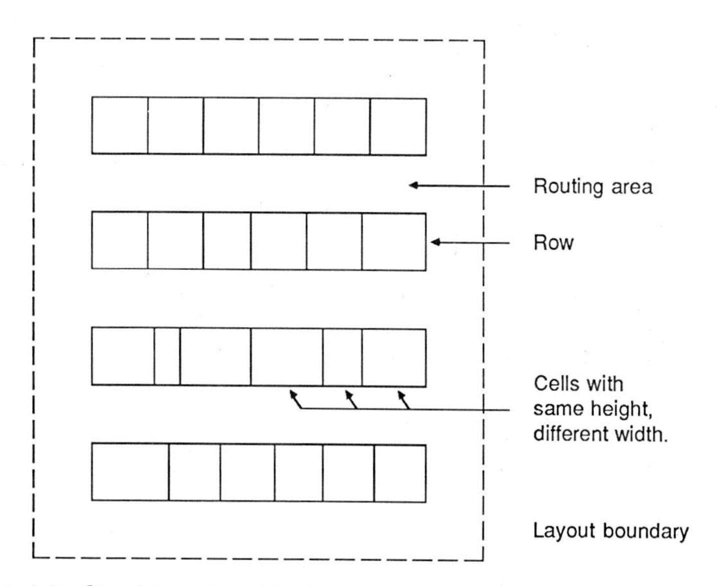

Figure 1.8 Standard-cell architecture.

[Hash71] and its variations [Deut76] [Yosh82], or to maximize the number of tracks available in the next column in the channel, as with the greedy channel router [RiFi82]. An excellent survey is given in [Sang87].

The advantages of a standard-cell architecture are its simplicity and popularity, which have led to the development of many placement and routing algorithms and tools. The disadvantages are poor area utilization and maintenance of the cell library. Because of the fixed sizes, pin positions, and functionality of standard cells, usually only 60–70 percent of the layout area is used by routing. Furthermore, any change in the fabrication process requires a redesign of the library since standard cells are usually designed for a particular process.

Another popular quadratic layout architecture is the *programmable logic array* (PLA), used in the implementation of sum-of-products Boolean functions. The architecture consists of AND and OR planes, as shown in Figure 1.9. For each input variable in the Boolean equation, there are two lines (the true and the complemented value of the variable) in the AND plane. The AND plane generates product terms by performing the AND operation on the selected set of input signals. In practice, AND and OR planes are implemented as NOR planes, as shown in Figure 1.9 for NMOS technology. The OR plane generates the outputs by performing the OR operation on product terms.

The area of a PLA is proportional to the product of the number of columns and the number of rows used, where the number of rows equals the number of product terms and the number of columns equals the sum of the number of output signals and twice the number of input signals. A PLA layout can be compacted by using logic minimization for reduction of the row number and folding for reduction of the column number.

Logic minimization consists of generating all prime implicants and selecting a minimal cover, that is, a minimal set of implicants that includes all terms for which the Boolean function is equal to 1. Although many optimal algorithms exist for a small number of variables, heuristic algorithms must be used for 10–15 input variables. Figure 1.9b shows a PLA implementation that uses a minimized set of Boolean functions taken from Figure 1.9a. In this case, only four rows are needed instead of six.

PLA folding achieves reduction of the layout area by sharing columns between two input or two output variables. In Figure 1.9c, variables X_1 and X_2 share the same column. That requires the row permutation <a,c,b,d>. On the other hand, two output variables, F_2 and F_3, can share the same column only if the row permutation is <a,b,c,d>. Thus, output signal folding prevents input signal folding and vice versa. Therefore, the objective of the PLA folding is to find the maximum number of pairs that can be folded at the same time. In addition to column folding, row folding, and segmented folding can be used if layout architecture is modified [DHNS85].

A storage logic array (SLA) is an extension of a PLA architecture [LePR84]. In contrast to the PLA architecture, where AND and OR

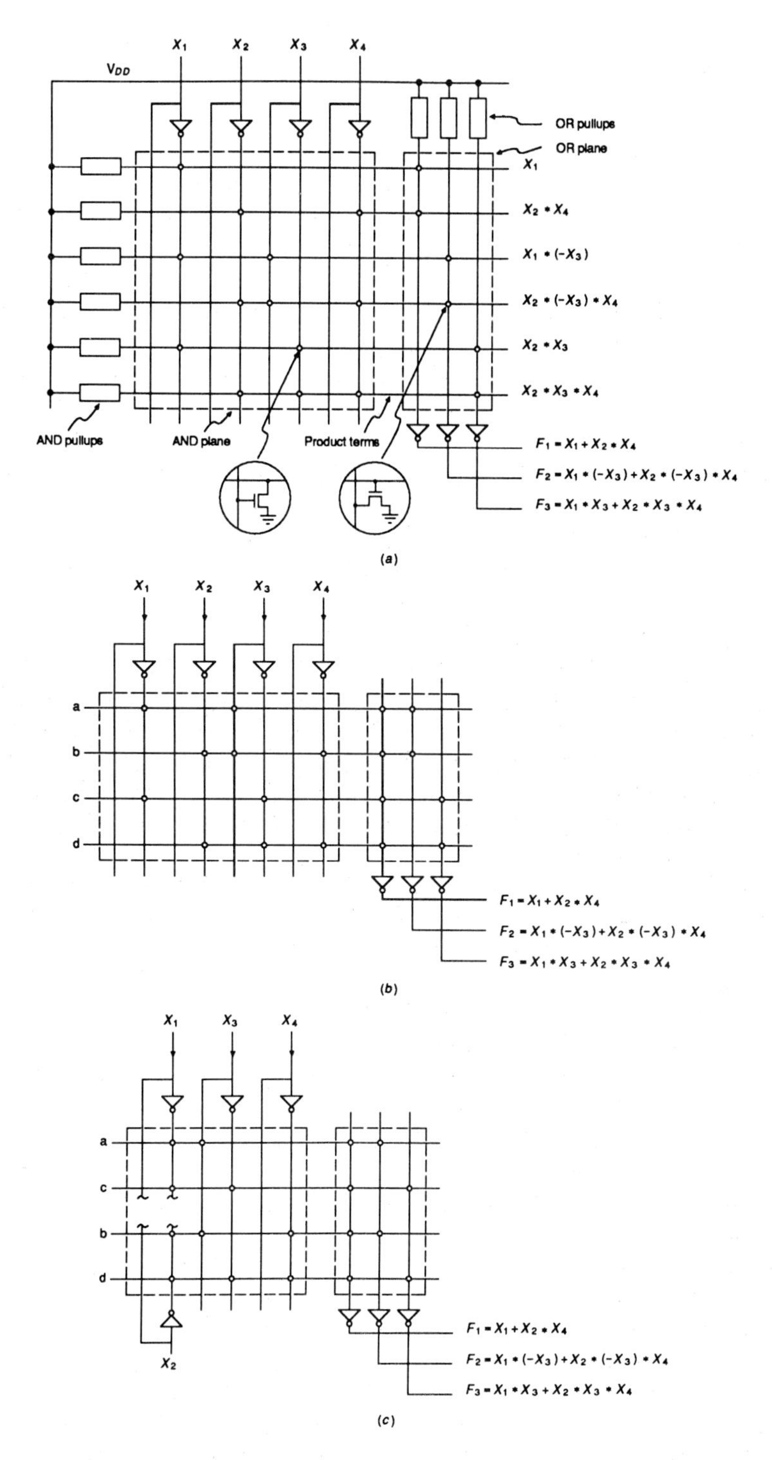

Figure 1.9 PLA layout architecture. (a) Nonminimized. (b) Minimized. (c) Folded.

operations are performed in different planes, SLA mixes both operations in a single plane. Furthermore, an SLA allows higher-level components such as flip-flops and inverters to be used at grid points, while a PLA architecture allows only single transistors.

1.3.2 Flexible Architectures

In the *flexible architecture* model, the layout of components is constrained very little or not at all. The layout shape and the pin positions are specified by the designer and not by the compiler. This extra flexibility makes the compilation a very complex process. One solution was to encode large quantities of IC design knowledge into the compiler. This led to the introduction of expert systems into the compilation process. The other solution, which led to an introduction of layout languages, was to give a designer more expressive power. In the latter case, the designer supplies part or all of the information for placement and routing of layout components.

The IC languages can be graphic, textual, or both. Since a graphic representation is much easier for designers to work with, some of the layout design systems provide a conversion from text to graphics and vice versa. Usually the layout is captured in symbolic form and converted to text. This textual representation can be converted to a compiler or procedural layout generator through parameterization. This is accomplished by the use of procedural constructs that allow instantiation of different layout versions for different sets or parameters. Thus, a procedural generator does not have a graphic representation.

In general, IC languages are based on three types of layout grid architectures: fixed, virtual, and relative grid architectures [Newt86]. In the *fixed grid architecture*, the layout area is divided into a grid uniformly spaced in both directions. The grid size is defined to satisfy the worst-case spacing requirement for a particular process. A symbol is defined for each combination of mask layers that may exist at a grid location. Each symbol may be thought of as a tile of unique color that a designer juxtaposes with other tiles to obtain a specific pattern. Figure 1.10a shows an example [West85] of a fixed-grid symbolic language and Figure 1.10 b shows the corresponding geometry.

The compilation process consists of selecting the proper tile and replacing each symbol with a corresponding layout. In Figure 1.10a, two different versions are used for the P symbol. The proper version is determined by looking at the neighboring symbols. Although the fixed-grid layout languages simplify layout specifications and compilation, they sacrifice silicon efficiency by using the worst-case grid spacing.

In the *virtual-grid architecture*, transistors, contacts, and wires are placed on a grid to facilitate easy design capture and interface to other tools, but the final geometric spacing between neighboring grid lines is determined by the worst-case spacing requirement between the layout components

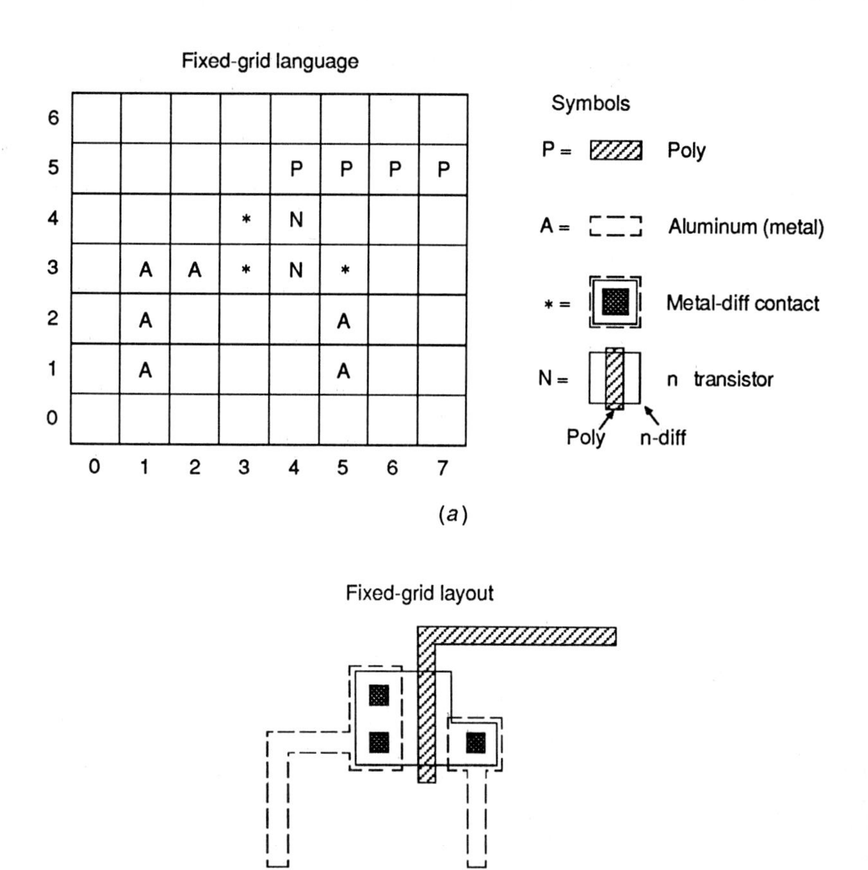

Figure 1.10 Fixed-grid architecture [West85]. (a) Language. (b) Layout.

on those grid lines. As an example, the layout of an inverter and its description are shown in Figure 1.11a and b. The specification consists of five parts: size, device, wire, contact, and pin specifications. Pins are used for interfacing to other cells. Note the concept of symbolic binding, which allows specification of points by name rather than by absolute value. Thus, the metal-to-diffusion contact at (2.2) is specified at nt.d, that is, the drain point of the n transistor. The compilation process consists of first compacting the virtual grid and then possibly stretching it for pitch matching with the neighboring cells.

The *relative-grid architectures* use a grid to indicate only relative placement of transistors, contacts, and wires and to determine the electrical connectivity of the circuit. The language based on the relative grid allows its user to design layouts at a conceptual level, at which neither sizes nor positions of the layout components need be specified. Mostly as a consequence of this, such layout language (a) makes the layout task more like programming than

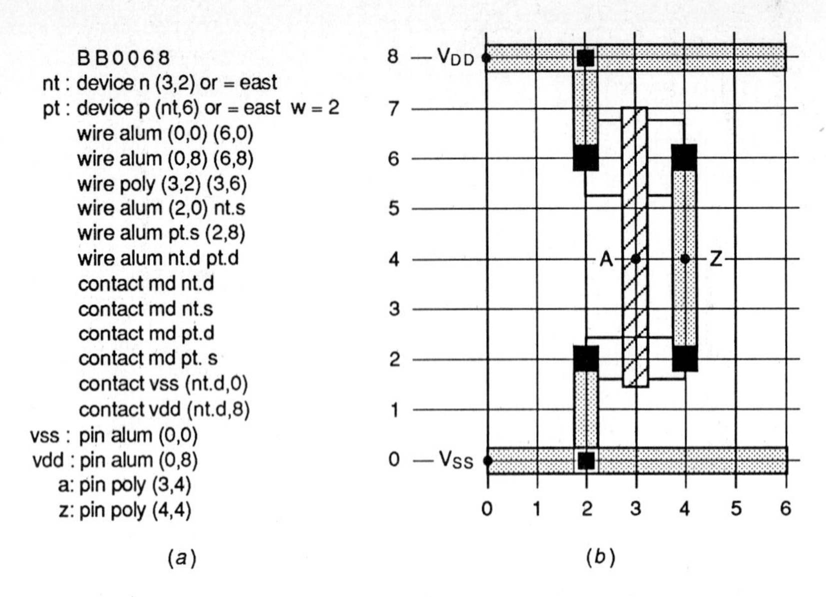

Figure 1.11 Virtual-grid architecture [West85]. (a) Language. (b) Layout.

editing, (b) eliminates the need for design rule checking after the layout is generated, (c) permits the creation of easy-to-use cell libraries, and (d) allows a hierarchical description, in which most of the detail at one level of the hierarchy is truly hidden from all higher levels.

Figure 1.12 shows a simple example using the procedural language ALI [Lipt82]. The ALI user specifies a layout by declaring the rectangles (also called boxes) of which it is composed and stating the relationships that hold among them. ALI then generates a minimal area layout that satisfies all the relationships among the boxes specified in the program. The program in Figure 1.12 consists of a declarative part and an executable part. To declare a box, the designer specifies its name (horizontal or vertical in this example) and its type (metal, for instance). The standard box types correspond to the layers of the physical layout. As the example also shows, the ALI user can define structured objects such as arrays.

The relationships among the rectangles are specified through calls to a set of primitive operations in the executable part. All such operations take boxes and possibly values of standard types (integers in this example) as arguments. The primitive "above" specifies that its first argument must appear above the second one in the final layout, the primitive "glueright" extends its first argument to the right to intersect its second argument, and "xmore" makes the size of its first argument along the x axis at least as large as the value of the second argument. In this example, ALI has determined the minimal separation between the horizontal elements as well as minimal sizes of boxes not specified by "xmore" (such as the height of horizontal metal lines) by accessing a table of design rules.

When an ALI program is executed, it generates two kinds of information.

```
chip simple;
    const
        hnumber = 10;
        length = 20;
        width = 6;
    boxtype
        htype : array [l..hnumber] ofmetal;
    var
        i : integer;
    box
        horizontal : htype;
        vertical : metal;
    begin
        for i := l to hnumber-l do begin
            above ( horizontal[i], horizontal [i + l] );
            glueright ( horizontal[i], vertical );
            xmore ( horizontal [i], length )
        end;
        glueright ( horizontal[hnumber], vertical);
        xmore ( horizontal[hnumber],length);
        xmore ( vertical, width )
    end.
```

(a) (b)

Figure 1.12 Relative-grid architecture [Lipt82]. (a) Program. (b) Layout.

The first is a set of linear inequalities involving the coordinates of the corners of the boxes in the layout as variables. These inequalities, which embody the relationships among the rectangles of the layout, are then solved to generate positions and sizes of the layout elements. The program also produces connectivity information about the rectangles in the program. This information is then used by a switch-level simulator that predicts the behavior of the circuit that has been laid out by the program's executable part.

A design system from Silicon Compilers Systems that is based on relative-grid architecture is described in Chapter 2. It uses a procedural language together with a graphic editor, a data base, and a set of support tools that allow the user to specify a variety of cell, module, and processor generators.

Since flexible architecture layouts represent a complex problem with many interacting subproblems and contradictory goals, knowledge-based systems seem to provide a good framework for solving these problems. However, we have to distinguish compilers that use just a rule-based programming paradigm from knowledge-based systems that achieve expert quality through lines of reasoning and explanation that parallel those of humans. Generally speaking, *knowledge-based* or *expert systems* consist of several knowledge sources, or "experts," working interactively on different aspects of the design. The knowledge is usually encoded in terms of if-then rules, where the *if* part recognizes a pattern and the *then* part modifies that pattern. All design data are stored in a data base usually called a "blackboard." The order of knowledge source application is determined by a control strategy usually called the "inference engine." In many early expert

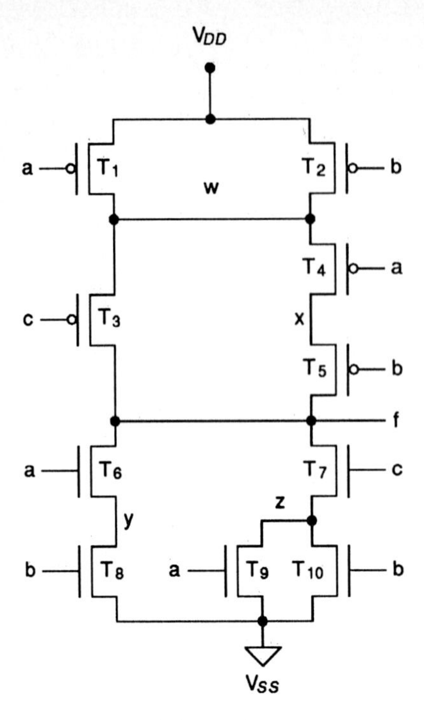

Figure 1.13 A CMOS gate schematic.

systems, the control strategy consisted of a fixed sequence of knowledge sources to be applied to the problem.

The Topologizer [Koll85] is a CMOS layout compiler that uses the rule-based programming paradigm. It follows a layout style in which all p transistors are set in rows parallel to the V_{DD} line and all n transistors are set in rows parallel to the V_{SS} line. The Topologizer accepts a transistor net list and environmental constraints as an input description. The transistor net list specifies transistor types and their interconnections. The environmental constraints include size or aspect ratio and pin constraints. Cell side, layer, location, and loading make up the pin constraints. The pin location is specified by an absolute position or a position relative to other pins. Output is a symbolic file, which makes the Topologizer free from technology-dependent design rule considerations. By using a virtual-grid language such as ABCD [Rose82] or ICDL [West81], the Topologizer's output can be translated into mask geometries.

The Topologizer synthesizes the CMOS cell layout by using a transistor placement expert and a routing expert. The placement experts simplifies the routing task by reducing the number of wires needed to connect transistors. In other words, the placement expert tries to position transistors into lines of diffusion if possible. Initially the placement expert generates a random transistor placement that satisfies the aspect ratio requirement. Next the placement expert improves the placement quality by rotating and swapping transistors to increase the number of pin matches. Two pins on a common net are matched if they are placed horizontally or vertically next to each other. Figure 1.14a shows an initial placement with a 5/2 aspect ratio of the schematic shown in Figure 1.13. The number of matches increases by 2 when transistors T_3 and T_4 are exchanged, as shown in Figure 1.14b. Then, if transistor T_{10} is rotated after being exchanged with T_9, a gain of one match will be achieved, as shown in Figure 1.14c. A further gain of two matches can be obtained by exchanging transistors T_4 and T_9 and rotating them, as shown in Figure 1.14d. This is the optimal placement for this example.

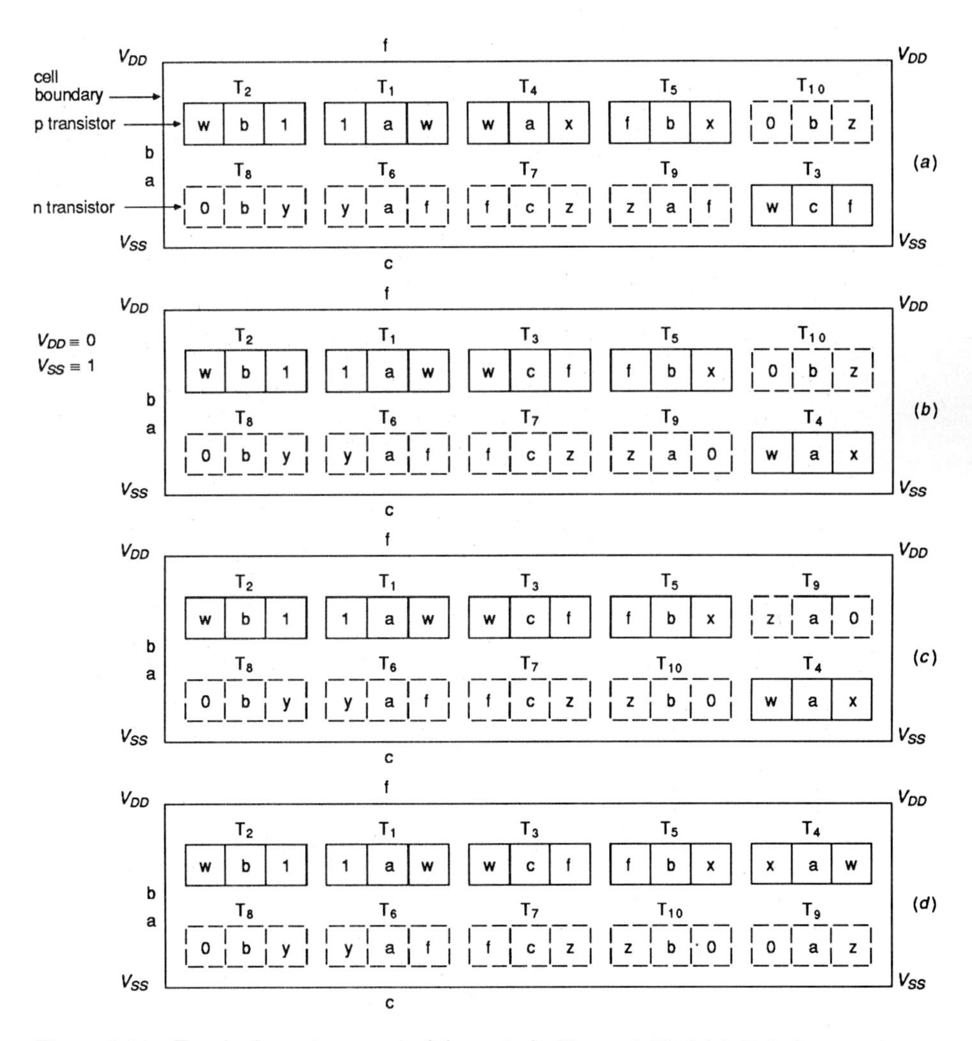

Figure 1.14 Topologizer placement of the gate in Figure 1.13. (a) Initial placement. (b) Placement after T_3 and T_4 exchange. (c) Placement after T_9 and T_{10} exchange. (d) Placement after T_4 and T_9 exchange.

After transistor positions have been determined, the Topologizer invokes its routing expert to connect the transistors. Initially the routing expert assigns a unique track to each pair of connected terminals. Figure 1.15a shows the result of this simple routing strategy. The routing expert then improves this initial routing by applying a set of optimization rules for track sharing and U-turn elimination. The final result is shown in Figure 1.15b.

The basic advantages of the Topologizer are its simplicity and its ability

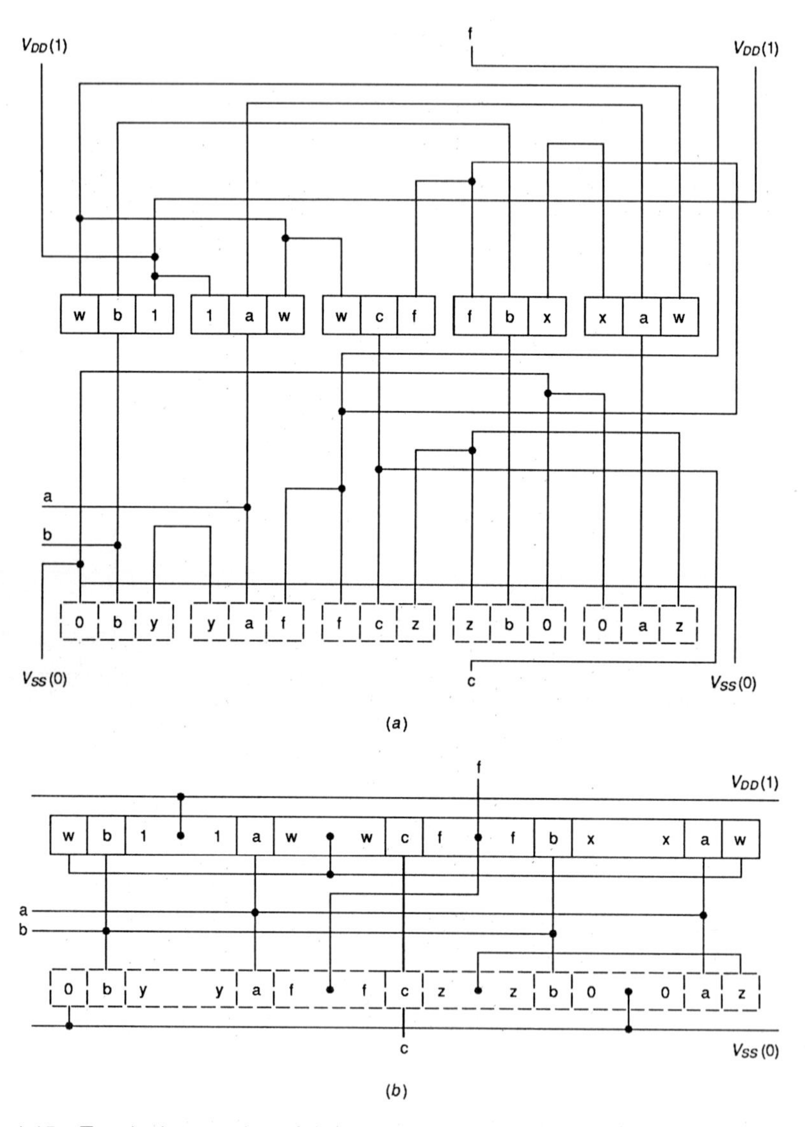

Figure 1.15 Topologizer routing of the gate in Figure 1.13. (a) Initial routing, one net per track. (b) Optimized routing. (All vertical lines are poly, and all horizontal lines are metal.)

to generate high-quality layouts for small circuits. Furthermore, it is easy to modify and new rules are easily added to its knowledge base. Finally, the Topologizer is highly interactive since the designer is allowed to modify the design after each phase of the design process. The main weakness of the Topologizer is its inability to deal with large circuits efficiently using logic schematic instead of transistor schematic LES [Lin87] is able to generate larger designs. Another expert system for cell compilation, called Talib, is described in Chapter 3.

1.4 Module Compilation

Modules can be roughly defined as microarchitectural entities that perform one or a few specific functions and consist of one or more arrays of cells or tiles of a specific type. Examples of modules are PLAs, ROMs, RAMs, register stacks, multipliers, ALUs, counters, and data paths.

Each modules is defined by a template, or frame, and a set of cells that occupy the template. For example, a generic PLA template [West85] is shown in Figure 1.16a. It consists of an AND array of cells of type A, an OR array of cells of type O, and such boundary cells as input drivers (IN), output drivers (OUT), the AND plane pullups (LA), and the OR plane pullups (TO). An example of the cell types for a dynamic CMOS PLA with a two-phase clock [West85] is shown in Figure 1.16b. A self-timed clock for precharging the OR plane is implemented using additional cells TL, TA, and TM.

The design of a module compiler consists of describing the template, tiles, interfaces, template personalization, and models used by other tools. The template can be described using a procedural language or an embedded language with some array constructs for the template assembly. The basic weakness of procedural languages is that they obscure geometrical relationships. Thus, some module compilers use layout editors for the tile specification and textual description for the template. More recent tools for module design use menus [Law85] or graphics [Mayo86] to specify the template. One such tool for module generation is described in Chapter 4. Another such tool, the Mocha Chip [Mayo86], uses annotated graphics, called *assembly diagrams*. Assembly diagrams show the relative position and orientation of subcells. Each subcell is either an assembly diagram or a leaf cell. In addition to displaying relative positions, assembly diagrams can take parameters, compute local variables, and pass new values down to subcells. Subcells may also access parameters and variables defined at the higher levels of the hierarchy. Two special cells make use of passed parameters. The *array cell* provides a form of two-dimensional iteration, and the *case cell* provides a form of conditional selection. These two building blocks can be combined to form more complex control structures similar to standard programming languages.

The interfaces between cells are usually by abutment, with the user specifying the bounding box and the alignment for each cell. Furthermore,

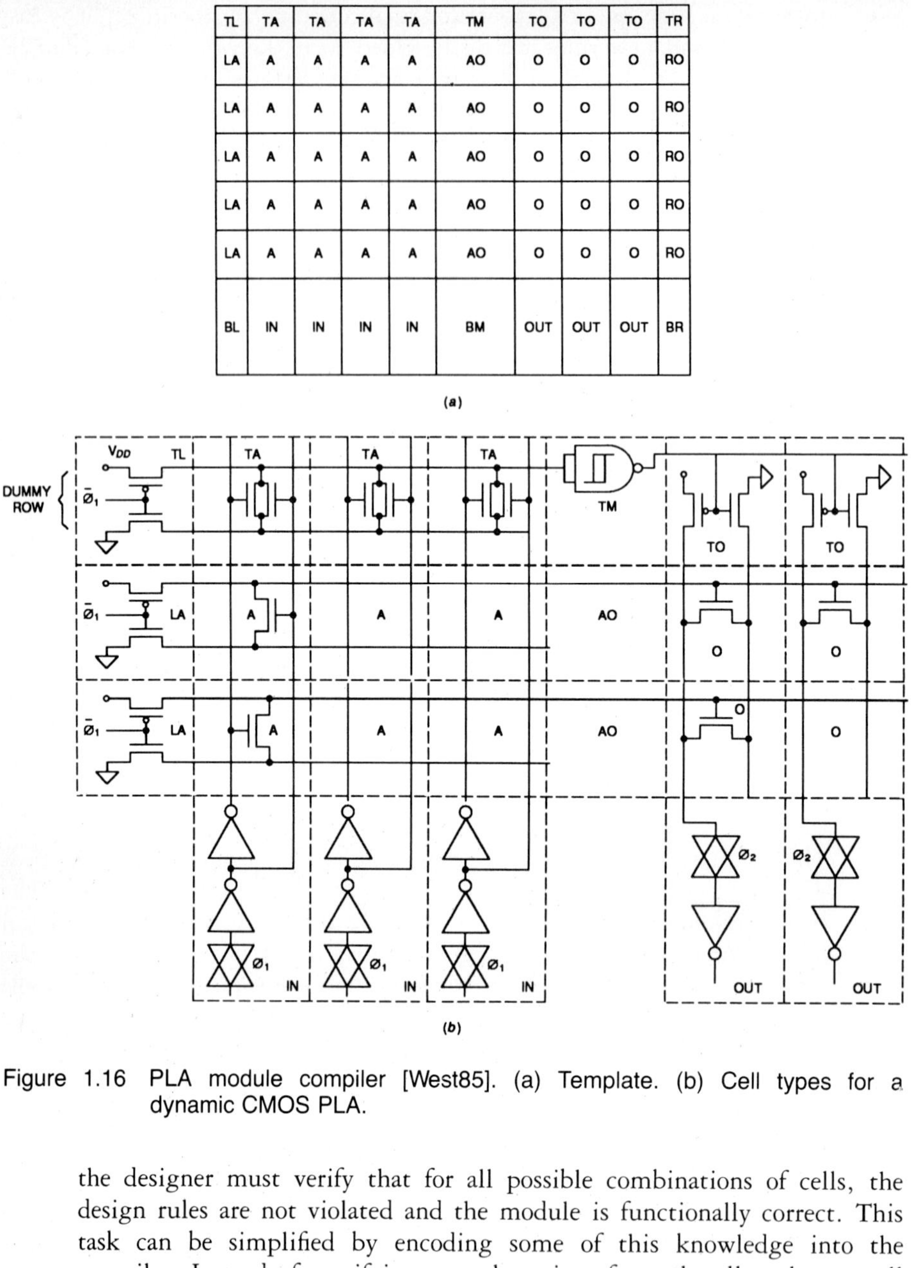
TL TA TA TA TA TM TO TO TO TR
LA A A A A AO O O O RO
LA A A A A AO O O O RO
LA A A A A AO O O O RO
LA A A A A AO O O O RO
LA A A A A AO O O O RO
BL IN IN IN IN BM OUT OUT OUT BR
(a)
DUMMY ROW
V_{DD}
TL
TA
TM
TO
$\bar{\varnothing}_1$
LA
A
AO
O
$\varnothing_1$
$\varnothing_2$
IN
OUT
(b)

Figure 1.16 PLA module compiler [West85]. (a) Template. (b) Cell types for a dynamic CMOS PLA.

the designer must verify that for all possible combinations of cells, the design rules are not violated and the module is functionally correct. This task can be simplified by encoding some of this knowledge into the compiler. Instead of specifying several versions for each cell, only one cell may be specified, and the compiler will perform pitch matching by stretching cells so that their I/O pins match on the common boundary. When pitch matching does not work, a compiler may use river routing or some other form of switch-box routing. Furthermore, a compiler may perform compaction by overlapping cells to share buses and power lines or may space them apart to avoid violation of geometrical design rules.

Personalization of a module is achieved by specifying a personalization matrix and defining a function that maps symbols in the personalization matrix onto a set of tiles. Some tools allow module optimization by selective replacement of tiles. As an example, consider the PLA in Figure 1.16. The tiles of type A in rows 2 and 3 and columns 3 and 4 can be simplified by removing the bottom horizontal wire passing through each cell. The regular-structure generator described in Chapter 4 allows optimization procedures to be associated with template description.

In addition to generating layouts, module generators must generate models to be used with other tools, such as a logic model to be used with a logic simulator or an interface model to be used with a floor planner.

1.5 Processor Compilation

The earlier sections of this chapter covered cell and module compilation and dealt predominantly with issues in structural and geometric domains. On the other hand, *processor compilation* is predominantly concerned with synthesis of behavioral description into a structure of microarchitectural components such as registers, ALUs, RAMs, and buses. Each structural component is instantiated by a module or cell compiler.

A typical behavioral description takes the form of a program in a programming language. Procedural statements are used to describe the conditions under which certain processing will occur. However, these procedures do not describe structural entities such as ALUs or registers. Rather, they describe the transformation of values that must occur in a system for a given set of inputs. Figure 1.2a is an example of behavioral description. Note that several statements are described but that no mention is made of the structural logic operators that will implement the arithmetic operations, for instance. This could be done with a central ALU, with separate incrementors, or in some other way.

Thus, we could describe the instruction set of a processor using a programlike language. From this description, we could propose a number of structural implementations from which we can derive a number of geometric designs all with the same behavior. Designing in the behavioral domain, then, involves transforming and then synthesizing the logical structure needed to implement the system's behavior.

1.5.1 Behavior Domain Transformations

The purpose of behavioral transformations is to transform the behavioral description into a form more suitable for implementation. At the system level of design, the transforms effect the control structure of the final implementation. There are two levels at which this occurs. At the higher level, we must decide on the number of controllers needed to implement the design. At the lower level, we must decide on the control sequence for each of the controllers.

A higher-level transform can be used to split a behavioral description into one or more processes. A *process* is a separate, independent thread of control, analogous to the usage of the term in the operation systems field. Each process is then implemented by a separate controller and data path. There are two advantages to this transform:

1. Each new controller and data path pair will be smaller than the original controller and data path. Thus, whereas the original system may not fit into a physical partition, the two smaller parts might fit their own partitions. With each process having a physical partition, all the critical paths of the process will be on the same partition, allowing for a faster clock rate for each.
2. With two controllers, one can respond to a critical timing relation while the other is doing more meaningful processing.

Of course, the splitting of a process must be undertaken with care. If the two processes are going to use two different physical partitionings, then it is wise (for timing reasons) that as little data as possible be transferred between them. In addition, synchronization of the two processes is needed to satisfy any mutual exclusion problems that may arise. As an example, consider the situation where a special-purpose processor must interface to a serial line. Three implementations come to mind:

1. The entire function of the serial line interface is programmed as part of the instruction set of the processor. Since the program for this function is small (on the order of 100 bytes), the cost is minimal. However, due to the asynchronous nature of the information coming in over the serial line, the processor must continually poll the interface looking for start bits and data. Thus, the performance of the processor will be severely curtailed.
2. The behavior is split into two processes, one consisting of the processor and the other consisting of the serial line interface. Some signaling mechanism, such as an interrupt, must be added to indicate when a value has been received by the serial line interface. Compared to the preceding case, there are two, smaller processes; the processor is smaller by the 100 bytes or so needed for the serial line function. However, the processor is able to perform calculations while the serial line is receiving information. Since both are on the same physical partition, there is little delay in communicating values between them.
3. We may postulate that the extra 100 bytes require just enough area so that the processor will not fit on a single chip. In this case, the serial line process is moved to a different physical partition (chip). Although the communications overhead between the two processes is greater due to the separate partitions, we could argue that the path is used so little that the effect is negligible. The alternative of having to split the processor onto separate chips would require a lower clock rate, reducing the performance for all the functions of the processor.

In summary, splitting a behavioral description into several processes can have far-reaching effects in terms of the physical space required for the system and the performance of the system.

The lower-level transformations are used to "clean up" a behavioral description. These transformations, originally suggested by Snow [Snow78], include ones often seen in optimizing compilers, such as constant folding, where constant expressions are evaluated at compile time and common subexpression elimination, where subexpressions are recognized and their results stored for later use without recalculation. Two more powerful transformations, in terms of hardware design, are inline subroutine expansion and select combination. The complete list can be found in Snow's thesis.

Inline expansion of subroutines allows a defined subroutine to be included at one or more sites in place of the call to it. In addition, the return is removed. Not only are the control steps for the call and return removed, but it is possible that some of the control steps in the subroutine can be interleaved with the surrounding statements when expanded in line. This would not be possible if the subroutine were kept separate.

The select combinations transformation allows for two or more sequentially specified IF or CASE statements to be combined into one CASE statement. Thus, instead of requiring several control steps to make several sequential decisions, they can all be implemented in one step. Of course, if the second (or *n*th) IF or CASE is dependent on values calculated from the preceding one, they cannot be combined.

It is instructive to consider why a compiler would be needed to make these decisions. The idea of designing from the architectural level of the behavioral domain is that a system-level architect can specify a behavior to be implemented, and the processor compiler can then assist by coming up with widely varying implementation alternatives. The theory is that the appropriate interface to a system-level architect is through software engineering, that is, a programming language. Experience with such languages to describe systems indicates that the descriptions are organized much like large programs. The description is modularized into procedures spanning many pages of text.

For instance, in the description of a processor the first page may state that the main control loop is a fetch-and-execute cycle. This would be modeled as an infinite repeat loop with calls to fetch-and-execute procedures. The page describing the execute procedure might state that the instructions fall into several groups and indicate how to determine the different groupings through CASE statements. Next the instructions in the groups would be decoded using another CASE statement. Finally, there would be a subroutine indicating how the instruction is executed. The point is that this is a nice human interface to the system-level architect who wants to specify the behavior of a system and use a set of CAD tools to evaluate and complete its design. But a high-performance implementation of the description would not call n subroutines in which there are m CASE statements because of the overhead in control steps. More typically, but not necessarily, there would be one CASE statement and no subroutines used to implement the situation.

These transformations allow the control structure of the behavior to be altered to a form more appropriate for implementation. Specifically, they can change the number of control steps needed, for instance, by reducing the number of subroutine calls and the number of CASE statements, and in some cases they simplify the data paths by reducing the number of common subexpressions. Thus, through a set of behavioral domain transformations, the capability of design space exploration is available to the system-level architect.

1.5.2 Behavioral-Level Design Decisions

The behavioral transformations discussed in the preceding section serve to mold the control structure of the final implementation of the system. The concurrency (number of processes) is specified, as are the basic blocks for the control sequence. The next step in the design process specifies what is commonly called the *register transfer* level of design. This synthesis step adds more detail to the logical structure and behavioral domain by specifying:

1. the control sequence,
2. the registers to hold the values indicated in the behavioral description,
3. the number and types of hardware operators for the operations described in the behavior, and
4. the busing and multiplexer interconnections of the registers and operators.

It should be emphasized that, given this list, there are many possible functional block implementations of the behavior. For instance, specifying the control sequence allows for implementations that differ in the parallelism of statement levels. Choice of design has ramifications on the performance and on the area required by the parallel data paths necessary to implement the design.

We will illustrate some of the possible design decisions through the use of a data flow graph. Early research in the area of automatic design at these levels [Snow78] suggested the similarity between the function of an optimizing programming language compiler and that of control step assignment in digital systems design. A data flow representation of the behavior domain description at the algorithm level, called the *value trace* (VT), was developed to provide a design representation capable of supporting the analyses, design transformations, and design decisions. Although the term *silicon compiler* up to now has not included such design decisions and transformations, we have enlarged the definition of silicon compilation to include these decisions, thus making it more analogous to the normal usage of the term *compilation*.

A data flow representation such as a VT can also be used as a data base with which the user can analyze the behavior and specify the structural information. Consider the simple ISPS description in Figure 1.17a and the associated VT shown in Figure 1.17b [Pang86]. Although this is a simple example, the main ideas behind using the VT as a representation are illustrated. The ISPS defines eight statements based on five input values (V_1, V_2, V_4, V_6, V_{10}). For this example we shall assume that the only values important beyond this basic block are these five values; the others are mere temporaries and are never used outside the basic block. Even though there are eight statements, the VT shows the behavior as a directed flow of information between operations. Names corresponding to the ISPS variable names have been kept in Figure 1.17b for ease of illustration.

The design decisions listed above can now be made. Consider first that several different control step assignments can be proposed. These different assignments arise from constraints on the underlying hardware, such as the functionality of operator modules in the database, as well as on timing constraints.

First we assign control steps in an "as soon as possible" manner. This approach packs operations into control steps when their input values are ready and typically corresponds to a direct translation from the data flow graph. Figure 1.18a shows the four control steps and assumes that all operations take the same amount of time. Logic suggests that there must be at least three operators to perform two additions and one division in step 3. To the extent that the other operations can also be implemented by these three operators, other operators may or may not be needed in the data path. Figure 1.18b shows an implementation using three operators.

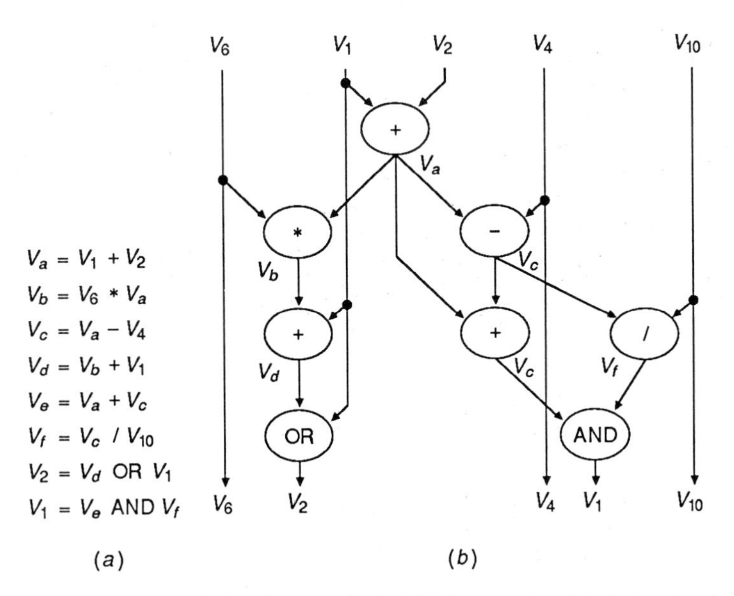

Figure 1.17 Behavioral Synthesis of straight-line code. (a) Behavioral description. (b) Data flow representation.

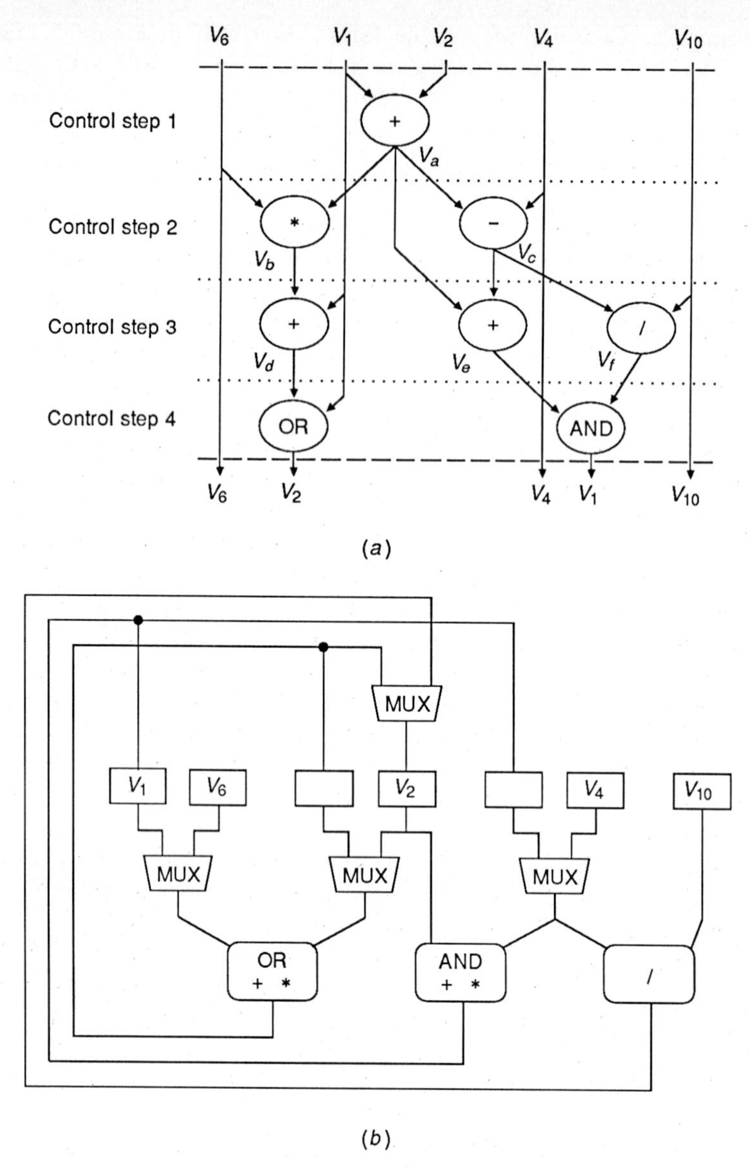

Figure 1.18 Synthesis of straight-line code. (a) Data flow representation with control sequence. (b) Data path.

However, the physical constraints of the system may indicate that three separate operators and supporting data paths use too much space. An alternate implementation would be the totally serial control sequence shown in Figure 1.19a. This figure shows the values being generated in the order V_a, V_b, V_c, V_d, V_e, V_f, V_2, V_1. This data path only requires one operator unit, but it must handle all the different operations. An example of the data paths for this approach is shown in Figure 1.19b. Although the data paths are simpler, the proposed implementation takes eight control steps rather than four.

Table 1.2 summarizes some of the differences between these two implementations. First note that the number of registers remains constant. This is a function both of the algorithm and of how the control steps are

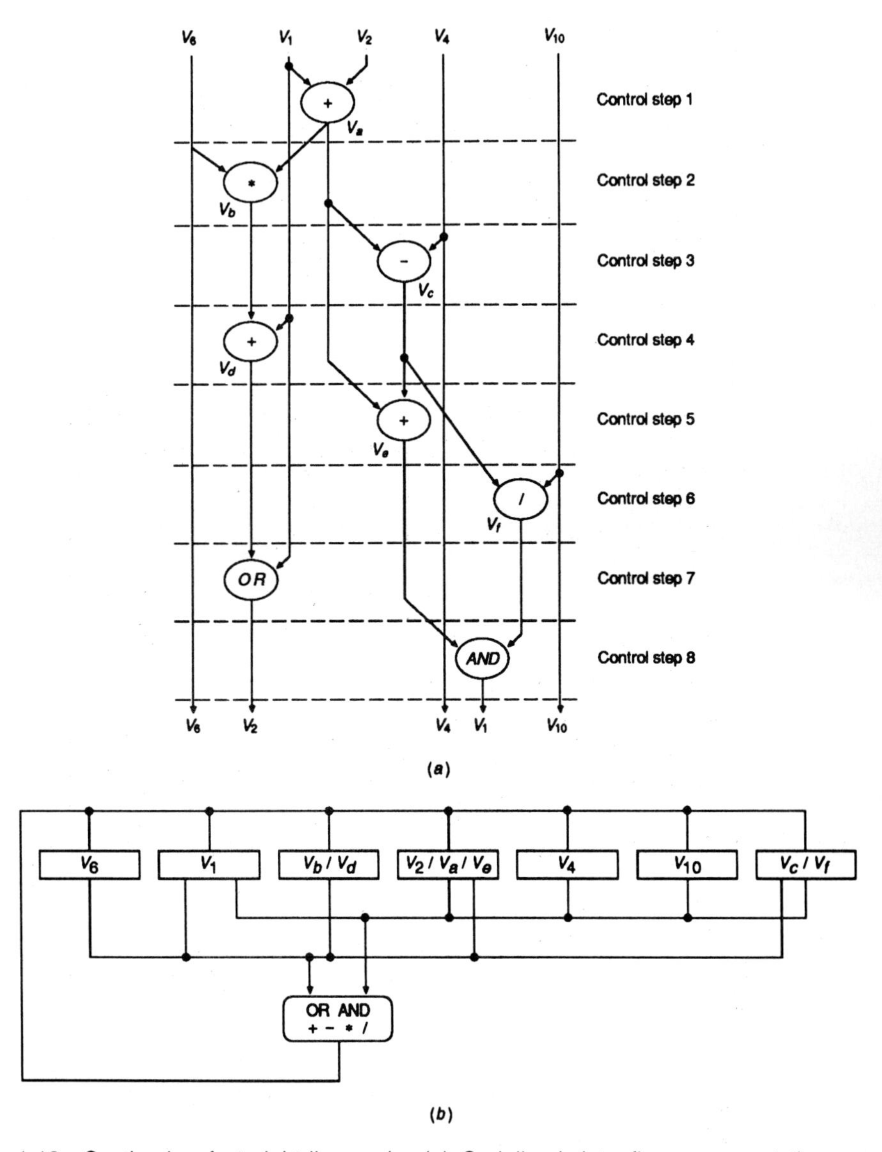

Figure 1.19 Synthesis of straight-line code. (a) Serialized data flow representation. (b) Serialized data path.

Table 1.2 Comparison of Parallel and Serial Implementations of Straight Line Code from Figure 1.11

Control Steps	*Registers*	*Statement-level Parallelism*	*Operators*	*Fan-in Buses*	*Fan-out Buses*
4	7	3	3	4	3
8	7	1	1	2	1

assigned. Any value in the data flow graph that crosses a control step line must be stored at a register. Thus, the maximum number of crossings in any control step of the basic block indicates the number of registers needed. Alternate control step assignments typically will change the number of registers required. The actual assignment of the values to registers is part of the data path synthesis task.

Statement level parallelism indicates how many register transfers involving separate operators are occurring. This can be found from the data flow graph and control sequence by counting the maximum number of operators in any step. This number is typically a guide of the complexity of the resulting data path since it implies multiple operators and busing.

The number of operators indicates the number of logical devices needed but does not specify their complexity. Three operators have been allocated for the behavior of Figure 1.18a, but a number of different operations have been bound to them. Thus, in Figure 1.18b, the OR or AND operations could be interchanged, but associated changes in the busing structure feeding the operators might have to be made. The assignment of operations to logical operators is a function of the data path synthesis program.

The number of fan-in buses is the sum of the multiplexers and destinations in a many-to-one busing structure. Fan-out buses indicate a one-to-many connection. These numbers give an idea of the complexity of the data path interconnections. Bits of input to fan-in buses and bits of output from fan-out buses can also be counted. However, since the structure of these busing devices tends to be very simple (several buses can run over a register stack and each register can either drive or take values from the bus), the important information is the number of separate buses needed and the number of connections to them. The number of buses that can run over a register stack cell and be accessed by the cell is a module-level technology parameter that should be considered at the behavioral level of design. Failure to take this into account might require redesign of the modules used to implement the data path.

There are several key points to note from these examples.

1. A number of different control sequence and data path implementations are possible for any one behavioral description.
2. The use of module-level information in behavioral-level decision making can help determine the appropriateness of the style of the resulting data paths and how well they will meet the constraints. The examples here

showed a parallel and serial style of data paths using modules typically found in a module database. If other modules were available, the design might be quite different. Thus, although the design process may be top-down, bottom-up module information is required.

3. The purpose of a data flow graph is to make apparent possible design alternatives, allowing the design programs or designer to make a decision after appropriate analysis. Using a data flow graph as the basic representation also allows the design system to verify that the changes being made are valid in the sense that the behavior of the system is not changed. One would expect no less from a software compiler.

1.5.3 Different Approaches to Control Sequence and Data Path

Different approaches to the automatic design of control sequences and data paths have been reported in the literature. In this section we review several of them.

The Elf [GiBK85] system performs control sequence assignment and data path synthesis simultaneously. A behavioral description in Ada is compiled into a flow graph representation. Operations are bound to states based on an urgency-scheduling algorithm. Node urgencies are calculated based on timing constraints on "as soon as possible" (ASAP) and "as late as possible" (ALAP) [Land80] schedulings. As a node approaches a timing constraint, its urgency weight is raised, thus raising the probability of the node being bound to the current state. Elf binds registers to arc in the flow graph sequentially from the top of the graph and chooses registers that are least costly to use at that time.

DAA [Kowa85a] [Kowa85b] performs control sequence assignment using an ASAP scheduling algorithm. This is a simple strategy that uses a maximally parallel implementation to bind operations to states. Such an algorithm effects fast parallel implementations that use suboptimally large amounts of hardware. DAA uses a two-pass expert system for data path synthesis. The first pass assigns components to elements in the flow graph. The second pass merges these components to form the actual structural components that implement the design. In both passes, decisions are made through a set of rules in a knowledge-based system. Chapter 5 describes the principles and detailed design process in DAA.

Emerald [Tsen84] is an algorithm-based system for performing data path synthesis from data flow graphs. The control sequence is based on ASAP scheduling. Its optimization strategy is based on a clique-partitioning algorithm. For instance, every operator is represented by a node on a graph. Two nodes are connected by an arc if the two operators they represent could be implemented by the same logical operator. Clique partitioning finds the smallest number of complete subgraphs, which correspond to the ALUs of the system. A similar approach is used for assigning values to registers.

MAHA [Park86] first determines the critical path through a behavioral description based on maximum time constraints and then schedules operations along the critical path by means of control steps. Data path hardware is first synthesized for the critical path operations. Then hardware is synthesized for the noncritical path operations using existing hardware where possible.

EMUCS [Thom83] is based on a greedy heuristic. It allocates hardware based on requirement estimates and iteratively binds data flow operations to hardware elements. The binding is based on a heuristic cost that reflects the impact of changes on the interconnection and data path logic previously allocated in the design. The cost function is parameterized, allowing the user to tune the operation.

The ISYN [Nest86] system performs control sequence assignment and data path synthesis for digital system interface. Maximum and minimum time constraints are placed on a behavioral description, and a control sequence assignment is made using a heuristic based on list scheduling [Fish81]. The constraints specify the maximum or minimum time between two operations. Based on data flow and timing constraints, other operations may be inserted between or moved before or after the constrained operations. The scheduling is similar to the urgency-scheduling algorithm in Elf. The EMUCS data path synthesis tool is then used.

1.5.4 Processor Compilation

The preceding sections described the essential issues in behavior-to-structure translation or synthesis. A complete processor compiler contains module compilers for each microarchitectural component specified in the synthesis part. Furthermore, each module compiler may call cell compilers to instantiate cells. In addition to these different types of compilers, a processor compiler also contains programs for placement and routing of components.

To simplify the compilation task, many processor compilers limit the language model, design style, floor plan style, or application domain. In Chapters 6, 7, and 8, three different processor compilers are described. The MacPitts silicon compiler [Sout83] uses an input language that describes each control step separately and generates a simple floor plan consisting of a linear data path and Weinberger array control. The Yorktown silicon compiler, described in Chapter 7, uses performance-driven synthesis on three levels: processor, module, and cell. The compiler also uses general placement and routing algorithms. The Cathedral compiler, described in Chapter 8, uses a multiprocessor architecture to solve signal processing problems.

1.6 Silicon Compilation–Based Design Systems

Traditional methodologies require a designer to build a structure and define its behavior with basic components such as gates, and then use it hierarchically to build higher-level structures. Once the design is finished, it is flattened into a structure of basic components for simulation, placement, and routing (see Figure 1.20a). This methodology doesn't efficiently exploit the hierarchical nature of the design, however, because the simulation,

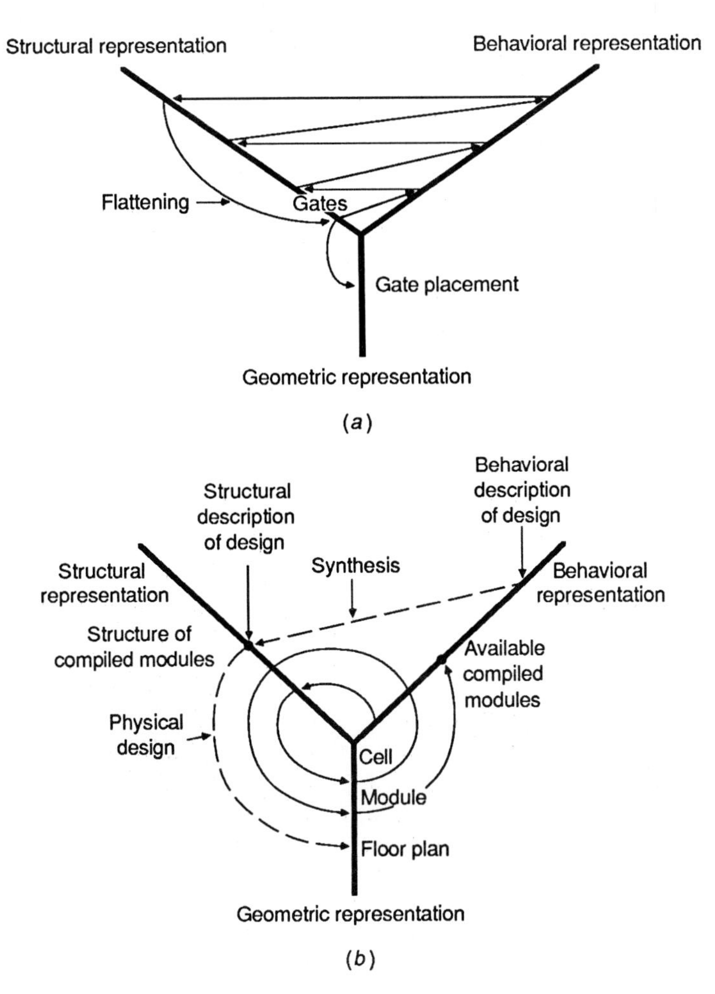

Figure 1.20 Design methodologies. (a) Bottom-up methodology. (b) Silicon compilation methodology.

placement, and routing are performed at the lowest level of abstraction (where they are the most expensive to perform). Furthermore, the fixed functionality as well as the fixed electrical and layout properties of the basic components leads to an inefficient layout. The silicon compilation methodology tries to overcome these deficiencies by providing basic components that are fine-tuned to the use's specification at a higher level of abstraction, such as ALUs, PLAs, RAMs, data paths, controllers, and core microcomputers. Furthermore, the methodology provides higher-level models for simulation, placement, and routing.

Silicon compilers generate layout descriptions together with higher-level models of those descriptions, such as functional, logic, timing, power, and testability models to be used by verification, analysis, and optimization tools such as functional and logic stimulators, timing and power analyzers, and layout compactors. The functional, electrical, and layout requirements for each basic component are passed as parameters to the corresponding silicon compiler. Silicon compilers as defined in this book include all possible compilers, that is, *a set of* cell, module, processor, and system compilers. A hierarchical methodology is supported by allowing silicon compilers to call other silicon compilers.

As with gate-array and standard-cell methodologies, silicon compilation requires two experts: the tool maker and the tool user. A silicon compiler writer (the tool maker) creates compilers for leaf cells and then uses these leaf cell compilers to construct module compilers, processor compilers, and other higher-level compilers. These compilers are parameterized, and parameters for different design components are stored in a table or a menu. Such a menu serves the role of a behavioral description for design components. The task of a silicon compiler writer is represented by the outward-going spiral in Figure 1.20b.

A system or application designer (the tool user) specifies the design using a behavioral or structural description. In the former case, five different tasks must be performed before the design is ready for fabrication:

1. The behavioral description is translated into a structural description (*synthesis*).
2. The layout of each structural component is instantiated by a silicon compiler (*compilation*).
3. All structural components are placed on silicon and routed (*physical design*).
4. The packaging is selected.
5. A test vector set is generated.

In the latter case, the translation task is not needed since the system designer specifies the design structure.

A ideal IC design system based on silicon compilation methodology is shown in Figure 1.21. The design structure is specified by the user or generated by the synthesizer from the behavioral description. In either case a menu/form package is used to capture the behavioral description of each component in the structure. The behavioral description is passed in the form of options or parameters to the corresponding silicon compilers. A technology file contains all of the process-relevant design rules used for generating the layout. In an interactive environment, layout or schematic editors are used to alter compiler outputs. However, in this case the "correct by construction" property of silicon compilers is lost. Similarly, timing, behavioral, or logic models are generated for each component in the structure. These models are linked together and passed to a timing analyzer or a simulator. Geometric models are linked together with placement and routing tools to form a chip composite. For interactive placement and routing, a composition editor or a silicon assembler, which allows preplacement of components and nets and evaluates the quality of the floor plan, may be used [Trim84]. A package editor is used to provide packaging information

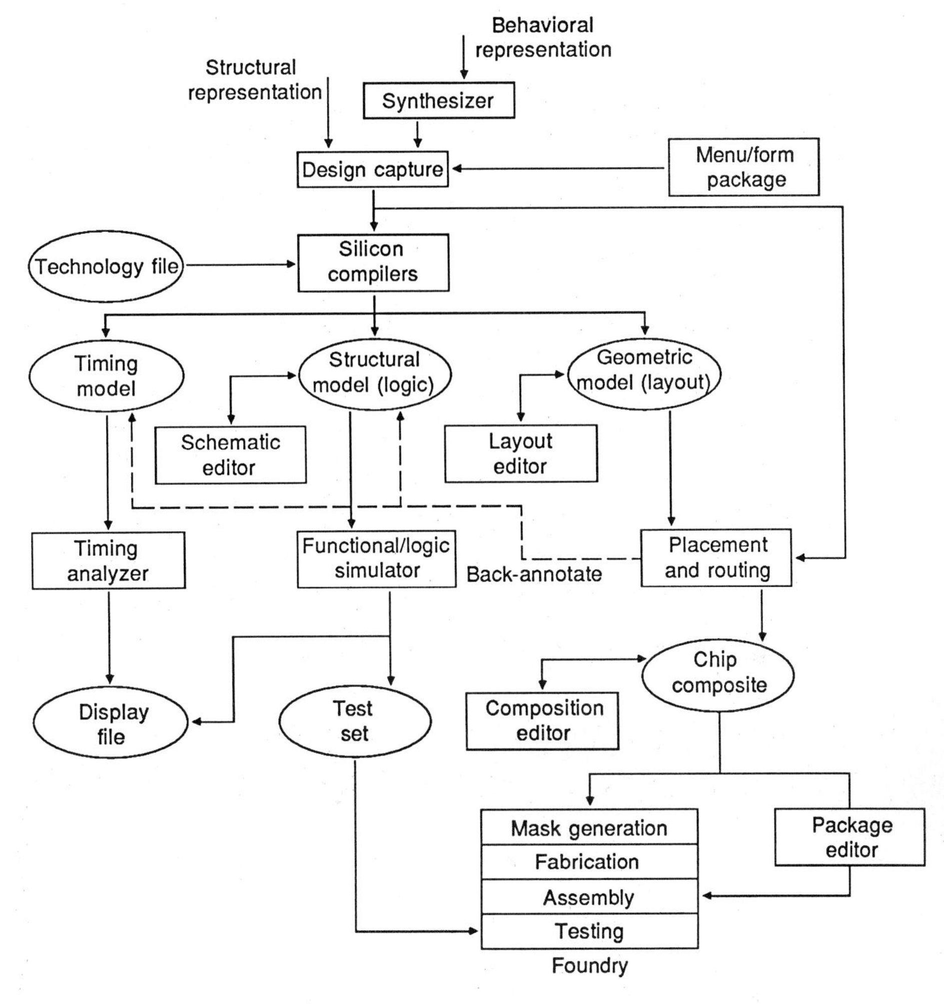

Figure 1.21 Silicon compiler-based design system.

to the foundry, and a simulator provides a test set for testing the assembled IC.

Most commercial silicon compiler–based systems fit well into the block diagram of Figure 1.21. A detailed exposition of the Genesil system from Silicon Compiler Systems. can be found in Chapter 9, and a description of the Seattle Silicon Technology Concorde system is given in Chapter 10. Commercial and research silicon compilers are classifiable according to several attributes, summarized in the following sections.

1.6.1 User Expertise

Figure 1.20b shows that the field is roughly divided into compiler writers that design silicon compilers and system designers that use silicon compilers to design application-specific integrated circuits (ASICs). Silicon compiler writers use IC languages to capture and parameterize layout and use other embedded languages to write simulation, timing, and power models for each silicon compiler. System designers specify systems as a structure of components where each component is: (a) a predefined library component, (b) a component instantiated by a silicon compiler, or (c) specified by the designer with a behavioral description. System designers are generally concerned with system issues independent of design rules and the fabrication process, and compiler writers are concerned with control over the design details and making compilers application-independent.

This crude division can be generalized by considering the design process as a generator of the final description on one of the design levels using a design language, such as the L language (described in Chapter 2) for IC description or VHDL [Shah86] for logic description. In the future we will see "silicon" compilers on different levels of design for gate-array, standard-cell, and custom design methodologies with different tradeoffs between the control over design and the expertise of the user.

1.6.2 Integrability

Silicon compiler systems come in two varieties:

1. They may be complete systems with their own set of proprietary tools, such as simulators, timing verifiers, and routers for complete IC design.
2. They may be integrated into a standard CAE workstation with most of the support tools provided by the host workstation environment.

The advantage of the first approach is enforcement of consistent design practices, a more efficient data base, and better interfaces between different tools. The second approach offers a smooth transition from existing design methodologies and the possibility of reusing parts already designed.

1.6.3 User Interface

The user interface specifies ways in which the designer may interact with the system during the design phase, in addition to specifying the initial input description. Some systems allow editing of the layout or schematic after it has been generated by a silicon compiler. However, the change must be reflected in all different representations or models of the compiler output. A layout editor is necessary during the compiler writing phase, since a cell layout is first created manually, then possibly converted into textual form and parameterized. Some systems allow preplacement and prerouting during the design composition.

Design documentation provides an auditing mechanism for the designer to keep track of the current state of the design. This includes simulation and timing analysis results at various levels, as well as several logic, schematic, and layout "views" that the silicon compiler systems produce. Some silicon compiler systems provide "browsers," facilitating extensive user monitoring of the design.

1.6.4 Richness

Richness is characterized by the number and types of module generators available in the system. Module types are divided into basic modules and sequential modules. Basic modules, such as SSI logic, registers, ALUs, PLAs, RAMs, ROMs, and data paths, use one control state for execution; sequential modules, such as counters, finite-state machines, controllers, and processors, require more than one control state for execution. Some companies offer several different types of controllers, such as interrupt, bus, DMA, and CRT controllers. Finite-state machine generators and other popular modules, such as bit-slice or standard microprocessors, are also available.

1.6.5 Process Independence

Silicon compilers are not technology-independent; that is, a silicon compiler written for a CMOS process cannot easily be derived from one written for a bipolar process. However, a compiler can be made process-independent by specifying design rules in a separate technology file and specifying the layout in terms of symbols and constraints between those symbols. During design instantiation, the symbols are replaced by their geometric representations, and the layout is compacted according to the constraint values in the technology file as described in Section 1.3.2 on flexible architecture. This method allows the compiler to cover a broad spectrum of processes, but its layout density depends on the sophistication of the compactor. An alternate strategy is to specify layout in generic lambda rules in which all constraints are a multiple of the basic lambda unit. Process independence is achieved by changing the value of lambda. The second method is simpler, but it may produce a less compact layout.

1.6.6 Quality Measures

Any design system based on silicon compilers can be evaluated according to three quality measures:

1. Transistor density (in square mil/transistor)
2. Compilation speed (in transistors/hour)
3. Design time (in transistors/person-hour)

These quality measures depend heavily on the complexity and the regularity of the design. Until standard benchmarks are established, these measures should be taken as indicative of silicon compilation capabilities and should not be used to compare different systems. Recently, Seattle Silicon Technology (SST) reported densities of 0.36 square mil/transistor, which is slightly better than the standard 80386 microprocessor. This demonstrates the viability of silicon compilation as a design methodology.

The density measure is similar to the computer performance measure of instructions/second, which is not very accurate but is a very popular approximation. A better density measure uses the design functionality instead of the transistor count as a normalizing factor, since the existence of each transistor may be difficult to justify. Compilation speed is defined as the compiler run time without plotting and should be normalized to the same machine. Reported compilation speeds vary from 10K to 25K transistors/hour for different chips and systems. The design time depends heavily on the design complexity as well as the abstraction level of the compiler. Design times range from 10 to 30 transistors/hour for various designs. Better design times (300–400 transistors/hour) are achieved by using high-level compilers such as processor compilers. These results indicate obvious tradeoffs between the design time, the compiler complexity, and the abstraction level of the design specification.

1.6.7 Support Tools

Tools for silicon compilation are divided into three groups: verification, analysis, and optimization tools.

The verification tools are used to verify the correctness of the input (behavioral or structural) description. Verification is accomplished by the designer's specifying a set of input-output pairs and observing for each pair whether the description produces the correct output for each input. Note that verification only proves that the design works for the given test set.

The most frequently used tools are behavioral, functional, logic, circuit, and fault simulators. Behavioral simulators are easier to write, particularly if the behavioral description language is embedded in other programming language. Functional and logic simulators are used in structural silicon compilation, where each module compiler generates a functional or logic module that is linked with other modules as specified in the input structural description. A fault simulator allows a user to specify a set of test vectors, and the simulator will determine the fault coverage of the given test set using an assumed fault model. Usually, a single "stuck-at" fault model is used, which assumes that each fault is caused by a wire being permanently connected to logic 0 or logic 1. Circuit simulation is used mostly on leaf cells or when a foreign module is imported at the circuit level.

Analysis tools are used to determine the quality, or "goodness," of the generated design. Very frequently, a timing analyzer is used to determine the delay from any input to any output and the delay between elements. The maximum delay between storage elements determines the clock period. Thus, a timing analyzer can be used to predict the performance of a design. Similarly, testability analysis tools calculate the controllability and observability figures, that is, the relative measures of difficulty in controlling and observing signal values. Together, these figures define the testability of a design.

Optimization tools improve the quality of the design without introducing any tradeoffs. Those tools make translation from behavior to structure or structure to geometry an easier task since translators have to produce only a correct, but not necessarily optimal, design. The most frequently used optimization tools at the layout level are programs for layout compaction and transistor sizing. At the topological level, many PLA folders minimize PLA area by sharing rows and columns between two or more input and output lines. At the logic level, synthesis programs such as SOCRATES [Greg86] and the Yorktown silicon compiler (described in Chapter 7) are capable of optimizing logic for a given input-output delay. Similarly, some behavioral level compilers are also capable of optimization for a given time delay. However, optimization at this level is a part of the synthesis and not a separate tool.

Presently, there are no commercial or research compilers that incorporate all of these tools. Most compilers include a simulator for checking the input description and a timing analyzer. Optimization tools are not readily available.

1.7 Conclusion

Silicon compilation has been presented as an evolving methodology that makes custom silicon affordable and removes the design time bottleneck. As this methodology evolves, the design level will rise from the circuit and logic level to the microarchitecture and system levels. This evolutionary process should blend well with present CAD tools, so more silicon compilation systems are expected to appear on standard workstations in the future.

Silicon compilers could be standard utility routines on future workstations, with the compiler layouts modifiable by the designer, and integrated with handcrafted custom and semicustom parts. Furthermore, a designer should be able to add his or her own compilers to the set, create new compilers by using existing ones, and add his or her own personal cells.

Present-day silicon compilers translate an input description into layout in a unique way. A designer working with a behavioral compiler must understand the translation process built into the compiler and modify the input description to force the compiler to produce the desired results. When working with structural compilers, the designer must be able to evaluate a design and choose a different style of a component if necessary. A future trend will be toward an intelligent silicon compiler that incorporates knowledge about the design process and uses this knowledge to guide the transformations of the input specification through possibly several design iterations until the specified set of constraints is met.

Intelligent compilation can be divided in three basic tasks: planning, design, and evaluation. Presently, only the design task is performed by silicon compilers; planning and evaluation are left to the designer. The planning task consists of selecting a style of design, formulating a strategy to obtain a design that meets the design goals and technology constraints, and specifying a proper set of parameters for each silicon compiler call. The evaluation task determines design quality, indicates whether the given constraints have been met, and performs a tradeoff analysis that sets new goals and constraints to drive the planning activity.

Once planning and evaluation are well understood, automation of the total design process will be possible.

References

1. [Asan82] T. Asano, "An Optimum Gate Placement Algorithm for MOS One-Dimensional Arrays," *Journal of Digital Systems*, vol. 5, no. 1, 1982, pp 1–27.
2. [Barb81] M. R. Barbacci, "Instruction Set Processor Specification (ISPS)," *IEEE Trans. on Computers*, vol. c-30, no. 1, January 1981, pp. 24–40.
3. [Camp86] R. Camposano and A. Kunzmann, "Considering Timing Constraints in Synthesis from a Behavioral Description," *Proc. ICCAD*, 1986, pp. 6–9.
4. [DeMi85] G. DeMicheli, M. Hoffman, A. R. Newton, and A. Sangiovanni-Vincentelli, "A Design System for PLA Based Digital Circuits," in *Advances in Computer-Aided Engineering Design* (A. Sangiovanni-Vincentelli, ed.), JAI Press, 1985, pp. 285–364.
5. [Deut76] D. N. Deutch, "Dogleg Channel Router," 13th Design Automation Conference, June 1976, pp. 285–292.
6. [Fish81] J. A. Fisher, "Trace Scheduling: A Technique for Global Microcode Compaction," *IEEE Trans. on Computers*, July 1981, pp. 478–490.
7. [Gajs83] D. D. Gajski and R. H. Kuhn, "New VLSI Tools," *Computer*, vol. 16, no. 12, December 1983, pp. 11–14.
8. [Greg86] D. Gregory, K. Bartlett, A. deGeus, and G. Hachtel, "SOCRATES: A System for Automatically Synthesizing and Optimizing Combinational Logic," *Proc. 23rd Design Automation Conf.*, 1986, pp. 78–85.
9. [Hash71] A. Hashimoto and J. Steven, "Wire Routing by Optimizing Channel Assignment within Large Apertures," *8th Design Automation Conference*, 1981, pp. 155–169.
10. [Joha79] D. Johannsen, "Bristle Blocks: A Silicon Compiler," *Proc. 16th Design Automation Conf.*, 1979, pp. 310–313.
11. [Kim86] J. Kim and J. McDermott, "Computer Aids for IC Design, *IEEE Software*, March 1986, pp. 38–47.
12. [Koll85] P. W. Kollaritsch and N. H. E. Weste, "TOPOLOGIZER: An Expert System Translator of Transistor Connectivity to Symbolic Cell Layout," *IEEE Journal of Solid-State Circuits*, June 1985.
13. [Kowa85a] T. J. Kowalski, *An Artificial Intelligence Approach to VLSI Design*, Kluwer Academic Publishers, Boston, 1985.
14. [Kowa85b] T. J. Kowalski, D. J. Geiger, W. Wolf, W. Fichtner, "The VLSI Design Automation Assistant: From Algorithms to Silicon," *IEEE Design and Test*, August 1985, pp. 33–43.
15. [Land80] D. Landskov, S. Davidson, B. Shriver, and P. Mallet, "Local Microcode Compaction Techniques," *Computing Surveys*, vol. 12, no. 3, September 1980.
16. [Law85] H-F. S. Law and J. D. Mosby, "An Intelligent Composition Tool for Regular and Semi-Regular VLSI Structures," *Proc. ICCAD-85*, 1985, pp. 169–172.

17. [Lin87] Y-L. S. Lin and D. D. Gajski, "LES: A Layout Expert System," *24th Design Automation Conference*, 1987, pp. 672–678.
18. [Lipt82] R. J. Lipton, S. C. North, R. Sedgewick, J. Valdes, and G. Vijayan, "ALI: A Procedural Language to Describe VLSI Layouts," *Proc. 19th Design Automation Conf.*, 1982, pp. 467–473.
19. [Lope80] A. Lopez and H. Law, "A Dense Gate Matrix Layout Method for MOS VLSI," *IEEE Trans. on Electronic Devices*, vol. ED-27, no. 8, August 1980, pp. 1671–1675.
20. [Mayo86] R. N. Mayo, "Mocha Chip: A System for the Graphic Design of VLSI Module Generators," *Proc. ICCAD*, 1986, pp. 74–77.
21. [Nest86] J. A Nestor and D. E. Thomas, "Behavioral Synthesis With Interfaces," *IEEE International Conference on Computer-Aided Design,* November 1986, pp. 112–115.
22. [Newt86] A. R. Newton, "Symbolic Layout and Procedural Design," in *Design Systems for VLSI Circuits: Logic Synthesis and silicon compilation* (DeMichelli, Sangoranni-Vincentelli, Antognetti, editors) Kluwer Academic Publishers, 1987.
23. [Pang86] B. M. Pangrle and D. D. Gajski, "State Synthesis and Connectivity Binding for Microarchitecture Compilation," Proc. ICCAD, 1986, pp 210–213.
24. [Park86] A. C. Parker, J. Pizarro, and M. Milnar, "MAHA: A Program for Data Path Synthesis," *Proc. 23rd Design Automation Conf.*, 1986, pp. 461–466.
25. [Rive82] R. L. Rivest and C. M. Fiduccia, "A Greedy Channel Router," *Proc. 19th Design Automation Conf.*, June 1982, pp. 418–424.
26. [Rose82] J. Rosenberg and N. Weste, "The ABCD Language," Tech. Report 82–01, Microelectronics Center of NC, 1982.
27. [Rude85] R. Rudell, A. L. Sangiovanni-Vincentelli, and G. DeMicheli, "A Finite State Machine Synthesis System," *Proc. ISCAS*, 1985, pp. 647–650.
28. [Sang87] A. Sangiovanni-Vincentelli, "Automatic Layout of Integrated circuits" in Design Systems for VLSI Circuits: Logic Synthesis and Silicon Compilation (DeMicheli, Sangovanni-Vincenteli, Antognetti, editors), Kluwer Academic Publishers, 1987.
29. [Shah86] M. Shahdad, "An Overview of VHDL Language and Technology," *Proc. 23rd Design Automation Conf.*, 1986, pp. 320–326.
30. [Snow78] E. A. Snow, "Automation of Module Set Independent Register-Transfer Level Design," Ph.D. diss., Carnegie-Melon University, April 1978.
31. [Sout83] J. R. Southard, "MacPitts: An Approach to Silicon Compilation," *Computer*, vol. 16, no. 12, December 1983, pp. 74–82.
32. [Thom83] D. E. Thomas, C. Y. Hitchcock III, T. S. Kowalski, S. V. Rajan, R. Walker, "Automatic Data Path Synthesis," *Computer*, vol. 16, no. 12, December 1983, pp. 59–70.
33. [Trim86] S. Trimberger, "VTIcompose—A Powerful Graphical Chip Assembly Tool," *Proc. ICCAD*, 1984, pp. 233–235.
34. [Tsen84] C. Tseng and D. P. Siewiorek, "Automatic Synthesis of Data Paths in Digital Systems," *IEEE Trans. on CAD*, vol. CAD-5, no. 3, July 1986, pp. 379–395.
35. [Ueha79] T. Uehara and W. M. vanCleemput, "Original Layout of CMOS Functional Arrays," *IEEE Trans. on Computers*, vol. C-30, no. 5, 1979, pp. 305–311.
36. [Wath85] R. A. Walker and D. E. Thomas, "A Model for Design Representation and Synthesis," *Proc. 22nd Design Automation Conf.*, 1985, pp. 453–459.

37. [Wein67] A. Weinberger, "Large Scale Integration of MOS Complex Logic: A Layout Method," *IEEE Journal of Solid-State Circuits*, vol. SC-2, no. 4, December 1967, pp. 182–190.
38. [West81] n. Weste, "MULGA—An Interactive Symbolic Layout System for the Design of Integrated Circuits," *Bell System Technical Journal*, vol. 60, no. 6, July– August 1981, pp. 823–858.
39. [West85] N. Weste and K. Eshraghian, *Principles of CMOS VLSI Design* Addison-Wesley, 1985.
40. [Yosh82] T. Yoshimura and E. S. Kuh, "Efficient Algorithms for Channel Routing," *IEEE Trans. on CAD*, vol. CAD-2, no. 1, January 1982, pp. 25–35.

CHAPTER 2

Design of Module Generators and Silicon Compilers

Misha R. Burich

2.1 Silicon Compilers and the Semiconductor Industry

Silicon compilation is a new branch of integrated circuit (IC) design automation. There are several reasons for its emergence:

- IC manufacturing technology offers capabilities for integrating highly complex circuits on a single chip: 100,000 to one million transistors on a chip are now common.
- Software technology offers highly sophisticated compiler and data base products.
- Due to high-quality heuristics, computational algorithms have advanced sufficiently to handle complex problems quickly. Placement and routing, logic minimization and synthesis, simulation, fault grading, and design for testability are examples of problems whose solution has been aided by good algorithms.
- Workstation technology offers powerful computing capabilities to engineers at ever-decreasing cost: 32-bit processors, virtual memory management, tens of megabytes of random access memory, hundreds of megabytes of hard disk memory, floating-point accelerators, and high resolution graphics are now standard features in workstations.

Due to these factors, a new type of IC product is gaining popularity. Instead of using standard, off-the-shelf chips, many engineers are designing their own application-specific integrated circuits (ASICs). Most often, they are integrating many functions onto a single chip and combining such devices with standard parts to complete their board, or system, design.

The system cost, the power consumption, the cost of the final product, and maintenance cost are lower with ASICs.

Also, manufacturers of standard parts have begun to offer many variations of their popular chips. Each variation includes some additional feature integrated on the chip. This trend has led to standard application-specific integrated circuits (SASICS). The manufacturers use new methods in design automation, including silicon compilation, to design these SASIC parts.

The relationship between the manufacturers of ASIC devices and their designer customers has rapidly evolved over the last few years. Gate-array and standard-cell layout styles have become the dominant technologies. The reasons for this are that the layout design is fully automated, the turnaround time from the design phase to prototypes is short, and the manufacturing reliability is very high.

Figure 2.1 shows a typical relationship between users and providers of ASIC products. The user takes advantage of workstations to perform conceptual design in schematic form. The design is verified by simulation, and its test procedure is developed by using test vector generation and fault grading. Physical layout has traditionally been done with gate arrays and standard cells, but a more general cell-based technology is emerging.

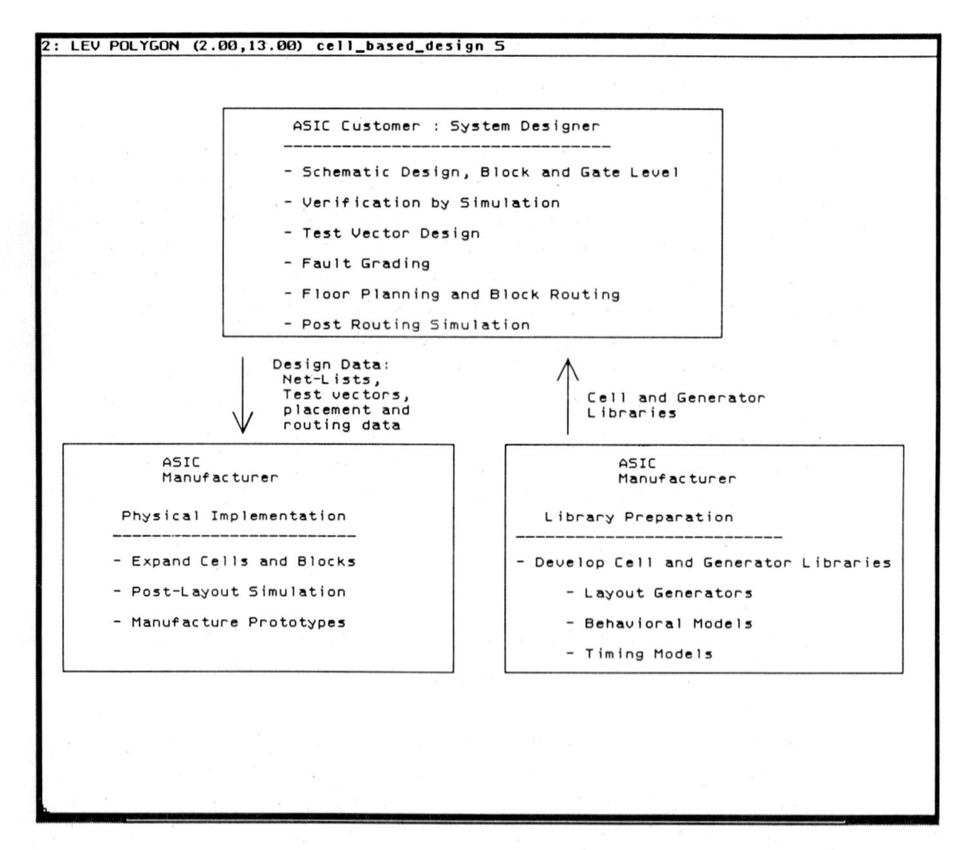

Figure 2.1 Relationship of ASIC companies and ASIC customers.

Cell-based technology relies on libraries of fixed or parameterized function blocks produced by module generators. It is characterized by larger function blocks mixed with standard cells in which some form of user-assisted floor planning may be employed.

2.2.1 The Design Process for Integrated Circuits

In any IC design, including traditional full-custom design, designers employ three major levels of abstraction: functional, structural, and geometric. The same levels of abstractions are needed for silicon compilation.

The functional description specifies system behavior, such as instruction set, logic functions, input/output behavior at pins, and timing relationships among signals. The functional description specifies *what* is to be built.

The structural description specifies the system's architectural components and their interconnections. The designer partitions the whole system into logic subsystems, data paths, memory subsystems, input/output blocks, and subsystem connectivity. The structural description specifies *how* the system is to be built.

The geometric description specifies the physical implementation of the system, including its floor plan, the placement of cells, the routing of the blocks, and the layouts of cells. The geometric description also specifies how the chip is built, but it is much more specific than the structural description because it is tied to a certain manufacturing process, such as CMOS, NMOS, bipolar, or GaAs. More than one geometric description of a system is possible for a given structural description. The designer's expertise is essential in forming an IC that meets the area and performance criteria.

A design system for custom ICs has to support all phases of the design process, as shown in Figures 2.2 and 2.3. The system designer starts by developing a functional model of the chip. The functional model is also called an *executable specification*. It is executable because it can be simulated. It is a specification because it spells out all details of the system; it is a precise blueprint of chip behavior. Functional models are developed in a functional description language. Examples include M (Silicon Compiler Systems), Helix (Silvar-Lisco), Verilog (Gateway Design Automation), and N.2 (Endot).

At this stage the designer develops test vectors to be applied to the chip. The test will be used throughout the design process to verify the correct behavior of all subsystems as they are developed.

As Figure 2.2 shows, the designer proceeds by developing a structural description of the chip. This is can be done by interactive graphical editors or by some structural description language, such as L (Silicon Compiler Systems) or Model (Lattice Logic). The system might be partitioned into a control logic section, an input/output section, the data path, and a ROM and RAM section. The interactive schematic editor replaces pencil and paper tools used in the past for this job. Simultaneously, the designer might start the geometric partitioning and floor planning.

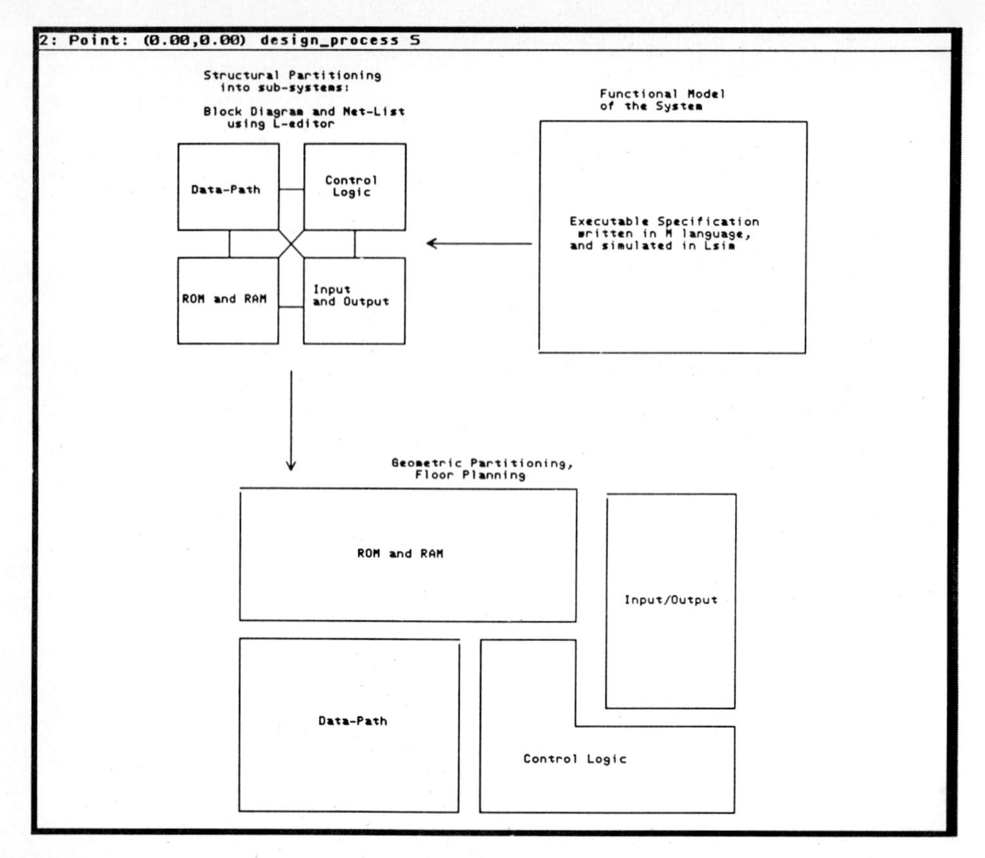

Figure 2.2 Different phases of the design process on the top level.

Figure 2.3 shows the next step, the design of the control logic subsystem through application of the exact same steps as before. The functional model of the control logic is developed in a functional description language and simulated by the simulator. The structural partitioning is described in terms of more primitive building blocks: PLA and gate-level logic. The schematic editor is again used to capture the design intent.

At this point the designer can verify that the detailed control logic block matches its functional model by using the simulator. The previously developed test vectors are applied to the whole chip, but the functional model is replaced by the structural model, the net list.

For geometric description, the module generator for the PLA can be used to provide information about the possible sizes and aspect ratios. The automatic standard-cell placement and routing software is used to evaluate the sizes of other parts of the control logic. This whole process can be executed rapidly if module generators and random-logic placement and routing are available. Such tools enable the designer to perform many iterations of the *what if* analysis. In many cases, custom layouts have to be developed for circuits for which there are no automatic layout tools.

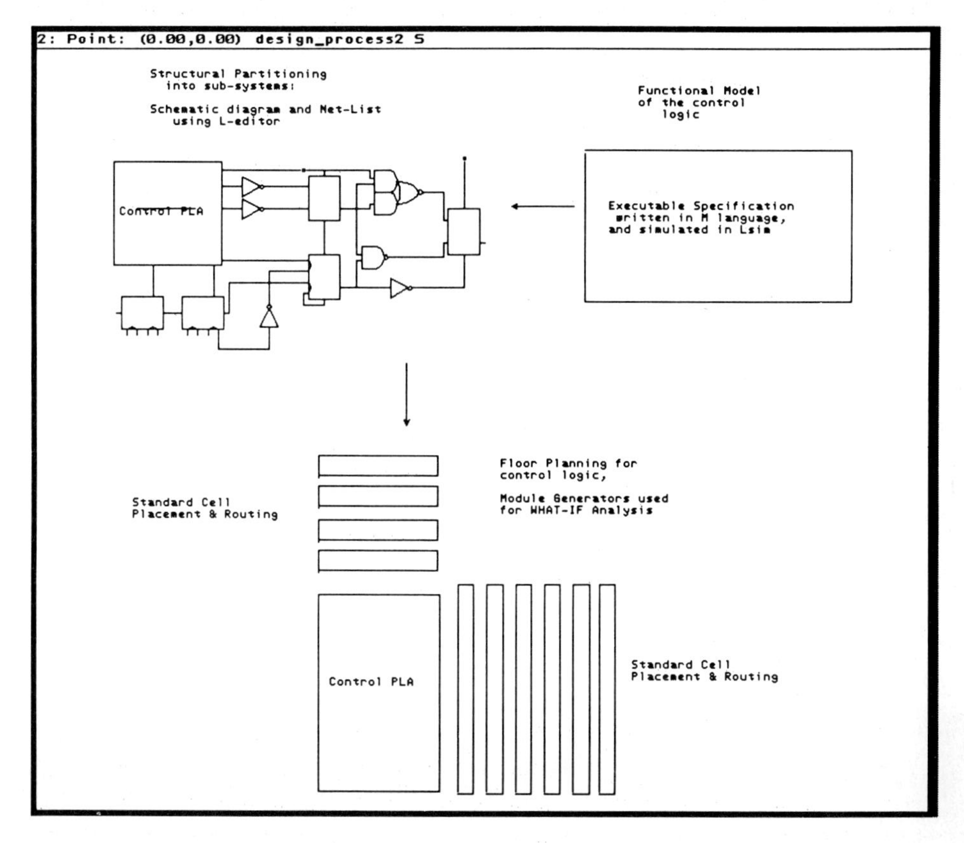

Figure 2.3 The design phases one level below the top level.

The same steps are performed on other sections of the chip. Verification is carried out by applying the test vectors at every stage of the design.

2.2 Developing Silicon Compilers to Automate the Design Process

Silicon compilation is a process of translating system design specifications into IC masks. The key problem in developing silicon compilers is how to capture and embed expert knowledge in some parts of the compiler, mix that knowledge with algorithms in other parts of the compiler, and provide a natural user interface for the IC designers.

Translation from functional models to structural models is most demanding. Design expertise that must be captured includes architectural knowledge, reasoning about system tradeoffs, understanding of technological tradeoffs, and the ability to experiment with different options. Today this translation is done by experienced architects and system designers. The most promising software technology for this component of the silicon compiler include knowledge-based systems, a methodology of capturing and using human expertise.

Translation from structural models to geometric and layout models requires a combination of expert knowledge, good algorithms, and module generators. The expertise to be captured in this component of the silicon compiler includes architecture design, logic design, circuit design, and layout design. The required algorithms include placement and routing, logic verification and simulation, circuit performance verification and simulation, mask layout verification, and fault grading and simulation for test purposes. Traditional software tools are used for this component of a silicon compiler: high-level languages, debuggers, and language parsers.

Module generators automate the translation from structural descriptions to geometric descriptions. Expert designers can use the module development tools to capture their knowledge. The generators may be as simple as logic gates in CMOS, NMOS, bipolar, or GaAs technologies, or they may be more complex parameterized function blocks, such as data paths and microcode sequencers. Alternatively, they may be very complex function blocks such as microprocessors. No matter what their complexity, the tools must be applied in the same consistent fashion, taking advantage of hierarchy and reusability of lower-level functions. The designers should not have to sacrifice layout density or circuit performance in building generators, and they should be able to build in mask-level process independence. There are several systems for developing module generators: Generator Development Tools or GDT (Silicon Compiler Systems), SILC tools (GTE), DPL language (MIT), and SLIC language (Seattle Silicon Technologies).

The developer of a general silicon compiler goes through the following steps:

1. Define the input behavioral/functional language. It has to be user-friendly and expressive, and it should not restrict the range of applications. Some graphical input could be useful. In case of interactive input, it is helpful to consider artificial intelligence (AI) techniques for a design assistant/advisor (DAA) approach. The DAA should have access to other parts of the compiler in order to provide viable design alternatives.
2. Develop techniques for mapping functional descriptions to structural descriptions. Knowledge-based systems with captured expert experience are probably the best alternative. The mapping can also be interactive, as in step 1. The mapping should also have access to estimates of chip size, performance, and power consumption. This is obtained from steps 3 and 4.
3. Develop mechanisms for structural-to-geometric translation. Here module generators, floor planning, placement, and routing play the major role. Also develop fast size and performance estimators to be used by procedures in step 2.
4. Develop simulation and verification models to be used in all phases

of the compilation. Functional modeling languages and simulators are intended to serve this function.

Fully general silicon compilers will emerge in the future. So far, the reported compilers always make assumptions about some aspects of the design. As shown in the following sections, the assumptions may concern the architecture or the layout implementation style. Most promising for the near future are compilers that start with a structural description of the design. Here the translation from functional to structural description is left to the experienced designers.

2.2.1 Compiling Functional Descriptions into Layout

Silicon compilation is a process of translating functional descriptions of systems to their structural descriptions, and then to their geometric descriptions, or mask layouts. There is no unique way to translate functional descriptions to structural descriptions, and no unique way to translate these to geometric descriptions. Therefore, every silicon compiler must incorporate some rules, constraints, and expertise in translating one representation to another.

One way to cope with this is to make assumptions about the underlying architecture (structural description) and layout style (geometric description). This simplifies the compilation task, can guarantee correctness by construction, and ease the verification problem.

Silicon Compilation with Fixed Layout Style

PLA design is one of the earliest examples of silicon compilation [1]. Functional descriptions can be expressed by finite-state machines. Structural description is assumed to be given by two-level AND-OR logic with registers at some inputs of the AND gates. Geometric description is assumed to be given by NOR-NOR planes with densely packed transistors that share common gate and drain wiring. As a result, automatic compilation from finite-state machine language to mask layouts is possible. In this approach, the PLA minimization algorithms are very important. They include logic minimization and PLA folding algorithms.

A different functional description language, retaining the same PLA layout style, has been reported [2]. Instead of using finite-state machines, the regular expressions were compiled into masks.

A silicon compiler developer could design a compiler of this type through the following steps:

1. Define an input language and write a translator for it. The translator produces an intermediate data representation suitable for required transformations. Recommended tools are yacc, Lex, and the C language, all standard in the UNIX system. Lex can be used to generate a lexical analyzer, and yacc is a parser generator program.
2. Develop algorithms and programs for transforming the intermediate representation to a PLA personality matrix.

3. Develop (or use available) logic minimization algorithms to remove redundant states and logic.
4. Develop (or use available) PLA folding algorithms and programs to minimize the size of the PLA.
5. If testability is needed, develop a PLA test strategy and algorithms for adding additional logic to the PLA.
6. Develop a PLA layout generator. Include options for folding, routing over the PLA, and user-specified locations for input and output terminals. This is done in a module generator development language, with the aid of the layout editor, design rule checker, and layout compactor. The verification and timing characterization is done with the simulator.
7. Develop a functional model for the PLAs produced by this compiler. This is done in a functional modeling language so that the user can simulate PLAs as a part of the larger system with the simulator.
8. Develop a program that will automatically invoke all the programs in steps 1–7 in the proper sequence. As an example, the designer can use the shell, the standard command interpreter in the UNIX environment. The various parts of the compiler could communicate through temporary files.

A silicon compiler of this type is easy to use if the input language reflects the most frequently needed constructs.

A different layout style, named storage logic arrays (SLA), has also been reported [4]. Once the architecture and layout style were fixed, the authors were able to develop automatic compilation from register-transfer-level (RTL) functional descriptions. To develop a silicon compiler of this kind, the developer would go through steps similar to those listed above. The algorithms would be different, as would the layout generator, but the principles are the same.

A different target layout style is exemplified by gate arrays and standard cells, which have been used extensively in automatic synthesis. The functional description can be a language such as MODEL [5]. Silicon compilation proceeds by translating functional specifications to gate-level interconnections (structural description). Finally, automatic placement and routing software translates this into layout. Simulation and verification programs monitor the success of the compilation.

If the generator development tools contain placement and routing for standard cells, the developer of the silicon compiler must pay attention to the higher-level steps, including the following:

1. Develop a functional design language. This might be a special-purpose language, such as for signal processing, communications protocols, or general-purpose RTL. Again, yacc, Lex, and the C language are used for this purpose.
2. Develop heuristics and programs for translation from functional to gate-level description. Translate these to the net list format by a standard-cell placement-and-routing package.

3. Develop the standard or special cells needed as primitive components. Use a module generator language or a layout editor.
4. Develop an interface to the placement-and-routing package to automatically synthesize layout.

Silicon Compilation with Fixed Architecture

Architectures based on data paths controlled by sequencers are popular among designers. The MacPitts silicon compiler uses such an architecture as its target [6]. A design is specified by describing the behavior of the system in a Lisp-like language. The designer controls the architectural resources by specifying the degree of parallelism desired in the circuitry to be output.

Bit-serial architecture was employed in FIRST, the silicon compiler intended for signal processing applications [7]. The specification language emulated the equations used by signal processing experts.

A complete parameterizable and programmable core microcomputer was the target of the Plex silicon compiler [8]. The architecture of the Plex microcomputer was fixed, but the user could specify all the major resources of the design. These included the number of registers in the data path, the size of the data memory, the size of the program memory, the stack depth, the number of interrupt lines, the size of the I/O buses, and the instruction set. The functional specification for the system consisted of the assembly language program that the resulting microcomputer was to execute. The compiler carried out the complete layout generation of the microcomputer, which could be augmented by other logic circuits on its periphery.

The generator development tools are well suited for developing silicon compilers of this type. The fixed architectures are constructed from well-defined but different components. These components are produced by module generators with many parameterized options. For example, the Plex compiler included a ROM generator, a RAM generator, a data path generator, a PLA generator, random-logic synthesis, and glue generators for binding the pieces together.

The steps in developing such silicon compiler tools are as follows:

1. Define the target architecture and the specification language for the users. This phase is very important because it sets the tone and range of applications for the silicon compiler.
2. Develop the language parser and the translation to the target structural description. The tools for this phase, again, are yacc, Lex, and the C language, all standard in the UNIX environment.
3. Develop module generators in a VLSI design language for all system components. Module generators should be hierarchical and should cover all needed options, both functional and geometric. The top-level generator interfaces to the input parser. It calls other generators and asks them to provide needed options. It completes the layout by including input/output pads. Floor planning, placement, and routing are done hierarchically and procedurally by generators on all levels of the hierarchy.

4. Develop the user interface, development aids, and debugging aids. This is done using the C language and perhaps some graphical utility programs. Develop functional simulation models. These are used in verification of the design.

SDL-2000 by Silicon Compiler Systems is such a silicon compiler for a microcontroller architecture. It is fully compatible with Intel's 80C51 microcontroller. It is to be used in applications requiring a processor, memory, and some user-specified control logic integrated on the same chip.

2.2.2 The Role of Module Generators in Silicon Compilation

Module generators are programs that produce functional, structural, and geometric descriptions from input specifications. They are an integral part of any silicon compiler. They can also be used in full-custom designs in a semi-automatic fashion. To be useful, the module generators should provide compact layouts and should take advantage of the distinctive features of the target manufacturing technology. They should be functionally parameterizable and geometrically flexible for easy integration with other modules.

Common functions provided by generator libraries include arithmetic functions (adders, counters, arithmetic-logic units (ALUs), shifters, and data paths), memory storage functions (register banks, RAMs, and ROMs), and control logic functions (PLAs, sequencers, and random-logic synthesizers). They may include complex systems, such as the SDL-2000 microcontroller.

Silicon compilers use generators when the structural description of the system is known. Once the design has been partitioned into subsystems and their interconnections, each of the subsystems may correspond to a generator from the library or to a random-logic block to be automatically synthesized. The silicon compiler calls the generators from the library, passes parameters to them, and assembles the system by placement and routing procedures.

Input specifications to module generators take many forms, such as property lists, bit patterns, and net lists. An ALU is an example of a system block that may be specified by a property list. The list might contain entries for the number of bits, the geometric pitch, the Boolean and arithmetic functions to be performed, and the type of output buffering. Tables of bit patterns are used to specify PLAs and ROMs.

In the example shown in Figure 2.4, the user fills out a form to specify an ALU. The form has spreadsheet capabilities for displaying real-time estimates for size and speed of the block being generated. Here the user has specified a 4-bit ALU with shift function, input latch, output latch, and the tristate output option.

Once the form is filled, the program constructs a call to the library generator to provide the user with two types of descriptions:

Simulation models: functional models for fast simulation; transistor or gate model for accurate timing simulation and fault simulation

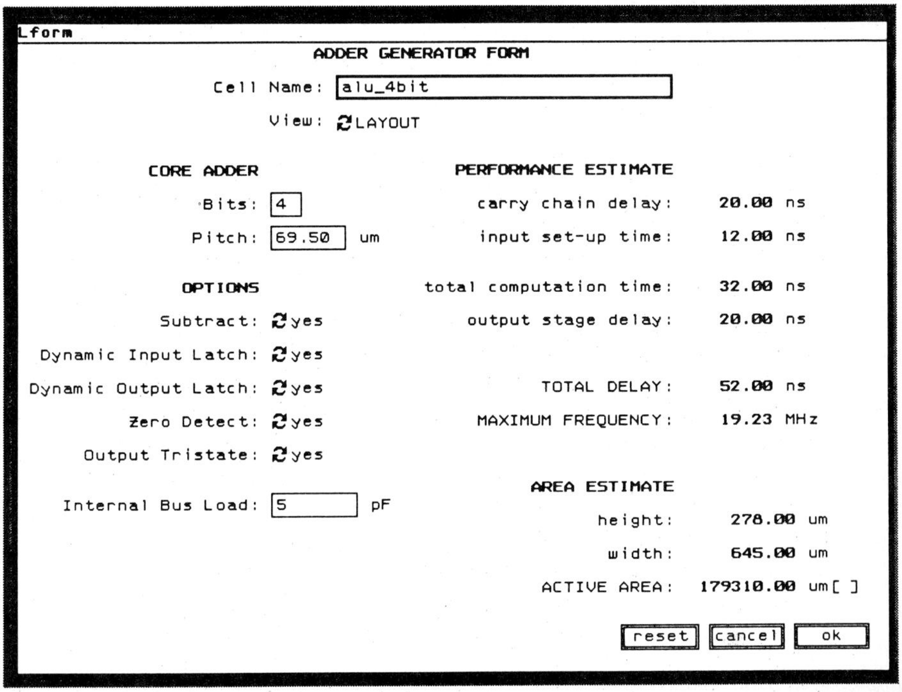

Lform

ADDER GENERATOR FORM

Cell Name: alu_4bit

View: LAYOUT

CORE ADDER

Bits: 4

Pitch: 69.50 um

OPTIONS

Subtract: yes

Dynamic Input Latch: yes

Dynamic Output Latch: yes

Zero Detect: yes

Output Tristate: yes

Internal Bus Load: 5 pF

PERFORMANCE ESTIMATE

carry chain delay: 20.00 ns

input set-up time: 12.00 ns

total computation time: 32.00 ns

output stage delay: 20.00 ns

TOTAL DELAY: 52.00 ns

MAXIMUM FREQUENCY: 19.23 MHz

AREA ESTIMATE

height: 278.00 um

width: 645.00 um

ACTIVE AREA: 179310.00 um[ ]

reset cancel ok

Figure 2.4 Form entry for an ALU.

Geometric models: bounding polygons and terminal positions for placement and routing; complete mask layout for final design verification and manufacturing

The layout of this ALU is shown in Figure 2.5.

A standard library of generators has to cover most commonly used system components. These can be divided into arithmetic functions (adder, counter, and ALU), memory functions (RAM, ROM, and latches), and control logic functions (PLA and standard cells with place-and-route package).

The standard cells in the generator library include inverter, tristate buffer, NAND, NOR, static latch, dynamic latch, latched multiplexers, OR-AND-INVERT, AND-OR-INVERT, exclusive-OR, and exclusive-NOR gates. The cells are used by the automatic placement-and-routing program in the synthesis of random-logic function blocks. Standard cells are either fixed cells or parameterized generators. If they are parameterized, they might include gates with a variable number of inputs, variable p-channel and n-channel sizes, and variable geometric pitch between power and ground buses. For example, the AND-OR-invert cell generator can produce 64 different logic gates. There can be one to four AND gates, whose outputs are connected to the NOR gate, and each AND gate can have one to four inputs. Figure 2.6 shows several standard cells produced by the generators.

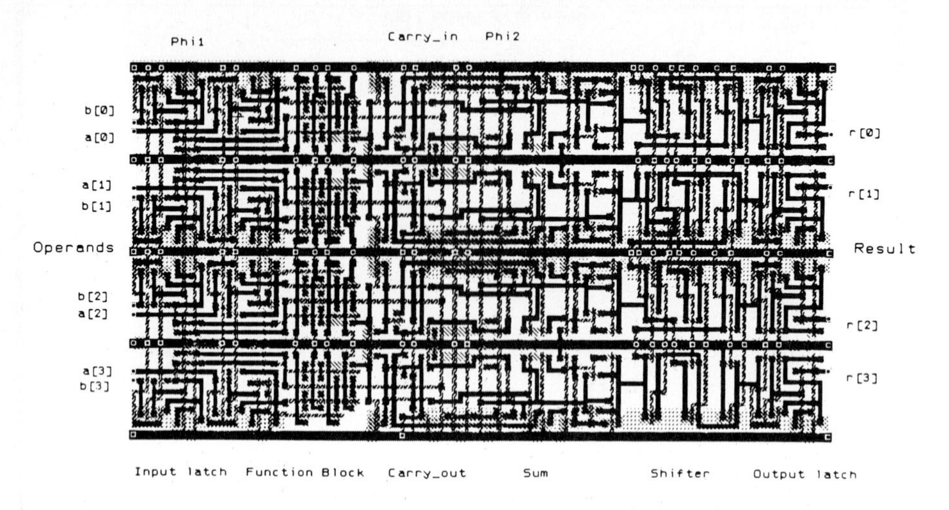

Figure 2.5 The ALU produced from the form entry.

ALU, adder, shifter, multiplier, and counter generators cover arithmetic functions found in digital designs. The function blocks can have high performance over a range of parameters if an expert designer develops specialized circuits for such a variable environment. Their layout can be very compact if an expert designer develops special leaf-level cells optimized for these functions.

Typically, an ALU generator produces a function block whose parameters include: the number of bits in the ALU, optional left/right shift function, optional registers at the ALU inputs, optional register at the ALU output, optional tristate buffer at the ALU output, and programmble geometric pitch between adjacent bits of the ALU. Programmable pitch is useful if it is desired to route ALU buses between the bits; thus, the number of buses can vary. The ALU performs add, subtract, bit-wise OR, bit-wise XOR, and bit-wise AND functions. If the shifter option is requested, the ALU can perform a shift in the same clock cycle in which an arithmetic operation is performed.

ROM, register file, and RAM generators produce memory storage functions, permanent and variable. They must be highly parameterized to provide flexibility to the designer using them. The parameters cover functional and geometric aspects of the circuits. Their circuit and layout design reflects the expertise of designers who developed them.

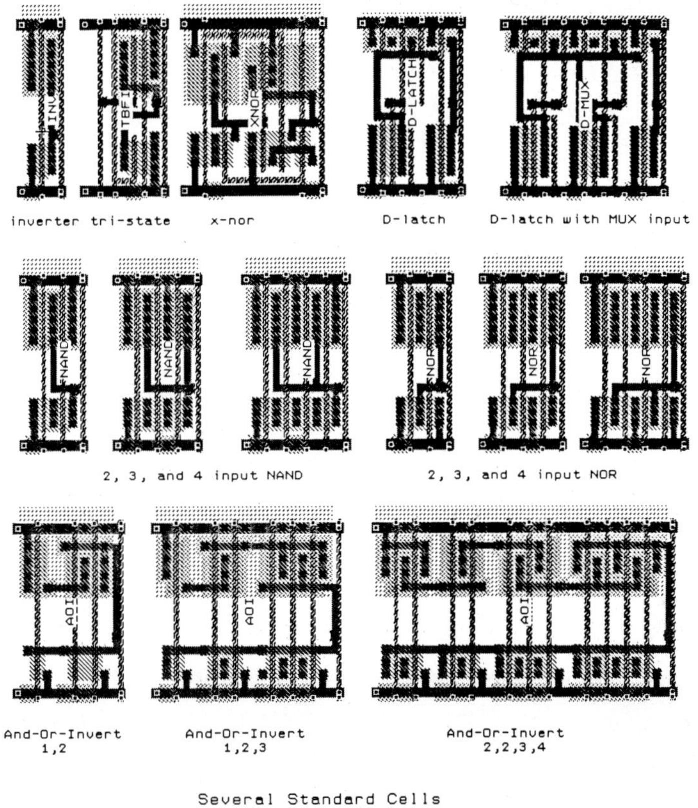

Figure 2.6 Several generated standard cells.

2.2.3 The Role of Standard-Cell Placement and Routing

Standard cells are relatively small circuits, typically simple gates. Their layout is designed so that they can fit together by abutment. Their height is uniform, and their terminals appear on the top and bottom. They are placed in rows, and the routing in one direction is accomplished in channels between the rows and in the other direction across the cells.

The versatility of standard cells stems from their small size. Placement and routing can be accomplished in such a way that different aspect ratios can be obtained for a given circuit. This simplifies the floor planning process, since standard-cell blocks can fit in regions between the rigid macro blocks.

Figure 2.7 shows an example of a CRT controller design. The layout consists of two data paths, obtained by module generation, and the control logic, obtained by standard-cell placement and routing.

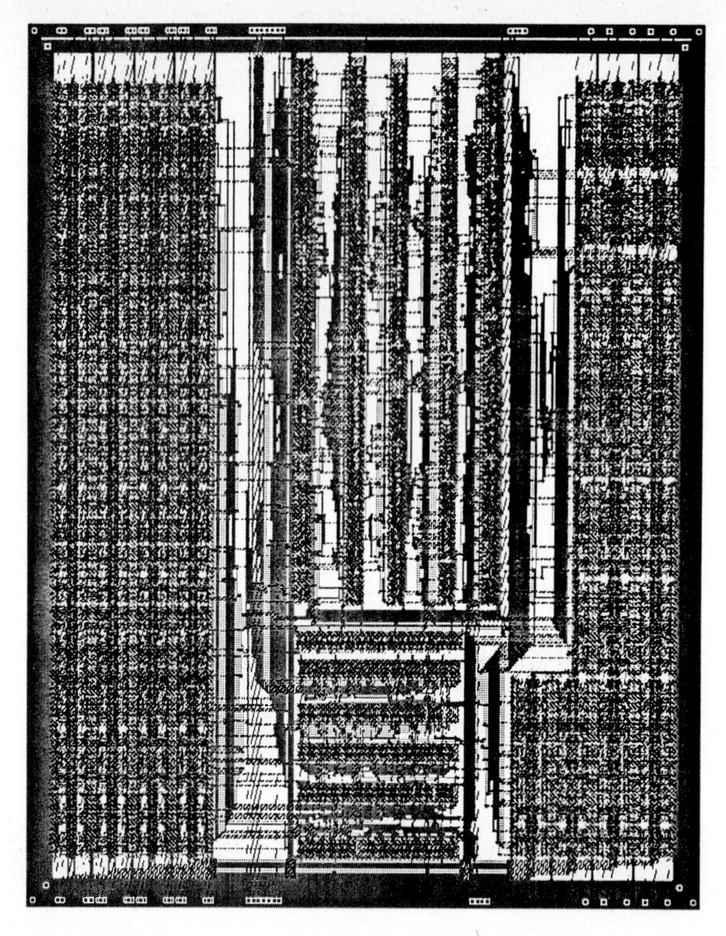

Figure 2.7 A CRT controller built out of data-path generators mixed with standard cell blocks.

2.2.4 Design with Libraries of Module Generators

A system designer can use libraries of module generators in an interactive environment consisting of the following components:

1. schematic capture
2. simulation
3. floor planning and placement and routing
4. library of parameterizable modules

For a complete chip design, the modules in the library must be supported by the following generators:

1. icon generator
2. functional simulation model

3. timing delay model
4. power consumption model
5. floor plan model generator
6. layout generator
7. detailed gate- and transistor-level simulation model

An icon generator produces a symbolic representation for the module to be used in schematic capture. The designer can quickly identify the function of a module by its icon in the schematic. The icon is parameterized and may change with the parameter changes entered by the designer in the schematic.

For example, the designer may develop a schematic as in Figure 2.8. Library components are accessed from the schematic editor and parameterized interactively. In this example the designer selected a ROM with 128 words by 8 bits, a RAM with 64 words by 8 bits, and 8-bit ALU, a PLA with 8 inputs and 20 outputs, and a block of random logic. The structure of the design is captured by interconnecting these blocks with buses.

The next step is to simulate the design. An interactive simulator with mixed-mode capabilities is especially useful for this application. Mixed-mode capability gives the designer the flexibility of using the same

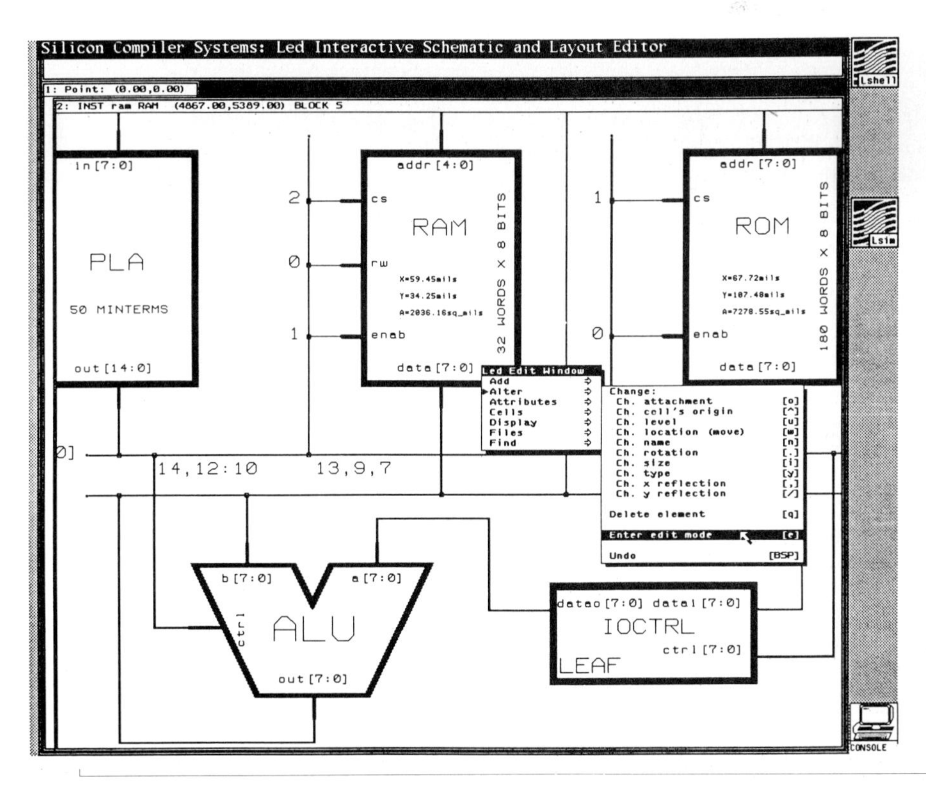

Figure 2.8 Schematic entry provides access to module generators.

simulator for functional simulation as well as for detailed gate- and transistor-level simulation. In fact, the designer may mix functional blocks with transistor-level blocks in the same simulation session. A functional simulation model exists in the library for every module. These models are used as the schematic is developed. Figure 2.9 shows an interactive simulation session during the schematic design.

Following simulation, it is necessary to experiment with floor plans. Floor planning is a process of positioning layout blocks for the best possible area utilization. The best results are obtained by combining automatic floor planning methods with interactive designer actions. The parameterized modules from the library contain floor plan model generators to assist the designer and the automatic tools in experimenting with floor plans. Figure 2.10 shows several alternatives for the design from Figure 2.8. A common method used to visually evaluate the floor plans consists of showing direct connections between the signal terminals on the blocks.

The next step in the design is to complete detailed placement and routing of the blocks based on the chosen floor plan. The floor planning step and the detailed placement and routing steps are usually applied iteratively until the best solution is found. It is important to use software tools that guarantee that this iteration will be as smooth as possible. The library of modules

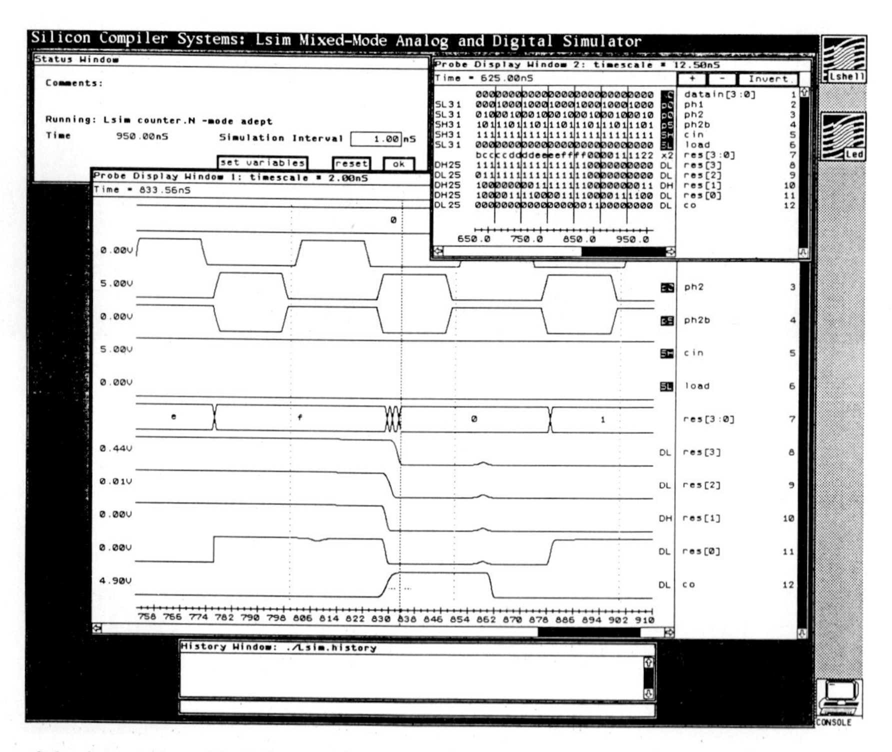

Figure 2.9 Interactive simulation environment.

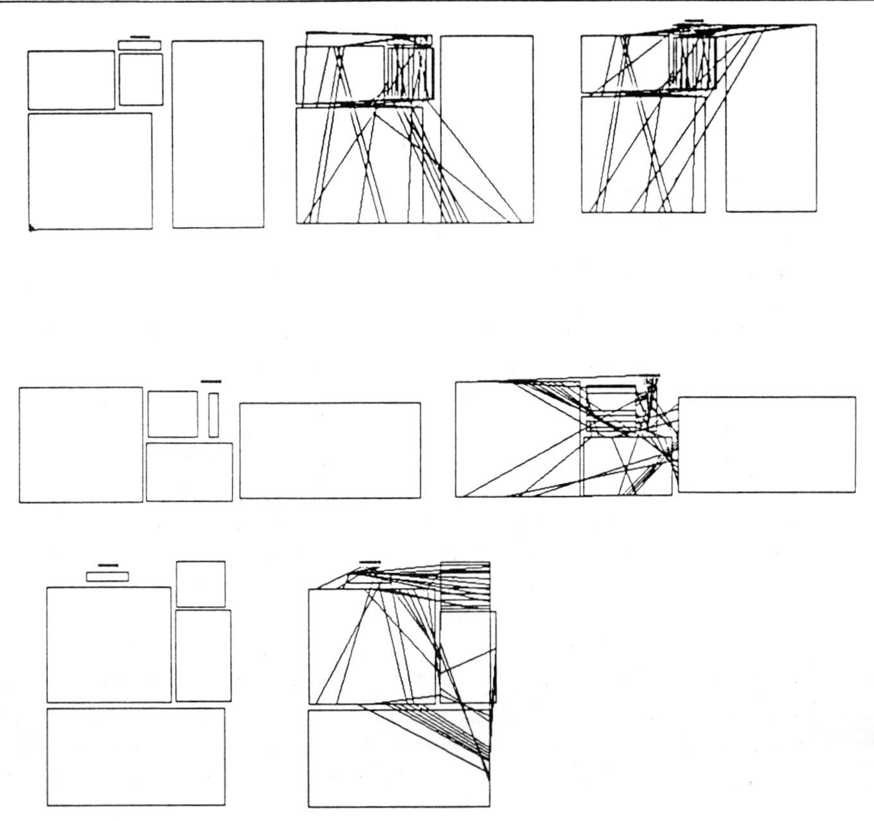

Figure 2.10

must provide layout generators and power consumption models for the final placement and routing. The power consumption models are used for sizing power and ground distribution lines on the chip. Figure 2.11 shows the detailed placement and routing for the design specified in the schematic from Figure 2.8.

2.3 Generator Development Tools and Their Use

The user of silicon compilers views the design process as top-down. The specification proceeds from functional to structural, and then from structural to geometric. With module generators, the geometric translation is fairly automatic.

However, in any design process there is a degree of bottom-up reasoning. Module generators offer the designer rather large primitive components for this phase. But what about primitives used by the designer of the module

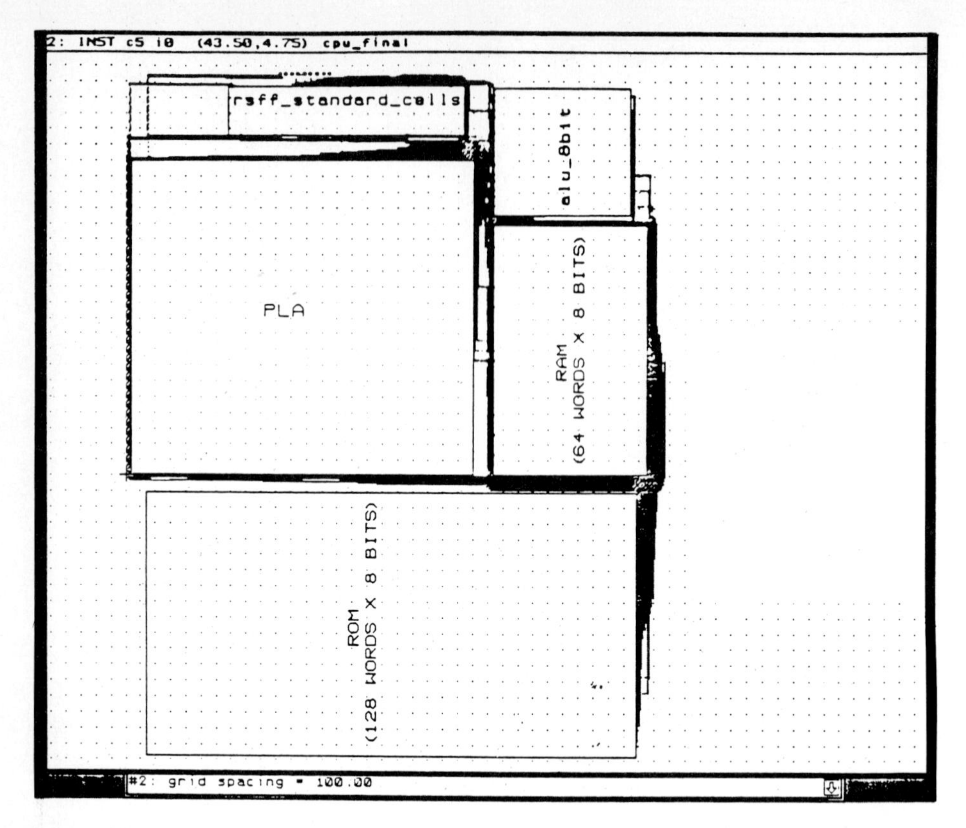

Figure 2.11 Floor-planning and block routing complete the physical synthesis.

generators, which might be as small as individual transistors? The tools for developing module generators have to take this into account.

The main challenge designers of generators face is creation of structural and layout generators. The difficulty arises because the nature of layouts is two-dimensional. Each component interacts with neighbors and has to satisfy manufacturing design rules. The layouts have to be dense, and they should be amenable to new design rules.

There have been several approaches to developing module generators. They are often written using general-purpose programming languages. Such generators simply output mask geometries either in geometric data formats, such as GDS-II or CIF, or in symbolic form, such as stick diagrams or character-symbolic representations. Some also output circuit net lists for simulation purposes. There are two distinct components in such an environment: the module generation component and the general-purpose design system. Module generation is an open-loop process; the programs don't have access to the design data base.

More powerful environments have been developed that use procedural languages to build module generators. These systems allow the program that produces the layout to be written in the same language as the description of the layout, improving the efficiency of the design capture process. Notable examples from the academic community are the Design Procedure Language (DPL), developed at MIT [9], and ALI and CLAY, developed at Princeton University [10].

The rest of this chapter will describe the Generator Development Tools (GDT) system from Silicon Compiler Systems (SCS). At its heart is a procedural VLSI design language called L. L can capture both the structural and the geometric aspects of a layout simultaneously. This allows a designer to cover two complete branches of the Y chart (described in Chapter 1) with one description language. The third, or functional, branch of the chart is covered by SCS's functional modeling language M, based on C, and the L-simulator. These capture the behavior of modules ranging from transistors to microcomputers.

2.3.1 The L Data Base

In contrast to other module generation systems, the L language for module generators and the L data base for design data constitute one uniform environment. During the IC design, the design data must be easily created, stored, accessed by application programs, and verified against different representations. The L data base (LDB) is a mechanism for satisfying these functions for all structural and geometric representations of a design. The L language is a procedural mechanism for building the design in the L data base. Libraries of L generator programs are stored hierarchically in UNIX file directories. Whereas L generators are general parameterized programs, LDB captures a particular design.

Figure 2.12 shows the interactions among the L program libraries, the L compiler, the L compiler's support programs (placement and routing and layout compaction), the interactive L-editor, and the L data base. The L compiler accesses L programs in the libraries as requested by procedures within the L programs. It compiles L programs into design data and stores these data in LDB. While compiling, the L compiler has access to LDB for retrieving information needed by L programs. It also has access to utility programs that perform specialized functions such as routing, layout compaction, and placement.

The L data base is implemented in such a way that very large designs can be handled easily and efficiently on modern workstations. The goal is to limit the size of the design only by the size of the disk attached to the workstation. The size of the main memory must not be a limitation to the efficiency of the design tools. In order to achieve this goal, LDB employs special handling of the design files at the top of the UNIX operating system.

LDB has a network structure. The design is represented by a hierarchy of *cells*. The cell types are schematic, layout, and icon. The scope of data is either *global* or *local*. Global information is visible to all parts of the design,

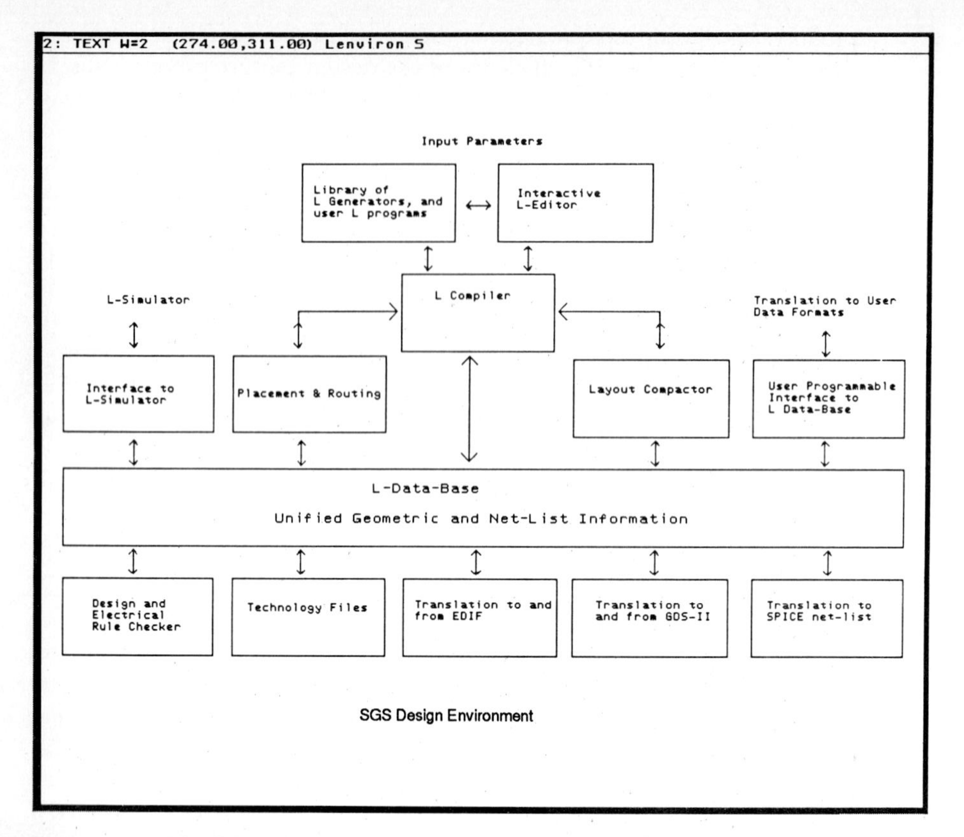

Figure 2.12 The GDT environment.

whereas local information is visible only within a cell. A cell is composed of the following atomic entities:

- cell properties and parameters
- transistors
- contacts
- terminals
- wires
- instances of other cells
- arrays of other cells
- polygons and rectangles
- properties attached to the atomic entities

Cell properties include cell type (icon, schematic, or layout), cell name, bounding polygon, generator parameters that produced the cell, file name of the generator, and technology name.

A transistor is a geometric and a circuit object. It is a four-terminal device. In MOS technologies it has a gate, source, drain, channel width and length. Geometrically, it consists of rectangles that define gate, source, drain, implants, well, etc. Every transistor in a cell has a name, and its electrical and geometric connection points are well-defined and also named. These names are taken from L programs to access the transistor and its connections.

Electrically, a contact is just a connection point for wires. Geometrically, it consists of rectangles that define contact regions, overlap regions of the layers being connected, and possible implant regions (buried contacts). Each contact has a name, which is used for accessing the contact from L programs.

Terminals are electrical entities without physical dimensions, on a certain layer (metal, polysilicon, diffusion, etc.). They are used to transfer signals into and out of cells to or from the next level of the hierarchy. Types of terminals are IN, OUT, INOUT, VDD, GND, and user-defined additional types. The types are used in electrical rule checking. In a schematic cell a terminal may be arrayed for bus connections. A terminal is named so it can be accessed from L programs.

Wires belong to a physical layer (polysilicon, metal, etc.). They may be named, they have width, and they always connect exactly two vertices. Wires are stretchable so that if one of the connection points moves, the wire remains attached. Wires can be attached to terminals, contacts, transistor terminals, instance terminals, array terminals, and other wires. In a schematic cell, a wire may be defined to be a bus.

An instance of a cell is a virtual copy of a cell, placed in a specific position and possibly reflected and rotated within the cell in which it is included. There may be many such copies of a cell in a design. If the original cell is changed, all copies will inherit the change.

The only connection points (electrical and geometric) to an instance are the terminal points declared in the original cell. An instance of a cell may be accessed from L programs by its name. In a schematic cell an instance may be arrayed so that its inputs and outputs form connections for buses.

Arrays of cells are similar to instances of cells in that they contain virtual copies of other cells. However, instead of only one copy of one cell, arrays contain many copies of one or more cells. These copies are arranged in a matrix, and the instances within the matrix are assumed to be connected to their neighbors.

Polygons are closed-plane figures with three or more vertices. They may be attached to other objects, in which case they are bound to them. If the object moves, its polygon maintains the relative placement. There is also electrical binding such that a polygon and its object share a common net identifier. Rectangles are special cases of polygons.

Properties are strings that can be declared and attached to objects (such as transistors, instances, and terminals). Properties are used for tagging objects with information that may be used by other L utility programs or user programs.

There are several built-in types of properties, but the user can define additional types. The built-in properties include SIG for naming signals, BUSCONN for specifying bus connections in schematics, VIEWNAME for displaying names of objects in the L-editor, TERMPLACE for specifying physical terminal locations for placement and routing, EQUIV for establishing equivalencies between schematic and layout terminals, SIMMODE for specifying simulation mode in the L-simulator, and TEXT as a general-purpose property.

2.3.2 Technology Files

Technology data for the L data base are obtained from a technology file. A technology file contains information specific to a particular manufacturing process. By abstracting technology information to a technology file, Generator Development Tools can be used across different technologies. The textual information from the technology files is stored as a special format in the L data base for fast access. Figure 2.13 shows a few of the technology file entries and their geometric interpretation.

Mask layer types are used to specify manufacturing details and other pseudolayers for schematic design. Each layer is defined by a name, such as POLY, NDIFF, PDIFF, and NWELL. A mask layer can also be a composite of other mask layers. Additional information attached to a layer definition includes minimum feature size, layer resistance, and layer capacitance.

Transistor types are also defined by the user in the technology file. A transistor entry describes the transistor type name, the general classification of the transistor (p-channel, n-channel, or depletion), the definition of the

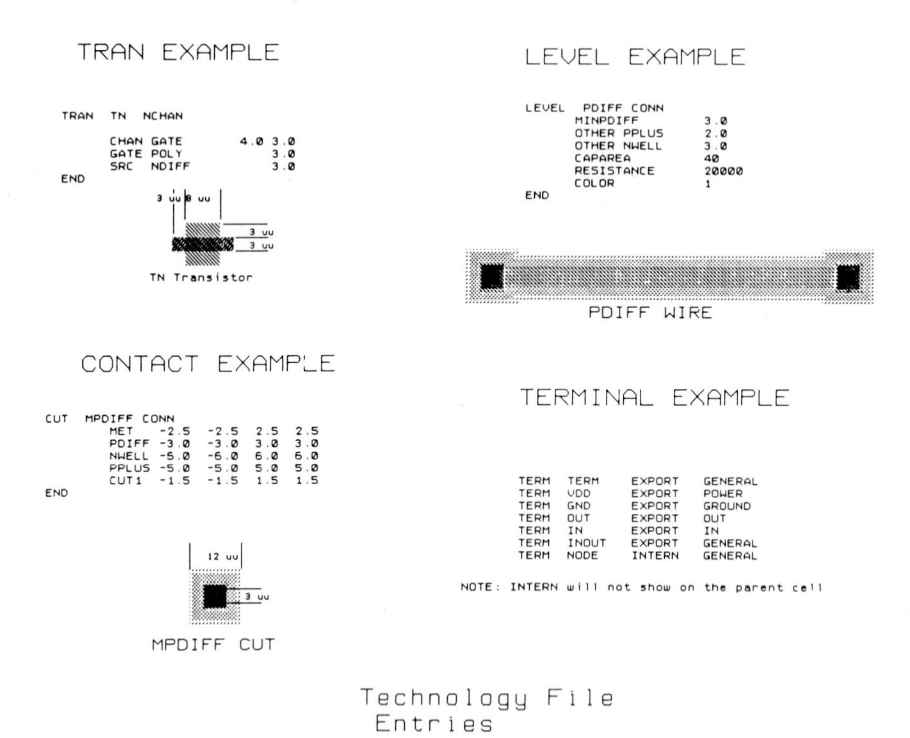

Figure 2.13 Some of the technology file entries.

channel, and layers making up the transistor. Channel length and width are parameters that are instantiated when the transistor is entered into the L data base.

Contacts are defined similarly in the technology file. Each contact type has a name and a corresponding collection of rectangles. Each contact serves the function of tying two objects that may be defined on different layers.

The design rules are an essential part of the technology file. They are needed in the L data base for the proper operation of the L compiler, routers, compaction, and net list computation. Some of the design rules are specified within layer definition, transistor definition, and contact definition. These include minimum feature sizes and overlap and overhang definitions.

The remaining design rules are defined as spacing rules. Since LDB contains connectivity information, it is possible to perform design rule checking using net list information. For example, two metal wires that intersect but have different net identifiers would be a design rule violation. If they have the same net identifier, then the intersection is allowed. With more traditional design tools, the intersection would not be considered illegal, but there would have to be a net list extraction followed by net list comparison against the schematic to verify the design.

2.3.3 L Language Description

L Basics

L programs reside in regular files. Basic units of design data are organized in cells. There may be any number of cell definitions per file.

The L compiler understands a sequence of files as if they constitute one large file. Each file is capable of including other L files. A large design with many L files may be initiated from a single L file.

L files are processed by L tools: Led, Lrc, and Lc. Each tool uses the L compiler as a front end. The L compiler interprets L cells and generators. It procedurally constructs LDB entries for variables and cells. Once the particular design is captured in LDB, the L compiler passes control to the particular tool that uses LDB for its application. Figure 2.12 describes the relationship.

Led is an interactive editor for examining, modifying, deleting, or adding new cells to LDB. Once an editing session is done, the contents of LDB are translated back to L language description or to a file with a special binary format. Lrc is a hierarchical electrical and design rule checker for L designs. Once it gets the design in LDB from the L compiler, it looks for potential violations of technology and electrical rules, such as shorts between electrical nets, incorrect or close mask features, and floating power supply terminals. Lc is a program that checks syntax of L files and translates LDB entries to L-simulator and other industry-standard data formats such as GDS-II and Spice.

L syntax is modeled after the C language. All declaration and expression statements are terminated by a semicolon (;). Space characters, tabs, new-line characters, and carriage return characters are ignored since they represent keyword delimiters. Comments start with a "sharp" (#) and end with a new-line character. Everything in between is ignored.

The symbols { and } are used to mark a block of L statements. Such blocks may be nested. { and } are used for cell specification and for WHILE, IF, and ELSE statements.

There are two types of keywords in L. The first group includes technology-independent keywords for functions, operators, and declarations. Some examples are IF, ELSE, WHILE, NUM, INT, GLOB, CALL, CELL, INST, WIRE, FOPENR, LEFT, and UP. The second group of keywords is defined by the user in the technology file. Each technology has its own set of naming conventions, so L supports technology-specific keywords. For example, MET, POLY, NDIFF, PDIFF, MET2, NWELL, PWELL are layer names for metal, polysilicon, n-diffusion, p-diffusion, second-layer metal, n-well, and p-well in CMOS technology. Typical transistor types are TN, TP, TD for n-channel, p-channel, and depletion transistors. Typical contact names are MPOLY, MNDIFF, MPDIFF, MNSUB, MPSUB, M1M2 for metal to poly, metal to n-diffusion, metal to p-diffusion, metal to n-substrate, metal to p-substrate, and metal layer 1 to metal layer 2.

Design spacing rules also have names. This feature allows L programs to compute physical distances based on symbolic expressions. The names of design rules are derived from the names of mask layers. For example, MET_MET, POLY_NDIFF, NDIFF_NWELL are used for rules governing layer spacing from metal to metal, poly to n-diffusion, and n-diffusion to n-well.

Names of Objects in L

All objects in L have names, including variables, cells, transistors, contacts, terminals, instances of cells, and arrays. The reason for this is that objects are related to each other geometrically and electrically, so the user should be able to express these relationships in L statements by referring to objects by their names.

Each object is assigned a name in a declaration statement. For example, a transistor might be declared by:

```
TP ptran W=expression ;
```

Here, a p-channel transistor is declared in a cell and given the name ptran. This name can used from this point on in the cell. Its channel width is defined by an arithmetic expression.

The scope of object names can be global or local. Global names are visible throughout a design in LDB both inside and outside of the cells. Objects that appear in the global name space are technology data, numerical variables, string variables, and cells. Global variables are used to communicate information between various parts of the design in the data base. One L procedure can set the values of some global variables, and another L procedure can read them and make decisions based on their values. Local names are visible within a cell. These are local numerical and string variables, transistors, contacts, terminals, instances of other cells, and arrays of other cells.

Variables and Expressions

Numerical variables hold numbers. Numbers can be either integers or floating-point numbers, but in LDB they are all kept in the same form.

String variables hold ASCII strings of arbitrary length. They are used to hold and communicate textual information among cells in the design hierarchy. Strings are supported by several operators and functions. Variables can be global or local.

The values assigned to global variables are visible in all cells and can be changed in all cells. They provide one way of passing information among cells. Their names are visible not only across the cells but also across the L files.

Arithmetic expressions in L follow the same rule as expressions in C. The operands are constant numbers, variables, geometric attributes of L objects (x and y coordinates, width, and length), and other arithmetic and logic expressions. The expressions always return a numerical value. The operators are the same as in C and use same precedence rules.

Logic expressions are also the same as in C; they return either 0 or 1. Logic expressions are used in conditional IF statements and looping WHILE statements. The logic operators are the same as in C.

Control Statements

One of the most important requirements for module generators is that the resulting modules be parameterized. This requires that the generator make decisions on how the circuit or layout should be created. IF and ELSE provide conditional execution of groups of statements. For example, an output transistor size may be determined by the value of the load it is driving:

```
IF ( load < 5 )
  TN driver W = 20;
ELSE
  TN driver W = 40;
```

In this example if the load is less than 5 pF, the channel width of the n transistor is chosen to be 20 microns; otherwise it is set to 40 microns.

Another important requirement in module generators is that certain operations be repeated a variable number of times. This is accomplished with the WHILE statement. In the following example, a group of n-channel transistors is declared within a WHILE statement:

```
NUM i = 1;
WHILE ( i < = 10 ) {
  TN tran[i];
  i+ +;
}
```

Declaring Objects

A typical declaration of an object (transistor, contact, cell instance, or array of cells) is:

```
OBJ_TYPE obj_name [obj_size] [orientation] [place_option];
```

OBJ_TYPE is one of the keywords defined in the technology file (transistors and contacts) or built into the L language (INST, ARRAY). obj_name is the name given to the object. All other components of the declaration are optional. Transistors are the only objects with optional size settings. The orientation option allows the object to be rotated or reflected. The orientation opinions are the same for transistors, contacts, instances of cells, and arrays of cells. L supports only 90°counterclockwise rotations. All objects rotate about their center. They may also be reflected about the *x* axis, the *y* axis, or both.

Location of the object is the optional part of the declaration. It can be declared with an AT statement such as

```
AT (expr1, expr2);
AT name;
AT name + (expr1, expr2);
AT name – (expr1, expr2);
```

Object placement is either absolute, as in the first statement, or relative with respect to some other object, as in the remaining three statements. If the AT location part of the declaration is omitted, the object is placed at (0,0), but it can be positioned later in the L program with other placement statements.

DECLARING CELLS Cells are the basic building blocks in L. They contain collections of transistors, contacts, terminals, instances and arrays of other cells, and interconnecting wires. Cells may be layout or schematic. Schematic cells obey the same rules as other cell types, except that the information related to physical layers (metal, poly, etc.) can be omitted from declarations.

If arguments are present, the cell is a generator. Arguments may be numerical or string variables. Numerical variables are specified simply by name or by NUM preceding the argument name. String arguments are declared by STR preceding the argument name. The L compiler checks the type of arguments in the CALL statement used to invoke the generator. When a generator is called, these will be initialized to numerical values or string values. Arguments to generators have the same properties as local variables.

Generator calls are performed by the following statement:

```
CALL gen_name CELL cell_name( arg1 , arg2, . . . .);
```

gen_name is the name of the generator file being called. Call arguments are arbitrary expressions or strings. The result of the call is a cell named cell_name. gen_name is a generator that resides either in the directory on the top of the stack of directories or in the public directory of generators.

A generator may call another generator, as in the following example:

```
CALL addergen adderx (bits,pitch);
```

This statement will call the generator whose name is addergen and will create the cell whose name is adderx. The arguments, bits and pitch, specify how many bits the adder should have and the distance (pitch) between adjacent bits. The generator addergen might look as follows:

```
CELL addergen( bits , pitch )
{
  NUM i = 0;
  CALL adderbit CELL bit( pitch );
  WHILE( ++i<=bits ) {
    INST bit bit[i];
    IF( i > 1 )
      bit[i].vdd AT bit[i-1].vdd;
  }
}
```

The adder generator calls the adder bit generator and assembles the bits into a complete adder.

DECLARING TRANSISTORS L directly supports rectangular transistors. These are formed by the overlap of two or more layers of material. The layers typically include polysilicon and diffusion, but there is nothing that precludes other materials and transistor types. Transistors are defined in the supporting technology file. The formats to declare a transistor are:

```
TRAN tran_name [size] [orientation] [place];
```

TRAN represents one of the keywords defined in the technology file. An NMOS technology file might contain an enhancement-type transistor called T and a depletion-type transistor called TD. A CMOS technology file might contain an n-type transistor called TN and a p-type transistor called TP. Other technology files might contain other transistor types and use other keywords. When a statement starts with a keyword defining a transistor, the statement declares a transistor.

The user provides a name for each transistor (tran_name). After a transistor has been declared, it can be referenced only by its name.

The size option specifies the transistor size. The size keywords

```
W=expr1
L=expr2
```

are used to specify the gate width (W) and gate length (l). Expressions expr1 expr2 can be any legal arithmetic expression. A transistor is defined to be symmetrical about its origin at the center of the gate region. Extensions of polysilicon and diffusion layers beyond the gate region are specified in the technology file. The technology file sets the minimum size so that the size option can be omitted.

A transistor has five coordinates that are known to the L compiler. The center, or origin, of the transistor is associated with its name. The other four coordinates are vertices where wires may connect. These vertices are called:

.d	drain located on the top (also .dn)
.gr	gate located on the right
.gl	gate located on the left
.s	source located on the bottom (also .dd)

Figure 2.14 shows an n-channel transistor named tran and its terminals.

DECLARING CONTACTS Contacts are used to change connectivity from one material layer to another. Like transistors, the contacts are defined in the technology file. The formats to declare a contact are:

```
CONTACT con_name [orientation] [place];
```

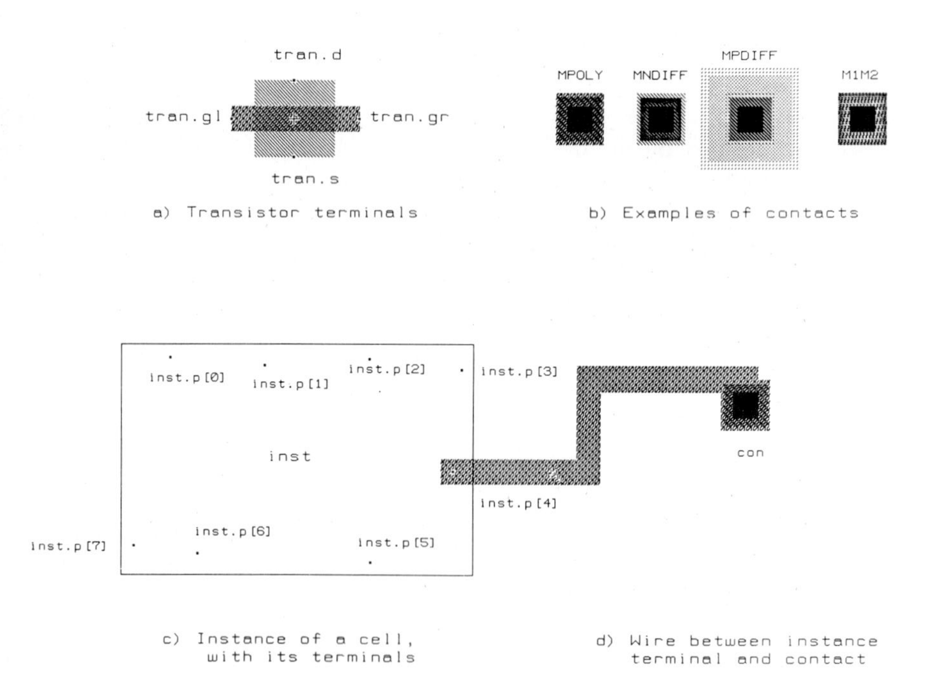

Figure 2.14 L declarations and their physical interpretations.

CONTACT stands for one of the keywords defined in the technology file. A COMS technology might contains the following types of contacts:

M1M2	metal-1 to metal-2
MNDIFF	metal-1 to n-type diffusion
MPDIFF	metal-1 to p-type diffusion
MPOLY	metal-1 to polysilicon
MNWELL	metal-1 to n-well
MPSUB	metal-1 to p-substrate

The orientation options are the same as described before. Most contacts are symmetrical so that rotation and reflection have no effect. However, some are assymmetrical, particularly the buried and butting contacts.

Contacts have only one vertex where wires can be connected. This vertex and the contact origin are the same coordinate. Figure 2.14b shows several contacts.

DECLARING TERMINALS Terminals are abstract circuit objects that are supported by L to help manage complexity. Their main purpose is to transport signals into and out of cells. Since cells are grouped together to make larger cells of greater functionality, one would like to hide as much of the detail of the smaller cells as possible. When an instance of a cell is used, the only places where wires can be connected are terminals.

One can think of hierarchical cells as layers of information. Each layer is connected to cells in a lower layer through terminals. A terminal can carry a signal only from one layer to the next. As cells are grouped into a new cell, all of the terminals that are needed externally must be propagated. It is not unusual for many layers of terminals to have the same coordinate. This is particularly true for clock, power, and ground terminals.

The formats for terminal declarations are:

```
TERM LEVEL term_name [location];
```

TERM stands for one of the terminal types defined in the technology file. There are three functionally different types of terminals: ground (GND), power (VDD), and signal (IN, OUT, INOUT). The VDD and GND terminals are special. The L design rule checker will check for shorts between these terminal types. It will also verify that none of these terminals are left unconnected. Once a cell has declared VDD or GND terminals, those terminals must be connected to other VDD or GND terminals within the new cell. In this way, the L rule checker ensures that all cells are connected to power and ground.

DECLARING INSTANCES OF OTHER CELLS Instances provide the mechanism for building the hierarchy because they allow cells to be used by other cells. An instance of a cell is not a copy of the cell, but rather a minimal description of what is needed to use the cell. Hence, it can be called a virtual copy of a cell, the only information needed for an instance of a cell is its location, reflection, rotation, and net information for its terminals.

The formats for an instance declaration are:

```
INST cell_name inst_name [orientation] [place];
```

INST is the keyword for an instance declaration. The cell_name is the name of the cell being using as an instance. The inst_name is the name given to the instance.

The names for the terminals of the instance are formed automatically by joining with a period (.) the inst_name and the terminal name from the cell:

```
inst_name.terminal_name.
```

Figure 2.14c shows an instance of a cell. Typically, only the bounding box and the terminals of the instance are shown, unless the user needs to look inside the instance.

DECLARING WIRES Wires in L refer to both geometry and connectivity. This is an important concept. There are checks in L that prevent errors due to unintentional connections or material overlaps. L will not assume that overlapping materials are connected or related unless there are statements that specify connection. Wire statements are needed whenever interconnection is needed between terminals, contact, transistors, and instance and array terminals.

Each wire connects exactly two vertices. A wire statement describes a path from the begin vertex to the end vertex. The path is a geometric description consisting of a number of segments. Each segment is either horizontal or vertical. The length of each segment can be set by an expression. The last two segments stretch to reach the end vertex. This allows objects to move and still maintain connectivity.

The formats for a wire statement are:

```
[WIRE] [level] [W=expr] begin_name path end_name;
```

The starting keyword for a wire statement can be WIRE, one of the LEVEL keywords defined in the technology file, or W=expr.

The L compiler determines the proper material level for the wire if possible. Usually the levels of the start vertex and the end vertex are such that only one level can be used. One can use the LEVEL keyword if the choice is ambiguous or to help with documentation.

One can set the physical width of a wire with W=expr. If it not specified or if the specified width is too small, the L compiler will use the minimum width from the technology file. The wire material is placed symmetrically around the segment, extending one-half the width all around the path.

The path is formed by a combination of one or more of the following:

```
RIGHT=expr
LEFT=expr
UP=expr
DOWN=expr
```

```
HOR=expr
VER=expr
```

Each keyword specifies the segment direction, and the value of expr is the length of the segment. If the path does not reach the end vertex, the last two segments will stretch to reach it. If the keywords HOR and VER are used for horizontal and vertical segments, the L compiler will automatically determine the direction needed to reach the end vertex.

Geometric Operators

Dot geometric operators provide information about an object. They are stated in the form:

```
object_name.operator
```

where name is the name of an object and operator is an x coordinate, a y coordinate, or the width or length of an object. These operators are used in two situations when the position of the object is known and when the position is unknown.

When an object's position is known, object_name.operator returns a numerical value. In these cases, dot geometric operators can be used in arithmetic and logical expressions. For example,

```
xxx=55+mdcontact.X
```

will set the variable xxx to the sum of 55 and the x coordinate of the object named mdcontact. When the position is unknown, dot operators may be used to specify the position of an object, its width, or its length using the statement

```
object_name.operator=expr;
```

Object Placement

One of the critical features of module generators is their ability to build two-dimensional layouts without knowing exact physical coordinates of subcomponents. All placements are relative to other objects. This makes layouts "elastic"; if some components change size, the others will adjust their locations.

The first group of object placement statements involves the AT operator. Its function is to specify the position for an object, but it also is loaded with certain side effects. The formats for using the AT operator are as follows:

```
object AT other_object;
object AT (expr,expr);
object AT other_object+(expr,expr);
object AT other_object-(expr,expr);
```

If an object is placed AT other_object, there are side effects. If other_object uses the same level as the object being placed, a wire on that level will automatically be added between the two objects. If the two objects are terminals of instances of cells, *all* other matching terminals of the two instances will be wired together on appropriate levels. (*Matching* means that the position and level are exactly the same.) All other AT statements place an object without automatic wiring.

The format +(expr,expr) or −(expr, expr) means that the coordinates of the object are the same as those of other_object, but modified by an (x,y) offset. Wire is added in this case also. It makes a horizontal jog first, followed by a vertical jog.

An example is shown in Figure 2.15a. The L program for the example is:

```
CELL abc()
{
  INST some_cell i0 ;
  INST some_cell i1 RX ;
  i1.p[0] AT i0.p[0];
}
```

The first statement declares an instance of cell some_cell. The second statement declares an instance of the same cell, but reflected around the x axis. Finally, the second instance is placed by the AT statement. Its terminal p[0] is placed at the terminal p[0] of the first instance. All terminals of the two instances that match (level and place) will get wires.

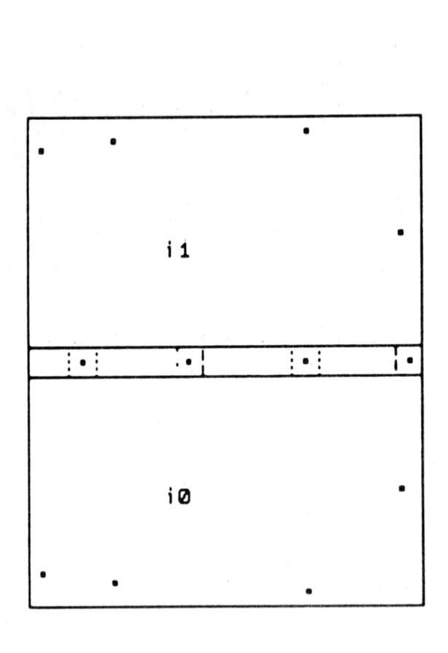

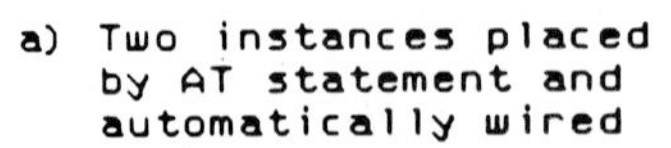

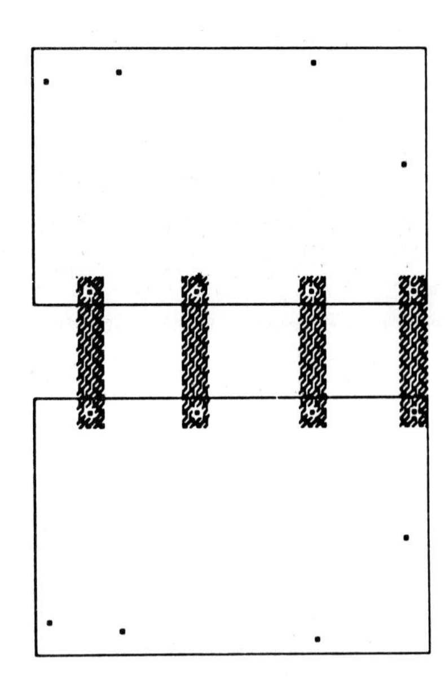

b) Instance i1 moved
UP by 14 microns

Figure 2.15 Relative placement of cell instances and their connectivity.

Another group of placement statements involves the MOVE function. MOVE repositions objects to a new location. This function is useful when two instances of cells are wired together by abutment and the need arises to create a wiring channel between them, as in Figure 2.15b. The move is accomplished by

```
MOVE i1 UP BY 14;
```

All wires that were created by the AT statement stretch to preserve connectivity between the instances.

The MOVE statement relocates an object according to the direction given (LEFT, RIGHT, UP, DOWN) by the value of the expression given. As an example, a description of an inverter and the corresponding state of the L data base is shown in Figure 2.16.

ROUTE Functions

L includes routing functions as part of the language. The L compiler prepares the routing specification from the description written with L statements and then makes a call to a separate routing program. There are four interconnection routines that can be invoked by the routing program through the L parser. The first two routines will wire objects within a specified routing channel. The rule-based greedy channel router generates two level connections between fixed circuit objects on either side of a channel and movable objects on either end of the channel. The river router handles only simple connections between objects and does not use side exits from the channel. Almost any routing problem can be decomposed into a series of simpler channel and river routes. This capability is exploited by the third type of router, a global router, which decomposes the region between the cells of the objects presented in the net list into routing channels. The global router automatically places the contacts to bridge routing channels during a loose routing phase and invokes the channel router for detailed wiring. The fourth type of router is the power and ground router. It is invoked through the global router and automatically sizes the power and ground buses connecting power and ground terminals.

The word *channel* as used in this section means an open area between circuit objects where wires and contacts can be freely placed. There are channels for horizontal routes and channels for vertical routes. A vertical channel route is used when one group of objects is located above another group of objects. The designer provides a list of NET statements that describe how the objects are connected together. The router assumes that wires can interconnect objects in a vertical direction without producing design rule violations. The router tries to connect all of the objects together according to the net specifications using a minimum of horizontal wire tracks.

On the other hand, a horizontal route is used when there are two groups of objects and one group is to the left of the other group. The router assumes that wires can connect objects in a horizontal direction without

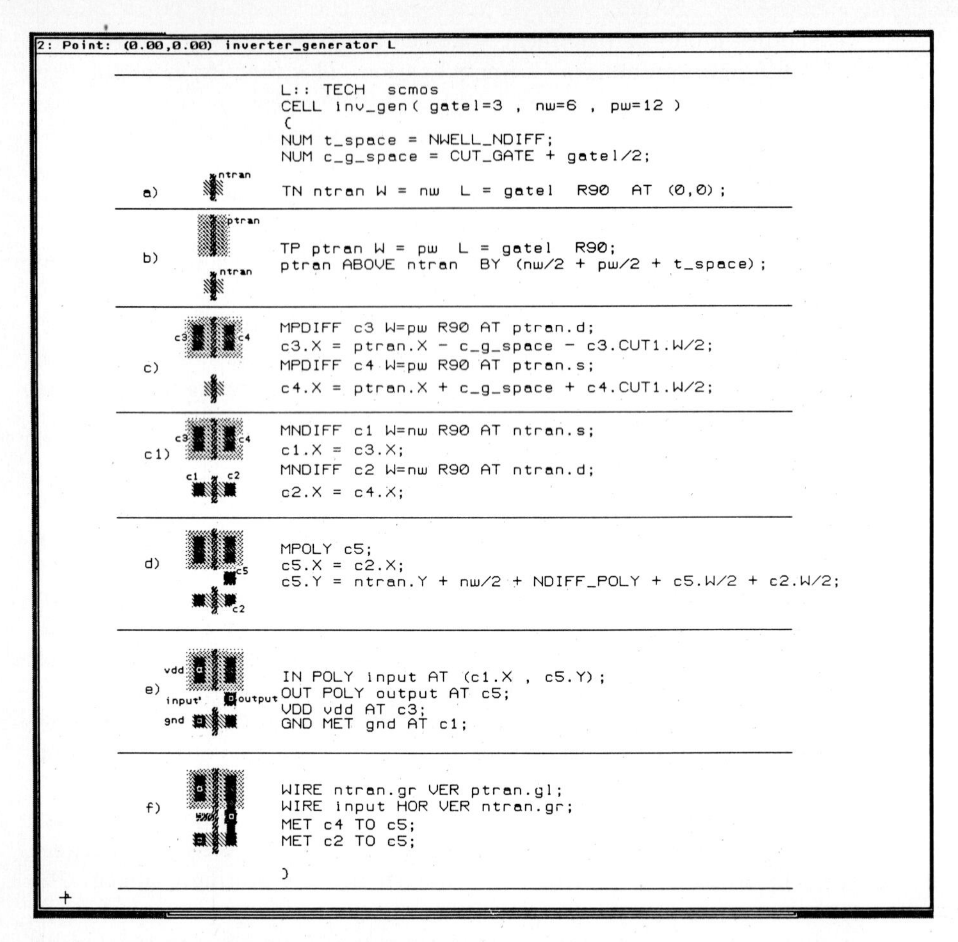

Figure 2.16 A complete step-by-step L generator. Shown are L statements and corresponding state of the L data base.

producing design rule violations. The router tries to connect the objects together using a minimum of vertical wire tracks.

Before starting a route, all of the circuit objects on either side of the channel must be positioned. The router will automatically move the objects apart if the channel is not wide enough. A vertical route can be specified with the following statement:

```
ROUTE VER (expr, expr) (expr,expr)
  VER LEVEL (v_pitch)
  HOR LEVEL (h_pitch);
  MOVE direction;
  LEFT EXIT name_l1, name_l2, . . . ;

  RIGHT EXIT name_r1, name_r2, . . . ;
  MOVE direction;
```

```
    NET name_n1a, name_n1b, . . . ;
    NET name_n2a, name_n2b, . . . ;
  REND
```

The keyword for a vertical route is ROUTE VER. The two (expr,expr) coordinates specify opposite corners of the rectangular routing channel. Objects (e.g., wires, contacts, nodes) that are generated by the router will be placed only within the top and bottom channel boundaries.

After the ROUTE statement, the statement starting with the VER keyword specifies the material level for the vertical wires. LEVEL stands for one of the levels in the technology file. VER POLY means that the vertical wires will be on the polysilicon level. The minimum space between the centers of adjacent vertical wires is specified by the value in parentheses (v_pitch). Likewise, the HOR MET keyword specifies that the horizontal wires will be on the metal level. The minimum space between the centers of adjacent horizontal wires is specified by h_pitch. If the wires between the top objects and the bottom objects do not cross, the horizontal and vertical wires can be on the same level. This kind of a route is known as a "river route," which is a sample case of the more general channel route.

The LEFT EXIT keyword is used to list the circuit objects to be placed to the left of the channel. The RIGHT EXIT keyword lists the circuit objects to be placed on the right side of the channel.

The NET statements are used to specify the objects to be connected together. All of the objects listed after the NET keyword and before the semicolon will be connected together. The user can use any of the L computation or flow control statements to construct the net lists. For example, the WHILE statement is useful for making bus nets where the terminals have indexed names.

The keyword REND is used to end the specification for the route. The objects are repositioned and the wires and contacts are added after the REND keyword is processed. All of this is finished before the next L statement is executed.

The following example shows a routing statement within the cell. Figure 2.17 shows the resulting routing.

```
CELL channelroute ()
{
  INT i;
  INT num;

  num = 10;
  CALL routecell CELL route(num);
  INST route up;
  INST route dn ;
  dn.top[1] AT (-20,-40);
  up.bot[1] ABOVE BY 40 dn.top[1];

  ROUTE VER (dn.URX,dn.URY) (up.LLX, up.LLY)
```

```
    VER POLY HOR MET;
    i=0;
    WHILE (++i<=num) {
    IF (i<=4)
    NET dn.top[i], up.bot[6+i];
    IF(i>4)
    NET dn.top[i], up.bot[i-4];
    }
    REND
}
```

First a cell called route is created from the generator routecell. Cells up and dn are then declared as instances of the cell route. Cell up is placed above cell dn. The routing statement calls for a vertical route with the channel corners placed relative to the corners of the instances. The wiring is in the metal and polysilicon layers. If the seed channel is too small, all of the objects above the channel are to be moved up—the default direction. NET statements specify the connections between the up and dn cells. The

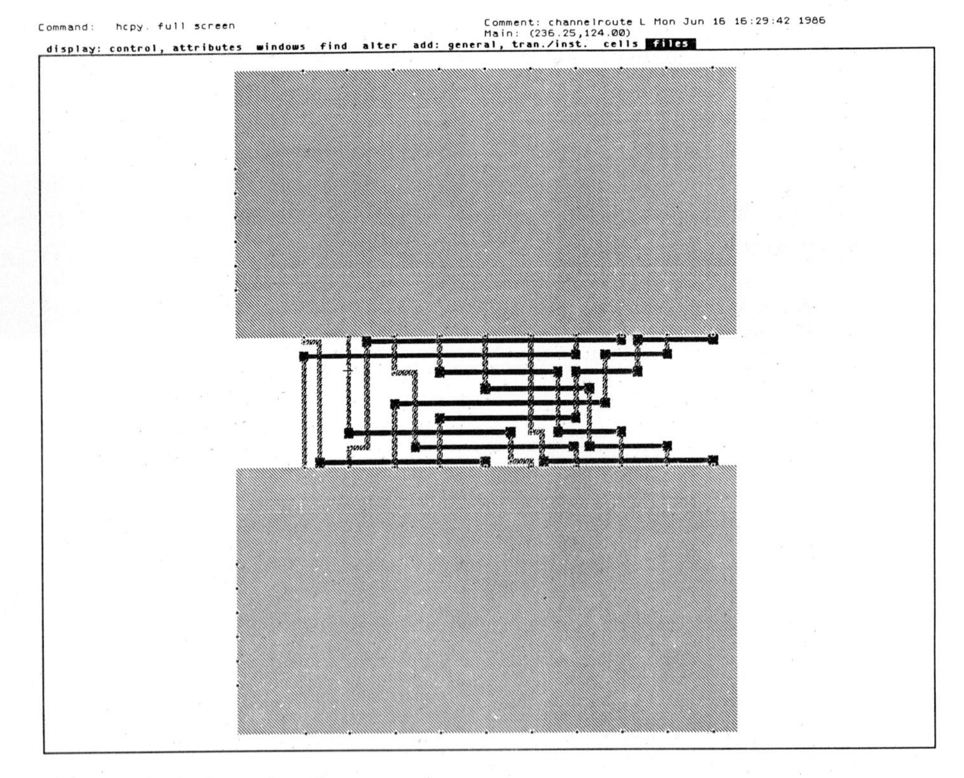

Figure 2.17 A channel routing example

routing is performed upon processing of the REND statement, and the up cell is moved up to make room for the wiring.

Global routing is the interconnection of objects whose signals span more than one routing channel. In effect, the global router must automate those design tasks necessary to perform multichannel routing. This includes accommodating power and ground wiring. The global router invoked by the L parser automatically decomposes the information derived from the routing specification block into five interconnection tasks. Driven by the net list specification, the global router will:

1. partition the area between cells into routing channels,
2. assign nets to channels,
3. order the channels for routing,
4. perform the detailed routing of channels, and
5. move objects (and cells) in a consistent manner if there is insufficient space to complete routing.

Exit contacts are automatically generated to bridge horizontal and vertical channels for signals that span more than one channel.

The area between routable objects is partitioned into T-shaped routing channels through a slicing algorithm. The algorithm uses bounding box and object placement information to derive the height and width of the channels. Objects are assigned to the closest available channel (shortest distance). To preserve the channel order for routing, the base (or trunk) channel of the T is routed before the crosspiece. By routing the trunk channels in the slicing hierarchy first, the exit contacts become fixed objects for routing in the adjacent crosspiece channels.

The greedy channel router is invoked to perform the detailed wiring. If the channel is undersized, repositioning of objects is then propagated through the hierarchy of slices.

Power and ground routing are invoked using the same NET keyword used for signal routing. The properties of the object, as defined in L, will alert the router that the object is part of a power or ground net. The net's contribution to the overall bus width is a function of the width of the wires connected to the object prior to the routing statements. If no prewiring is done, the width of a cut is used. The width of the bus will never exceed ten times the width of a contact. In constructing the power bus tree, a branch will contribute a default value of 75 percent of its width to the trunk. This value can be modified by declaring a global or local variable called BRANCH_FACTOR, which must be a positive integer. A value of 60 indicates that the branch will contribute 60 percent of its width to the trunk.

Standard-Cell Placement and Routing

The Lpar package in GDT generates a standard-cell layout directly (without manual intervention) from a schematic drawing by translating a circuit logic description into a standard-cell layout with the specified aspect ratio and terminal positions. The aspect ratio is controlled by defining the number of rows. The location of terminals can be specified with varying degrees of accuracy by entering the side, group of sides, or position within a side for each terminal. Lpar also provides a special facility for handling clock signals.

A standard cell is assumed to have two power and ground terminals located at its corners. The Lpar package uses the power and ground terminals as cell abutment points. A cell provides one equivalent physical terminal on the top and bottom for each logical terminal.

Terminal positioning assists Lpar in producing a better layout and minimizes the wiring required to interface with the layout Lpar created. The layout created by Lpar is interfaced to the circuit through channel routing. Unless otherwise specified, Lpar will not place all terminals where they were initially indicated. The Lpar package places terminals where the best compromise between external interface overhead and internal area requirements is achieved. If direct abutment rather than routing is used to interface to the circuit, Lpar can position top and bottom terminals exactly as specified.

Terminal positions can be defined with different degrees of accuracy by specifying the side (top, bottom, left, right), vertical (top, bottom), or horizontal (left, right) positions. When a side has been specified, the designer can further define the position within the side by specifying a single position or a range of potential positions for the terminal. The absolute position and range for top and bottom terminals represent the offset from the left side. The absolute position and range for left and right terminals represent row numbers (counting from 1). When a single position is defined for a vertical terminal, Lpar can place the terminal exactly where specified if a top channel (or bottom channel, as appropriate) is also defined.

When a top (bottom) channel is requested, Lpar places the top (bottom) terminals at the exact positions requested. Terminals for which only one position is specified are placed as close as possible to their requested positions. Finally, the terminals specified with a range of positions are placed.

The output produced by Lpar is an L file that contains the instance, net, terminal, and exit statements relevant to the circuit layout.

2.4 Methodology for Developing Module Generators

The development of a module generator is similar to the design of a custom IC. However, there is the additional goal of developing a parameterized and technology-adaptable function block as opposed to a fixed cell layout. The following steps must be performed by the designer.

- Design the architecture and circuit for the module.
- Develop the floor plan with variations for all parameter values.
- Design the generators for the leaf cells.

- Develop hierarchical composition cells by placement and routing of leaf-level cells and other composition cells. Placement and routing may be either simple abutment or more complex relative placement and automatic routing.

There are basically two types of module generators: leaf cell generators and complex generators.

2.4.1 Development of Leaf Cell Generators

There are two ways to develop a simple-cell generator:

- The basic cell is designed with an interactive layout editor. The designer maintains technology process independence with the help of the layout compactor.
- The basic cell is described in an L program in which the designer describes the relative placement of transistors, contacts, and wires. The designer uses layout compaction to ensure process independence.

The designer may choose the first method when the cell structure does not change as parameters change. Examples include exclusive OR or NOR circuits, some fixed latches, and signal-buffering circuits.

The second method is useful when the circuit structure changes with the parameters. The simplest example is an AND gate generator where one of the parameters is the number of inputs. As the number of inputs changes, so does the number of the transistors and their connectivity.

Of the standard cells shown in Figure 2.6, some were developed by the first method, such as XNOR and latch cells. Others were developed by using the second method. These are variable-input NAND, NOR, and AND-OR-INVERT cells.

2.4.2 Development of Complex Generators

Complex generators are composed of many cells that are the same or similar. All subcells are typically interconnected by abutment. A memory array in a RAM is a good example. The developer of such generators has to design the hierarchy and a floor plan such that user options such as size of the array and number of bits can be easily included.

Often, the basic cells that make up the regular structure are best developed by using the interactive editor. These cells are called *leaf* cells. The *composition* cells are the ones in which leaf cells are abutted or routed together. The composition cells that are fixed may be edited in the interactive editor. However, if the composition cells are variable and parameterized, the designer can use L programs to piece them together.

Figure 2.18 shows an example of a two-bit adder produced by the adder generator. The generator offers the following options:

- number of bits in the adder
- zero-detect circuit

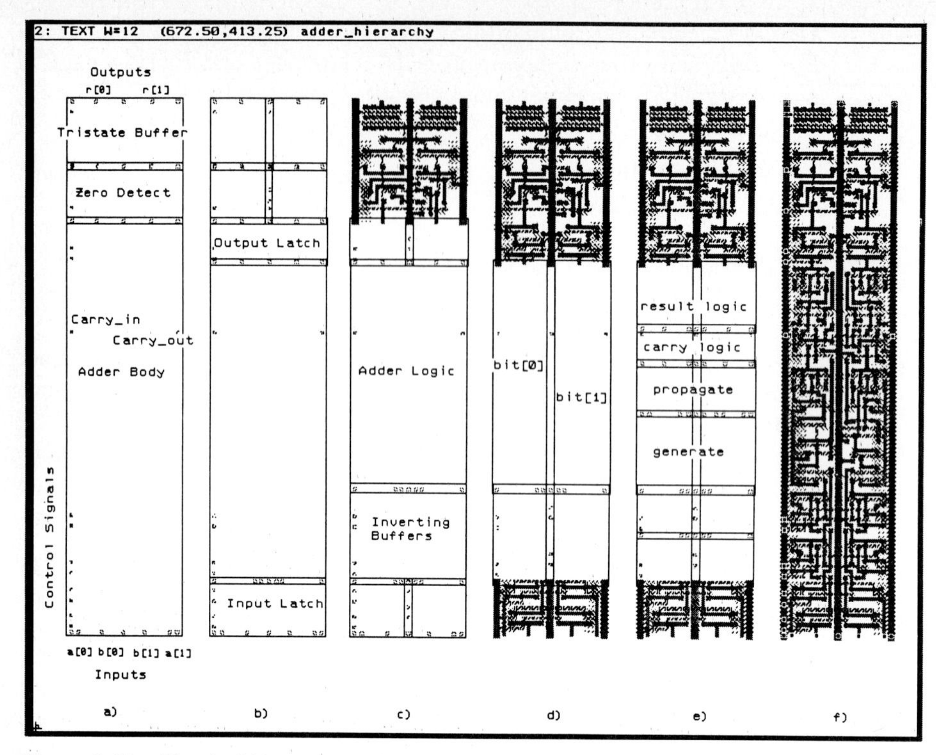

Figure 2.18 Physical hierarchy of a generated adder.

- Output tristate buffer
- Output result latch
- Input operand latch

The optional circuits are attached to the main body of the adder only if needed. The figure shows the physical hierarchy of the adder by displaying the same circuit at successively deeper levels from left to right.

Figure 2.18a shows the view of the adder at the top of the hierarchy. It shows three components: adder body, zero-detect circuit, and tristate buffer. These circuits are attached because the parameters specified these options. If the parameters had not asked for these options, the zero-detect and tristate buffer would be absent.

The next view, in Figure 2.18b, shows the circuit one level deeper in the hierarchy. The internal structure of the adder body is shown to consist of the main section, input latch, and output latch.

Figure 2.18c shows the next level down the hierarchy. It reveals the layout of the zero-detect and tristate buffer circuits. It also shows that the main section of the adder consists of the inverting buffers and the adder logic.

Figure 2.18e shows that the adder logic contains generate, propagate, carry logic, and result logic circuits. Finally, Figure 2.18f shows the complete layout of the 2-bit adder. It is important to partition the design into manageable pieces such as these. The layout in Figure 2.18f seemingly has no structure, but the analysis of the floor plan shows otherwise.

The generator that produces the layout follows the structure of the floor plan. A top-level L program (composition cell) to compose the adder is:

```
CELL adder( bits , zero_detect , tristate , inlatch , outlatch )
  #number of bits, and 1 or 0 flags for options
{
  CALL add_body_gen adder_body ( bits, inlatch, outlatch );
  INST adder_body abody;
  IF( tristate= =1) {
    CALL tri_gen tbuf( bits );
    inst tbuf tb;
}
IF( zero_detect= =1) {
  CALL zdet_gen zdet( bits );
  INST zdet zd;
}
IF( zero_detect ) {
  zd.vddl AT abody.vddu; #paste zero_detect to body
  IF( tristate )
    tb.vddl AT zd.vddu; #paste tristate to zero-detect
}
ELSE IF( tristate )
  tb.vddl AT abody.vddu; #paste tristate to body
}
```

The cell generator "adder" has arguments that describe user options. Most of the options are yes/no options handled by 1 or 0 choices. The generator calls three other generators, two of them only if needed. The generators are:

- add_body_gen, which generates the basic adder,
- tri_gen, which generates tristate buffer. This generator is called only if the tristate option is asked for.
- zdet_gen, which generates the zero-detect circuit. This generator is called only if the zero-detect circuit is required.

The decisions are made by IF and ELSE statements. They get all the arguments from the top-level generator. These generators create cells that are installed in the L data base by the L compiler at the time the program is run.

Next in the program, instances of these cells are created if needed. Finally, placement of these instances is completed by the AT statement, which ensures correct placement of the cells with respect to each other. The points used for connection are vddl and vddu, the lower and upper VDD terminals on the cells. The AT function serves a dual purpose. Firstly, it places the objects by assigning new locations to specified objects. Secondly, it looks for all coincident terminals of the instances being placed and wires them together.

The same procedure is applied to develop other generators in this example. Leaf-level cells (the ones with transistors, contacts, and wires) can be developed graphically with the L-editor. Their process independence is maintained by the layout compactor.

2.4.3 Layout Compaction

The goal of compaction in GDT is to provide technology adaptability to layout generators. Chip layouts consist of leaf cells and composition cells. In a well-structured design, the goal is to confine technology independence to the leaf cells, since composition cells combine leaf cells by abutment or by automatic routing. In case the designer uses abutment, the compactor enforces the pitch matching to ensure that the cell boundaries interconnect correctly.

Compaction types include the following:

- compaction of leaf-level cells
- pitch matching of composition cells
- compaction of a flattened hierarchy of cells

The designer can achieve technology independence of layout designs by programming the generators in the L language. This is done by describing geometric relationships of objects in terms of the symbolic equations involving design rules. This method is useful but may require considerable time for verifying many possible combinations of the design rules. Compaction is an alternative method that requires much less engineering time.

A compactor can be used in two ways:

- interactively, from an interactive editor
- procedurally, as invoked from generators

The designer can use the compact statement in many different ways. The simplest way is given in the following example:

```
CELL abc()
{
    terminal declarations;
    transistor declarations;
    contact declarations;
    wire declarations;
    COMPACTXY;
}
```

In this program the cell content is specified first. At the end of the program, the compactor is invoked to complete the cell.

Compaction provides design rule adaptability by automatically moving objects closer together or further apart based on the design rules found in the technology file. There are four directions of compaction available: *x* compaction (horizontal), *y* compaction (vertical), *xy* compaction (horizontal followed by vertical), and *yx* compaction (vertical followed by horizontal).

There are a number of different compaction algorithms. One of these is a virtual-grid method, described here. During the compaction of a leaf-level cell, a compactor performs the following operations:

- It constructs a list of distinct vertical and horizontal "virtual" grid lines and assigns objects to them. Objects remain on these lines throughout compaction.
- During *x* compaction it compresses the distance between vertical grid lines so that objects satisfy design rules relative to their neighbors to the left.
- During *y* compaction it compresses the distance between horizontal grid lines so that objects satisfy design rules relative to their neighbors below.

Figure 2.19 shows a leaf-level cell, a CMOS inverter. It is shown before compaction (part a), after the *x* compaction (part b), and, finally, after the *y* compaction (part c).

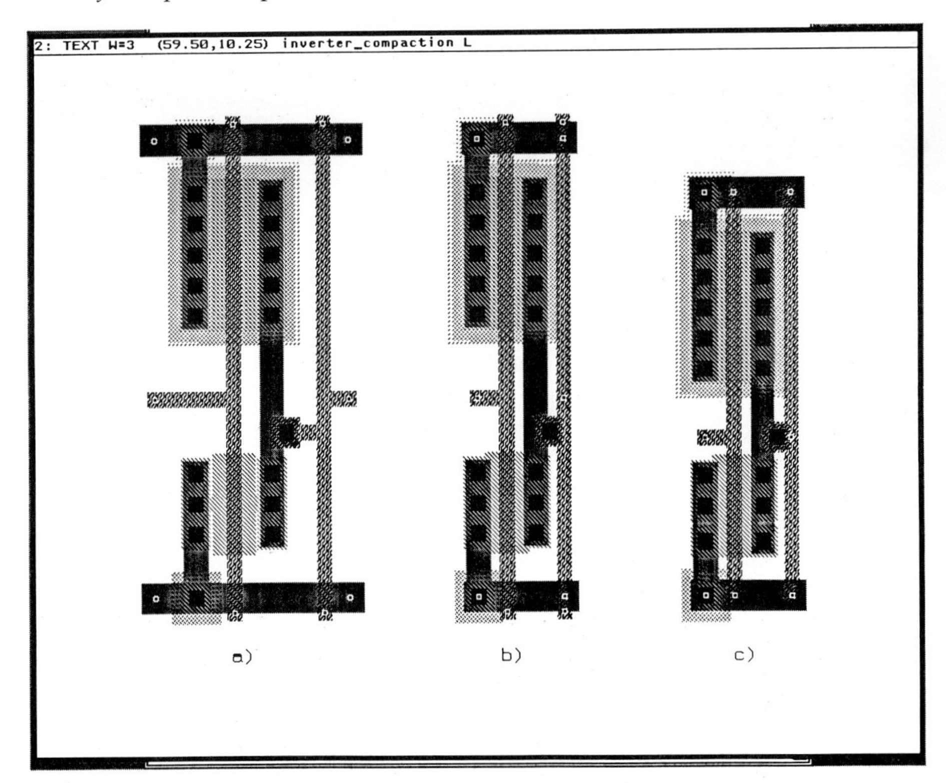

Figure 2.19 Compaction of an inverter. (Layout-to-layout compaction.) (a) Original cell. (b) After X compaction. (c) After Y compaction.

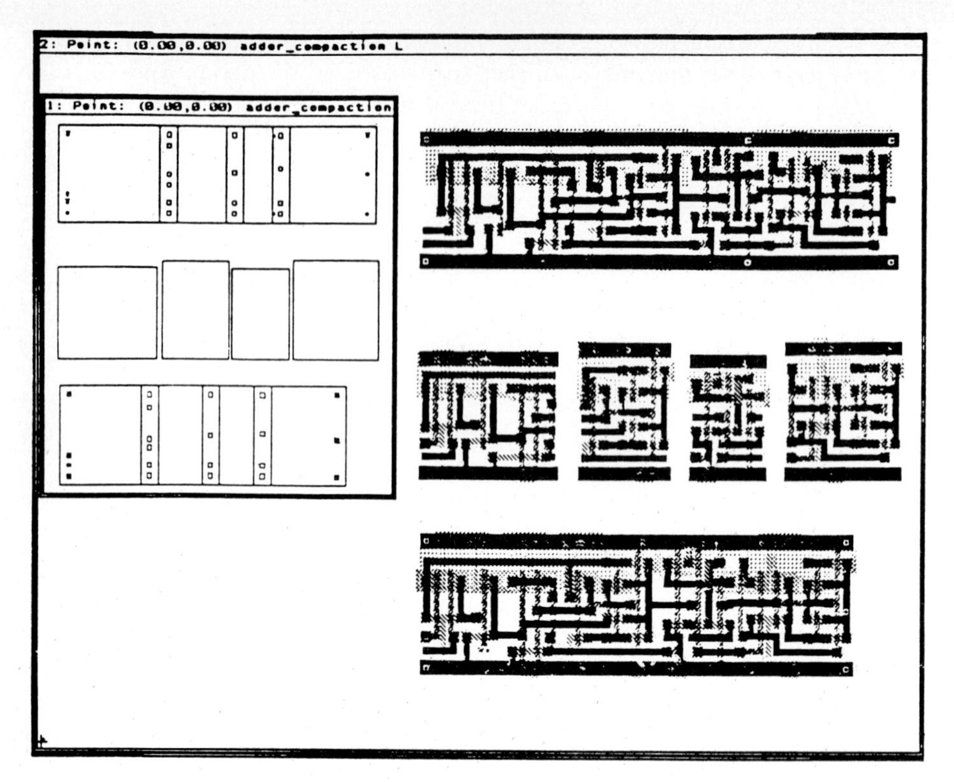

Figure 2.20

Compaction with pitch matching of abutted cells is a process of minimizing the layout area while still keeping the cells well-connected. The compactor does pitch matching by creating a flattened version of the template cell, compacting the flattened version, and then modifying the individual cells to match the result. The flattened version is then thrown away.

Figure 2.20 shows an adder bit undergoing compaction. The insert at the left of the figure shows the floor plan of the adder. The figure shows the result of compaction it if is applied separately to the individual leaf-level cells that the adder uses. It is evident that the individual cells no longer match after the separate compaction. Finally, the figure shows the adder after compaction with pitch matching, which has preserved the abutment of the leaf-level cells.

The complete adder is assembled by the following program, which invokes the compactor at the end:

```
CELL adder_bit()
{
      #instantiate desired cells
   INST generate_cell gen;
   INST propagate_cell prop;
   INST carry_cell carry;
   INST sum_cell sum;
```

```
    #abut them together
  prop.vddl AT gen.vddr;
  carry.vddl AT prop.vddr;
  sum.vddl AT carry.vddr;

    #compact
  COMPACTXY;
}
```

2.4.4 Verification

To aid in the design of complex integrated circuits, the M modeling language was developed to describe subcircuits at the behavioral, logic, and transistor levels. The behavioral level uses an algorithmic model to describe the circuit being simulated. It also can describe a register transfer language, which is useful for verifying an architecture. The logic level can be used for SSI and MSI components, which function as basic building blocks. The term *functional module* describes modeling at the behavioral, functional, and logic levels because the language and compiler are identical for all three levels.

Transistor-level models are provided as a part of the L-simulator. The mixing of M modules with switch- and circuit-level simulation, where their correct interaction is verified within the unified L-simulator environment, allows designers to detect and correct flaws early in the design cycle. L-simulator is an environment for networking functional models of system components. First, it connects inputs and outputs of modules according to the net list. Then it initializes the circuit and inspects it for illegal connections. For example, it reports transistor gates that are not connected to any signals and power supplies that are disconnected or shorted. Finally, it facilitates simulation by monitoring activity on nodes, maintaining queues of active nodes, scheduling activation of functional models, and arbitrating signal values on the nodes. One of the key concepts in the L-simulator is that each functional model is written as a stand-alone unit. It does not have to contain any knowledge about its environment or how the concurrency of events is handled. This simplifies the development of functional models.

A module written in M language conceptually represents a finite-state machine with the outputs dependent on the inputs and past history. A module will typically read its inputs and, after performing the necessary computation, set the desired output signals (after a specified delay, in the timing mode). The distinct interface between the nodal arbitration scheme and the functional module simplifies the development of the functional module.

A module compiler compiles functional modules. In addition to creating the appropriate object code, the compiler generates additional information to permit interactive viewing, debugging, and state editing within a simulation session. To circumvent the preliminary linkage phase required by behavioral simulators, M object modules are dynamically loaded into the simulator on demand, using a unidirectional linkage mechanism.

Conclusion

Silicon compilation is an emerging computer-aided design discipline. Its evolution depends on module generation techniques. Module generation provides the link between structural and physical representation of the systems on a chip.

References

1. J. L. Hennessy, "SLIM: A Simulation and Implementation Language for VLSI Microcode," *Lambda*, April 1981.
2. H. Trickey and J. Ullman, "A Regular Expression Compiler," *IEEE COMPCON*, 1982.
3. Carver Mead and Lynn Conway, *Introduction to VLSI Systems*, Addison-Wesley, 1980.
4. Z. Navabi et al., "Storage Logic Array Realization of RTL Descriptions," *IEEE CHDL Conference*, June 1983.
5. *Design Specification Using the MODEL Language*, Lattice Logic, 1982.
6. J. Southard, "MacPitts: An Approach to Silicon Compilation," *Computer*, December 1983.
7. N. Bergmann, "A Case Study for the First Silicon Compiler," *Proc. Third Caltech Conference on VLSI*, Computer Science Press, 1983.
8. M. R. Burich, T. G. Matheson, C. Christensen, "The Plex Project: VLSI Layouts of Microcomputers Generated by a Computer Program," *IEEE International Conference on Computer-Aided Design*, September 1983.
9. *VLSI Design System*, VLSI Technology, Inc., 1983.
10. J. Batali and Anne Hartheimer, *The Design Procedure Language*, A.I. memo no. 598, MIT, September 1980.
11. R. Lipton et al., "ALI: A Procedural Language to Describe VLSI Layouts," *Proc. Nineteenth Design Automation Conference*, 1982.

CHAPTER 3

Knowledge-Based System for IC Design

Jin H. Kim

3.1 Introduction

The high cost of designing a complex IC chip is due to the large volume of detail that must be managed. Designing an IC chip involves specifying the logical and physical characteristics of a large number of functional subsystems relative to one another. Each subsystem performs a particular function, such as address decoding. For each subsystem, the designer must develop the associated circuitry, which takes into account the characteristics of the particular technology and the physical realization of the circuitry for a specific manufacturing process. This design process must ensure that the overall chip performs the desired function while satisfying all constraints on performance, power, testability and area. To handle the complexity associated with designing large-scale IC chips, methodologies based on abstraction hierarchies and maximization of regularities in layout structures have been developed. For a computer aid to be effective, it must operate within a hierarchical design environment while enforcing structural regularities. The research described in this chapter examines the issues involved in constructing such a computer aid for IC layout that relies on the use of large amounts of domain-specific knowledge for its problem-solving power.

The IC chip design process is composed of three closely coupled phases. One phase involves the functional design, in which the behavioral and performance requirements of the system are specified. In the second phase, consisting of the structural design, a network of basic logical units is defined to realize the behavior specified during the functional design phase. The network is designed to take advantage of the capabilities provided by the specific technology and to regularize the interconnections between logical blocks. Finally, during the physical design phase, the functional network is mapped onto the semiconductor surface through determination of the precise geometry and position of each constituent unit and the interconnecting wires. The objective during the physical design phase is to find good or near-optimal shapes and shape placements to minimize the area used and still satisfy the performance requirements.

Physical design of a complex IC chip is usually carried out in a hierarchical manner. The complexity of chip layout is managed by partitioning the original problem into less complex subproblems at lower levels in the abstraction hierarchy. Typically, a chip floor plan, which partitions the layout into areas for functional units, such as ALUs and shifters, is initially developed at one level of abstraction. Using the chip floor plan as a guide, each functional area is in turn partitioned into areas for cells, such as a bit slice of ALU, in a hierarchical manner. At this point the cells describe small circuits, typically containing less than 100 transistors. Once the layouts for the cells are designed, the cells and their functional areas are merged, and fine adjustments are made to produce a satisfactory layout. By using hierarchy and introducing regularity into the layout problem, complexity is reduced as subunits are replicated many times and the interconnections between the subunits are simplified.

In recent years there has been a growing interest in developing systems that automatically synthesize layouts for cells from functional specifications. An automatic cell layout system can

1. reduce design costs by eliminating errors
2. allow efficient exploration of the design space, and
3. reduce overall design time by letting designers conceptualize at a higher level of abstraction.

It also eliminates the need for costly verification and design-checking steps because it substitutes correctness by construction for correctness through checking.

The complexity of cell layout, however, makes it difficult to develop a purely algorithmic solution, and most existing computer aids suffer from one of two shortcomings. Some design tools can accept problem descriptions at a high level of abstraction but produce layouts that do not compare favorably to those generated manually. Conversely, tools capable of synthesizing area-efficient layouts usually require input descriptions at a lower abstraction level.

An alternative approach is to incorporate heuristic knowledge specific to a domain, used by human designers, to complement existing algorithmic techniques. The set of heuristic knowledge can generate hints based on contextual cues to guide the algorithmic subsystem and react to exceptional conditions. The algorithmic subsystem can handle those subproblems for which efficient techniques have been discovered.

This chapter covers the design and evaluation of one such system, TALIB, a knowledge-based system for cell layout design.

3.2 Overview of Artificial Intelligence

One way to view the cell layout design process is as a search through a problem space. Each state in the problem space represents either a partial or a complete solution. The cell layout designer's task is to find a path from an initial problem state to a state that qualifies as a solution. The search of the problem space is carried out by utilizing a set of design or problem-solving operators; application of an operator transforms a current state into a new state. The objective of the cell layout designer is to find and apply an operator that always results in a state that is in the solution path. Knowledge that will cut down the amount of search required can be expressed as a set of rules; systems with knowledge encoded in rule form are called *rule-based systems*.

A rule-based system can be characterized as pattern-directed and consists of three major components:

1. a collection of rules that can be activated by specific patterns in the problem space,
2. one or more memories that represent the problem space and can be examined and modified by the rules, and
3. an interpreter that controls the selection and activation of the rules.

Each rule represents a chunk of domain knowledge in the form of *if*-condition-*then*-action. Rule-based systems work by applying rules, noting the results, and applying new rules based on the changed situation. The flexibility of rule-based systems stems from the explicit separation of the rules from the interpreter or the inference engine.

The organization of rule-based systems allows the representation of knowledge in a highly uniform and modular way. The organization of knowledge into separate modular units facilitates the incremental expansion of the system. Each rule is designed to be an independent chunk of knowledge with its own criteria of relevance. Since a rule can be chosen only if its criteria of relevance have been met, the system behavior remains reasonable even as rules are successively deleted or added. Hence, the modularity of the knowledge representation allows a rule-based system to gradually acquire expertise by incorporating a large body of knowledge over an extended period of time.

3.3 TALIB Overview

The TALIB system creates layouts by augmenting some algorithmic techniques with large amounts of domain-specific knowledge encoded as rules to develop design plans at several abstraction levels for the purpose of controlling the problem-solving behavior of the system. The design plans are successively refined until a satisfactory layout is obtained. The design plans let TALIB identify and tackle the most critical subproblems first.

Development of the problem-solving architecture for a knowledge-based system usually involves resolving issues in representing and using knowledge. Because the automatic cell layout system must be built incrementally, a representational framework that supports a modular structure has been favored. Thus, the development of a system such as this exhibits an incremental improvement in performance as more knowledge covering exceptional conditions is added. The large body of collective experience in using a rule-based formalism influenced the selection of OPS5 [1], a general-purpose rule-based language, for this project.

The version of TALIB described here is geared toward cell layouts in NMOS technology. The fabrication technology is essentially the single-metal, single-polysilicon, NMOS technology described by Mead and Conway [2] but with buried contacts. Although this research describes layout designs in NMOS, the concepts are applicable to other semiconductor technologies. Finally, to make the research more tractable, some simplifying assumptions have been made. All layouts produced by TALIB are based on Manhattan geometry. This means that the edges of geometric shapes of every mask element are parallel to either the x or the y axis. The orthogonal geometry results in many simplifications in the internal data structures of TALIB without compromising much of the final layout's quality.

3.4 Input/Output

A cell layout problem is presented to TALIB in two parts. The first part describes the circuit components and their interconnections. This involves a list of transistors and logic gates along with the signal nets to which the terminals of the transistors, the logic gates, and the I/O pins are connected. The circuit's performance requirements are expressed by specifying the sizes of the various transistors. If transistor size is not explicitly specified, a default value is generated by TALIB based on minimal tolerances of the fabrication technology.

The second part of the input provides the topological and geometric requirements around the outside boundary of the circuit layout. The topological information describes the order in which external connections must appear on the cell boundary. This information can vary in detail from just a specification of the cell side on which each signal must appear, to a specification of the ordering of each signal relative to the other signals on that side, to a specification of the location of each signal relative to the cell's reference point. The topological information can also describe the layer on which the external connections are to be implemented.

Definition 1. A *topological level layout* specifies the layout in terms of geometric primitives for the transistors and contacts, the relative locations of the geometric primitives, the orientation of each geometric primitive, the creation of wires to satisfy the interconnection requirements, and the layer assignments of the wires without regard to spacing tolerances.

The geometric information describes the physical parameters of the cell layout. This can be stated in terms of the upper bounds on the dimension of the layout or in terms of the cell's aspect ratio. The combination of the topological and geometric layout descriptions make up the boundary constraints of the layout problem.

3.5 Task-specific Knowledge

To carry out its problem-solving tasks, TALIB relies on four types of knowledge. First, it has analysis knowledge to identify situations in the design state as instances of more general design situations. Second, it has synthesis knowledge to convert design situations at one abstraction level into more specific designs at a lower abstraction level. Third, it has axiomatic knowledge to govern how design decisions affecting one part of the design state affect other parts. Finally, it has control knowledge to determine the order in which subtasks must be carried out by the system. All four types of knowledge are represented as rules in TALIB's OPS5 rule memory.

In addition to the domain-specific knowledge, there is certain amount of knowledge required to control the basic behavior of TALIB. For example, the knowledge that makes up the inference engine is required to input problem descriptions, output results, run through the basic task sequence for planning, and carry out the compaction. Since this knowledge determines how the domain-specific knowledge is subsequently used, it is somewhat task-independent. This basic set of knowledge can be used in other domains that require problem-solving behavior similar to that used for cell layout.

Definition 2. The *inference engine* of TALIB describes that knowledge requiring to run through a sequence of a predetermined set of actions and general bookkeeping functions in order to properly apply the four types of domain-specific knowledge.

3.5.1 Analysis Knowledge

An important phase of any design deals with the subtasks that analyze the current design state. The goal of this analysis phase is to identify familiar design situations. For example, given the design state shown in Figure 3.1, an experienced layout designer can identify an instance of the crossover interconnection pattern between logic devices d_1 and d_2.

In another form, the analysis knowledge identifies equivalent circuits in the input specification and regroups circuit components to simplify the subsequent wiring. This type of analysis knowledge includes deducing the presence of a NOT gate when a pullup-type transistor is connected to a single pulldown-type transistor through the drain or source terminal.

Recognizing instances of a known design situation resolves problems such as determining what subgoal should be accomplished next and accessing synthesis knowledge associated with a particular class of design situations.

3.5.2 Synthesis Knowledge

Once instances of design situations are identified at one abstraction level, synthesis knowledge is needed to generate more detailed designs at a lower abstraction level. For the example in Figure 3.1, once the instance of a crossover interconnection pattern has been identified, a good designer will plan ahead to reserve sufficient layout space to insert a metal wire. This plan is based on the fact that crossovers represent a locally nonplanar wiring situation; therefore, one of several cell floor plans for resolving nonplanar wiring is invoked. The most commonly used technique is to implement some of the wires on the metal layer and others on the diffusion and polysilicon layers. In TALIB, the synthesis knowledge is used to generate design objects, relationships between objects, and constraints on the objects. Synthesis knowledge is local in context between a specific pair of abstraction levels: synthesis knowledge handles each design situation separately. This provides the needed modularity in the knowledge base: the use of a piece of synthesis knowledge for handling a particular design situation only on the presence of that design situation and the current design state, not on any other design situations.

3.5.3 Axiomatic Knowledge

Axiomatic knowledge covers the rules of interaction between design objects in accordance with the physical laws governing the problem. This knowledge is needed for three reasons. First, TALIB needs knowledge to handle interactions between subproblems. Complex design problems are usually solved through decomposition techniques that produce a number of simpler, interrelated subproblems; design decisions affecting one subproblem also affect others. For example, because the synthesis knowledge in

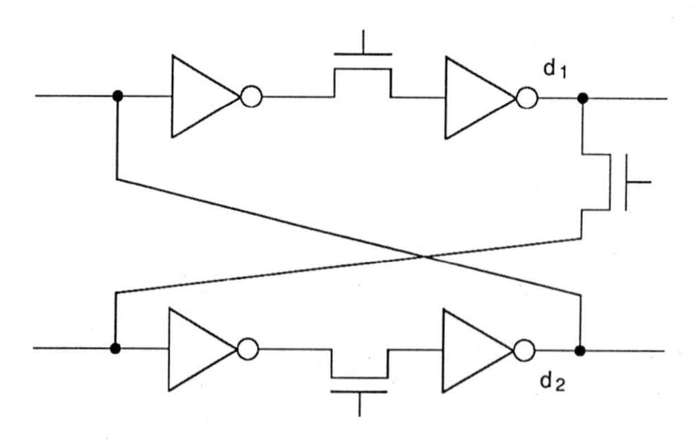

Figure 3.1 Instance of crossover design situation.

TALIB are organized around design situations, the subplans generated through synthesis knowledge often overlap or conflict with each other. Axiomatic knowledge is used to handle the necessary interactions between the subplans. Second, axiomatic knowledge is required to take advantage of local opportunities. The synthesis knowledge produces a design in a top-down fashion by taking into account the global relationships between design objects. However, to produce high-quality layouts, TALIB must exploit opportunities to introduce technology-specific design decisions. Finally, a good design system must recognize inconsistencies in the design state. TALIB's knowledge base includes knowledge to detect inconsistencies and undo design decisions associated with the inconsistencies.

3.5.4 Control Knowledge

The problem-solving behavior of TALIB is built around recognizing occurrences of familiar design situations and synthesizing design plans in response to them. An important component of the knowledge dealing with the cell layout task involves generating the order in which subtasks are invoked. At one level, this involves recognizing when to carry out the analysis subtask rather than the subtask for inputting the problem specification or the subtask for synthesizing more detailed design plans. At another level, the ordering of subtasks corresponds to determining which design situation to evaluate next. This ordering ensures that the more critical situations are handled before the less constrained ones.

For example, in a set of design situations describing wiring requirements between transistors, a topologically impossible situation can result if the wrong problem is resolved first. Such a situation can result from allowance of insufficient layout space to introduce the necessary metal wires. This domain-specific control knowledge is distinct from the domain-independent knowledge embedded in the inference engine for selecting rules.

3.6 Design State

Implementing hierarchical planning in a cell layout system involves representing the state of the layout at several abstraction levels, the subgoals to be achieved at different abstraction levels, the relations between subgoals, and the design actions that cause changes in the design state. The description of the layout state consists of design objects and their interrelations.

An example of a design object is a layout primitive representing an enhancement-mode transistor, shown in Figure 3.2. Figure 3.3 shows an example of a design relation between a transistor primitive and a wire primitive. This design relation indicates that the termination point for one end of the wire and the drain terminal of the transistor primitive occupy the same location. The subgoals are represented through abstract design objects in the layout state and design requirements on these objects.

```
primitive
      status:          { in or out }
      id:              { identification tag }
      type:            { transistor or wiring primitive }
      is-a:            {   }
      orientation:     { x + or x - or y + or y - }
      drain-x:         { x axis grid of drain }
      drain-y:         { y axis grid of drain }
      gate-x:          { x axis grid of gate }
      gate-y:          { y axis grid of gate }
      source-x:        { x axis grid of source }
      source-y:        { y axis grid of source }
                       :
                       :
```

Figure 3.2 Example of a design object: part of a transistor primitive description.

For example, Figure 3.4 shows an abstract design representing a wiring requirement on the metal layer. The channel objects describe an abstraction of routing tracts available on the different layers. The *merge* requirement on the newly generated channel describes a subgoal to sort the new objects in the existing ordering of channel objects.

3.6.1 Control Mechanism

TALIB's primary mechanism for governing problem-solving behavior is a combination of match and planning. In many of the layout subtasks, there is sufficient information at the completion of a subtask for TALIB to recognize precisely what to do next. For example, a control rule to detect

```
Primitive
   status:          in
   id:              g00032
   type:            transistor
   is-a:            etnx-1
   orientation:     y+
   drain-x:         g00049
   drain-y:         g00078
   gate-x:          g00092
   gate-y:          g00134
   source-x:        g00179
   source-y:        g00204

Primitive
   status:          in
   id:              g00041
   type:            wire-primitive
   is-a:            diffusion-wire
   orientation:     x+
   terminal-a-x:    g00049
   terminal-b-y:    g00073
```

Figure 3.3 Example of a design relation: connection relations between the diffusion wire and drain terminal of a transistor.

an opportunity to share contacts between two wiring problems can initiate the exact sequence of subtasks to be activated. In cases where match cannot be applied, TALIB relies on planning based on recognition of commonly occurring situations and generates tentative subplans.

Planning in TALIB has both top-down and bottom-up characteristics. During the top-down process, it generally follows a strategy known as *least-commitment* [3], in which the planner does not commit itself to a plan step or to an ordering between plan steps that might have to be abandoned later. For example, in Figure 3.5, two design situations each describe a point-to-point wiring requirement. Implementation of one wiring requirement on either the diffusion or the polysilicon layer will present a wiring barrier to the other requirement that can only be resolved through a metal-layer or depletion-mode transistor. At this point in the design, there is insufficient information to unambiguously develop an ordering between plan steps to handle these conflicting requirements.

In TALIB, the rules are organized to skip such situations and handle other design situations first. In many cases, a refinement of another design situation will generate information that is relevant to the earlier, unsolved problem. The problem of Figure 3.5 is solved by implementing the signal net associated with wiring requirement s_{10} on the metal layer as a result of solving another design situation. This type of additional information is propagated from one part of the design state to another through constraints. In general, this style of planning is not sufficient for layout subtasks because deadlocks can occur in which all the subproblems associated with the planning can be locally underconstrained. In these cases, TALIB relies on arbitrary selection of solutions with a limited form of backtracking in the event that inconsistencies develop.

TALIB has a bottom-up quality in that no one can predict the consequence of each planning step in the layout domain. The goals associated with each planning step describe only the critical relations that must be satisfied and not the potential side effects that might occur. For example, placing a transistor primitive adjacent to a particular contact may prohibit a reduction in the x axis because the transistor primitive now becomes part of a critical path on the x axis. Because of the many potential side effects, TALIB has many rules that deduce new facts in the design state based on

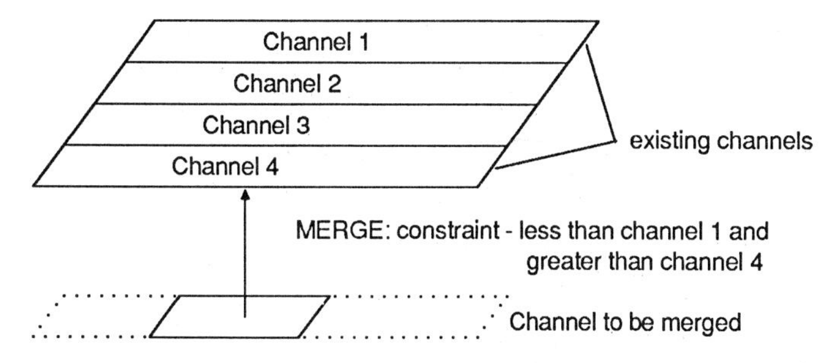

Figure 3.4 Representation of a subgoal.

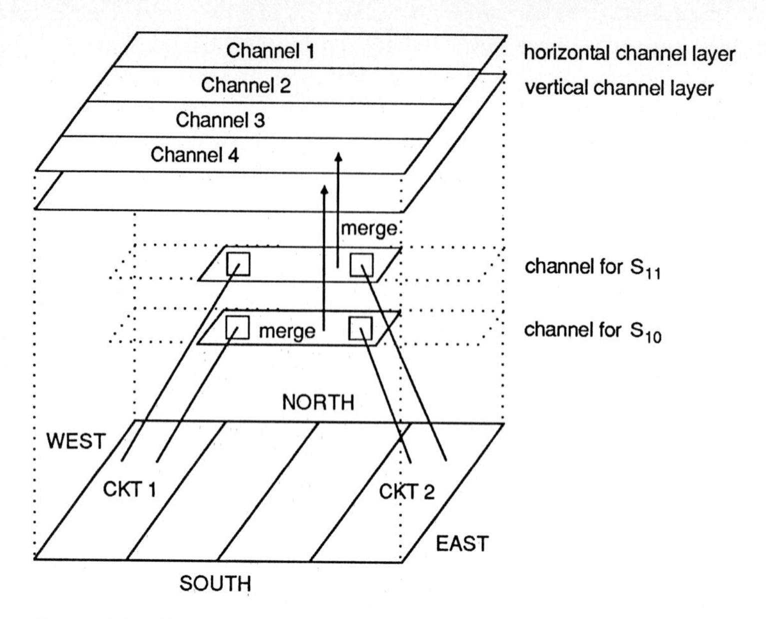

Figure 3.5 Representation of a design plan.

axioms about cell layout. In many instances, a newly deduced fact presents a design opportunity that should be taken advantage of. TALIB then alters the design's focus of attention, generates separate planning islands that can later be merged into the main plan, and modifies the existing plan.

Using Match

TALIB relies on match to carry out four types of actions. These correspond to actions that can be taken on the basis of information available locally in the design state.

First, match is the principal problem-solving technique for detecting and correcting inconsistencies in the design state. Once a particular type of inconsistency (for example, lack of space for implementing metal wire) is detected, a sequence of steps must be taken to correct the error. The exact order in which the steps are taken depends on the local context in which the inconsistency was first detected. However, because it is impossible to specify a course of action for each type of inconsistency, match is an appropriate technique to control this behavior.

Second, the series of steps used to take advantage of local opportunities can also be based on match. For example, when an opportunity develops to introduce a jog into the design state, the sequence of abstract steps for breaking an existing wire, creating a jog, and regrouping the circuit components and wires into new compaction primitives (for example, *x*-group and *y*-group objects) must be taken. The exact sequence of primitive actions to be taken depends on the composition and cannot be prespecified. Because a set of actions that take advantage of design opportunities can be partially ordered, match is an appropriate technique.

Third, once a design situation has been detected, a sequence of steps must be carried out to generate a subplan to accommodate the details related to the situation. In all cases, the scripts for these local design situations have at least a partial ordering among the steps. Although the exact primitive actions required cannot be determined before the design script is instantiated, they can be partitioned into subtasks. This type of knowledge application is easily controlled through match.

Finally, the basic problem-solving behavior of invoking abstract steps to analyze the design state, propagating constraints, and refining and merging subplans can be carried out through match. Again, this is possible because there is sufficient information available at an abstract level to invoke the next abstract subtask without the need to undo the decision for invoking the subtask at some later time. The element of uncertainty becomes a part of carrying out the specific subtask.

Using Planning

The overall strategy in planning relies on identifying design situations that are familiar to TALIB through an analysis phase. For each design situation, TALIB has rules in its knowledge base to refine constituent design objects and relations into more specific objects and relations based on the current state of the layout. Once all of the design situations at one abstraction level have been examined, the layout state at a more detailed level is examined. This process is repeated until the lowest abstraction level is reached and an acceptable solution is generated. Match cannot be used in these instances because the knowledge for developing and ordering the design situations is uncertain. The rules for analyzing the design state generate or update a world model.

Polya suggested a problem-solving strategy that relies on identifying aspects of the new problem that make it similar to problems that have been solved before [4]. Cell layout experts use this strategy to approach a given design problem in terms of past, familiar design situations. This repertoire of design situations describes commonly occurring relations between design objects at each abstraction level and forms a metalanguage with which experts reason about the layout domain.

Similarly, TALIB devotes a substantial amount of its resources to identifying instances of known situations in the layout state.

ANALYSIS The analysis phase has three parts. In the first part, a set of all plausible design situations is developed. A design situation is considered plausible if certain evidence is present in the layout state. During the second part, other evidence is collected to support or refute design situations considered in the first part. In the final part of the analysis phase, all those design situations with support weight above a threshold value are activated. Also activated are those design situations describing objects in the layout state not covered by any active design situation, even if their support weights are below the threshold. This can occur when TALIB uses default knowledge to develop a particular design situation.

SUBPLANS For each subplan, TALIB applies its axiomatic knowledge to patch subplan elements into the layout state at a lower abstraction level. Since the subplans represent solutions to different subproblems to satisfy subgoals, they do interact with each other. In some cases the interactions uncover technology-specific design opportunities, for example, opportunities to share electrical contacts. In other cases the elements of subplans can produce inconsistencies. This is usually the result of refining individual design situations without considering their interaction. When inconsistencies are detected, TALIB uses its limited knowledge of plan patching to backtrack and modify erroneous subplans. The rules for monitoring these conflicts are independent of the major subtasks mentioned earlier. These control rules constantly monitor the design state for a recognizable conflict and invoke corresponding subtasks to resolve the conflict.

CONSTRAINTS TALIB operates with four types of constraints. First, there are constraints governing the ordering of design situations. These constraints determine precedence relations among design situations based on their relative importance. Since design situations define subgoals to be achieved, these constraints govern the ordering of subgoals. Second, there are constraints at the topology level. These constraints govern the relative location of components such as wires, transistors, and circuit clusters. Third, in addition to constraints describing relative placement, there are those that govern the layout characteristics of each component. Constraints in this category describe things such as the layer on which a particular wire must be realized, the type of layout model to be used in refining a transistor, and how circuit clusters should be refined in terms of primitives. Finally, there are constraints governing the spacing between compaction primitives.

Not all subplan interactions can be handled through constraint posting. In certain instances, TALIB will reach a point where all of the subproblems associated with refining a design situation are underconstrained. To advance the design state, TALIB will make a reasonable guess by selecting the least specific rule within a rule group. Unfortunately, whenever TALIB takes a guess, there is the possibility that it will block the resolution of another subproblem.

3.6.2 Modifying Design Plans

Inconsistencies in the design state are introduced for three reasons. First, design knowledge is based on information available at one level of abstraction that constrains subsequent design activities at a lower level. It is not possible to enumerate all the conditions under which a particular rule is applicable because the consequence of a rule does not become apparent until the rule has been applied. So, although TALIB follows the least-commitment principle of applying the most general knowledge first and then successively applying more specific knowledge as information becomes available, it will need to backtrack because of the uncertainty inherent in the rules themselves. Second, at certain points in the design, there will be insufficient information to apply any design rule with certainty. In these cases TALIB must use a default rule and take a guess in order to move the design forward. Finally, modifications of the plan structure based on design opportunities are local in context. For example, in refining the wiring requirements, a decision to share a contact between two separate but adjacent subcircuits reflects only the opportunities at the local level and does not take into account the global parameters, such as whether the contact is part of a critical path in either the x or the y axis.

TALIB can modify its plans by explicitly building and maintaining a link between each design situation and the design objects created as a result. Any time a new assertion is introduced into the working memory, whether it describes a design object or a constraint, a link is introduced between it and the design situation that caused it. For example, if a constraint on the relative location of two circuit clusters is formulated as part of refining a design situation that describes the interconnection of circuit clusters, TALIB creates a link between the constraint and the design situation to indicate dependency. The dependency link also maintains consistency in the plan. For example, relative placement of subcircuits is based on a specific subnet between them, which is in turn based on a specific layout model for the circuit cluster, and if some other planning step alters the layout model of subcircuit sufficiently to invalidate the particular subnet, TALIB uses the dependency link to undo any portion of the plan that was the consequence of the particular subnet. In addition to the dependency link, TALIB has a collection of rules about how to patch a portion of the plan when constraints are violated. To a large extent, backtracking is avoided through the least-commitment strategy, but there are instances when TALIB backtracks and tries alternatives.

3.6.3 Compaction Subsystem

Once TALIB develops a topological-level layout, it relies on an existing technique based on critical-path analysis to remove the unused space between layout components. Compaction steps on the x and y axes are carried out separately under the design plan. A set of primitives are defined and are moved as one during the compaction process. By conducting compaction on a subset of components at a time, two purposes are accomplished. First, the time spent on compaction analysis is reduced since only a small number of components are involved at any one time. Second, by compacting subcircuits and operating on the result as a new primitive in subsequent compaction analysis, global information on connectivity is introduced at the compaction analysis level. From the perspective of planning, the compaction operation is just another primitive operator.

The constraints imposed by the fabrication-specific rules are of the lower-bound type. In addition, there are upper-bound constraints introduced by TALIB during topological-level design. Using these lower- and upper-bound constraints, compaction proceeds by constructing a constraint graph plotting G_x and G_y for a specific set of subcircuits to reflect spacing constraints in the x and y axes, respectively, for the underlying primitives. A graph algorithm for finding the longest path is applied to G_x and G_y to determine the coordinates of the mask elements after compaction. The algorithm used is very similar to the critical-path analysis approach used in the CABBAGE system [5].

TALIB cannot evaluate the merit of its topological plan without executing it by invoking the compaction subsystem. The result of such evaluation might indicate that the cell layout resulting from a particular topological plan (or partial plan) does not meet boundary constraints. The type of feedback that TALIB receives from the compaction or critical-path analysis on a particular axis includes:

1. what groupings (and, hence, the individual components making up each grouping) are part of the critical path in either the x or y axis,
2. by what amount the x or y axis bound was exceeded (this is reported in terms of "slack" in the direction of the compaction; if the bound was exceeded, then the slack is negative),
3. what the slack is in the axis orthogonal to the compaction axis for components that are not part of the critical path.

These types of information are used by TALIB in making local changes to the design plan.

3.7 Effect of Knowledge on Performance

The problem-solving behavior of a system can be characterized by either of two extremes. At one extreme, it can be based on search with minimal knowledge about the task domain. For reasonably complex domains, this can result in a combinatorial explosion as large numbers of alternatives are generated at each decision point. This type of system will require large amounts of search to discover a solution. At the other extreme, the availability of large amounts of knowledge about a task can result in a problem-solving behavior based purely on match. If match can be used for a particular task, then essentially no search is required. Between the two extremes, knowledge about the task can be used to reduce the number of alternatives to be examined at each decision point, hence reducing the amount of required search.

TALIB exemplifies a problem-solving strategy based on augmenting search at multiple abstraction levels with large amounts of domain knowledge to guide the search process at the various levels. Because of the complexity of the task domain, there is a large body of domain knowledge that can be brought to bear during a cell layout. In order to understand how large amounts of domain knowledge can be used to enhance the problem-solving capabilities of design systems such as TALIB, it is necessary to evaluate the role of domain knowledge. Ideally, the problem-solving system should be able to make a graceful transition from a knowledge-intensive problem-solving behavior to one based on search with causal or deep knowledge [6]. Although TALIB does not possess the deep knowledge to attack the cell layout problem from first principles, its knowledge base is built in a layered fashion, with more general knowledge at the lower levels and more specific knowledge at the higher levels. This architecture is suitable for evaluating the effects of different types of knowledge.

When design is viewed as a search problem, there are two ways in which the domain knowledge can be used in developing a design tool. One way is through simplification of the problem space by taking advantage of the task's characteristics. Another way is to use knowledge to assist in the actual search process.

3.7.1 The Problem Space

The problem space for a task consists of states that describe partial solutions, operators to move from one state to another, a state designated as the initial state, and a set of states designated as goal states. A search within a problem space is represented by the problem solver's attempt to reach one of the goal states from the initial state. The issues in search control include detecting when a problem has been solved, determining when a dead end has been reached, selecting an operator, and selecting the next state upon which to apply an operator. To be useful, the information needed for controlling the search must be directly available to the problem solver. Otherwise, the problem of infinite recursion, involving searching an infinite number of problem spaces, can result.

The Basic Problem Space

In its simplest form, the cell layout problem can be formulated as determining the rectangular shapes for different circuit components and their locations in each of the various layers of semiconductor materials such that the resulting layout exhibits the required function. The spacing and overlapping of the rectangles are constrained by the specific design rules for the fabrication process. Within this formulation, the states in the cell layout problem space consist of partial or complete placement of rectangles that form the transistors, contacts, and wires such that none of the design rules are violated. A solution is any state that has a properly defined set of rectangles for each transistor in the input specification, a proper set of rectangles to realize the required interconnections, and a placement of rectangles that satisfies the boundary constraints on layout size and ordering of I/O terminals. This formulation defines a problem space that provides the greatest amount of flexibility in designing layouts within the framework of Manhattan geometry.

The problem space for the cell layout task has two characteristics that encourage the incorporation of domain knowledge for simplifying the search. First, the problem space associated with the cell layout task is very large. This implies that an exhaustive search of this problem space is impractical for all but the most trivial situations. Second, most of the cell layout problems are initially underconstrained, and as a result there are typically numerous acceptable solutions. Thus, the cell layout process may be halted after discovery of any solution that satisfies the initial boundary constraints, rather than the solution that uses minimal layout area. Although a tradeoff can be made between the amount of time spent on search and the effectiveness of the solution in terms of layout area, in most situations it is not warranted. This is because the cell layout problem is only a part of the larger chip layout problem. The most area-efficient layout for one cell can result in a larger chip layout due to the extra area required to implement a more complex intercell wiring. Therefore, the most important criterion in carrying out the cell layout task is minimization of the time required to discover any of many solutions so that a variety of cell layout alternatives can be explored by altering the boundary constraints.

There are several approaches for using domain knowledge to simplify the problem space and reduce the required search. TALIB relies on two techniques to reduce the basic problem space. One way is to recast the problem by defining new objects, relations, and operators.

Within TALIB the layout problem has been reformulated to simplify the most primitive problem space with minimal sacrifice in layout efficiency. This simplification consists of restricting the number of possible layouts for individual transistors and contacts. Thus, TALIB is constrained to design layouts from a standard set of primitives for transistors and contacts. This simplification has the primary effect of allowing the knowledge in TALIB to be focused on developing good placements for primitives to minimize the area needed to realize the interconnection requirements. Since the bulk of the layout area in a cell is affected by how transistor primitives are placed and not by the individual layout of the transistors, this simplification is not restrictive.

Another approach to reducing the amount of required search is to rely on abstractions of the basic problem space. Unlike the problem reformulation technique, the use of abstraction levels only temporarily hides the unimportant details; they are subsequently reintroduced into the design. However, by properly defining and using abstraction levels, TALIB can highlight important parameters at different levels and conduct a search through a simpler problem space. A solution obtained at one abstraction level is then used to guide the search at a lower abstraction level.

3.7.2 The Domain Knowledge

The domain knowledge contained in TALIB to reduce the amount of search falls into four categories. First, there is knowledge involved in identifying the instances of known design situations in the layout state. Second, in response to the development of design situations at one level, there is knowledge to refine these situations into more specific designs at a lower abstraction level. Third, there is axiomatic knowledge about the cell layout domain for maintaining consistency in the design state and propagating effects of local design decisions to other parts. Finally, there is control knowledge that determines on which part of the design and at what abstraction level the system should focus next.

These four categories of knowledge account for 76 percent of the more than 2000 rules in the knowledge base of TALIB. Of the remaining 24 percent, 14 percent are used for design rule checking and compacting the layout. The remaining 10 percent are responsible for inputting problem specifications, setting up internal data structures, generating diagnostic information, and outputting the layout. In this subsection, each type of knowledge, its representation in the knowledge base, and its effect on the search process is described.

3.8 Summary

The performance of TALIB can be evaluated from two perspectives. First, the efficiency of the layouts designed by TALIB can be compared to those obtained through other means. Second, the efficiency of knowledge utilization by TALIB must be considered. In terms of layout efficiency, TALIB has designed layouts that are within 5 percent of manually generated layouts that are similarly constrained (same technology and Manhattan geometry). The effect of knowledge utilization can be measured in terms of number of rule activations in different knowledge categories as compared to the complexity of the design. In general, the run time performance of TALIB is more dependent on the boundary constraints for the layout than on the number of transistors in the circuit. Typical run times are on the order of 5 CPU minutes on VAX-11/780 running VMS for a circuit containing about 30 transistors. A detailed study of TALIB's performance is reported in a Ph.D. thesis [8].

Advances in IC fabrication technology offer the potential for implementing a variety of exotic, specialized systems through lowered manufacturing costs. This in turn has led to a great deal of pressure toward improving the efficiency of the IC design process. Currently, the improvements in design tools seriously lag advances in fabrication technology. As a result, the design phase constitutes a major bottleneck in the generation of IC layouts. One way of enhancing the effectiveness of design tools is to supplement them with large amounts of domain-specific knowledge. The knowledge can be used to detect and handle exceptional conditions that arise in the design state. This paper has examined the issues of using domain-specific knowledge for developing a more effective design aid for the cell layout task.

Acknowledgment

The research work at Carnegie-Mellon University that produced the results described in this chapter was conducted through a fellowship from XEROX-PARC and grants from the Semiconductor Research Corporation. I owe a special debt to Professor Daniel Siewiorek and Dr. John McDermott for their advice and suggestions on this work at Carnegie-Mellon University. My special thanks to Ms. Kristin Buono for helping me put together this chapter.

References

1. C. L. Forgy, *OPS5 User Manual*, Tech. report CMU-CS-81-135, Carnegie-Mellon University, 1981.
2. C. Mead and L. Conway, *Introduction to VLSI Systems*, Addison-Wesley, New York, 1979.
3. E. D. Sacerdotti, *A Structure for Plans and Behavior*, Elsevier, New York, 1977.
4. G. Polya, *How to Solve It*, Doubleday Anchor, New York, 1957.
5. Min-Yu Hseuh, *Symbolic Layout and Compaction of Integrated Circuits*, Ph.D. diss., University of California, Berkeley, 1979.
6. P. S. Rosenbloom, J. E. Laird, J. McDermott, A. Newell, and E. Orciuch, "R1-Soar, An Experiment in Knowledge-intensive Programming in a Problem-solving Architecture," *IEEE Workshop on Principles of Knowledge-based Systems*, IEEE Computer Society, 1984.
7. B. Hayes-Roth and F. Hayes-Roth, "A Cognitive Model of Planning," *Cognitive Science*, vol. 3, 1979.
8. Jin H. Kim, "Use of Domain Knowledge in Computer Aid for IC Cell Layout Design," Ph.D. diss., Carnegie-Mellon University, 1985.

CHAPTER 4

An Intelligent Composition Tool for Regular and Semiregular VLSI Structures

Hung-fai Stephen Law
Graham Wood
Mindy Lam

4.1 Introduction

Regular and semiregular structures such as ROMS, RAMs, PLAs, ALUs, and register files are being used more and more widely in integrated circuit design. Because of their regular structures, these components can be assembled automatically by programs called *module generators*, which shortens the design time. In addition, module generators allow quick changes, yielding more efficient designs.

First-generation module generators[1,2,3] have been written for a variety of components. These generators are often built individually and are fairly rigid, so that changes to the structure or to the set of cells used imply a rewrite of the software, which can only be done by a software expert.

Bamji et al.[4] proposed a "design by example" approach for defining the legal combination of cells in a library. All legal relative placements between any two cells must be defined by sample layouts before any composition using these cells can begin. Variations of legal placements between two cells can be easily specified by providing the corresponding sample placement layout files. Cell overlapping is allowed. To create a module, the composition must be described in a procedural design file using a language embedded in LISP.

In this paper, we present a new tool called the *structure compiler*, which is used to build *module generators*. This concept is similar to the notion of a compiler-compiler in the software literature, in that modifying a module generator amounts to only changing the "rules" used by the structure compiler to produce the module generator. The basic information needed by the structure compiler to produce a module generator consists of (1) a parameterized block diagram or floor plan of the array, called the *array structure template,* and (2) a library of cells to be assembled. Placement information for all the major blocks in a module is interpreted from the array structure template. Textual procedures are used only to provide extra optimization of the module as well as friendly user interface. Because of the generality of this tool, generators for both regular arrays such as ROMs and semiregular arrays such as PLAs can be obtained.

4.2 Module Generation with the Structure Compiler

The structure compiler is a tool that automatically generates the final layout of a module from an array structure template, a library of master cells, and an optional *personality* file. The module to be generated may contain multiple blocks of arrays. There are two types of arrays: homogeneous arrays and heterogeneous arrays.

A *homogeneous array* is an array using only one master cell. Only the number of rows and columns together with the master cell are needed for the generation of a homogeneous array. An array of input buffers to a ROM is an example of a homogeneous array.

A *heterogeneous array* is an array of many different master cells. The structure compiler uses a personality file specified by the user to determine which cell is to be placed at a given location. For example, the layout of a ROM generally contains two different cells, one with a transistor and one without. The personality file specifies where to insert one cell versus the other.

In the following sections, we explain how to specify the structure of a module by the array structure template and how to specify the cell library.

4.2.1 Structural Specification of a Module

The array structure template is a block diagram representation of the module to be generated. All generic structural information that is not technology-dependent is stored in the array structure template. Figure 4.1 shows a template for a programmable logic array (PLA) structure. Each array structure template is composed of one or more rectangles, called *array blocks*. Each array block signifies a single array of one or more master cells from the user-provided library. The specification of which cell is to be placed at each location is provided in the personality file.

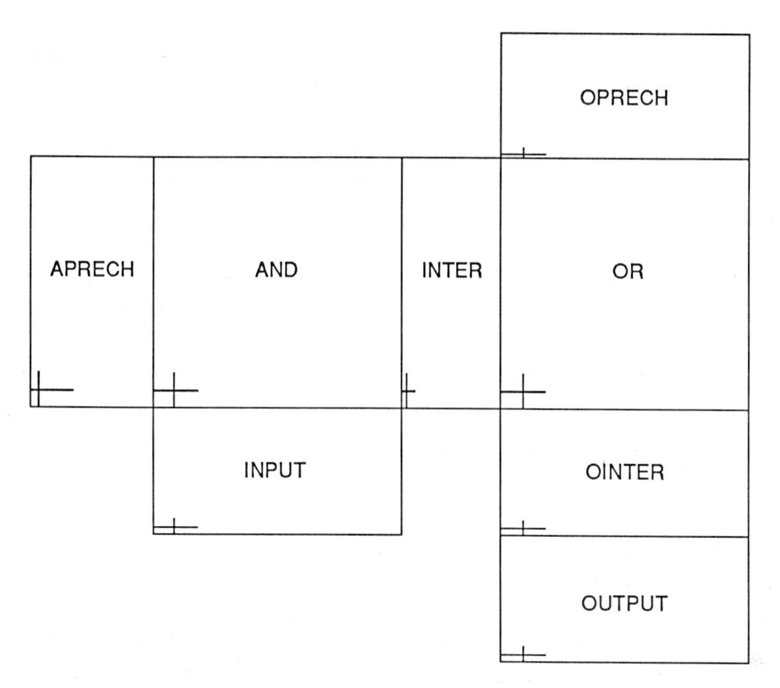

Figure 4.1 Array structure template for a PLA.

The relative positioning of the array blocks in a module is specified through matching corners of neighboring array blocks. The template in Figure 4.1 has eight array blocks, each representing a different array with a specific functionality. Each array block also has a list of properties or parameters, including:

- which section of the personality file to use
- the type of array (homogeneous or heterogeneous)
- the mapping of single-character symbols in the personality file to other single-character symbols (For example, the symbol "O" is mapped onto two symbols: "N" followed by "T")
- the number of columns and rows, if needed (If this property is omitted, the array will be sized automatically by the personality file; the number of columns or rows can also be set to equal the number of columns or rows of another array block)
- the ways to translate symbols in the personality file to cell names in the library
- a two-dimensional *orientation pattern* to be repeated throughout the entire array
- rules for promoting "pins" from lower-level leaf cells to the module level
- a procedure to further manipulate the array

Figure 4.2 shows the property list from the AND array block in Figure 4.1. In Figure 4.2, the property *personality* specifies a block named "pO"

```
name: AND
personality: p0
type: HETEROGENEOUS
set:
map: 1:TN; 0:NT; -:NN;
masters: N:plaTran0 layout; T:plaTran1 layout;
orientation pattern: 2 4
                     R180 MY
                     MX R0
                     MY R180
                     R0 MX
pin:
procedure: OptimizeAnd( )
```

Figure 4.2 Properties of an array block.

in the personality file for the placement of different masters. The *type* is heterogeneoous. Since a block of the personality file is going to be used, the *set* property is not specified; the size of the array block will be automatically set based on the personality file. The *map* property specifies that all the "1" symbols in the personality be replaced by the symbol "T" followed by the symbol "N," all the "O" symbols be replaced by the symbol "N" followed by the symbol "T," and all the "-" symbols be replaced by two "N" symbols. Once these replacements are made, only the symbols "N" and "T" will remain in the personality. The *masters* property further specifies the actual cells to be used for the corresponding symbols in the personality. The *orientation pattern* specifies a 2-by-4 pattern of orientation transformation to be used throughout the entire array block. Since the AND block of the PLA module specified in Figure 4.1 does not directly communicate with external circuitry, the *pin* property is empty. A procedure named "OptimizeAnd" is specified for evaluation by the structure compiler after this block has been synthesized.

It is sometimes necessary to set the size of an array to be the same as that of another array. For example, the *set* property for the INPUT array block is specified as:

```
set: columns = columns(AND)
```

This causes the size of INPUT to track the size of the AND block, which in turn tracks the size of the "pO" section of the personality file.

The array structure template is used as a graphical procedure for the structure compiler to generate the module.

4.2.2 Master Cell Library

There is no limitation on the kind of information in the master cells. The master cells are typically entered using a graphics editor. The only requirement for the cell is a rectangle drawn on a special layer. This rectangle provides the size of the master cell to the structure compiler. Since the dimension information is stored at this level, the structure compiler can assemble arrays that have different spacing for different columns and rows.

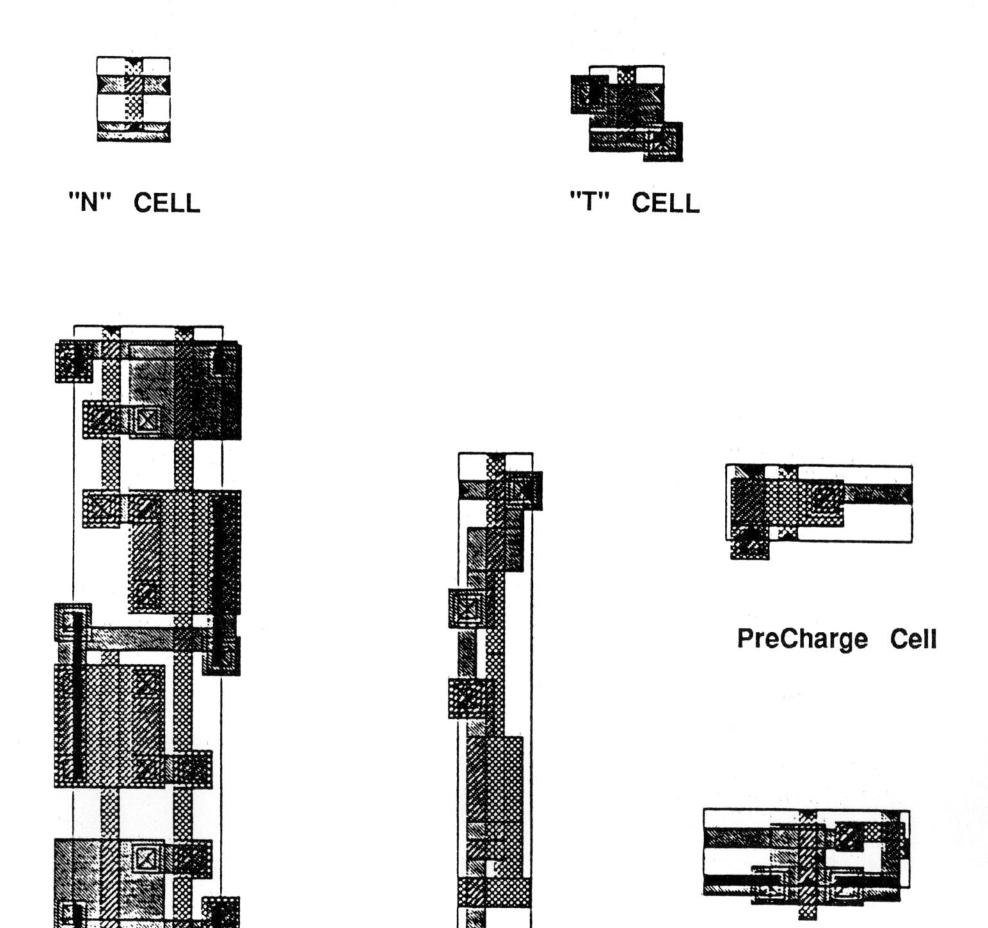

Figure 4.3 Master cells for a PLA.

This is an important feature in supporting heterogeneous arrays. Figure 4.3 shows the master cells used by the PLA structure shown in Figure 4.1.

4.2.3 Personality File

To generate a module that contains heterogeneous arrays, a personality file may be provided. A personality file contains connectivity information and is used to assemble the parts for the overall structure. This file can be the truth table for a PLA if the module is a PLA. It can be the character symbolic layout file for a fixed-grid character-based symbolic layout system. It can be a relative placement file for the different blocks in a data path or array multiplier. Each symbol in the personality is a single character that can be

```
.name alustate
.pname p0 p1
------00000   0000000000000000000
------00001   1100000000000000100
0000-0001-   0010110101011000000
0001-0001-   0010110101111000000
0010-0001-   0000110101011000000
0011-0001-   0000110101111000000
0100-0001-   0010001111011000000
0101-0001-   0011110111011000000
0110-0001-   0010110111011000000
0111-0001-   0011001111011000000
1000-0001-   0011000111011000000
1001-0001-   0010111111011000000
1010-0001-   0010011111011000010
1011-0001-   0010011111011000001
1100-0001-   0000110101010000000
110110001-   0000110101001010110
110100001-   0000110101000010110
------10010-   0010110101001011110
------00010-   0010110101000011110
------10101-   0010110101001000110
------00101-   0010110101000000110
------10110-   0010110101001001110
------00110-   0010110101000001110
------11001-   0010110101001110110
------01001-   0010110101000110110
------11010-   0010110101001111110
------01010-   0010110101000111110
------11101-   0010110101001100110
------01101-   0010110101000100110
------11110-   0010110101011101000
------01110-   0010110101010101000
```

Figure 4.4 Personality file for a PLA.

further mapped to one or more single-character symbols by specifying the mapping function in the array block property list. Each of these symbols is then translated to one of the cells in the library. Figure 4.4 shows a personality file that is the truth table for a PLA named "alustate."

4.2.4 Array Procedure

To allow the user more freedom and power to further optimize the arrays, a user-provided procedure can be accepted by the structure compiler to manipulate symbols in an array after it has been generated from a template and a personality file. The procedural language permits the user to inquire about the contents of a certain row and column location of the personality and can conditionally specify the following operations:

- insert a column or row
- replace a symbol at a certain location with another symbol
- embed a new symbol at a certain location and thus increase the row/column size

- set new masters
- set new orientation patterns

The array procedure can be used to generate a new personality based on properties provided by the user. The language used is a functional language. The syntax of this language is very close to the programming language C. It has all the control constructs, such as *if*, *for*, *while*, and *case*, as well as *procedure* and *macro* capabilities commonly found in programming

```
;
; Function for stripping unused poly from product
; terms in the OR plane of a PLA.
; Name of the function is "OptimizeOr"
procedure(OptimizeOr( )
        ;
        ; Initialize variables.
        ;
        no_poly = "M"
        no_transistor = "0"
        ;
        ; Start at the right end of each row of the OR plane,
        ; and work towards the left end, replacing
        ; "no-transistor" cells with "no-poly" cells,
        ; until a "transistor" cell is encountered.
        ;
        ; Variables rows and columns are defined by the system
        ; for array, representing the number of rows and
        ; columns in the current array.
        ;
        for (i 0 rows
                stopP = nil
                ;
                ; Iterate over the current row, replacing cells.
                ;
                j = columns - 1
                while ( ( (j > = 0) && ( not stopP) )
                        cell =  geticon (i j)
                        ;
                        ; If the cell at location (i j) is
                        ; a "no_transistor" cell, replace it
                        ; with a "no_poly" cell.
                        ;
                        if ( (head (cell) = = no_transistor) then
                                replacetile  ( (i j) no_poly )
                        else
                                ;
                                ; otherwise, we are done for the current row.
                                ;
                                stopP = t
                        )
                        j = j - 1
                )
        )
)
```

Figure 4.5 A procedure example.

languages such as C and LISP. A detailed description of this language and its capability is presented in a separate paper.[5] Figure 4.5 shows a procedure for stripping unused polysilicon from product terms in the OR array block of the PLA structure shown in Figure 4.1. The semicolon (;) is used to mark the beginning of comments in each line.

4.3 Steps in Module Synthesis

The structure compiler first generates a list of array blocks to be processed based upon the array structure template. Then, for every array block, an array is synthesized according to the properties of the array block. The user-provided procedure is then executed. After all the arrays are synthesized, the structure compiler places them according to how the corners of the array

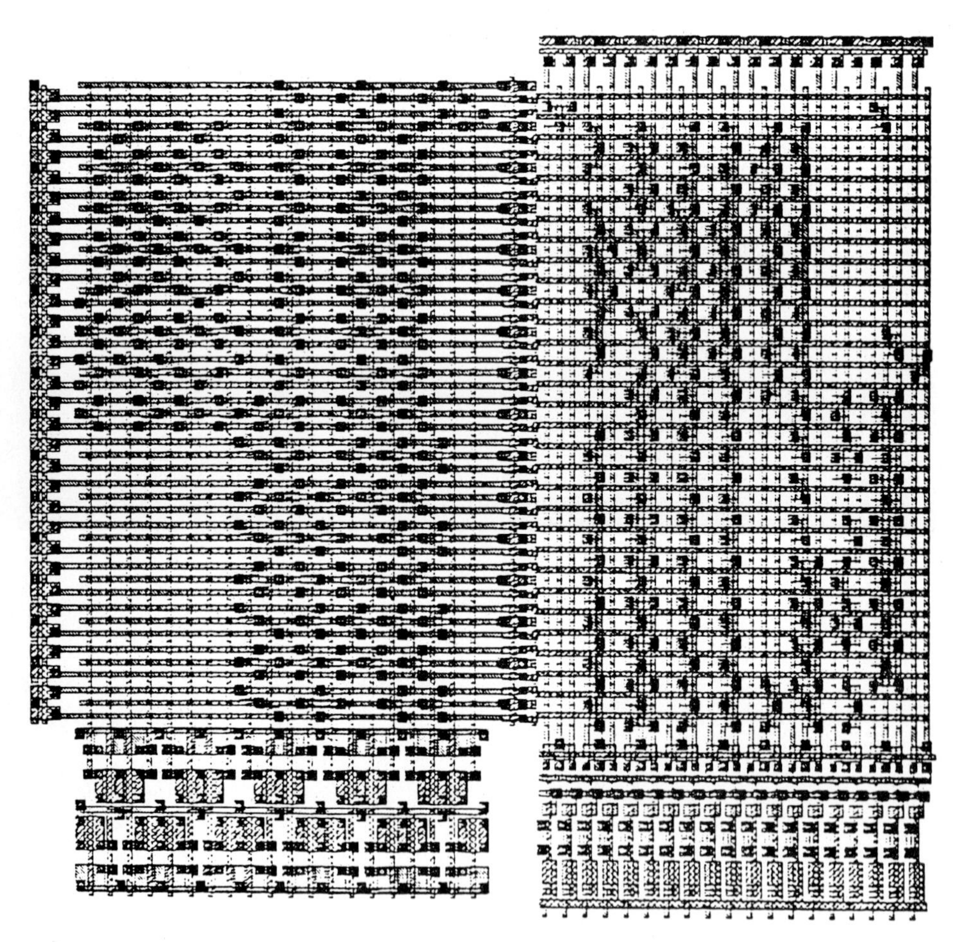

Figure 4.6 The generated layout for a PLA.

blocks in the template are matched. For example, in Figure 4.1, since the lower left corner of the AND array block matches the lower right corner of the APRECH array block, the structure compiler will place the synthesized arrays corresponding to these array blocks with the same corners matched. If the upper corners are also matched, the structure compiler will check to make sure that the corresponding synthesized arrays are the same height. This corner-matching operation is carried out for all corners of each array block until all the arrays are placed. Throughout the synthesis process, error messages as well as graphical feedback are returned if the master cells or the arrays do not fit or connect properly. Figure 4.6 shows the final layout of a PLA using the template in Figure 4.1, the library of cells in Figure 4.3, and the personality file in Figure 4.4.

4.4 Conclusion

The structure compiler provides the basic functionality of many module generators, such as PLA and ROM generators. However, instead of modifying a program in order to change the module structure, which can be error-prone, time-consuming, and may not be an option available to designers, the user of the structure compiler has only to change the array structure template, which is also a block diagram for the module. No programming is needed. Furthermore, since the array structure template is entirely symbolic and has no absolute dimensions, it is totally technology-independent and can be used for any other technology through provision of a new library of master cells.

References

1. M. W. Stebnisky, M. J. McGinnis, J. C. Werbickas, R. N. Putatunda, and A. Feller, "A Fully Automatic, Technology-independent PLA Macrocell Generator," *Proc. IEEE International Conf. on Circuits and Computers*, September 1982, pp. 156–161.
2. K.-C. Chu and R. Sharma, "A Technology Independent MOS Multiplier Generator," *Proc. 21st Design Automation Conf.* June 1984, pp. 90–97.
3. H.-F. S. Law and M. Shoji, "PLA Design for the BELLMAC-32A Microprocessor," *Proc. IEEE International Conf. on Circuits and Computers*, September 1982, pp. 161–164.
4. C. S. Bamji, C. E. Hauck, and A. Jonathan, "A Design by Example Regular Structure Generator," *Proc. 22nd Design Automation Conf.*, June 1985, pp. 16–22.
5. G. Wood and H.-F. S. Law, "SKILL—An Interactive Procedural Design Environment," *Proc. IEEE 1986 Custom Integrated Circuits Conf.*, May 1986, pp. 544–547.

CHAPTER 5

The VLSI Design Automation Assistant: An Architecture Compiler

*Thaddeus J. Kowalski**

5.1 Introduction

Recent advances in integrated circuit fabrication technology have allowed larger and more complex designs to form complete systems[1] on single VLSI chips. These chips use 1-micron to 5-micron features to achieve complexities equivalent to 100,000 to 250,000 transistors. This level of design complexity has created a combinatorial explosion of details that are a major limitation in realizing cost-effective, low-volume, special-purpose VLSI systems. To overcome this limitation, design tools and methodologies capable of automating more of the digital synthesis process must be built.

We have been developing just such synthesis tools at AT&T Bell Laboratories[2] and Carnegie-Mellon University.[3] These tools help the designer develop the algorithmic description of the system and interactively add the details required to produce a finished design. Our approach is aimed at aiding the designer by producing data paths and control sequences that implement the algorithmic system description within supplied constraints. Thus the designer can consider many alternatives before deciding on a final

* Reprinted with permission from T. Kowalski: *An Artificial Intelligence Approach to VLSI Design.* Kluwer Academic Publishers: Norwell, Mass., 1985.

design. This structured approach can decrease the time it takes to design a chip, automatically provide multi-level documentation for the finished design, and create reliable and testable designs.

A series of acquisition interviews[4] and an initial prototype system[5] have been used to bootstrap a system that generates a technology-independent list of hardware components such as operators, registers, data paths, and control signals from an algorithmic description and a list of design constraints. The Design Automation Assistant, (DAA)[6] has been used to design an IBM System/370 and was favorably evaluated by an IBM System/370 designer.[7] We are currently pursuing the design of Digital Signal Processing (DSP) and pipeline processors at AT&T Bell Laboratories.[8]

This chapter focuses on the synthesis, or allocation, of the implementation design space as it advances from an algorithmic description of a VLSI system to a list of technology-independent registers, operators, data paths, and control signals. The following discussion illustrates how the DAA uses large amounts of expert knowledge to design an architecture with little backtracking. This chapter discusses the development of the DAA's knowledge base by providing several of the MOS Technology Incorporated MCS6502 microcomputer designs along with the additions to the knowledge base. It shows the generality of design knowledge in the DAA by comparing and contrasting an IBM System/370 bipolar gate array microprocessor chip, μ370,[9] designed by an expert human designer, Claud Davis, against the IBM System/370 design produced by the DAA, D370. This chapter takes a retrospective look at that codified knowledge base, examining what has been learned about VLSI design. It discusses both the major steps in the implementation design process and the extent to which each rule embodies domain knowledge. It provides an English translation of several rules and a brief discussion of the translated rules. These translations must be taken with a grain of salt because rules do not stand by themselves; they are part of a larger network of rules connected by the working memory and the inference engine. Finally, the chapter provides an example design with typical rules from each of the major steps in the implementation design process.

This synthesis task has inspired a variety of approaches, ranging from the most simplistic backtracking methods through the most complicated constraint propagation methods.[10–14] Owing to the complexity of design synthesis, simplistic back-tracking schemes consume large amounts of CPU time, and the constraint propagation method is too cumbersome for large designs. Because of the combinatorial explosion of details and implicit dynamic constraints involved in choosing an implementation, this problem does not lend itself to these algorithmic solutions. An alternate approach to design synthesis uses a large amount of design knowledge to eliminate backtracking; whenever possible, the focus is on specific design details and constraints. Artificial intelligence researchers have called systems developed under this heuristic approach knowledge-based expert systems (KBESs).[15]

5.2 Conception

KBESs are generally developed in several stages. First, "book knowledge" of the problem is codified as a set of situation action rules; interviews with experts then fill in knowledge gaps and refine current knowledge. Then, many example problems are given to the KBES, and experts closely examine and validate the results. Often, errors are found through the examples, and new rules are added to the system to correct the error situations.

This iterative process is necessary because experts are often unaware of exactly how they go about designing a chip and are inexperienced at articulating the procedure. Furthermore, the knowledge base is not an exact codification of the expert's knowledge, but a compilation of what is understood by the knowledge engineer.

After gathering current book knowledge about synthesis of the architectural design space, [10–12] we interviewed four designers of varied experience: one was a novice, two were moderately experienced, and one was an expert. The interviews, which lasted about an hour each, started with a determination of the designer's background, including years of experience, logic families used, and designs created. Most of the time was spent discussing the design process, with some time given to a discussion of the DAA system. Our interview method was designed to allow the interviewees as much freedom as possible in generating ideas; we emphasized such questions as "What do you do next?" and "Could you elaborate?"

The designers discussed the global picture, partitioning, selection, and allocation tasks. They began with a high-level overview of the hardware, which listed inputs and outputs to the outside world, the functions the hardware should provide, general constraints, and design feasibility with consideration of the target technology. They generally partitioned the global picture into smaller blocks and emphasized minimizing connections among blocks, selecting blocks that operate as parallel or serial units, and grouping according to similarity of function. Partitions were chosen for allocation in a decreasing order of difficulty or degree of constraint. The designers reasoned that if the most difficult part could be designed, the rest of the design was feasible.

Once a partition was selected for allocation, it was carried out either in parallel or in series. A parallel design made thinking of the control logic much simpler, while a serial design minimized the design area. The constraints of the parallel design were examined for size violations to determine the parts to be serialized by adding data paths, registers, and control logic to the initial parallel design. The constraints of the serial design were examined for speed violations to determine the parts to be reimplemented in parallel. If the designers recognized a part of the design as similar to a part of a previous design, they used what they knew had worked in the past. Within each partition, designers allocated clock phases, operators, registers, data paths, and control logic. The order was interesting because once registers and data paths were allocated, they were not changed.

The control was changed because it was the hardest thing to think about and because it depended on a constant structure for the data path elements.

The designers described the iteration process as a step-by-step refinement to meet violated constraints. They looked for a technology change to meet a constraint before making a design change. This could be as simple as finding a new chip in the TTL data book or as complicated as a design rule shrink. Next, they would sacrifice functionality to meet a constraint. One designer summed it up best by saying, "An engineer's training teaches him when constraints can be swept under the rug."

The relative importance of constraints is application dependent. The designers mentioned the constraints of speed, area, power, schedule, cost, drive capabilities, and bit width. Other design changes consisted of global improvements not recognized until the design neared completion. This suggests that the general choice of partitions and the initial design style selections approached optimum and that designers do not seem to use much backtracking in their designs.

5.3 Birth

Even though many details were missing, enough book knowledge had been gathered to put together a prototype version of the DAA system using the OPS5[16] KBES writing system. While the DAA system was far from perfect at this point, it stimulated further elicitation sessions with expert designers. We now turn our attention to the flow of control in the prototype system and how the KBES approach formulates the problem.

5.3.1 Flow of Control

The DAA starts with a data-flow representation extracted from the algorithmic description. This representation resembles the internal description used by most optimizing compilers,[17] but computer programs manipulate it more easily, and it is felt to be less sensitive when the same algorithm appears in a variety of writing styles.

The DAA produces a technology-independent hardware network description. This description is composed of modules, ports, links, and a symbolic microcode. The modules can be registers, operators, memories, and buses or multiplexers with input, output, and bidirectional ports. The ports are connected by links and are controlled by the symbolic microcode.

The DAA uses a set of four temporally ordered subtasks to complete the synthesis task. First, the base-variable storage elements—constants, architectural registers, and memories with their input, output, and address registers—are allocated to hardware modules and ports. Second, a data-flow BEGIN/END block is picked, and the synthesis operation assigns minimum delay information to develop a parallel design. Third, it maps all data-flow operator outputs not bound to base-variable storage elements to register modules. Fourth, it maps each data-flow operator, with its inputs and outputs, to modules, ports, and links. In doing so, the DAA avoids multiple assignments of hardware links; it supplies multiplexers when necessary. The last two mapping steps place the algorithmic description in a uniform notation for the expert analysis phase that follows.

The expert analysis subtask first removes registers from those data-flow outputs where the sources of the data-flow operator are stable. Operators are combined, according to cost and partitioning information across the allocated design, to create ALUs. The DAA also examines the possibility of sharing non-architectural registers. When possible, it performs increment, decrement, and shift operations in existing registers. When appropriate, it places registers, memories, and ALUs on buses. Throughout this subtask, constraint violations require tradeoffs between the number of modules and the partitioning of control steps. The process is repeated for the next data flow BEGIN/END block.

5.3.2 The OPS5 Writing System

The DAA is implemented as a production system via the OPS5 KBES writing system. The KBES tool is based on the premise that humans solve problems by recognizing familiar patterns and by applying their knowledge in the current situation. The tool formulated a problem by using three major components: a working memory, a rule memory, and a rule interpreter.

Working Memory

The working memory is a collection of elements that describe the current situation. The elements resemble the records in conventional programming languages:

```
literalize module
id: adder.0
type: operator
atype: two's complement
bit-left: 17
bit-right: 0
attribute: +
```

This working-memory element describes an operator module *adder.0,* which can do two's complement addition on 18 bits of binary data.

Rule Memory

The rule memory is a collection of conditional statements that operate on elements stored in the working-memory. The statements resemble the conditional statements of conventional programming languages:

```
IF:
    the most current active context is to create a link
    and the link should go from a source port to a destination port
    and the module of the source port is not a multiplexer
```

and there is a link from another module to the same destination port
and this other module is not a multiplexer

THEN:
create a multiplexer module
and connect the multiplexer to the destination port
and connect the source port and destination port link to the multiplexer
and move the other link from the destination port to the multiplexer

This rule recognizes situations in which a multiplexer needs to be created to connect one port to another.

Each subtask in the DAA is associated with a set of rules for carrying out the subtask. The previous IF-THEN statements provide an example of a rule for the fourth subtask. Most of the rules, like this example, define situations in which a partial design should be extended in some particular way. These rules enable the DAA to synthesize an acceptable design by determining, at each step, whether a certain design extension respects constraints.

The Rule Interpreter

The rule interpreter pattern matches the working-memory elements against the rule memory, to decide what rules apply to the given situation. The rule selection process is data driven; the rule interpreter looks through the rule memory for a rule whose antecedents match elements in the working memory. This is also called *forward chaining* or *antecedent reasoning.* The consequences of the rule are applied, and the process is repeated until no more rules apply or until a rule explicitly stops the process. If more than one rule applies, the rule dealing with the most current working memory is selected first. If multiple rules are still applicable, the most specific rule is selected. This selection mimics following a train of thought, as far as possible, and uses special-case knowledge before general-purpose knowledge. The separation of expert knowledge from the reasoning mechanism makes the incremental addition of new rules and the refinement of old ones easy because the rules have minimal interaction with one another.

5.4 First Steps

The prototype DAA system had about 70 rules and could design a MOS Technology Incorporated MCS6502 microcomputer in about three hours of VAX 11/750 CPU time. We asked many expert designers at INTEL and AT&T Bell Laboratories to critique the design by explaining what was wrong, why it was wrong, and how to fix it. After each critique, rules were modified, new rules were added, and the MCS6502 was redesigned. Based on the critiques, the development DAA system now has over 300 rules and has designed a much better MCS6502 microcomputer in about five hours of

VAX 11/750 CPU time. In retrospect, clearly much of what we learned was common sense design knowledge, the same things human designers learn through apprenticeship. The DAA has undergone many improvements and produced many designs of the MCS6502 microcomputer. We will now illustrate a few of these changes.

Each knowledge acquisition interview began by giving the designer a drawing of the design with a sheet of clear plastic over it. Before the designer started the critique, pieces of cardboard were placed over the design. As the designer proceeded, a piece of cardboard had to be lifted, the correction had to be written on plastic, and a new layer of plastic placed over the design. This provided a complete record of where the designer was focusing attention and what was corrected. The designers found this elicitation procedure compatible with their normal spatial mode of operation.

The first prototype DAA system was used to produce the design summarized in column 1 of Table 5.1. Each row shows the bits of the specified operator or register type found in the design. (An illustration is not provided as a complement to column 1 of this table because such a figure is totally inscrutable). The expert criticism is summarized in four points:

- Operators of different types and sizes should be combined into ALUs.
- Any 1-bit operators within the same block should not be combined, because multiplexers are more expensive than most 1-bit modules.

Table 5.1 MCS6502—Three Designs

Designs	*1*	*2*	*3*
And	20	20	20
Cmp	177	1	1
Minus	64	0	0
Or	9	9	9
Not	21	21	21
Plus	540	0	0
Shifts	35	1	1
Xor	9	9	9
Alu	0	35	35
Dreg	450	281	210
Treg	1,227	0	62
Mux In	2,122	2,657	473
Mux Out	293	377	84
Bus In	0	0	769
Bus Out	0	0	210

- Registers should increment, decrement, and shift their values internally, whenever possible.
- Temporary registers to the controller should be eliminated, and one latched register should be placed in front of the controller.

The rules were changed to produce the design summarized in column 2 of Table 5.1 and illustrated in Figure 5.1. To produce this design, partitioning information was added, based on connectivity of data paths and similarity of operators among blocks. This simplified the decision about which modules to combine when hardware operators are shared among abstract operations detailed in the algorithmic description. Rules were also added to combine modules of different sizes and types. As column 2 shows, the ALU number

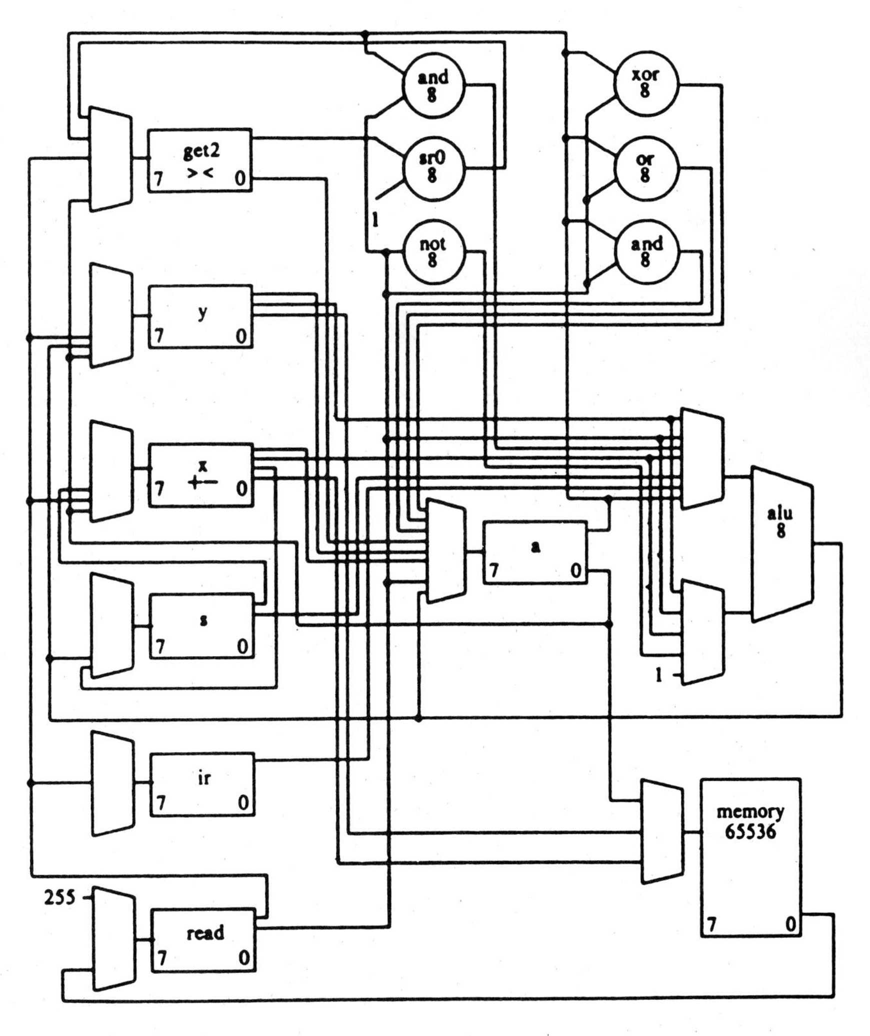

Figure 5.1 MCS6502—8-bit MUX data paths.

increased, decreasing the plus, minus, shift, and compare numbers. Rules were also added to decrease the amount of temporary register storage.

Figure 5.1 shows the 8-bit data paths of the MCS6502. The 1-bit and 16-bit data paths were omitted for clarity. Each of the symbols represents a module. The circles are single-function ALU modules to AND, SHIFT, NOT, XOR, and OR data; the small trapezoids are multiplexers that gate one of their inputs to the output, the small rectangles are registers; the large rectangle is the memory; the large trapezoid is a multifunction ALU; and each of the lines represents a link between the modules. Where the links join with the modules, a port is defined. An obvious problem, pointed out by our experts, was overuse of multiplexers. They suggested ways of distributing the multiplexer hardware to form buses.

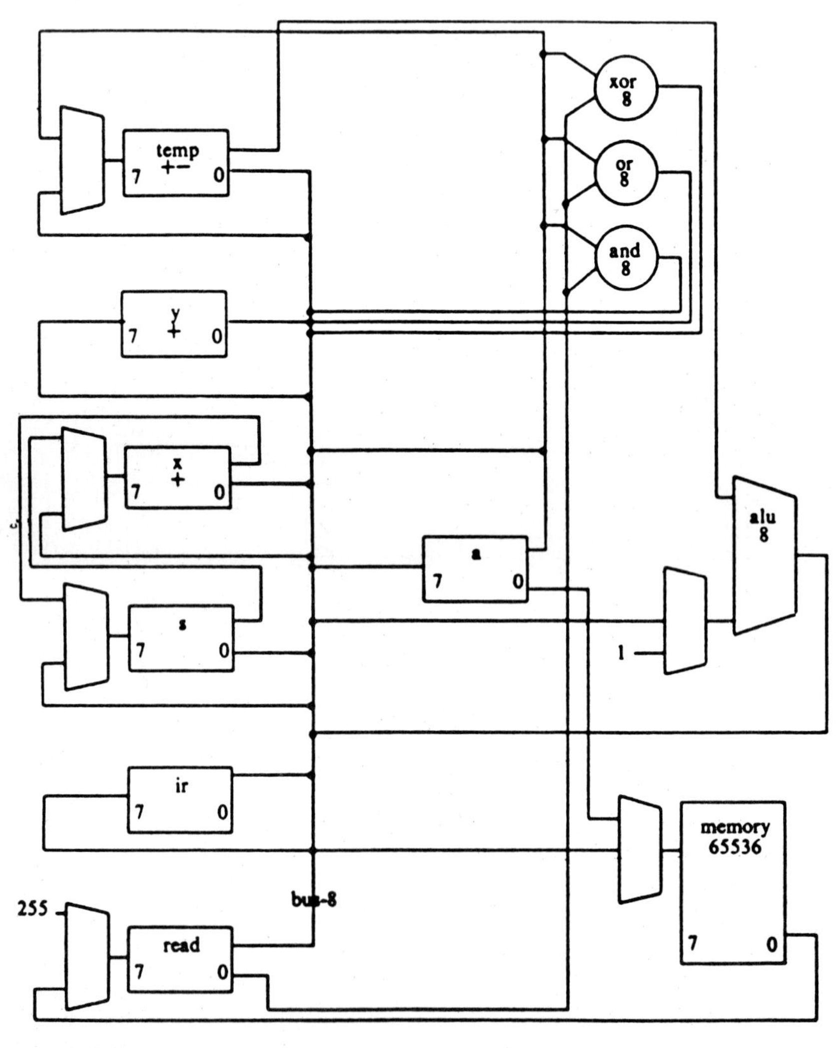

Figure 5.2 MCS6502—8-bit BUS data paths.

The rules were changed in response to these critiques, resulting in the design shown in column 3 of Table 5.1 and in Figure 5.2. To produce this design, rules were added to recognize when a multiplexer should be converted into a bus and how to share that bus with other distributed multiplexers. In addition, new rules decreased the amount of declared register storage. Specifically, registers were not needed to multiplex information into the data-flow BEGIN/END blocks. As column 3 shows, the multiplexer numbers decreased, increasing the bus numbers. The declared register number also dropped.

Although this design was acceptable to our experts, it was not perfect. Further changes led to improvements such as multiple buses of different widths. However, these changes did not affect the MCS6502, because it did not require multiple buses. This brings up an interesting point about expert systems: they are never totally finished. Like human designers, the DAA becomes a better designer as its rule memory expands. Until all possible world knowledge about designing microprocessors has been codified in the DAA's rules, there will always be room for improvement in its designs.

5.5 The IBM System/370 Experiment

After the DAA successfully designed a MCS6502 microprocessor, it had to be determined whether the system had also acquired knowledge about processor design in general. In this regard, an experiment was designed to see whether the DAA could design a processor substantially different and more complex than the MCS6502.

An ISPS description was chosen for the complete IBM System/370 from the descriptions maintained at Carnegie-Mellon University. This description included memory management operations, channel controller I/O instructions, and all the 370 instructions, except the extended-precision floating point, the characters under mask, the edit and mark, and the packed-decimal instructions. The unmodified System/370 description, missing only a small percentage of the total 370, is more than 10 times larger than that of the MCS6502, and it had not been used to build the DAA. (The next bigger description, the Digital Equipment Corporation VAX 11/780, would not compile through the VT compiler in a 6-megabyte address space.) Important benefits of this choice are that a single-chip design of the 370 had been made at IBM, information was publicly available, and Claud Davis, the design team manager and a key designer, was willing to critique the design. Thus the experiment was a fair and convenient way to test the generality of the DAA's design knowledge.

5.5.1 The DAA 370 Design

The DAA designed the D370, its version of the System/370, in 47 hours of CPU time on a VAX 11/780 with 6 megabytes of memory and 2 memory controllers. The D370 was designed without rule modifications or design iterations of any type.

The D370 is an IBM System/370 data-flow design using a 50x clock, where x is some scaled unit of time (e.g., μ seconds) with multiplexer and bus-style data paths. The DAA's constraints were set to produce a high-performance machine—that is, it could use as much hardware as required to allocate the data paths and retain maximum parallel operator usage. To meet this performance constraint, the D370 has 8-bit, 24-bit, and 64-bit buses, 32-bit, 64-bit, and 68-bit ALUs, a few discrete components, 6 memory arrays, and a great many architectural registers.

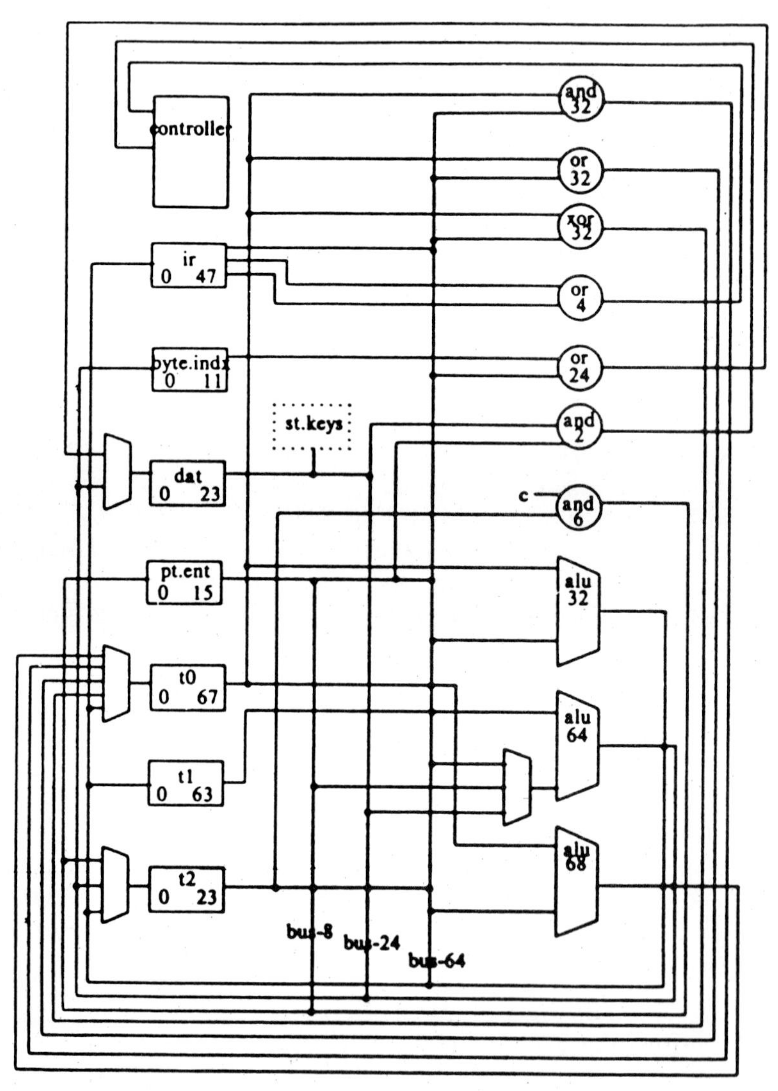

Figure 5.3 The D370 design—part 1.

The Three ALUs

The three ALUs and bus sizes arose from three different groups of data operations and major transfer widths in the IBM System/370. The basic busing style placed a temporary register before an input to each ALU and assumed the ALU latches its result so it can be read on the next clock phase transition. Thus a two-phase clock set up the inputs to an ALU in one clock cycle and stored the result on the next clock cycle.

For clarity, the data path of the D370 is drawn in two separate figures (see Figures 5.3 and 5.4.) The connection between the two figures is through the 8-bit, 24-bit, and 64-bit buses. Figure 5.3 contains the arithmetic

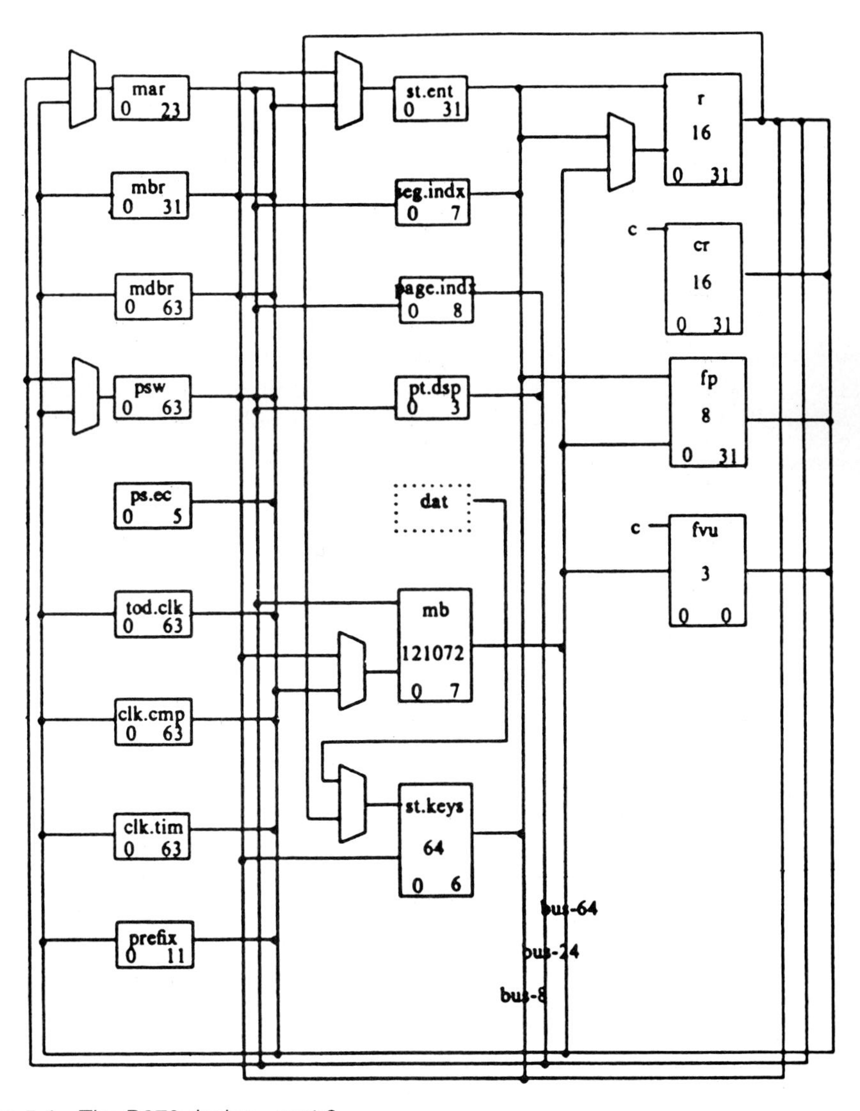

Figure 5.4 The D370 design—part 2.

portion, the temporary registers, and the controller; Figure 5.4 shows the architectural registers, including the register arrays, such as the 16 general-purpose registers R.

The 32-bit ALU is used for most of the arithmetic operations in the System/370 architecture. It can ADD, SUBTRACT, and COMPARE two binary numbers from the T0 temporary register and the 64-bit bus. It gates a result out on this bus.

The 64-bit ALU is used for most of the address calculation operations and a few low-frequency operations, such as MULTIPLY and MODULUS, in the System/370 architecture. It can ADD, SUBTRACT, COMPARE, MULTIPLY, MODULUS, and SHIFT RIGHT two binary numbers from the T1 temporary register and a bus. It can gate a result out on either the 24-bit or 64-bit bus.

The 68-bit ALU is used for most of the floating-point operations in the System/370 architecture. This ALU can ADD, SUBTRACT, COMPARE, and SHIFT LEFT two binary numbers from the T0 register and the 64-bit bus. Its result is gated onto the 64-bit bus.

The Discrete Components

Not all data manipulation is done in the ALUs. To aid debugging, one of our expert designers from INTEL keeps single-function logic outside the ALU. Thus the logic instructions are implemented with separate distributed logic elements: the 32-bit AND, OR, and XOR.

Smaller logic elements are provided for a variety of functions. The 4-bit OR takes input from two fields of the instruction register (IR) and feeds the result to the microcontroller. This aids in instruction decoding. The virtual storage system uses three discrete components. The 24-bit OR takes input from the byte index (BYTE.INDX) register and the 64-bit bus and places the result in the dynamic address translation (DAT) register. The two-bit AND takes input from the 24-bit bus and the page table entries (PT.ENT) and feeds the result to the microcontroller. The 6-bit AND takes input from a group of constants and T2 and places the result on the 8-bit bus.

The Memory Arrays

The D370 architecturally defines the primary memory (MB), the storage keys (ST.KEYS), the general-purpose registers (R), the control registers (CR), the floating-point registers (FP), and the floating-point error registers (FVU). Table 5.2 lists the bit width of each memory array, the number of words in the array, and what buses *bb* or registers connect to the address, input, and output ports. Thus there are 64 storage keys, each 7 bits wide, with their address ports connected to the general-purpose registers and the dynamic address translation register. The input and output ports are connected to the 8-bit bus.

The Architectural and Temporary Registers

The D370 architecturally defines many registers. Table 5.3 lists the bit width of each register and tells what buses or registers connect to the input

Table 5.2 Memory Arrays in the D370 Design

Abbreviation	*Bits*	*Words*	*Address*	*Inputs*	*Outputs*
MB	8	121072	24bb	8bb, 64bb	64bb
ST.KEYS	7	64	R, DAT	8bb	8bb
R	32	16	8bb	8bb, 64bb	ST.KEYS, 8bb, 24bb, 64bb
CR	32	16	Constants		64bb
FP	32	8	8bb	64bb	64bb
FVU	1	3	Constants	64bb	64bb

and output ports. Thus the instruction register is 48 bits wide, with its input connected to the 64-bit bus and its output connected to the 64-bit bus and the 4-bit OR described previously.

The Control Specification

A symbolic microcode word controls the D370. A microcode word is required for each cycle of the machine. The generation of either a PLA-based or ROM-based microengine is possible in later phases of the design synthesis task. A sample sequence from the dynamic address translation BEGIN/END has the following activities occurring simultaneously:

1. The 24-bit bus gates MAR to the MB address port.
2. The 64-bit ALU adds the MAR from the 24-bit bus and the temporary register T1. The result is gated on the 64-bit bus to the MB input port.
3. The 6-bit AND operates on the temporary register T2 and a constant, gating the result on the 8-bit bus to a field in the PT.ENT.

This illustrates the high degree of parallelism possible in the D370.

5.5.2 The μ370 Design

The μ370 is an IBM System/370 microprocessor data-flow on a single bipolar gate-array masterslice chip. It uses a 100-nanosecond cycle clock and is capable of executing 200,000 instructions per second. The physical chip is 7×7 mm and dissipates 2.3 watts. The plan was to use no more than 5,000 wired circuits, 3 watts of power, and 200 pins. To meet these size and power constraints, the problem was divided into on-chip and off-chip sections. This section discusses the functional blocks of the μ370.

The On-Chip Functional Block

The on-chip functional block has an 8-bit ALU, a 24-bit incrementer/decrementer, I/D, a 24-bit shifter, two 9-bit parity generators, 17 8-bit working registers, two 8-bit buffer registers, a 16-bit status register, a 24-bit

Table 5.3 Registers in the D370 Design

Abbreviation	*Name*	*Bits*	*Inputs*	*Outputs*
MAR	Memory address	24	24bb, 64bb	24bb, 64bb
MBR	Memory buffer	32	64bb	8bb, 64bb
MDBR	Memory double buffer	64	64bb	8bb, 64bb
PSW	Processor status word	64	24bb, 64bb	8bb, 24bb, 64bb
PS.EC	Extended code	6		64bb
TOD.CLK	Clock	64	64bb	64bb
CLK.CMP	Clock comparator	64	64bb	64bb
CPU.TIM	CPU timer	64	64bb	64bb
PREFIX	Prefix	12	64bb	64bb
IR	Instruction register	48	64bb	64bb, 4-bit, OR
ST.ENT	Segment table entry	32	8bb, 64bb	8bb
PT.ENT	Page table entry	16	8bb	8bb, 64bb 2-bit AND
SEG.INDX	Segment index	8	24bb	8bb
PAGE.INDX	Page index	9	24bb	24bb
PT.DSP	Page table displacement	4	24bb	24bb
BYTE.INDX	Byte index	12	24bb	24-bit OR
DAT	Dynamic address translation	24	24bb, 24-bit OR	24bb, ST.KEYS
T0	Temporary 0	64	64bb, 32-bit AND, OR, XOR	64bb, 32-bit ALU, 68-bit ALU, 32-bit AND, OR, XOR
T1	Temporary 1	64	64bb	64bb, 64-bit ALU
T2	Temporary 2	24	8bb, 24bb, 64bb	8bb, 24bb, 64bb, 6-bit AND

register and hardware to calculate the next microcode address. These components are wired together with two fan-in 8-bit buses, a fan-out 8-bit bus, a bidirectional 16-bit bus, a fan-in 24-bit bus, and a fan-out 24-bit bus. These are shown in Figure 5.5.

The 8-bit ALU can ADD, SUBTRACT, OR, AND, and XOR either binary or packed-decimal numbers. The arithmetic operations of the ALU can be controlled directly from a microcode field or indirectly through a

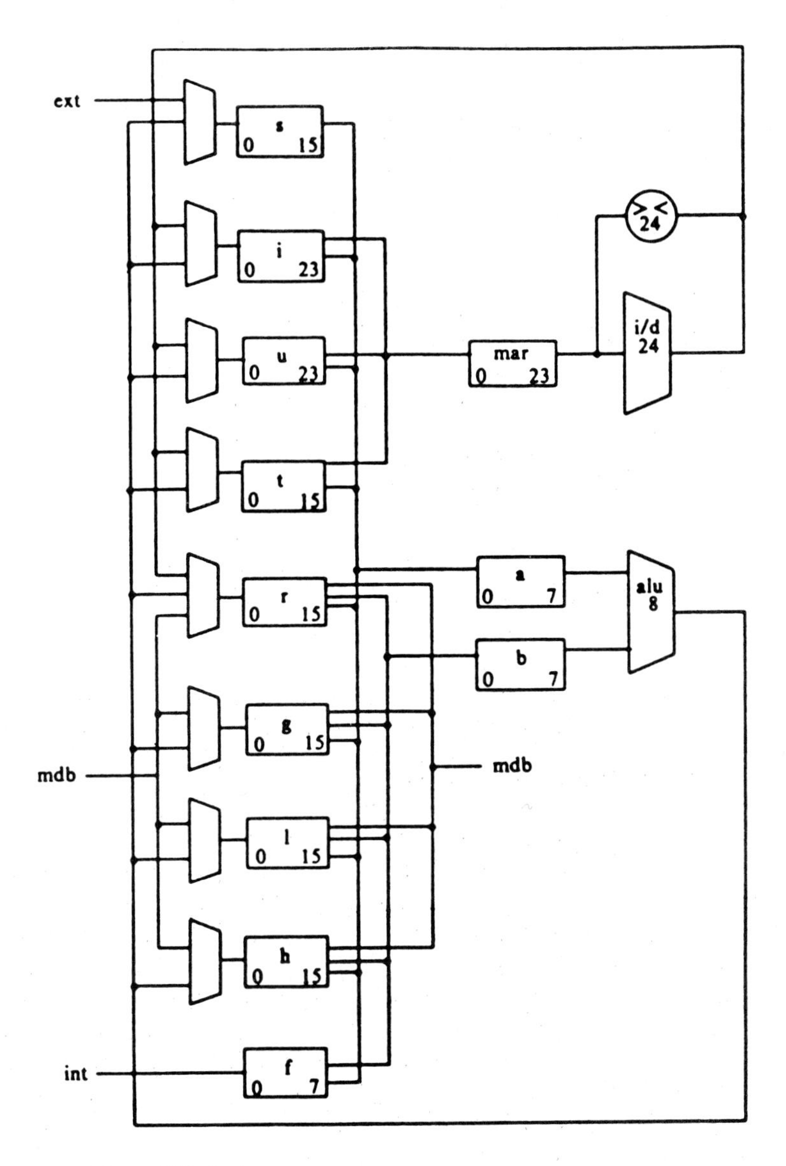

Figure 5.5 The μ370 design.

status bit located in register S. This indirect control feature allows sharing of microprogramming routines for the ADD and SUBTRACT operations. Two 8-bit buses feed two 8-bit buffer registers, A and B, which feed the ALU. These two registers can selectively gate groups of four bits that correspond to hex digits within the byte or pass the complete byte to the ALU. This gating is used for decimal operations. The A register can also pass its eight bits, rotated by four bits to reverse its two hex digits. This is used by the pack and unpack instruction. The output of the ALU is placed on an 8-bit bus gated to all the working registers.

The 24-bit I/D is a special-purpose adder that can add a 24-bit binary number with the constants 0, 1, 2, 3, -1, -2, or -3. The constant input is directly controlled from a microcode field. This I/D is dedicated to address calculations. An address is gated from a set of three working registers onto the 24-bit bus feeding the memory address register (MAR). Any value present in this register is gated to the memory address bus (MAB), the input of the I/D, and a shifter. The output of the I/D is placed on the 24-bit fan-out bus and gated back to the same three registers that fed the MAR.

The 24-bit shifter can do a few complex SHIFT operations to produce a 16-bit result. This shifter is dedicated to handling the 12-bit address field used for page addressing in the virtual storage system. Input is gated from the MAR, and output is gated onto the 24-bit fan-out bus.

Two 9-bit parity generators check the parity of each data byte arriving at the chip and place it on the 16-bit memory data bus, MDB. They can also affix a parity bit to each data byte leaving the chip from the MDB.

The μ370 has 17 8-bit registers, two 8-bit ALU buffer registers, a 16-bit status register, and a 24-bit MAR. The 17 8-bit registers are grouped in functional pairs and triples. The R, G, L, and H register pairs are primarily the memory data registers, (MDRs). Because the G register pair has special microcode branching capabilities, it is the OP code register. The I and U register triples are the program counter and the operand register, respectively. The T register pair is the local-store address. The I, U, T, and R register groups can pass two bytes of data to one another with or without a displacement. This feature is part of the virtual storage management of the IBM System/370 architecture. The S register pair is the CPU status register and serves as input to the microcontroller. S1 can be set and reset by external inputs. The interrupt register F can also be set by external conditions.

A 54-bit microcode word controls the μ370. A microcode word is required for each cycle of the machine and is fetched during the last 75 nanoseconds of each cycle from the read only store, (ROS). To select the next ROS word, a 16-bit address is generated in the first 25 nanoseconds of each cycle. Six bits of the ROS address are taken directly from the ROS word. The low order two bits are extracted from conditions within the chip. ROS fields dictate the internal conditions to be examined. The remaining bits are taken from the previous ROS address. However, if an external trap bit is raised, the ROS address is forced to a specific value for a trap handler. Possible traps are parity errors, IPL request, page overflow, storage wrap, memory protect violation, stop request, and I/O control.

The Off-Chip Functional Block

The off-chip functional block has the architectural registers, two external memories, an I/O port, and the ROS. The μ370 chip uses 512 bytes of architectural registers. They are kept in a local store that can be accessed in 60 nanoseconds. The local store, which is limited to 64K bytes, is addressed by the T0 and T1 registers. The μ370 uses up to 16 megabytes of memory, which is addressed by the MAB, while data are gated on the MDB. Read, write, memory-1, memory-2, and ready signals allow up to two memories of any speed to be interfaced to the chip. If neither memory line is asserted, the MDB is connected to an I/O bus that uses the MAB to choose the I/O device being serviced.

5.5.3 The D370 and μ370 Design Comparison

The previous two sections have discussed the individual attributes of the two designs. This section brings those designs together by comparing and contrasting their differences. Claud Davis compared the two designs at IBM Poughkeepsie. During his career of over 25 years at IBM, Davis has worked on designs and managed teams of designers for the 701, 702, 7074 MA, 360/50, FAA, 360/67, and the μ370. His vast experience with the higher-performance processors and the μ370 made his critique valuable in two ways. First, we could determine what was needed for a single-chip IBM System/370 architecture; second, we could determine what was needed for a higher-performance processor. Davis summarized his comparison.[18]

> The 370 data-flow we reviewed exhibited the quality I would expect from one of our better designers. The level of detail was what we call second level design. This encompasses all "architected" registers, status latches and sufficient working registers to implement the functions defined by the instruction set. This level of design is independent of implementing technology.
>
> The review included a test for "architected" registers, data path widths, latches for exceptional conditions, signs, and latches for temporary information in multi-cycle instructions.
>
> The assumptions for clocking and controls were examined and found to be consistent.

The complete transcript[19] is also available.

In the following discussion, differences are grouped by objectives and functional blocks including: ALUs, buses, memories, and registers, which are summarized in Table 5.4. For each difference, possible changes in the CMU/DA and the DAA systems are discussed.

The Objectives

The objectives and testing of the two designs differed. The μ370's objective was to place a fully functional System/370 on a single chip, while observing such technology constraints as number of wired circuits, power, and I/O pins. The D370's objective was to design a high-performance System/370 sensitive to technology constraints but independent of power and number of I/O pins. The μ370 was produced as a working chip, whereas the D370 was only a paper design.

The ALUs

The number, size, and type of functions supported by the ALUs in the two designs differed. The D370's design had one extra ALU that can be directly traced to the implementation of floating-point operations, while

Table 5.4 IBM System/370—Design Differences

Design	*D370*	*μ370*
Objectives	High performance, technology sensitive, independent of power and I/O pins; paper design	Strict observance of technology criteria such as number of wired circuits, power, and I/O pins; working chips
ALUs	32-bit, 64-bit, and 68-bit; Binary numbers; hardware for virtual memory, floating point, and multiply	8-bit and 24-bit; Binary and packed-decimal numbers; microcode for virtual memory and multiply
Buses	8-bit, 24-bit, and 64-bit; bidirectional	Three 8-bit, a 16-bit, two 24-bit; fan-in, fan-out, and bidirectional
Memories	12-byte buffer; single ported	8-byte buffer; single ported
Registors	Discrete	Memory array

the μ370 planned a separate floating-point chip. In addition, the D370 implemented the dynamic address translation hardware, while the μ370 supplied a shifter that had a few complex shift patterns to aid in calculating the virtual address by using microcode.

The DAA does a high-level floor layout to help decide how to partition the algorithmic description, but this does not currently allow exclusion of functionality; the whole algorithmic description is implemented. Changes that modify the algorithmic description by including or excluding functionality are best made by changing the initial description or by having a postprocessor feed size constraints to the DAA.

The ALUs also differed in size. The μ370 serialized the 32-bit and 64-bit operations of the System/370 architecture into four or eight cycles through an 8-bit ALU. The DAA's constraints were set to design a high-performance processor, and thus the data paths were not serialized. Less than a dozen rules could be added to the DAA to allow it to serialize on ALU width. However, this change would be better made by adding a transformation to the CMU/DA system that removes a single abstract data-flow operation and replaces it with several smaller ones.

Finally, the ALUs differed in the functions they provided. The μ370 has an ALU that can ADD and SUBTRACT packed-decimal numbers; the D370 performed these operations by adding hardware and microcode. The D370 has an ALU that can MULTIPLY, while the μ370 MULTIPLIED by SHIFTING and ADDING. Davis felt the choice of an ALU that could MULTIPLY was reasonable and consistent with the constraints used by the DAA.

The Buses

The designs differed in the number, size, and type of buses used. The μ370's 8-bit buses and 16-bit bus serve the same purpose as D370's 8-bit and 64-bit buses. The size difference is accounted for by the μ370 serializing the 32-bit and 64-bit operations of the System/370 architecture down to 8-bit operations, as discussed previously. Also, the D370 uses bidirectional buses, whereas the μ370 used separate fan-in and fan-out buses. Davis felt the design choices made by the DAA were reasonable for the higher-performance D370 design, citing the IBM System/370 model 158 as an example of this style of busing.

The Memories

The memory functional blocks differed only slightly. The D370 uses an 8-byte buffer with a 4-byte memory data register, while the μ370 uses 8 bytes of memory data registers. Davis felt this and even more elaborate cache schemes suited the higher-performance processors. He suggested that the D370 use dual-ported memories for its general-purpose registers to allow the use of two registers during the same cycle. Dual-ported memories would require a few rule changes but would allow up to two memory array accesses during the same clock cycle. The rules for finding the address, the input, and the output ports of memories would have to be enhanced to check for idle ports. All told, about 20 rules would have to be modified or extended to effect this change.

The Registers

Both descriptions have about the same number of bytes of architectural and temporary registers. However, the μ370 groups all the architectural registers off-chip in a fast local store, which can be thought of as memory. This would be a major change in the structure of the DAA. However, it could be done simply, as a postprocessor pass by the CMU/DA system when other technology-specific hardware is bound to the modules.

5.6 DAA Knowledge

The DAA is implemented as a production system using the OPS5[16] knowledge-based expert system (KBES) writing system. One merit of codifying knowledge in a KBES is that it can be easily quantified and qualified. An important goal of the research has been to understand how VLSI designers choose computer implementations. The problem division they use, shown in Figure 5.6, consists of general service functions, global implementation allocation, data-flow allocation, structural allocation, and global improvements to the implementation. A summary of these subtasks and the number of rule firings, or activations, for the complete design of the structured

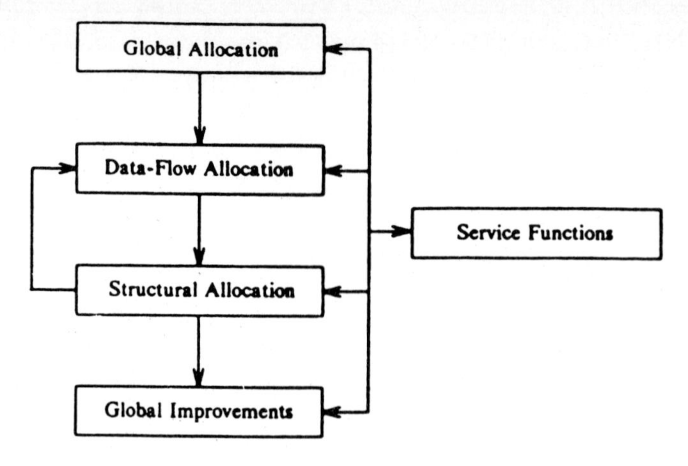

Figure 5.6 DAA Subtasks.

control flow processor,[20] SCF3, is provided in Table 5.5*. These subtasks, implemented in the DAA, are discussed by functional sections along with two procedures that provide high-level floor planning information.

5.6.1 DAA Subtasks

The first functional section is the set of service rules. These 88 rules control the ordering of subtasks throughout the implementation, minimize the connectivity between modules, and perform the simple bookkeeping to maintain and express the design. Thus these rules both control and are controlled by the other four functional sections.

The next functional section is the set of global allocation rules. These 78 rules (1) allocate hardware that cannot be optimized by the DAA, (2) provide default constraints and timing information where none is supplied, and (3) initialize the bookkeeping information used by the service rules.

Table 5.5 Rules by Function

Section	*Rules*	*Firings*
Service Functions	88	4901
Global Allocation	78	104
Data-Flow Allocation	55	1468
Structural Allocation	47	1610
Global Improvements	46	460
Total DAA	314	8543

* The SCF3 is a RISC architecture being designed and studied at Carnegie-Mellon University.

The hardware that cannot be optimized by the DAA consists of architectural components such as memories, architectural registers, and controllers. Figure 5.6 shows that this set of rules is fired only for the first partition allocated.

Once the DAA has allocated the architectural hardware, the next task is to partition the whole design into smaller blocks and select a partition. Within each partition, the DAA creates clock phases, operators, registers, data paths, and control logic in two subtasks—data-flow allocation and structural allocation. This allows the DAA to gather all the information about register usage in the data-flow allocation and then create registers and modules in the structural allocation.

The 55 data-flow allocation rules assign operators to control phases, determine the minimum size needed to represent an operation, allocate temporary registers and ALU modules, and allocate multiplexers. This creation of register and ALU modules is not allocating hardware, but rather mapping the data-flow representation into the structural and control representation to gather timing, connectivity, and functionality information.

After data-flow allocation, the 47 structural allocation rules (1) remove registers that are not required to maintain testability, (2) combine other registers, (3) use fan-out from existing ALU modules, and (4) create ALU modules. Figure 5.6 shows that the data flow and structural allocation rules are repeated for each of the partitions created by the service function rules.

As a design nears completion, the DAA starts examining it for components that are no longer needed or could be better shared. This cleanup includes removing unused ALU modules and registers, reducing multiplexer trees, combining ALU modules by fan-out, and allocating buses in high traffic areas. This set of 46 rules, like the global allocation set of rules described previously, is activated only once per design.

5.6.2 Estimators

As the DAA evolved from the prototype system, it became clear some things could not be handled well in rules. Primarily, these were actions that required summing, counting, or checking for set membership. Represented in OPS5 rules, these calculations were terribly inefficient; they could be much better represented as algorithms in a language like C. The decision making connected with the results of these estimators was kept in rules, but the calculation was moved into a C program. Through the interviews with designers, these calculations evolved to focus attention on similarities of functions and connections in data-flow graphs and modules. They mimic the back-of-the-envelope floor plan on which designers base many of their decisions. These estimators have closely matched expert designer performance in partitioning and cost analysis experiments.

Primarily there are two estimators. One bridges the algorithmic to fabrication-dependent hardware-network level as a partitioner, while the other bridges the technology-independent to the technology-dependent hardware-network level as a cost function.

Partitioning

The first estimator is a partitioner that bridges the algorithmic to fabrication-dependent hardware-network levels. It is unlike most current partitioners of digital hardware[21–24] because it does not require the specification of physical modules and their interconnections. That is, in our terms, it uses only the abstract and imprecise information contained in the ISPS[25] description to make predictions of how a data path may be laid out. This layout guess is often referred to by designers as the *floor plan* of a chip. The floor plan partitioner attempts to share hardware effectively and minimize the interconnections between partitions.[26]

Cost Estimator

The second estimator is a hardware pricer that bridges the technology-independent to technology-dependent hardware-network levels. It is unlike the partition estimator because it requires that the system be specified by technology-independent modules and their interconnections. It provides information about what percentage of the required hardware for a new module already exists in another module. That is, in our terms, it uses only the abstract and imprecise information contained in the structural and control description to make predictions of how much it would cost to upgrade the functions and interconnects of an existing module to contain a new module. Whereas the partition estimator gives a high-level floor plan, this estimator gives a much more local view, thus augmenting the high-level floor plan with more detailed information.[27]

5.7 A Sample Design

To help the reader better understand the relationship between the design steps, a simple ISPS example description (see Figure 5.7) is examined throughout the implementation design process. This ISPS fragment is from the description of the Digital Equipment Corporation PDP-8. It first defines the current page and the instruction carriers. It then labels specific fields of the instruction carrier as the page-zero bit and page-address carriers. The last part of the ISPS fragment decodes the page-zero bit of the instruction carrier and sets the effective-address carrier equal to the page-address carrier concatenated with either zero or the current-page carrier. Sample rules are also provided to highlight various areas of knowledge in the DAA. These rules are discussed and translated in the following section. The reader must be warned that these translations must be taken with a grain of salt because rules do not stand by themselves, but are part of a larger network of rules connected by the working memory and the inference engine.

5.7.1 Sample Global Allocation

Figure 5.8 shows the design of the decoding loop of Figure 5.7 after global allocation has been done. Each of the symbols represents a module with

```
Example :=
Begin

** Storage.Declaration **

cpage \current.page<0:4>,
i \instruction<0:11>
  pb \page.0.bit<> :=i<4>,
  pa \page.address<0:6> :=i<5:11>

** Address.Calculation **

Global eadd \effective.address<0:11> :=
  Begin
  Decode pb => 
    Begin
    0 := eadd ='00000 @ pa,
    1 := eadd =cpage @ pa
    End
  End
End
```

Figure 5.7 Sample ISPS description.

the bit width given as the bottom pair of numbers. The small rectangles are registers, the small squares are constants, and the large rectangle is the controller. The registers r2.reg, r3.reg, and v6.reg are allocated because of the ISPS definitions for cpage, i, and eadd, respectively. A controller and four constants, one for each of the constants shown in Figure 5.7, are also allocated. Although unseen in the figure, the technology data base and constraints are initialized. The sample rule in Figure 5.9 is used during this

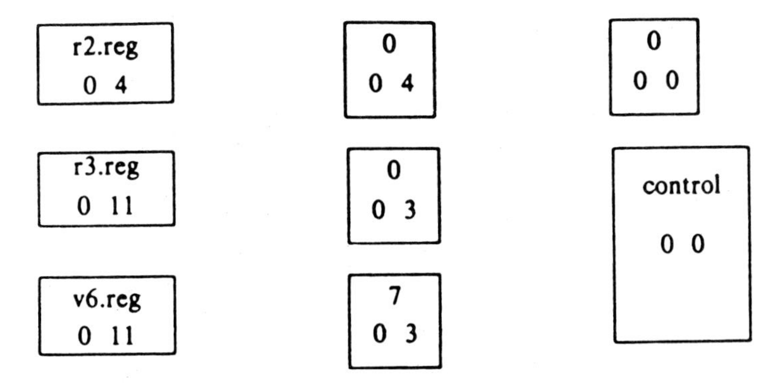

Figure 5.8 Design after global allocation.

IF:

the most current active context is declared variable allocation
and there is a VT-body that is either a carrier, VT-body, or section list
and it is not an array
and it has a nonzero width
and the parent VT-body is a section list

THEN:

create a register module
and create an output port
and create an input port

Figure 5.9 Find registers.

subtask to allocate registers. It finds data-flow entities, called VT-bodies, that are not arrays and allocates a register module with input and output ports.

5.7.2 Sample Data-Flow Allocation

Figure 5.10 shows the design of the decoding loop of Figure 5.7 after data-flow allocation has been done. Each of the symbols represents a module. The circles are single function wiring modules that bring together or concatenate two sets of signals, the trapezoid is a multiplexer that gates one of its two inputs to its output, the small rectangles are registers, the small squares are constants, the large rectangle is the controller, and each of the lines represents a link between the modules. Where the links join with the modules, a port is defined. The state of the design shows the creation of

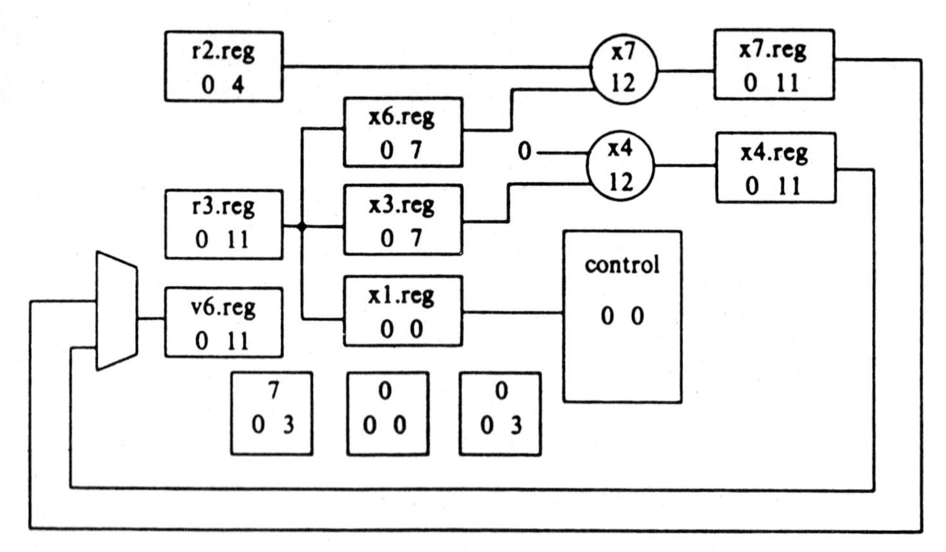

Figure 5.10 Design after data-flow allocation.

IF:

the most current active context is temporary variable allocation
and there is a produced value outnode
and the outnode is not associated with an architectural register

THEN:

create a temporary register module
and create an output port
and create an input port

Figure 5.11 Find temporary registers.

the temporary registers (x1.reg, x3.reg, x4.reg, x6.reg, and x7.reg), the concatenation modules (x7 and x4), a multiplexer, and many connections. Along with each connection, the operator assignment to control steps and the data-flow references are specified, so that a control specification can be generated. The temporary registers are created to latch values for testability reasons. The concatenation modules are created, rather than assigned as register attributes, because they are wiring instructions and cost nothing to create. The multiplexer is created because the output of either x7 or x4 needs to be stored in v6.reg. Finally, a constant has been redrawn as an input to x4. An example of a rule to allocate temporary registers is provided in Figure 5.11. This rule finds output values, called *outnodes,* from data-flow operators that are not associated with architectural registers and creates temporary registers with input and output ports for them.

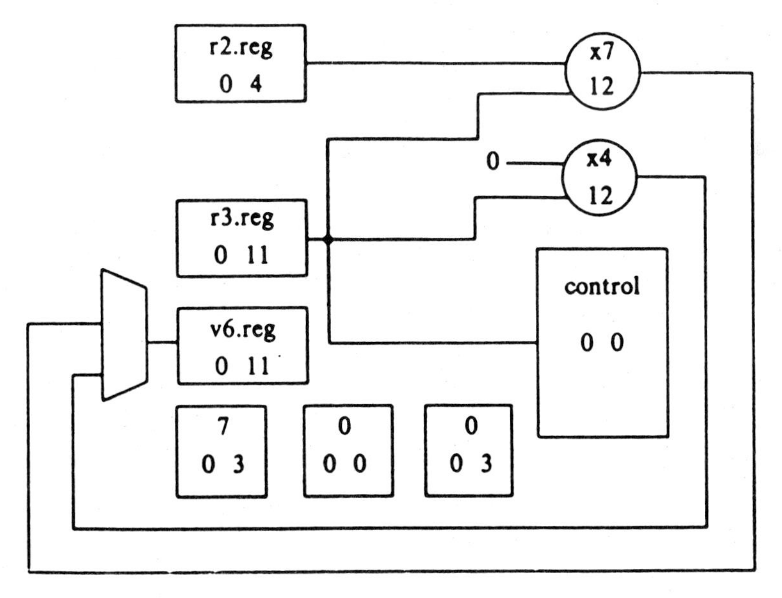

Figure 5.12 Design after structural allocation.

IF:

the most current active context is fold allocation
and there is a temporary register
and there is a link to the temporary register
and there is not a link from the temporary register

THEN:

remove the temporary register
and remove its link

Figure 5.13 Fold no output register.

5.7.3 Sample Structural Allocation

Figure 5.12 shows the design of the decoding loop of Figure 5.7 after structural allocation has been done. The state of the design shows the removal of the temporary registers, x1.reg, x3.reg, x4.reg, x6.reg, and x7.reg, and many connections. For this simple design, it turned out that all the registers are stable and not needed to keep the design testable. Thus they were not made permanent. Also, the two wiring operations cannot be combined because they do not have the same inputs. Had the operators not been wiring operators, they would have been examined using the partition and cost estimators shown in earlier sections and possibly combined. An example of a rule to remove stable registers is provided in Figure 5.13. If a temporary register has an input value, but its output is not used anywhere, the register is found to be stable and it is removed.

5.7.4 Sample Global Improvements

Figure 5.14 shows the design of the decoding loop of Figure 5.7 after global improvements have been made. The state of the design shows the removal

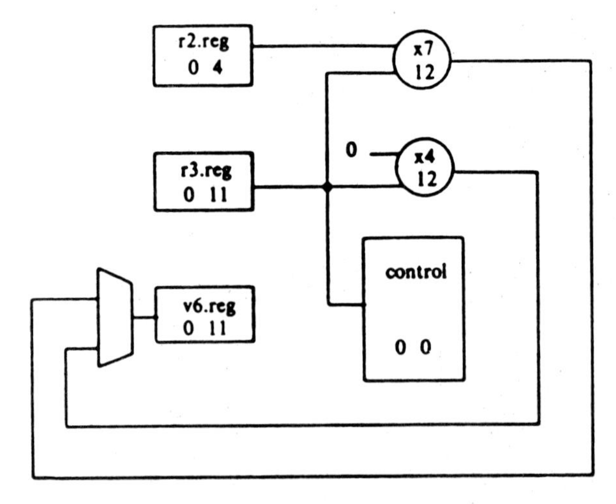

Figure 5.14 Design after global improvements.

IF:

the most current active context is bus allocation
and there is a module that is a multiplexer
and there is another module that is also a multiplexer
and there is a link from a non-bus module to the first multiplexer
and there is a link from that module to the second multiplexer
and there is a link from another non-bus module to the
first multiplexer
and there is a link from that module to the second multiplexer

THEN:

place these connections on an idle bus

Figure 5.15 Convert MUX inputs to bus.

of the unused constants and the input ports to r2.reg and r3.reg. Because the initial ISPS was so simple, no multiplexer cleanup or bus allocation is required. An example of a rule to allocate bus structure is provided in Figure 5.15. It finds modules that are connected to the same multiplexers and places them on a bus that is not transferring data during this time step.

5.8 Domain Knowledge

This section analyzes the knowledge in the DAA by examining the extent to which each rule embodies domain knowledge. Knowledge directly related to the implementation design task is domain-specific, while knowledge of a more general type is domain-independent. An example of domain-specific knowledge is how to make particular design decisions and an example of domain-independent knowledge is how to count things. Distributed between these two categories, DAA contains seven distinct rule types. Rules that extend the partial design, control the context or focus of attention, and maintain the technology-independent data base are domain-specific. Rules that remove unneeded working-memory elements, transform the input description, simulate list and set operations, and simulate procedure calls or calculations in antecedents are domain-independent.

Table 5.6 summarizes these categories by listing the number of rules in each category and the number of rule firings for the complete design of the SCF3 processor. Table 5.6 shows that three-quarters of the rules and two-thirds of the rule firings are domain-specific. Most of the domain-specific rules and firings are used to extend the partial design (see the Design row in Table 5.6). The remaining rules and rule firings are domain-independent or overhead rules. Most of these rules and firings are used to clean up unneeded working-memory elements and simulate list and set operations (see the Cleanup and List rows in Table 5.6).

Table 5.6 Rules by Knowledge Type

Domain specific			*Domain Independent*		
Type	*Rules*	*Firing*	*Type*	*Rules*	*Firing*
Design	151	3768	Cleanup	33	1586
Context	18	1562	Input	9	64
Setup	70	70	List	14	1290
			Test	19	203
Total	239	5400	Total	75	3143
Total DAA	314	8543			

5.9 Summary

This chapter has shown how expert VLSI designers choose a MOS microcomputer's implementation and how a knowledge-based expert system, the DAA, can mimic their results. It begins by showing that the KBES technique is a promising approach to design synthesis. Using a KBES approach has sped the development of the DAA by providing a framework that allows incremental addition of common sense modular design knowledge and queries about the knowledge during the design task. This framework has increased the flexibility of the final system and reduced the execution time by replacing backtracking techniques with match techniques. We have seen how the DAA, like a human designer, has become a better designer as its rule memory expands. The extraction, codification, and testing of the expert designers' knowledge has facilitated a better understanding of VLSI design synthesis, while providing another KBES system for computer scientists and knowledge engineers to examine.

This chapter has explored the generality of design knowledge in the DAA by comparing and contrasting an IBM System/370 designed by an expert human designer, Claud Davis, against the design produced by the DAA. This is the first large design, automatically generated from an algorithmic description and constraints, that has exhibited the quality expected from one of IBM's better designers. Furthermore, the design required 47 hours of CPU time, which with some work can be reduced by a factor of 12 to about 4 hours of CPU time. This clearly shows the dramatic improvement in CPU time for large designs obtained using methods that replace backtracking by match techniques.

This chapter also has shown how designers partition the design task into four major subtasks. These subtasks involve global allocation of architectural modules and registers; local allocation of control steps, operators, registers, data paths, and control; global allocation of operators and registers; and global improvements to the design. Within these subtasks, a dominant goal is to partition the design into smaller, more manageable pieces.

While making the design more manageable, it is also important to retain a global view of the hardware. Designers retain this global view by using high-level floor plans that are filled in as the design proceeds. This floor planning mechanism is composed of estimators for partitioning and pricing hardware. These estimators deal with connectivity and functionality of the hardware networks. At the same time, the designers feel it is important to design testable synchronous designs. This constraint is even more important than making the least expensive connections or optimally sharing hardware. After testability, the most important consideration is how the design will lay out or how to minimize connectivity. Thus, creating a good design is not just minimizing components, but also paying careful attention to testability and connectivity.

In retrospect, much of what we learned was common sense design knowledge, the same things human designers learn through apprenticeship. This learning process is not complete, nor will it ever be complete. Like human designers, the DAA becomes a better designer as its knowledge base expands. Until all possible world knowledge about designing microprocessors has been codified in the DAA's knowledge base, there will always be room for improvement in its designs.

This KBES approach has opened the door to a whole new class of intelligent computer-aided design tools.

References

1. C. Mead and L. Conway, *Introduction to VLSI systems*, Addison-Wesley, Reading, Massachusetts, 1980.
2. T. J. Kowalski, D. J. Geiger, W. H. Wolf, and W. Fichtner, "The VLSI Design Automation Assistant: From Algorithms to Silicon," *Design and Test of Computers*, vol. 2, no. 4, August 1985, pp. 33–43.
3. D. E. Thomas, C. Y. Hitchcock III, T. J. Kowalski, J. V. Rajan, and R. Walker, "Automatic Data Path Synthesis," *Computer*, vol. 16, no. 12, December 1983, pp. 59–70.
4. T. J. Kowalski and D. E. Thomas, "The VLSI Design Automation Assistant: First Steps," *26th IEEE Computer Society International Conf.*, February 28, 1983, pp. 126–130.
5. T. J. Kowalski and D. E. Thomas, "The VLSI Design Automation Assistant: A Prototype System," *Proc. of 20th Design Automation Conf.*, June 27, 1983, pp. 479–483.
6. T. J. Kowalski, *An Artificial Intelligence Approach to VLSI Design*, Kluwer, Boston, Mass., 1985.
7. T. J. Kowalski and D. E. Thomas, "The VLSI Design Automation Assistant: An IBM System/370 Design," *Design and Test of Computers*, vol. 1, no. 1, February 1984, pp. 60–69.
8. T. J. Kowalski, D. J. Geiger, W. H. Wolf, and W. Fichtner, "The VLSI

Design Automation Assistant: A Birth in Industry," *IEEE International Symposium on Circuits and Systems*, June 5, 1985.

9. C. Davis, G. Maley, R. Simmons, H. Stroller, R. Warren, and T. Wohr, "IBM System/370 Bipolar Gate Array Micro-Processor Chip," *Proc. of the International Conf. on Circuits and Computers*, October, 1980, pp.669–673.
10. P. Marwedel and G. Zimmermann, *MIMOLA Software System User Manual*, vol. 1, Institut Für Informatik und Praktische Mathematik, Christian-Albrechts-Universität Kiel, 1979.
11. L. J. Hafer, *Automated Data-memory Synthesis: A Format Method for the Specification, Analysis, and Design of Register-Transfer Level Digital Logic*, PhD Thesis, Department of Electrical Engineering, Carnegie-Mellon University, June, 1981. Also in Design Research Center DRC-02-05-81.
12. L. Hafer, *Memory Allocation in the Distributed Logic Design Style*, Master's Thesis, Carnegie-Mellon University, December 21, 1977.
13. C. Y. Hitchcock III, Automated Synthesis of Data Paths, CMUCAD-83-4, SRC-CMU Center for Computer-Aided Design, Carnegie-Mellon University, January 1983.
14. C. J. Tseng and D. P. Siewiorek, "Facet: A Procedure for the Automated Synthesis of Digital Systems," *Proc. 20th Design Automation Conf.*, June 27, 1983, pp.490–496.
15. E. A. Feigenbaum, *Knowledge Engineering: The Applied Side of Artificial Intelligence*, Computer Science Department, Stanford University, 1980..
16. C. L. Forgy, *OPS5 User's Manual*, Department of Computer Science, Carnegie-Mellon University, 1981.
17. A. V. Aho and J. D. Ullman, *Principles of Compiler Design*, Addison-Wesley, Reading, Mass., 1979.
18. C. Davis, Personal letter to Dr. D. E. Thomas, August 12, 1983.
19. T. J. Kowalski, The VLSI Design Automation Assistant: The IBM/370 Critique, AT&T Bell Laboratories Internal Memorandum, August 1, 1984.
20. M. A. Rose, *Structured Control Flow: An Architectural Technique for Improving Control Flow Performance*, Master's thesis, Department of Electrical Engineering, Carnegie-Mellon University, November 1983.
21. B. W. Kernighan and S. Lin, "An Efficient Heuristic Procedure for Partitioning Graphs," *Bell Systems Technical Journal*, vol. 49, no. 2, (1970), pp. 291–308.
22. D. G. Schweikert and B. W. Kernighan, "A Proper Model for the Partitioning of Electrical Circuits," *Proc. 9th Design Automation Workshop*, 1972, pp. 57–62.
23. M. A. Breuer, "A Class of Min-Cut Placement Algorithms," *Proc. 13th Design Automation Workshop*, 1976, pp. 284–290.
24. T. S. Payne and W. M. vanCleemput, "Automated Partitioning of Hierarchically Specified Digital Systems," *Proc. 19th Design Automation Conf.*, 1982, pp. 182–192.
25. M. R. Barbacci, G. E. Barnes, R. G. Cattell, and D. Siewiorek, *The ISPS Computer Description Language*, Department of Computer Science, Carnegie-Mellon University, 1979.
26. M. C. McFarland, "Computer-Aided Partitioning of Behavioral Hardware," *Proc. 20th Design Automation Conf.*, June 1983, pp. 472–478.
27. T. J. Kowalski, Computer-Aided Cost Estimation from Implementation Specifications, (unpublished).

CHAPTER 6

Algorithmic System Compilation: Silicon Compilation for Systems Designers

Jay R. Southard

6.1 Introduction

The silicon compiler described in this chapter takes a program-like, or algorithmic, description of the behavior of a digital system and implements it as an integrated circuit. Hence, it is an example of an Algorithmic System Compiler. We will describe the concepts and syntax of the language compiled by this Algorithmic System Compiler. Some simple examples will be designed to demonstrate the implementation, in this approach, of such digital systems concepts as pipelining, parallelism, and interface protocols. Finally, some implementation details of the Algorithmic System Compiler will be considered.

6.2 User Background

Far too often the typical user is implicitly assumed to be a hardware-oriented designer. Given that 75 percent of today's system design dollar goes to the implementation of system functionality through *programming*, any implicit assumptions should identify the user as a programmer. To be more explicit, the optimal background of an Algorithmic System Compiler user is presented in Table 6.1. This shows a spectrum of "knowledge areas" that spans a digital system from its basis of implementation (physics) to

Table 6.1 Spectrum of Knowledge Areas of the Algorithmic System Compiler User

Score	*Knowledge Areas*
	Semiconductor physics and manufacturing
*	IC layout
*	IC circuit
**	Gate-level design: TTL logic, gate array, standard-cell
***	Digital system: behavior-level design: FSM, pipelining, systolic arrays, bit-serial arithmetic, block-level design
****	Hardware manipulation programming: bit-slice, array processor, μ-code, multiprocessors
***	Application algorithms in some area
**	Basis of computation: abstract models of computability, NP completeness; numerical methods: roundoff, stability, effect of word length

Key

*	Usually knows when to ask questions and usually understands answers
**	Basic familiarity with concepts and principles
***	Competent professional
****	Primary skill

its applications. The asterisks indicate a combination score of knowledge, interest, and capability.

It is impossible to be an expert in one area and know *nothing* about adjacent areas. Thus, there will be a spectrum of knowledge areas for each individual. The "peak" score will determine what kind of tools the reader will be most interested in. For example, an engineer who peaks at the level of application algorithms will be more interested in the Algorithmic System Compiler approach, while the logic gate-level designer will be more interested in a Structural Silicon Compiler.

6.3 The Nature of "Algorithm"

There is a great deal of confusion surrounding algorithms, hardware, and software. This confusion stems from viewing design as comprising a succession of "structural" levels, with the addition of an "algorithmic" design level. To counteract this confusion, let us begin *tabula rasa* with the nature of design.

6.3.1 Algorithm versus Structure

"Algorithm" is one of the few basic design paradigms. Every instance of such paradigms must have *primitives* and *composition methods* for combining those primitives. Another basic paradigm, for example, is "structural composition," for which the composition method is "connection"; primitive blocks can be combined by connecting the outputs of one to the inputs of another.

Sometimes hierarchy is allowed, which means simply that a given group of interconnected blocks may be treated elsewhere as a single block, subject to further connection to blocks external to the given group. Alternatively, the connection mechanism may be limited to a two-dimensional grid or array (i.e., data paths).

There are two basic (and two optional) algorithmic composition mechanisms:

sequence: (step 1) A = A + 1
(step 2) B = B + A
selection: *if* A *then* B = B + 1
else C = C + 1
cascade (optional): A = B + C − D
parallelization (optional): (par A = A + 1
B = B + A)

Each of these will be discussed in detail in this chapter. The results of the two different composition paradigms can be encapsulated by saying that algorithms are *temporally* local, while structures are *topologically* local. In other words, in a program, statements ordered sequentially on the coding sheet are likely to be executed in sequence. On the other hand, components on the same schematic diagram sheet are likely to be interconnected.

Notice that the complexity (sometimes called the "conceptual" or "abstraction" level) of the primitives was not addressed in the preceding discussion. The adjective *higher-level* can only compare primitives' abstraction levels within a single design paradigm. This means that a structural and an algorithmic method may use similar levels of primitives and still be fundamentally different. Therefore, we use the orthogonal axes of the Y chart rather than a single axis to compare design methods. This is not merely an academic exercise. Lower levels of primitives are more flexible, whereas higher levels allow quicker design along a single axis, *but not necessarily between axes*.

We will argue that very high level primitives are probably generally less effective than intermediate-level primitives, irrespective of combination mechanisms. Thus, the combination mechanisms are more critical for determining design cost and implementation effectiveness. Which method, algorithmic or structural, is "better" is primarily determined by the demands of a specific problem, somewhat modified by taste. Sometimes the ability to have connections on two or more dimensions outweighs the utility of sequence and selection. Algorithmic design has always been preferred when its *physical implementation* is of satisfactory performance. One might say that algorithmic system compilation is a method that extends the implementation performance levels of the algorithmic system design approach.

The reader should be aware that some observers have confused the term *algorithm* with the "highest" level on a single axis of design levels. This might be represented as a pyramid, as in Figure 6.1. Such terms as *behavioral*, *functional*, and, in other contexts, *architectural* have all been used as synonyms for the top level. Although *behavioral* and *functional* have been

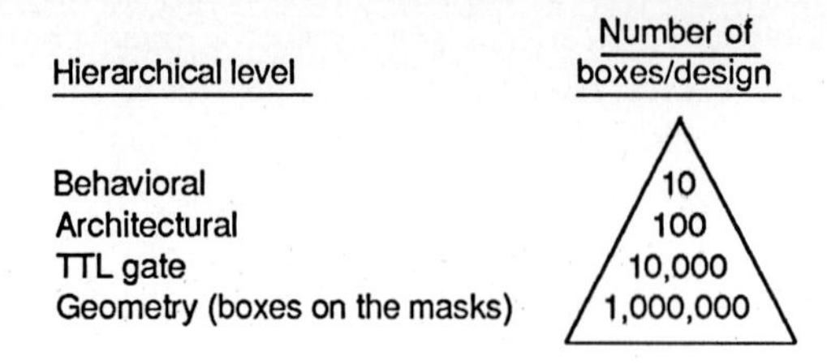

Figure 6.1 Hierarchical levels of complexity.

abused to the point of being meaningless and *architectural* has at least three different, precise meanings in as many subdisciplines of digital engineering, the overall meaning is quite clear: each level is composed of interconnections of subunits from the next lower level. This is fundamentally, therefore, a "structural" view of design.

6.3.2 Design versus Implementation, Hardware versus Software

We are accustomed to thinking about hardware as structure and software as algorithm (in this context, *algorithm* is synonomous with *procedure* or *program*). They are, however, totally independent: one is a design method, and the other is an implementation mechanism (see Figure 6.2).

There are actually several algorithmic design–hardware implementation combinations extant that implement finite-state machines (FSMs) in so-called programmable logic arrays (PLAs). Since FSMs are at the level of Boolean operations, they represent a low level in the algorithmic design hierarchy. One might argue, however, that since they customize rather than create hardware, they do not constitute a "true" combination of algorithmic and hardware methods. The Algorithmic System Compiler we are about to describe, however, unquestionably fills the empty box in Figure 6.2, since it creates custom hardware (rather than customizing predesigned PLAs) and is designed at a higher algorithmic level than FSMs.

	Hardware	Software
Structure	Schematic capture	UNIX "pipes" -OR- Object-oriented programming
Algorithm		Standard "structured" [sic] programming

Figure 6.2 Design method versus implementation (prior to algorithmic system compilation).

6.4 Road Map to the Chapter

This chapter will cover elements from the knowledge "trees" of Figure 6.3. Items in bold will be presented in greater detail. The central tree concerns algorithmic system compilation. Just as no human can be expert in a branch of engineering without some lesser knowledge of adjacent areas, so too is it impossible to present one tree properly without some recognition of concepts in "nearby" knowledge trees.

The reader will observe a minor subtree in Figure 6.3 labeled "silicon compilation techniques." Because MacPitts and its offspring (notably MetaSyn and SILC) as well as Bristle Blocks and its offspring have all been

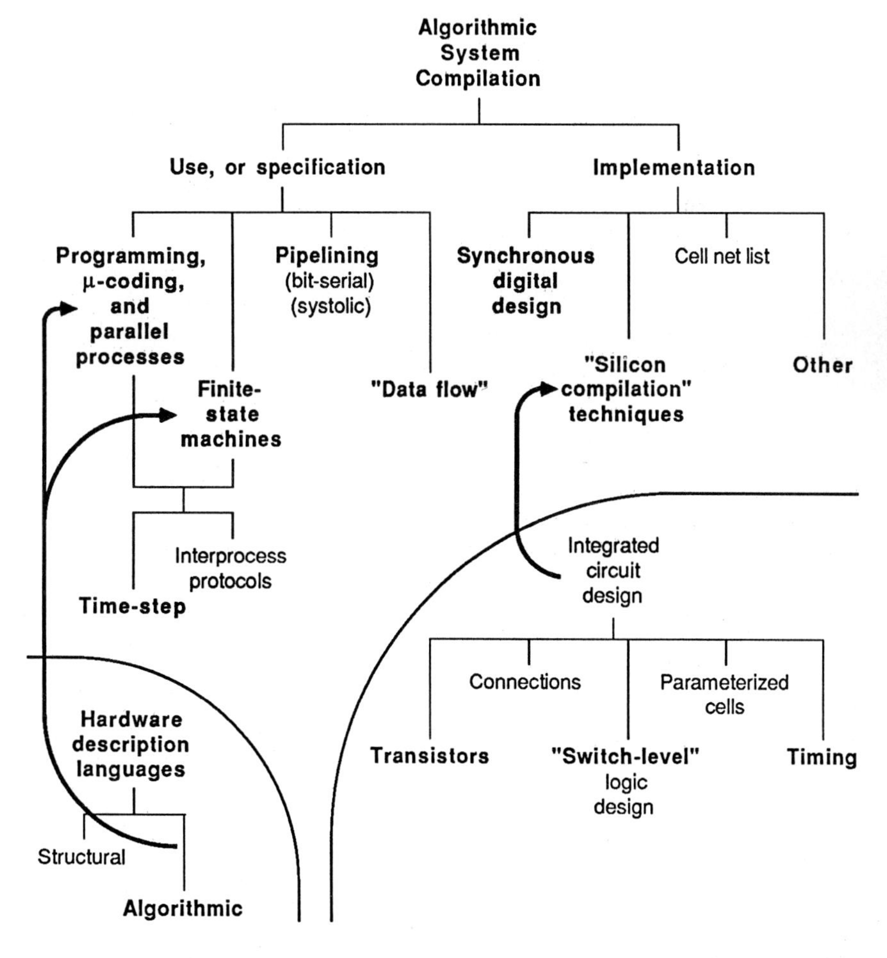

Figure 6.3 Algorithmic System Compilation knowledge tree, with some details of adjacent trees: "hardware description languages" and "integrated circuit design".

referred to as "silicon compilers," there has been more than a little confusion as to what constitutes "silicon compilation." In the past, proponents of the MacPitts style have described Bristle Blocks–style compilers as "silicon assemblers." However, in the interests of simplification, the term *silicon compilation* will be restricted here to describe those techniques pioneered by Dave Johannsen in Bristle Blocks. The term *algorithmic system compilation* will be used to describes MacPitts-like logic synthesis combined with any method of physical implementation.

MetaSyn, a commercial derivation of MacPitts (which is accepted as the earliest algorithmic system compiler), will be used to illustrate examples and concepts throughout this chapter.

6.5 Use of the Algorithmic System Compiler

The four concepts of Algorithmic System Compiler, programming, finite state machines, pipelining, and data flow, have guided digital system design for the past 15 years. For example, in designing a computer, the microprogramming (bit-slice) style of design is most likely to be employed to generate the initial specification, and it is often the implementation mechanism of choice as well. In designing a controller, the FSM style is most likely to be employed. In digital signal processing, data flow diagrams are often the first expression of the digital design, although these are usually preceded by equations. Pipelining is often called upon to improve performance.

In other words, most system design proceeds from an understanding of the system requirements and the general design techniques of FSMs, programming, pipelining, and data flow. These techniques are applied to the problem until a satisfactory solution is reached.

The algorithmic system compiler ideally is a tool that helps to implement a system in the most cost-effective and highest-performance medium, and that also allows the system designer to think and design in terms of familiar conceptual frameworks. Such a compiler should allow simple specification of systolic systems (bit-serial and pipelining are subsets of systolic systems), as well as microprogramming, FSMs, and data flow. When MacPitts was being developed, its authors were not ready to tackle all the issues inherent in specifying and synthesizing systolic systems. Moreover, pipelining is usually perceived as a "performance booster" rather than as an initial and necessary system requirement, so that it is perhaps less important than the other concepts. Also, pipelining can be implemented in terms of data flow and microprogramming, so that pipelined systems can still be rather easily designed even if no direct specifications are possible.

The final antecedent of the system compiler comes from "hardware description languages" (HDLs). It might seem that an Algorithmic System Compiler specification language is just one instance of an HDL. However, the Algorithmic System Compiler is intended to *synthesize* hardware, whereas HDLs are typically intended to describe hardware that already exists (and that invariably was designed structurally). There are thus major differences. As Buckminster Fuller once said, "Form follows function."

6.5.1 Hardware Description Languages

In theory, hardware descriptions can be any combination of:

$$\begin{Bmatrix} \text{textural} \\ \text{graphical} \end{Bmatrix} \times \begin{Bmatrix} \text{algorithm} \\ \text{structure} \end{Bmatrix}$$

The graphical-structure approach is epitomized by today's CAE workstations. The graphical-algorithm approach has not yet caught on, although there has been constant interest, especially now that graphics-capable hardware is becoming less expensive. The textural-structure approach to HDLs is the "path most traveled." It combines inexpensive data entry hardware with the structural mindset. Not surprisingly, it has been used mostly for postdesign description and simulation, since structural designers still tend to design graphically. Also not surprisingly, it has a reputation for being academic. The fourth choice for hardware descriptions is textural-algorithm. The two seminal algorithmic hardware description languages are AHPL and ISPS.

ISPS is a derivative of ISP, which was first described by Bell and Newell in their book *Computer Structures*. ISPS differs from MetaSyn and AHPL mainly in that it does not employ a single, implicit view of time (the "time step"). Due to this characteristic, ISPS has been fruitful for research which uses AI techniques to assign instruction states for efficient machines. In a later section, we will compare an ISPS description with a MetaSyn description of the "instruction decode" section of a CPU.

AHPL, like MetaSyn, was designed as a hardware synthesis language and, therefore, is time-step oriented. However, its operator set is limited to Boolean operations on bits and words. In theory, this set could be extended to include arithmetic operators; however, no one has come up with the synthesis and practical optimization routines for them, and the APL character set is always a practical nuisance. We will compare an AHPL specification with a MetaSyn specification in a simple example.

First, however, we will discuss the critical "time step", concept from the programming viewpoint. We also discuss this concept from the standpoint of FSMs.

6.5.2 Programming, Time Step, and Parallelism

In this section we will dissect a single programming concept, the "instruction." We will extract two primitive concepts from it (the more important being "time step"), and by exploring all their combinations, we will derive most of the algorithmic, time-step language principles.

Consider the instructions:

```
A = A + 1
B = B + A
```

It is a tribute to the spread of programming that these two instructions are thought to be intuitively obvious, when in fact, to a precisionist, their meaning is quite different from the common interpretation. To interpret these precisely, we must examine A and B *between* the instructions:

Before first instruction	A has value v_1 B has value v_2
First instruction	A = A + 1
Between instructions	A has value $v_1 + 1$ B has value v_2
Second instruction	B = B + A
After second instruction	A has value $v_1 + 1$ B has value $v_2 + v_1 + 1$

The precise interpretation, therefore, requires an alternating sequence of instruction and interinstruction times. If a single computer is to interpret this program, it will do so in a series of "cycles," with one pair of instruction-interinstruction times being a cycle or a *time step*.

Let us generalize this by separating the arithmetic operations from the cycle. This separation will allow us to do more than one arithmetic operation in a cycle, thus fostering *parallelism*. One way of combining operations and cycles is sequentially, as in the original example. However, we can also specify that both additions are to be executed in the same cycle:

```
(par A = A + 1
     B = B + A)
```

We call this a *form*, defined as a collection of operations that all happen in one cycle:

Before form	A has value v_1 B has value v_2
Form	(par A = A + 1 B = B + A)
Before next form	A has value $v_1 + 1$ B has value $v_1 + v_2$

Note the difference in the final value of B. Note also that since two additions happened in the same cycle, compiler-synthesized hardware will have at least two adders. Note also that, all other things being equal, this will result in a higher-performance ASIC. Incidentally, a single operation always executes in a single cycle.

Bit-Slice μ-Coding

What we have just described can be said to form the basis for a theory of bit-slice μ-coding or array processor programming. In either case, there are a number of operations that can take place in each time step. Indeed, the hardware is created or selected just for that capability. μ-coding must then be employed to realize the performance advantages for the specific application. As we will see in more complex examples, the idea of μ-coding is a useful way of looking at the Algorithmic System Compiler-based design process, with the interesting twist that instead of programming in the maximal parallelism a fixed set of base hardware will tolerate, the Algorithmic System Compiler user trusts the compiler to create the minimal hardware needed to implement the parallel program that forms the specification.

With this bit-slice concept in mind, we can model the Algorithmic System Compiler design process as programming a general, ultimately flexible machine. The programmer's model of the general "MetaSyn machine" is depicted in Figure 6.4. The MetaSyn machine programmer may use any number of any of the primitive operations supported in the vocabulary (+, −, increment, decrement, etc.) and any number of integer (word-length) or Boolean variables. These can be combined by the four algorithmic combination mechanisms described in Section 6.3.1:

sequence
cascade
selection
parallelization.

In any specific Algorithmic System Compiler program, the designer will use neither infinite operations nor a full cross-bar network of the implied units. The compiler thus implements only that fraction of the infinitely general machine necessary for the particular program. It is this implementation which is constructed in hardware.

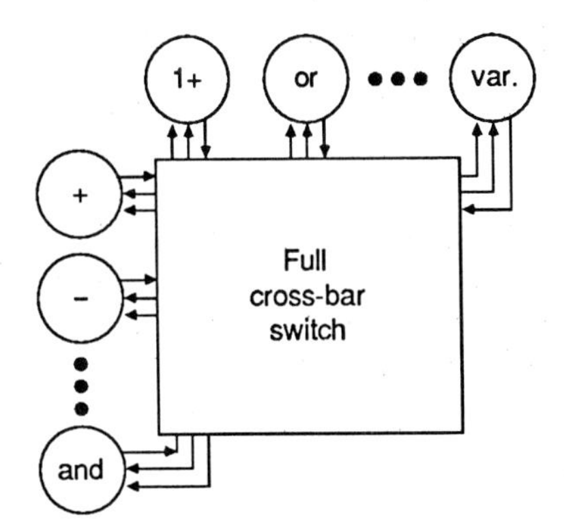

Figure 6.4 General "MetaSyn machine"—programmer's model. The result of any operation can be switched on every time step to be the argument (input) of any other operation.

Synchronous Digital Design
We often refer to the operations/cycle interpretation as a "master-slave" relationship. Note that an implementation of the A and B variables as singly clocked, master-slave registers, along with combinational logic for the addition operations, is precisely what is required (see Figure 6.5). In phase 0, the slave registers are stable, and their outputs are driving the combinational logic block, which implements the operations. The logic block eventually settles, presenting stable values at the inputs of the master registers. In phases 1, the master latches these values. Notice that the slave outputs have not changed, and therefore the masters' inputs remain stable. Once the masters have latched, they are isolated from any further changes in the combinational logic; in phase 2, their new values are propagated to the inputs of the slaves. Finally, in phase 3, the slaves latch their new values, which then begin to affect the combinational logic, and the cycle repeats. Although this cycle seems persnickity, it completely avoids "race" conditions, which often confound designs using less exacting disciplines. This basic cycle is often called "two-phase" clocking, since in some technologies, phases 1 and 3 can safely take place on clock edges, and in other methodologies, the four phases are "encoded" by the signals of two nonoverlapping clocks. Later we will see how MetaSyn implements the master-slave cycle in NMOS circuitry.

Non-Time-Step Operations (Syntactic Convenience)
The combination of stored variables or states with combinational operations is not really tractable for specification. Human beings generally want to

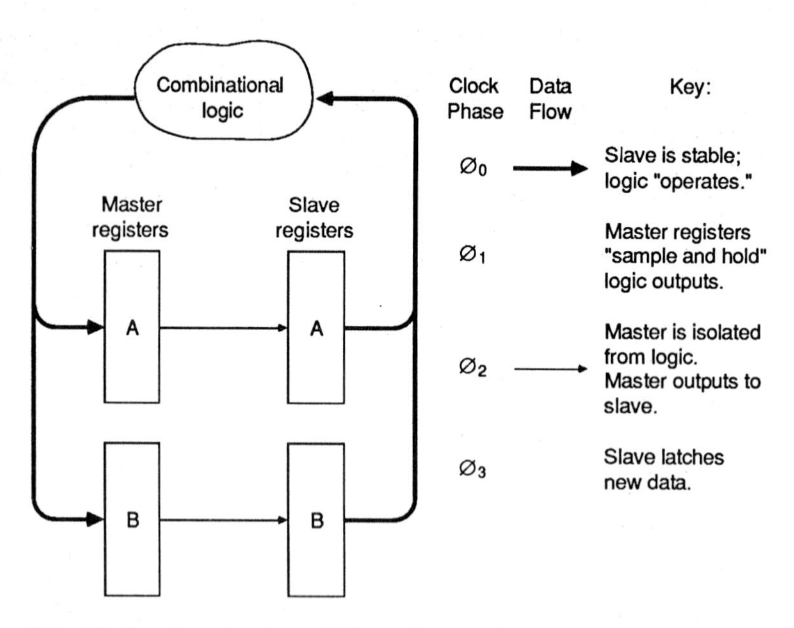

Figure 6.5 Master-slave implementation of operations cycle in synchronous digital hardware.

break up their operations. For example, consider the following computer program:

```
C = A + B
B = C
A = C - E
```

This is equivalent to:

```
B = A + B
A = A + B - E
```

in which the variable C does not appear at all. Intuitively, C may be thought of as a "named bus" or a "temporary variable." Since all practical Algorithmic System Compilers use such variables, the remainder of this section discusses this concept in more detail.

A recent concept in computer science (applicative programming) goes so far as to declare the first program expression "bad" and the second program expression "better." Such a declaration implies that the first expression conforms to a natural human tendency to break up a computation into smaller pieces. For a "natural" programmer, of course, the response would be, "Who cares?" If anything, the first expression is superior to the second since an optimizing compiler would translate the second expression into the first to avoid an extra addition operation. If the first expression is "natural" and results in at least as good machine code, why worry?

Irrespective of one's position in this controversy, the two program expressions above are significantly different if each is to be executed in a single time step:

```
(par C = A + B
     B = C
     A = C - E)
```

and

```
(par B = A + B
     A = A + B - E)
```

An optimizing algorithmic system compiler will, in each case, generate the same combinational operation components (i.e., one adder and one subtractor). A little investigation, however, will show that the behavior of both expressions is very different. In the first expression, the next value of A depends on the current value of C, which depends on the *previous* values of A and B. On the other hand, in the second expression, the next value of A depends on the *current* values of A and B. How can we specify things so that both expressions will perform identically?

First we recognize that the instruction C = A + B is subtly different from B = A + B. The first instruction does not contain "feedback." Its "natural" interpretation is as valid as the more precise one we have developed. In the above program expression, C is used as a place holder: the result of A + B is going to be used twice, so the programmer gave the result a name and used the name twice rather than the expression twice. Considered in another light, C could be considered a "fan-out node." As a result of this, C does not have to be a register; in fact, it can be a "bus." Unlike a register, which is single-valued only between instructions, a bus is valid only during the instruction (the corresponding jargon for a Boolean variable might be "flag" and "wire"). Thus, if C is declared a "bus," both program expressions will perform identically. Both AHPL and MetaSyn have this definition capability.

A Synchronous Example

We are now ready to observe the differences between AHPL and MetaSyn by creating a simple programmable modulus counter. This counter, when enabled, will clock through the sequence:

0, 1, 2, ..., *limit*–1

where *limit* is the programmable modulus and, hence, an input of the chip. A *reset* input signal will override the counting operation and return the counter to 0 (see Figure 6.6). Both languages require the buses, wires, and registers to be predefined.

EXAMPLE CODED IN METASYN The specification code for the actual operation of the chip in MetaSyn is as follows:

```
(always                         ; every clock cycle
  (if reset
    (setq count 0)                                    ; reset
    (if enable
      (if (= limit (1+ count))
        (setq count 0)          ; reset if overflow
        (setq count (1+ count)))                      ; or increment count
      (par)))                   ; nop if not enabled
  (setq out count))                                   ; in any case output
```

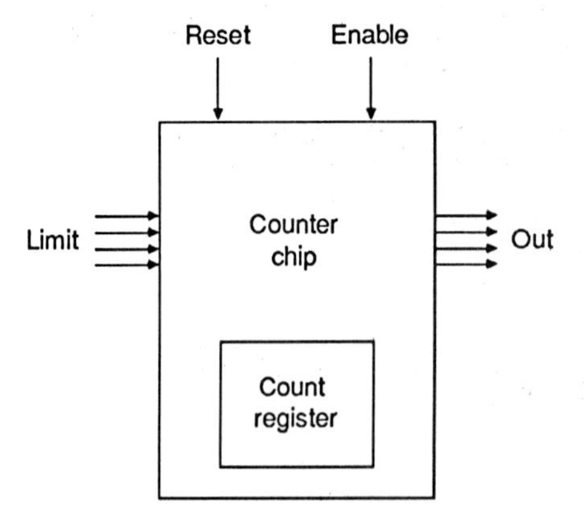

Figure 6.6 Counter Block diagram.

MetaSyn is a prefix, parenthesized language. This means that the first element following an opening parenthesis is the name of an operation. The second and following elements are arguments or operands of the operation. The arguments, of course, can be the results of other operations. Operation statements are ended by a closing parenthesis matching the operation's opening parenthesis.

The *always* keyword specifies that every operation within its scope (i.e., up to the matching closing parenthesis) is to be executed every clock cycle.

The first operation is an *if*. The syntax of an *if* is as follows:

```
(if condition consequent alternate)
```

This generates hardware with the effect that if the condition is true, then the consequent will be executed, and if the condition is false, the alternate will be executed. In this case, if *reset* is true, then the operation

```
(setq count 0)
```

will be executed. Otherwise, the second complex *if* operation will take effect.

The *setq* operation is technically called "assignment." Its syntax is as follows:

```
(setq variable value)
```

This generates hardware that sets the variable to the value (which might be a constant or the result of some other operation). The operation

```
(setq count 0)
```

sets the register *count* to 0 (between this time step and the next).

Let us now continue with the next *if* operation (which will take effect only if *reset* is not true). If *enable* is true, then we need to check if the next *count* will overflow the limit. We do this by making the condition of the next *if*

```
(= limit (1+ count))
```

Parenthesized notation, of course, must be interpreted inside out. The operation

```
(1+ count)
```

results in a value of 1 plus the current value of *count*. This value immediately becomes one of the arguments of the = operation. The other argument is the current value of *limit*.

The = operation results in a Boolean value of "true" if and only if its two word-length arguments are equal; otherwise its result is "false." As we have said, the result of the = operation becomes the condition of the *if*. When the equality is "true," this means that the counter is about to overflow and the count should be set to 0. Otherwise, the count should be incremented by assigning the result of

```
(1+ count)
```

to the variable *count*. Note that just writing

```
(1+ count)
```

does not in itself modify *count*. The result of the operation can either be fed into another operation (the intermediate result is "anonymous"), or it can be assigned to a variable, or, as in this case, both.

Since our edit program has auto-indented the text as we typed it based on the parenthesis nesting level, we see that we have not completed the syntax of the *if*, which started off as

```
(if enable
```

Since no operations should take place if both *reset* and *enable* are "false," we complete this *if* by creating a "no op" alternate. A *par* with no operations in its scope denotes nothing in parallel, that is, "no op."

Finally, since we are still within the scope of the *always*, whatever the outcome of the various nested *if*s on the next cycle's value of *count*, the current count will be output on the bus *out* every clock cycle.

This example demonstrates two characteristics of MetaSyn specifications. First, in comparison with a computer running a similar program, the MetaSyn-generated chip will operate much faster. A standard computer such as a Motorola 68000 or even a reduced instruction set computer (RISC) would take a minimum of one cycle to set condition codes based on the value of *reset*, and another to execute the associated branch. Another two cycles would be used to test *enable*, and then there would be a cycle to increment *count*, another for a check on equality between the increment and *limit*, another for the associated branch, and so on. This ignores the fact that most μ computers use several cycles per instruction. In short, for similar technologies, the MetaSyn chip will operate about 50 times faster than the same algorithm programmed on a μ computer.

Second, this MetaSyn specification clearly demonstrates the difference between a structural or logical hardware description and an algorithmic specification that will *synthesize* logic. Note that a 1+ operation appears in two places. In one statement the result is fed into the = operation, and in the other it is assigned to a variable. If two 1+s appeared in a structural hardware description, the resulting hardware would be two increment operators. In the algorithmic specification, however, the compiler must perform logic synthesis. Thus, the compiler will recognize that in this case, and more general cases, the operations can share a single operator. The compiler creates the appropriate multiplexers, control, and interconnection so that the compiler-generated hardware will implement the user-specified algorithm.

EXAMPLE CODED IN AHPL

A "top-level" AHPL specification for the counter is as follows:

```
1. COUNTINC ← BUSINC (COUNT)
   → (RESET)/(5)

2. NO DELAY
   → (UP)/(6)

3. NO DELAY
   → (BUSEQ (LIMIT ; COUNTINC))/(5)

4. NO DELAY
   COUNT ← COUNTINC
   → (6)

5. NO DELAY
   COUNT ← 0

6. NO DELAY
   OUT ← COUNT
   → (1)
```

Statement 1: A bus labeled COUNTINC is set to BUSINC (COUNT). We are immediately caught up in a complexity of AHPL due to the lack of integer arithmetic operations. BUSINC is a user-defined combinational *operator* which inputs a word value and outputs 1 plus that value. (Note the use of the word *operator* as opposed to *operation*; the significance of this distinction was explained in the preceding section.) The AHPL hardware compiler should "optimize" and synthesize logic for Boolean operations; however, it does not do so for arithmetic ones. Thus, it is up to the designer to realize that only the BUSINC of COUNT is required, and that since it is required in two statements (3 and 4), the value should immediately be computed and broadcast on a bus. (The theory of internal bus values and their relationship to the time step was considered in Section 6.5.2.) The second line of the statement says that in parallel with BUSINC, if RESET is "true," go to statement 5. Otherwise go on to statement 2.

Statement 2: This statement starts out with NO DELAY. Normally every numbered statement is a new time step. However, in this case (and indeed throughout this specification) we want everything to take place in the same time step. Although there is no time step delay, statement 2 will not be executed if statement 1 calls for a "go to 5." Statement 2's action causes a "go to statement 6" if UP is "false." Otherwise go on to statement 3.

Statement 3: BUSEQ is another user-defined combinational logic operator. This one signals "true" if its two word-length inputs are equal in all bit positions and "false" otherwise. In the former case, the machine will go to statement 5; otherwise it will continue with statement 4.

Statement 4: The register COUNT will be assigned the value of COUNTINC for the next time step, and the machine will skip over statement 5.

Statement 5: We get to statement 5 either from statement 1 (if RESET) or from statement 3 (if we are about to overflow LIMIT). In either case we want to set COUNT to 0.

Statement 6: Connect the current value of COUNT to the OUT bus, and return to statement 1. Since statement 1 *does not* start with NO DELAY, this will cause a time step to occur.

So far AHPL does not seem terribly punishing because arithmetic operations are second-class verbs. We have yet to design the user-defined operators, however. BUSINC would be a parallel machine or process also described in AHPL. (The same facility is available, but less often necessary, in MetaSyn and ISPS.) A simple specification might be as follows, with BIN the bus input, BUSINC the bus output, and C0 through C2 defined as "wires":

1. NO DELAY
   $\mathrm{BUSINC[0]} \leftarrow \mathrm{BIN[0]}$
   $\mathrm{C0} \leftarrow \mathrm{BIN[0]}$

2. NO DELAY
   $\mathrm{BUSINC[1]} \leftarrow (\overline{\mathrm{BIN[1]}} \wedge \mathrm{C0}) \vee (\mathrm{BIN[1]} \wedge \overline{\mathrm{C0}})$
   $\mathrm{C1} \leftarrow \overline{\mathrm{BIN[1]}} \wedge \mathrm{C0})$

3. NO DELAY
   $\mathrm{BUSINC[2]} \leftarrow (\overline{\mathrm{BIN[2]}} \wedge \mathrm{C1}) \vee (\mathrm{BIN[2]} \wedge \overline{\mathrm{C1}})$
   $\mathrm{C2} \leftarrow \mathrm{BIN[2]} \wedge \mathrm{C1})$

4. NO DELAY
   $\mathrm{BUSINC[3]} \leftarrow (\overline{\mathrm{BIN[3]}} \wedge \mathrm{C2}) \wedge (\mathrm{BIN[3]} \wedge \overline{\mathrm{C2}})$
   $\rightarrow (1)$

The reader might attempt the BUSEQ design. In more complex examples, it is not the need for user-designed integer operators that makes AHPL more difficult to use (the most useful ones will be in a library anyway), but rather the fact that the designer must manually create the logic *structure* for arithmetic operators to implement the underlying algorithm. MetaSyn's automatic synthesis and optimization of arithmetic *operator* elements turns out to be just as valuable as the synthesis and optimization of Boolean operators.

DEFINING VARIABLES IN METASYN As we have said, most HDLs and Algorithmic System Compiler languages require variables to be predefined as to their type. In computer science jargon, these languages are "strongly typed." In contrast, LISP, with which MetaSyn is often confused, is an example of a language that not only is not strongly typed, but is not even statically typed; that is, a single variable can be a number, a list, a symbol, and a function all within the space of a single program. The MetaSyn definition section for the counter would read something like this:

```
(program counter 4
   (def reset signal input 5)
   (def enable signal input *)
   (def limit port input (7 8 9 10))
   (def out port output *)
   (def count register)
```

The line

```
(program counter 4
```

specifies that *counter* is the name of the chip and that all arithmetic on the chip is to take place in 4 bits (i.e., modulo 16). Each of the next five lines defines a variable.

The line)

```
(def reset signal input 5)
```

defines the variable named *reset* to be a signal of type "input" that will be coming in on pad number 5. A "signal" is a single-bit, unstored, unclocked variable, which in other nomenclatures might be called a "wire." We have specified that this signal is to be input from pad 5, which means that the compiler must create a pad and the appropriate lightening protection, level shifting, etc. Any attempt to set or write to this signal will result in an error during the compilation phase. Other signal types are "output," "i/o," and "tri-state." An output signal could never be used as the argument of a computation or operation. If an output signal is not written to during a time step, it defaults to an output of "false" or 0. An i/o signal can be either input or output during a time step (but not both at the same time). Its default is "input" (which is high impedance as far as the external circuitry is concerned). The last signal type, tri-state, also defaults to high impedance, although it cannot be used as an input, only as an output.

The line

```
(def enable signal input *)
```

is similar to the previous definition; however, by using the * symbol, the designer indicates indifference to the actual location of the pad.

The line

```
(def limit port input (7 8 9 l0))
```

defines a bus (often called a "port") that is input from pads 7, 8, 9, and l0. Pad 7 is the most significant bit, pad 8 the next, and so on. Like signals, ports may be of type input, output, i/o, or tri-state, with similar properties. An output port, however, will be output an undefined value if it is not written to during a time step.

The line

```
(def out port output *)
```

defines an output bus named *out*. This time, however, we have not specified the pads for the bus, leaving it up to the compiler to generate their locations.

Finally we define the *count* variable as a register.

6.5.3 Finite-State Machines

The master-slave cycle is not only the basis of computers and programming, it is also the basis of FSMs. The FSM has no stored data except its so-called state. From its current state and inputs, the combinational logic of the FSM determines what its outputs should be and what its next state should be. An FSM is commonly implemented as a PLA with a master-slave register (see Figure 6.7). Note the similarity between Figures 6.5 and 6.7. This indicates that possibly all computation could be viewed and implemented as an FSM or a combination of FSMs.

The FSM has been a very powerful paradigm for controllers. From a theoretical standpoint, integer and other multibit data can be encoded as "states." On a practical basis, however, FSMs have been useful only with Boolean or single-bit variables and Boolean or logic operations. This is because all bits of an FSM state are expected to contribute equally to the next-state mapping function. In arithmetic operations, this is not efficient either in conception or in implementation. Consider the specification and implementation of a standard integer addition operation if each bit of the result must be described as a Boolean function of the input bits.

FSMs combine the qualities of a formal (and theoretically tractable)

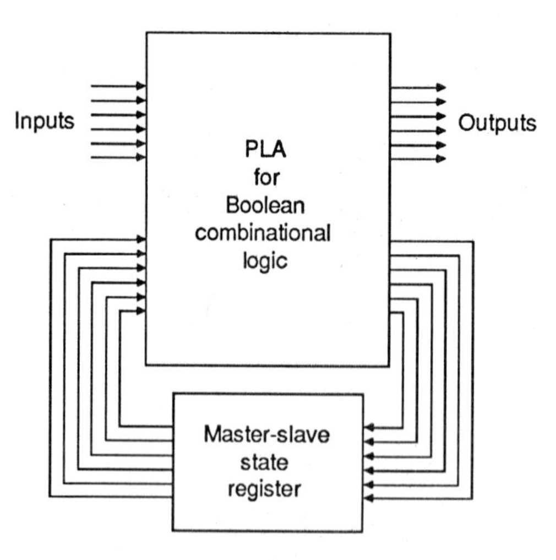

Figure 6.7 PLA and master-slave implementation of an FSM.

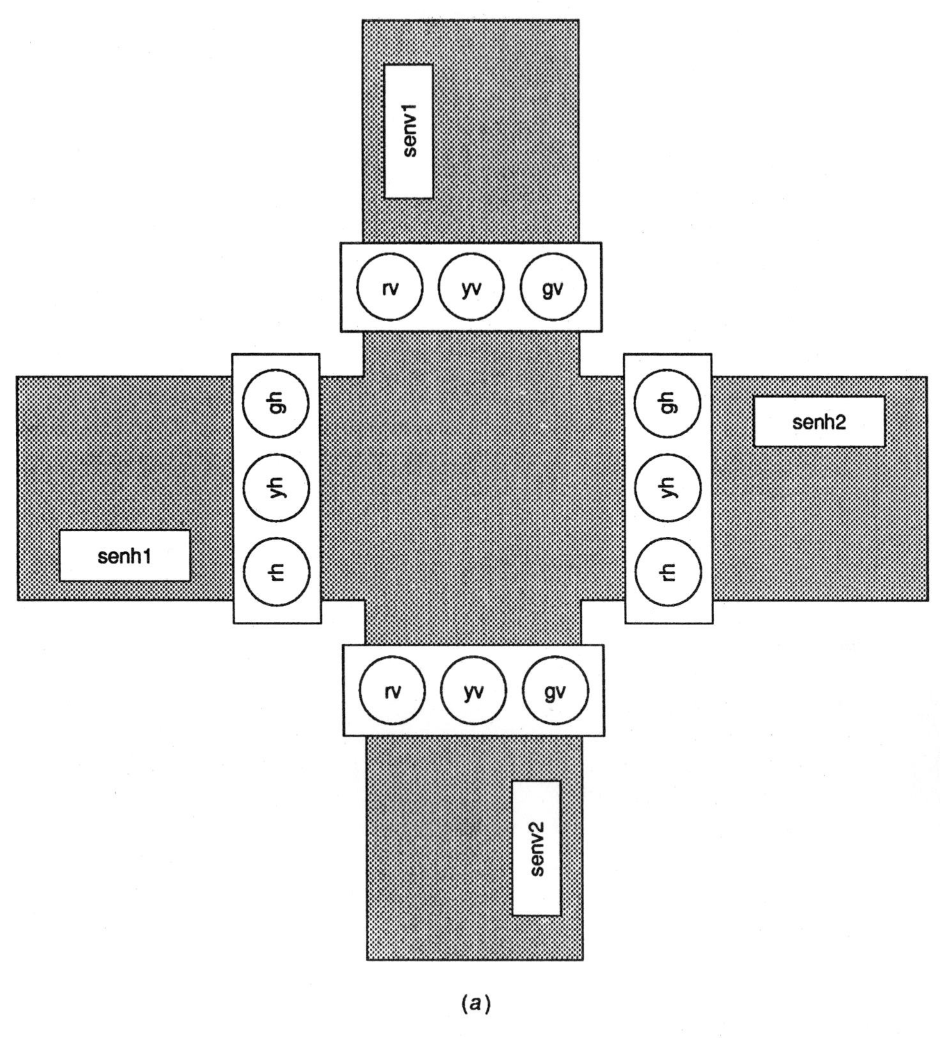

Figure 6.8 Traffic light controller. (a) Intersection layer.

mathematical definition, practical usefulness as controllers, and implementability, and this has resulted in much work to aid, automate, and "optimize" the design of FSMs. One such mechanism is AHPL. The companies that manufacture programmable logic arrays of various types have developed their own languages similar to AHPL, although perhaps more reasonable to use. The MetaSyn language is also exceptionally useful for such implementation, maintaining the FSM concepts of parallelism and providing several added conceptual benefits, including:

arithmetic as well as Boolean operations
storage of information as variables as well as pure-state
submachines that are analogous to subroutines

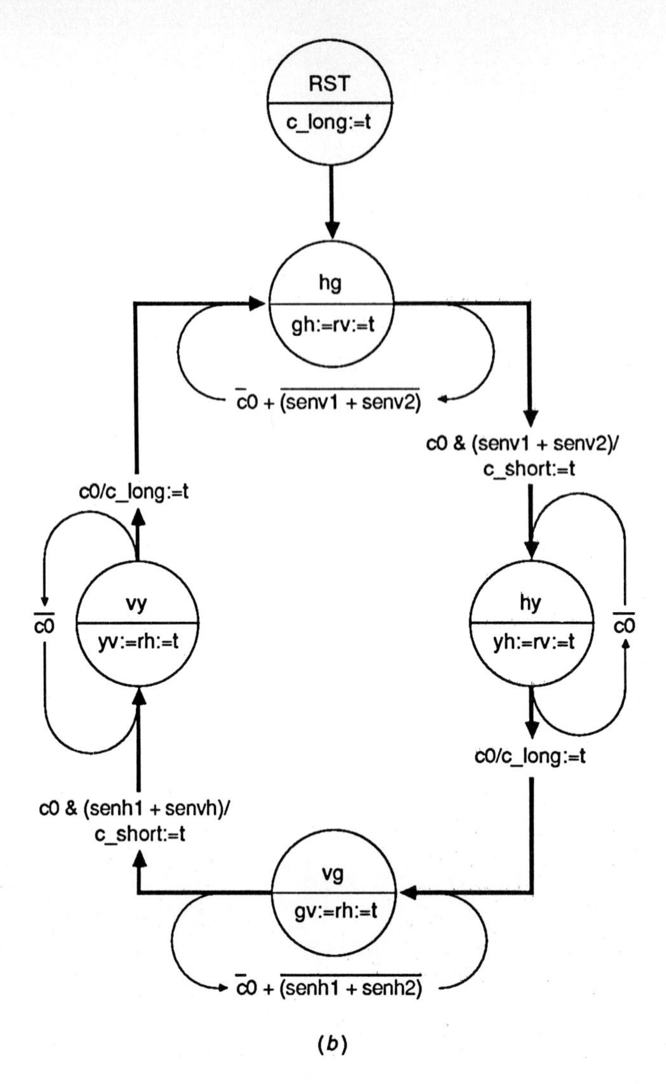

Figure 6.8 Traffic light controller. (b) FSM diagram.

As a practical matter, MetaSyn is also useful in providing automatic state code generation and logic reduction (minimization). It is also customary to provide a *reset* signal, which causes any state of the FSM to transit to a fixed reset state. This allows initialization of the FSM; otherwise it will begin operating in a random state, or worse, in a random state encoding that may be unassigned. This reset action is extremely awkward to specify in pure FSM notation; however, MetaSyn covers all the messy details implicitly rather than explicitly. In MetaSyn, each FSM is known as a *process*. Multiple processes may be defined, and each has a name and a *stack depth*, the latter for implementation of the sub-subroutine concept.

Figure 6.8a (see page 171) depicts a simple traffic intersection of two streets: H and V. Each street has a pair of the traditional sets of red, yellow, and green signal lamps and a pair of sensors indicating the presence of an oncoming vehicle. Figure 6.8b (see page 172) diagrams the FSM that will implement the traffic light controller.

Note that the lights have been labeled by two letter combinations with the "color" initial first, whereas FSM states are labeled with direction initials first.

The FSM must control not only the lights (labeled gh, yh, rh and gv, yv, rv for green, yellow, and red horizontal and vertical lights, respectively), but also a down counter, which may be preset with either a long or a short count. The down counter signals the FSM when it reaches 0. This counter is used to set the minimal time between various light changes. *Thus, the FSM diagram is not sufficient. We must also create a counter with the appropriate behavior* (see Figure 6.8c). Some of the elements of this behavior, especially the detailed timing relationships between c0 (controlled by the counter and signifying that it has counted down to 0) and c_long and c_short (controlled by the FSM to start long or short time delay countdowns) are quite subtle.

There are several commercially available FSM generators that would handle this FSM satisfactorily, and MetaSyn is one of them. In fact, unlike the others, MetaSyn could describe the down counter as well as the FSM. However, MetaSyn is even more powerful, since the entire system can be described in a single arithmetic extended FSM description. This is just as easy to write as the initial FSM description and much more understandable, since it is self-contained and intuitively complete. Figure 6.8d shows the extended FSM hy (horizontal yellow) state, redrawn to include the countdown.

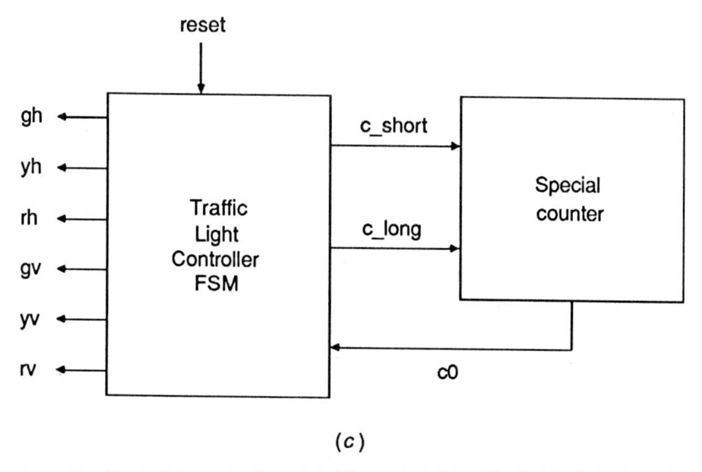

(c)

Figure 6.8 Traffic light controller. (c) "Complete" traffic light digital system.

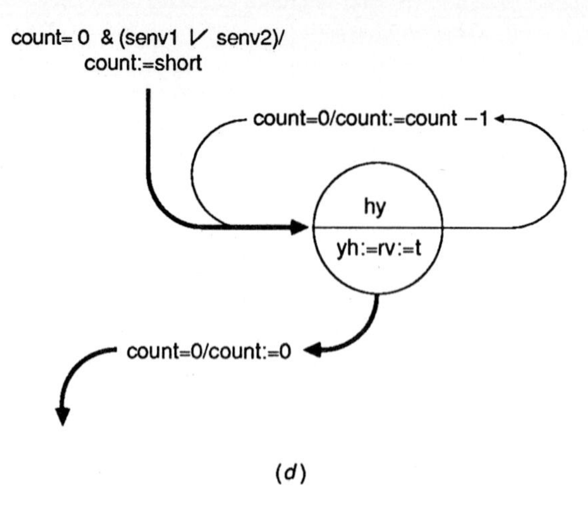

Figure 6.8 Traffic light controller. (d) Extended FSM diagram for **hy** state in traffic light controller. Note the use of arithmetic operations.

The MetaSyn code for the RST, hg, and hy states is as follows:

```
(process lights 0          ; FSM named lights with no subroutines
RST
   (setq count 0)          ; if reset start in some nominal state
hg
   (par (setq rh (setq yh (setq yv (setq gv f))))
        (setq gh (setq rv t))                    ; steady lights
        (if (=0 count)
           (if (or senv1 senv2)                  ; counter ran out
              (par (setq count short)            ; time for V street
                   (go hy)) ; first yellow on H
              (go hg))                           ; no cars on V
           (par (setq count (1-count))           ; run counter
                (go hg)))) ; and wait for runout
hy
   (par (setq rh (setq gh (setq yv (setq gv f))))
        (setq yh (setq rv t))                    ; steady lights
        (if (=0 count)
           (par (setq count long)(go vg))        ; time for V green
           (par (setq count (1-count))           ; or run count
                (go hy))))
```

Recall from the previous example that an *always* indicates that its argument operations are to be executed in a single time step. In this code we use a *process* instead, to indicate that the operations which follow it are to be executed one at a time. Of course, within any of these operations, more parallelism may, and usually will, be specified. The name of this process is "lights" (which fact will be useful in a simulation). It has no subroutine stack. The first state in the code is labeled RST. (Labels are easily distinguished from state operations since operations must begin with an opening parenthesis.)

Most FSMs have a built-in transition to some nominal reset state upon receipt of some global signal. In MetaSyn the reset signal, when "true," will implicitly cause such a transition for each process to the first state in that process. The fact that it is implicit saves a great deal of typing, and more importantly, prevents the designer from accidentally leaving it out. In this case, the reset state is labeled RST, and its sole function is to set the count register to 0. Another common human error in designing FSMs is to leave out a transition case. MetaSyn assumes a *program* orientation, so if no transition is specified or if some of the transition cases are unspecified, it creates the hardware to transit to the next state in lexical order.

We are in the **hg** (horizontal green) state, and we must maintain the traffic light signals in a meaningful fashion. The green horizontal and the red vertical lights must be turned on and all the others turned off. Then there are basically three significant transition cases. First, the count may not have run down, in which case **hg** will decrement *count* and loop back to itself. Alternatively, the count may have run down, but there are no cars indicated by either of the vertical sensors, in which case *count* will not be touched, but **hg** will continue to loop to itself. Finally, there is the case where *count* will equal 0 and one of the vertical sensors will be active. In this case (indicated in the code by the arrow), *count* will be reinitialized to *short* and the FSM state will transit to **hy**.

Since the first action in the **hy** state is a check for a zero value in *count*, it is vital that both the FSM state and the variable *count* be synchronized in the same "master-slave" time step. Fortunately they are. The reader should be able to correctly interpret the **hg** and **hy** code and see that it corresponds to the extended FSM diagram.

It will also be an instructive exercise to program the **vg** and **vy** states. For added interest, try to write a program that will return to green from yellow if the appropriate sensor is no longer active (right on red after stop).

Note that the benefits obtained over simple FSMs are derived from two sources. First, we are able to do arithmetic directly in the "machine." Second, we separate the "data state" from the "control state." In an FSM there is only one kind of state: control. The traffic light case has a *count* data state of which only "*count* equals 0" is significant for control. Imagine the encoding boredom (to say nothing of induced errors) if we needed to describe (for 8 bits of *count* data) all 1024 states! These options together make MetaSyn much more widely applicable than traditional FSMs.

6.5.4 Data Flow

One might say that FSMs are at one end of the digital system spectrum in that they exhibit only control flow and no data flow. In the digital signal processing (DSP) area especially, we often encounter the exact opposite: algorithms with no control flow, only data flow. Consider the typical flow diagram for a second-order filter shown in Figure 6.9.

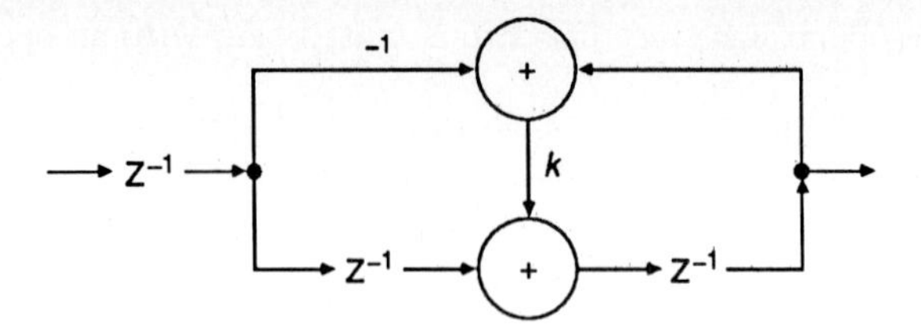

Figure 6.9 Second-order IIR filter.

The Z^{-1} notation implies a one-time-period delay; -1 and k are multiplicative constants. MetaSyn does not have a multiplier; however, this is remedied by implementing the multiply operation as a combination of shifts and adds. For the sake of illustration let us assume that $k = 0.6875$ (.1011 binary).

```
        (program filter 16
        (def xin port input *)
        (def rs port output *)
        (def * phia)(def * phib)(def * phic)
        (def rx register)(def rv register)(def rw register)
        (def i1 port internal)(def i2 port internal)(def i3 port internal)

        (always
        (setq rx xin)
        (setq rv rx)                ; these three are data transfers without
        (setq rs rw)                ; any calculations
        (setq i1 (−rw rx)
; constant is .1011
        (setq i2 (+(a>> i1)(a>> i1 3)(bit 2 i1)))
; li2= .5*i1 + .125*i1 with rounding
        (setq i3(+ i2(a>> i1 4)(bit 3 i1)))
; etc for other constants
        (setq rw (+ rv i3))))
```

The addition operation

```
(+ word1 word 2 [boolean])
```

has two (optionally three) operands. The two word arguments are added together; however, if the Boolean operand is specified, it becomes the “carry-in” for the least significant bit of the addition. Similar options are available for the subtract, increment, and decrement operations. The operation

```
(a>> word number)
```

shifts the word right by the specified number of places, maintaining the sign bit—in short, an arithmetic right shift. The operation

(bit number word)

creates a Boolean value which is equal to the numbered bit in the word.

It may be appropriate at this point to demonstrate how one debugs a MetaSyn design. The input/output relationship we expect for a second-order filter of this type is shown in Figure 6.10.

Among its other features, MetaSyn includes a simulator, which will be described in more detail in a later section. One simulator mode allows the user to interact with the design through the use of "windows." Figure 6.11 shows several MetaSyn simulation windows superimposed over each other. In the first frame (at time cycle = 0), the user has set RX to be 10. RS is still 0 since the system has not yet been clocked. In the next frame (time cycle = 1), RX is still 10; however, RS is now −7. In subsequent frames, RX has been returned to a value of 0, and by observing the values of RS we can verify that this MetaSyn design did implement a second-order filter.

6.5.5 Pipeline, Cascade, and Sequential Designs

The preceding DSP example uses a single word delay (Z^{-1}) as part of the algorithm for a second-order filter. This delay mechanism also forms the basis for pipelining. There are many algorithms that can be implemented variously in pipeline, cascade, or sequential form.

In the sequential form, the algorithm "unfolds in time," using the same hardware operation units and returning a result some number of clock cycles after the input is presented. In the cascade form, the algorithm "unfolds in space." A block of hardware operation units is repeated on the chip, and data "cascade" out of one block and into the next, culminating in the result

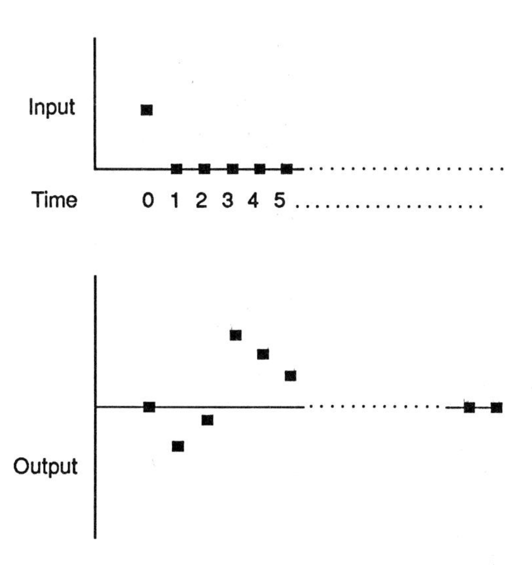

Figure 6.10 IIR I/O relationship.

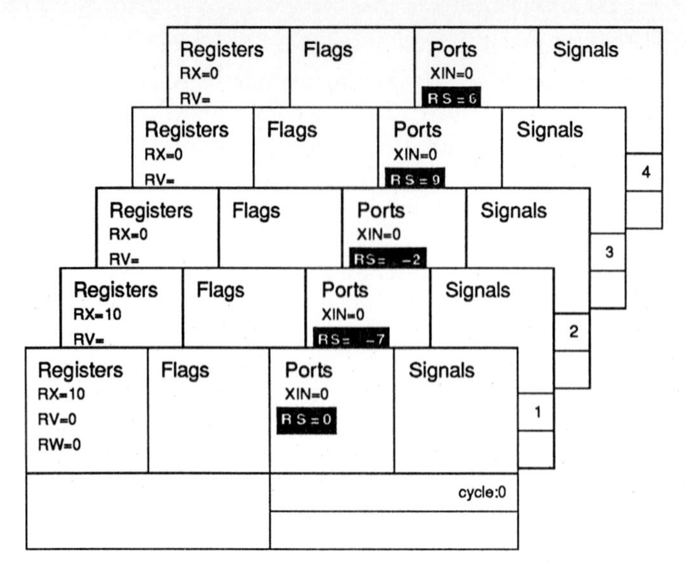

Figure 6.11 Several frames of IIR filter simulation.

being returned in the same cycle as the inputs are presented. Of course, this clock cycle will be longer than the sequential clock cycle since the critical path flows through more arithmetic physical hardware units. However, since the cascade version executes in only one cycle instead of many shorter ones, it will usually have higher system throughput. The cascade chip is also generally bigger than the sequential version because it has more arithmetic hardware units. Finally, the cascade form can be turned into a pipeline form by replacing the data cascade wires with pipeline registers. The pipeline version combines the short clock cycle of the sequential form with the one result per clock cycle of the cascade form. In compensation, however, the extra pipeline registers result in the largest area of all. Figure 6.12 shows all three possibilities in general block diagram form.

As we will see, it is relatively trivial to move from one mode of parallelism to another. Indeed, it is eminently reasonable to mix schemes, such as alternating pipeline and cascade stages. One of the advantages of using MetaSyn is that the system designer can examine the various cost/performance tradeoffs of each approach, or hybrids of various approaches, with very little extra design cost.

Let us consider the example of a multiplier design. MetaSyn is well suited to implementing Booth's encoding multiply algorithm or the "array" multiply method; however, for simplicity we will use a standard add-and-shift method. Figure 6.13 represents one stage of this algorithm pictorially.

This method uses the standard assembly language programmer's trick to minimize register usage. Each step yields a bit of the result, which will never again be modified, and also "uses up" a bit of the multiplicand, which

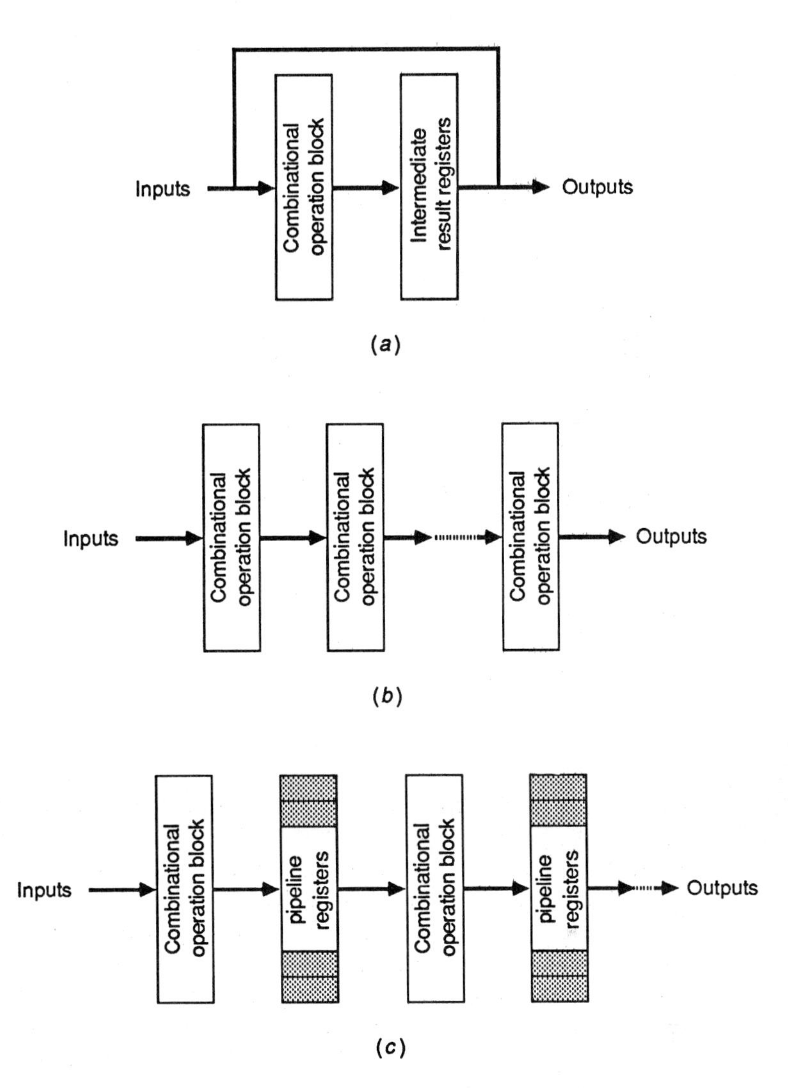

Figure 6.12 Alternative parallelism for the same algorithm. (a) Sequential implementation: intermediate results are fed back to the operation block, and after a number of cycles the final result is output. (b) Cascade implementation the output of each operation block is fed into the next. The critical path includes all the operation blocks from input to output. (c) Pipeline implementation: the output of each operation block is fed into the next through pipeline registers. The critical path is only as long as one operation block.

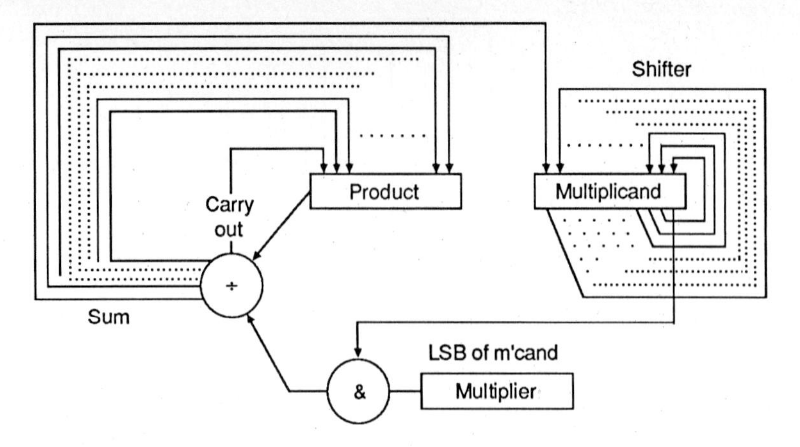

Figure 6.13 Add-and-shift multiply stage.

will never need to be tested again. Hence, we shift the used-up bit out of the multiplicand (mcan) at the same time we shift the least significant result bit in. In the end, mcan contains the low-order word of the result, and the product (prod) contains the high-order result word. Here is an encoding of the sequential version:

```
(program mult–seq 8
   (def mplr–in port input *) (def mcan–in port input *)
   (def res–hi–out port output *) (def res–lo–out port output *)
   (def * phia) (def * phib) (def * phic)

   (def mplr register) (def mcan register) (def prod register)

   (process mult 0
loop
      (par (setq mplr mplr–in)                              ; input mplr and
           (if (bit 0 mcan–in)
              (par (setq prod (>> mplr–in))                  ; calculate first
stage
                   (setq mcan (>> mcan–in 1 (bit 0 mplr–in))))
              (par (setq prod 0)                         ; of product and mcan
                   (setq mcan (>> mcan–in)))))                ;from inputs.
      (if (bit 0 mcan)
         (par (setq prod (>> (+prod mplr) 1 (+carry? prod mplr)))
              (setq mcan (>> mcan 1 (bit 0 (+ prod mplr)))))
; NOTE – the MSB shifted into prod is the carry out of the sum, which
;        extends the range of the algorithm
;      – the MSB shifted into mcan is the LSB of the sum
; the compiler recognizes that all these sums can be carried out by one physical
; adder unit (and two shift units).
```

```
(par (setq prod (>> prod))
     (setq mcan (>> mcan 1 (bit 0 prod)))))
                                          ;repeat 5 more times

(if (bit 0 mcan)
   (par (setq res-hi-out (>> (+prod mplr) 1 (+carry? prod mplr)))
        (setq res-lo-out (>> mcan 1 (bit 0 (+ prod mplr))))
        (go loop))
   (par (setq res-hi-out (>> prod))
        (setq res-lo-out (>> mcan 1 (bit 0 prod)))
        (go loop)))))
```

The >> operation is, of course, a right-shift operation. Its general syntax is

```
(>> word [number [Boolean]])
```

The word operand will be shifted the specified number of positions to the right (the default number is 1). The optional Boolean operand specifies the bit value to be shifted into the most significant bit of the word (the default Boolean value is "false" or 0).

This code also uses an interesting operation:

```
(+carry? word word [Boolean])
```

This operation yields the value of the final carry output of the indicated addition. There are similar operations for subtract, increment, and decrement operations. They are typically used (as here) for multiprecision arithmetic. As a very simple example, a single-time-step, double-precision addition can be performed as follows:

```
(par(setq a-lo (+ b-lo c-lo))
    (setq a-hi (+ b-hi c-hi (+carry? b-lo c-lo))))
```

The compiler will recognize that both (+ b−lo c−lo) and (+carry? b−lo c−lo) can utilize the same hardware.

After a little study, the question may arise: Why not write this as a loop, using a single program statement and an explicit loop count variable? In software, this would result in less compiled code. However, the algorithmic system compiler does not create "code" in the same sense. Moreover, the hardware for an explicit loop will be no more efficient than for an "inline" specification, and if the loop is not programmed carefully, it will be less efficient. Also, the inline expression is easier to modify for the cascade and pipeline versions.

To make a cascade version, we declare three series of internal ports:

```
prod0, prod1, ..., prod6
mplr0, mplr1, ..., mplr6
mcan0, mcan1, ..., mcan6
```

Next we remove all the registers. Finally, we modify each stage *i* to input from prodi, mplr, and mcani, and to output to prodi + 1, mplr + 1, and mcani + 1. Since the entire calculation is to happen in one clock cycle, the process turns into an *always*. The code section now becomes:

```
(always
  (setq mplr0 mplr-in)      (setq mplr3 mplr2)      (set mplr6 mplr5)
  (setq mplr1 mplr0)        (setq mplr4 mplr3)
  (setq mplr2 mplr1)        (setq mplr5 mplr4))
(always
  (if (bit 0 mcan-in)
    (par (setq prod0 (>> mplr-in))                        ;calculate first stage
         (setq mcan0 (>> mcan-in 1 (bit 0 mplr-in))))
    (par (setq prod 0)                                    ; of product and mcan
      (setq mcan (>> mcan-in))))                          ; from inputs
  (if (bit 0 mcan0)
    (par (setq prod1 (>> (+ prod0 mplr0) 1 (+carry?prod0 mplr0)))
         (setq mcan1 (>> mcan0 1 (bit 0 (+ prod0 mplr0)))))
    (par (setq prod1 (>> prod0))
         (setq mcan1 (>> mcan0 1 (bit 0 prod0)))))
  (if (bit 0 mcan 1)
    (par (setq prod2 (>> (+ prod1 mplr1) 1 (+carry? prod1 mplr1)))
         (setq mcan2 (>> mcan1 1 (bit 0 (+prod1 mplr1)))))
    (par (setq prod2 (>> prod1))
         (setq mcan2 (>> mcan1 1 (bit 0 prod1)))))
                          .
                          .                               ;etc. 4 more times
                          .
  (if (bit 0 mcan6)
    (par (setq res-hi-out (>> (+ prod6 mplr6) 1 (+carry? prod6 mplr6)))
         (setq res-lo-out (>> mcan6 1 (bit 0(+prod6 mplr6)))))
    (par (setq res-hi-out (>> prod6))
         (setq res-lo-out (>> mcan6 1 (bit 0 prod6)))))
```

The two *always* sections are strictly for convenience.

Turning a cascade version into a pipeline version is the ultimate in simplicity: Simply redefine all the internal ports as registers, and recompile!

The user cannot always predict the relative merits of different versions, especially between the cascade and the sequential. There is always an overhead in any sequential process that will require registers and extra logic in the control section. For examples in this range of complexity, the cascade version is usually about the same size, if not smaller, and has about the same power drain as the sequential version. The pipeline version usually has twice the power drain (registers draw power) and is 30–50 percent larger.

6.5.6 Interface Protocols

Protocols fall into two general categories: hardware and architecture. MetaSyn has been used to implement both types of protocols for applications ranging from coprocessors to debouncing switches. As an example, we will discuss the interface between a MetaSyn chip and a standard memory (see Figure 6.14). We will assume that the memory can read or write faster than the MetaSyn chip can clock.

Electrical Protocols

The typical memory chip has $\overline{\text{READ}}$ and $\overline{\text{WRITE}}$ activation lines. For various electrical reasons, these are usually "wired NOR" lines, which means that they are terminated with pullup resistors and wired to several potential memory masters (such as coprocessors and DMA controllers that must "pull down" the appropriate line. A nonselected master must leave its activation lines high. This convention is in opposition to the MetaSyn standard, in which a single not asserted is "false." Thus, somewhere in the definition section there will be:

```
(def read signal internal)
(def write signal internal)
(def readbar signal output *)
(def writebar signal output *)
```

and in an *always* section there will be:

```
(setq readbar (not read))
(setq writebar (not write))
```

Address and data buses, on the other hand are usually "tristate" buses. This means that nonselected masters must present a high impedance to the bus, whereas the selected master drives it with both high and low values. This high-impedence state is the default state of MetaSyn i/o and tri-state ports if they are not set to any value in an instruction state. Thus,

```
(def address port tri-state *)
(def data port i/o *)
```

Architectural Protocols

We have now handled the two most common hardware interface protocols. Architecturally there is a much greater variation. Typically the input from

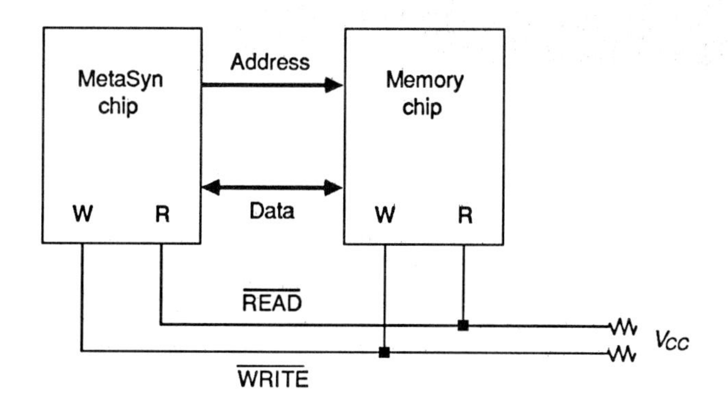

Figure 6.14 MetaSyn chip–memory chip system.

the memory to the master device is straightforward. The address port and the read signal must be valid before ("setup") and after ("hold") the data port can be safely read from the memory (see Figure 6.15). Thus, the safest read protocol would be a multi-time-step sequence:

```
(par (setq address x)(setq read t))
(par (setq address x)(setq read t) (setq y data))
(par (setq address x)(setq read t))
```

All MetaSyn ports and signals settle to their values *before* any registers are latched; therefore, a single "data read" time step includes a setup time, which may be sufficient to eliminate the need for a separate setup time step. Also, the read hold time is usually small, zero, or negative (depending on the memory chip); therefore, the third time step can also usually be deleted, resulting in a single-time-step protocol. We will use this one-cycle protocol in the example of Section 6.5.7.

Write protocols are not as standard. Using the Intel 2115 RAM as an example, we see that the sequence indicated in Figure 6.16 is necessary. For a write, the address and data must be stable before and after the write signal is given. Again, a three-time-step MetaSyn sequence is safest:

```
(par (setq address x)(setq data x))
(par (setq address x)(setq data x)(setq write t))
(par (setq address x)(setq data x))
```

To safely reduce this to a single-clock-cycle protocol, the write pulse may be ANDed with the phia clock signal, or the memory itself may have a clock for latching the data input, in which case it should share the MetaSyn chip's clock.

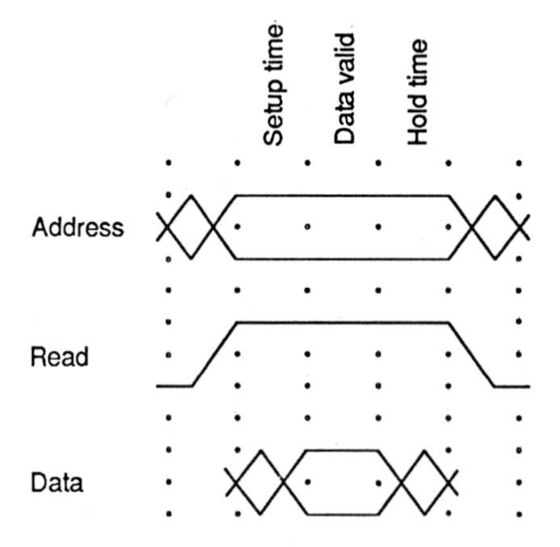

Figure 6.15 Read timing sequence.

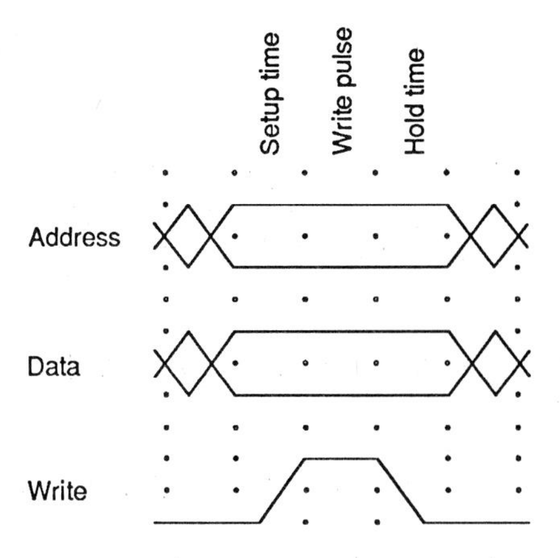

Figure 6.16 Write timing sequence.

Designers using these simple techniques will be able to concentrate on the application aspects, such as, for a μ-processor, pipelining every instruction in such a way as to fully utilize every clock cycle for a memory access.

6.5.7 Comparison with Non-Time-Step Descriptions: Instruction Decode Example

We will now present an example that will demonstrate the difference between MetaSyn, which has an implicit time step, and ISPS, which does not. Both MetaSyn and ISPS have a "decode" operation. This operation is a multi-way branch (or *case* statement) with remarkably similar properties between the two formats. We will use the operation decode to describe the "instruction decode" phase of the PDP-8.

ISPS Description

As usual, buses and registers (such as op, AC, and pc) have been predeclared.

```
decode op = >
  Begin
  #0\and := AC = AC And M[eadd()],
                  .
                  .
                  .
  #2\isz := Begin
            M[eadd] = M[eadd()] + 1 next
            If M[eadd] Eql 0 = > pc = pc+1
            End,
```

```
#3\dca := Begin
              M[eadd()] = AC next
              AC = 0
              End,
              .
              .
              .
```

The code is shown only for the **and**, **isz** ("increment and skip if zero"), and **dca** ("deposit and clear accumulator") instructions.

The ISPS instruction "decode" checks the op bus, which in this case will have been defined as a "field" or subset of bits of the instruction register. If the field is 0, then the **and** branch is taken; if it is 2, then the **isz** branch is taken; and so on.

If the **and** branch is taken, the effect should be to perform a logical AND of the AC register with the data value in the "effectively addressed" location in memory M. The "effective address" is the memory address pointed to by the current instruction. This pointing can be one of the various functions (usually called "memory address modes") of program counter, index registers, immediate operands, and so on, depending on the instruction architecture of the machine. The functions eadd() and eadd are used to compute the effective address of the data for the current instruction; they are defined by the user in ISPS elsewhere in the code. Unlike user-defined functions in AHPL, these functions may take more than one clock cycle. The result of the **and** is stored in the AC register.

If the **isz** branch is taken, the effectively addressed memory location is incremented. The keyword *next* is an ISPS instruction that states that the previous instructions must be completed before the following ones can be begun. In this case it is important that the memory be updated before it is tested. If the new value of the memory is 0, then the pc is incremented an extra time to skip the next instruction.

Note that *next* is not identical to a time step; it does not imply that the previous instructions are to be executed in parallel. One might say it only creates a "partial time order," whereas a time step orientation creates a "strict time order." With most of today's memories, the code

```
M[eadd]= M[eadd()]+1
```

will take at least two clock cycles over and above the effective address calculation, if only because separate cycles are required for the memory read and the memory write.

Note that this two-cycle requirement was qualified as being for "today's memories." When the PDP-8 was first designed, the standard memory access cycle was "read-modify-write." The then-standard "core" memories used destructive read techniques, so that even if the intent were to simply read from the memory, the unchanged value would have to be rewritten. In any case, the ISPS code simply does not give enough information to actually interface to any real memory. Compare this with the more complete information specified by the MetaSyn code in Section 6.5.6. *In general, this is a major problem with a non-time-step-oriented specification: It is very difficult to interface to an outside world that often requires strict time sequencing.*

Finally, if the **dca** branch is taken, the AC is stored in the effectively addressed memory location; once that has taken place, the AC is cleared. Note that if the description were time-step-oriented, both actions could take place in the same cycle.

One immediately obvious effect of the lack of time step orientation is that we have no idea how many clock cycles the decode operation itseslf is going to take. The possibilities correspond roughly to how "vertical" or "horizontal" the underlying μ-code engine is. For example, the hardware might be so restricted that the test for each branch might take another cycle, very much as a standard computer would. Alternatively, there might be the capacity for a multi-way μ-code branch in one cycle, but nothing else could be done in parallel; the actual operations would be done in succeeding μ-instructions addressed by the branch. Finally, both the branch and the initial operation could be done simultaneously in a very horizontally μ-coded machine. All of these options and more could be described by the same ISPS program. In order to synthesize hardware, therefore, a "compiler" must investigate the alternatives and make a decision among them. Automatic investigation (usually involving "artificial intelligence" techniques) has therefore been the hallmark of "compiler" work based on ISPS as an input specification.

MetaSyn Description

Unlike the designer working with ISPS, the MetaSyn designer must make a parallelism decision before coding. It is relatively simple to change one's mind afterward, so we will show two versions of the decode instruction. In the first version, the multi-way branch is executed in a single time step, but the instruction is executed in separate time step. In the second variation, as much as possible will be done in the first time step.

LESS PARALLEL

```
(decode i (11 10 9)     ; the op field is composed of the 3 msbs
  (go AND) ; op = 0 => And
  (go TAD) ; op = 1 => Add
  (go ISZ)  ; op = 2 => Increment and skip if zero
  (go DCA) ; op = 3 => Deposit and Clear Accumulator
     .
     .    )  ; end decode
AND               ; the label of another time step
  (call EADD-CALC)    ; first: call "subroutine" to calculate effective
  (par (setq addr eadd)     ;    address, then address that location
       (setq read t)     ;   read from it (onto the "data" bus)
       (setq AC (and AC data))     ;    and do the AND in 1 time step.
       (go INSTRUCTION-FETCH))
```

```
ISZ
  (call EADD-CALC)
  (par (setq addr eadd)
       (setq read t)   ;   I+M[eadd] is loaded into
       (setq dataout (1+data)))   ;   the scratch register "dataout."
  (par (setq addr eadd)
       (setq data dataout)   ;   dataout is sent out on the data bus
       (setq write t)   ;   which is written into the addressed location
       (go INSTRUCTION-FETCH)
       (if (= 0 dataout)
          (setq pc (1+pc))   ;   if dataout = 0 then increment the pc
          (par)))
CMA
  (call EADD-CALC)
  (par (setq addr eadd)
       (setq data AC)
       (setq write t)
       (setq AC 0)
       (go INSTRUCTION-FETCH))
```

The MetaSyn *decode* is a selection type combiner; in fact, it macro-expands (see Section 6.6.1) into a nested tree of *if* statements. It operates on a field of bits in a word (either a register or a bus). The bits of the field are specified in the operation; thus if "i" is the "instruction register," then the three most significant bits (numbered 11, 10, and 9 in a 12-bit word) are the "opcode" field. Each of the $2^3 = 8$ possibilities will have an operation among the remaining arguments of the *decode*. In this case, each of the operations is a *go* statement. A *go* transfers control to a labeled time step. Thus, since we see a

```
(go AND)
```

we expect a time step labeled AND somewhere in the code. As with any MetaSyn form, and unlike its ISPS namesake, the *decode* must unambiguously be executed in a single clock cycle.

The AND time step contains a

```
(call EADD-CALC)
```

If the designer treats the MetaSyn source code specification as μ-code, then this statement corresponds to a μ-code subroutine call to the time step labeled EADD-CALC. The EADD-CALC time step (or series of time steps) will perform the effective address calculation and return to the time step immediately following AND.

The time step immediately following the AND-labeled step does several things in parallel. We have checked our memory chip catalog and found that to read from memory, the proper memory address (eadd) must be placed on the addr bus and the read signal must be asserted to be "true."

Under these conditions, our memory will output the addressed location's value on the data bus.

MORE PARALLEL

```
(decode i (11 10 9)   ;   the op field is composed of the 3 msbs
  (par (setq ac (and ac data))   ;   op = 0
       (setq addr eadd)   ;   other code calculates effective address
       (setq read t))   ;   trigger read from m[addr]-> data
  (...)   ;   op = 1
  (par (setq dout (1+data))   ;   increment data read, save result for output
       (setq addr eadd)   ;   other code calculates effective address
       (setq read t))   ;   trigger read from m[addr]-> data
       (if (= (1+data) 0)   ;   why wait to calculate the pc
       (setq pc (1 = pc))
       (par))
  (go WRITE-DOUT))   ;   must go to another time step to write out.
(par (setq data ac)
     (setq addr eadd)   ;   we can write out in this time step since
       (setq write t)   ;   we aren't inputting also.
       (setq ac 0))
       .
       .
       .
```

Incidentally, several industry standard as well as custom processors have been designed using MetaSyn. They typically take 1–2 person-weeks to design, are 20% to 100% larger than hand-done designs, and have superior speed performance due to architectural improvements (see Section 6.6.5).

6.6 Implementation of the MetaSyn Algorithmic System Compiler

Figure 6.17 shows the MetaSyn software components and program flow. Note that in the case of MetaSyn, the physical layout procedure is by "silicon compilation." Although other chapters in this book have covered these silicon compilation techniques, we will cover what are, perhaps, some unique aspects of MetaSyn, in which "IC," "hardware," and "electrical" considerations all take a back seat to "system" considerations.

6.6.1 Prepass

The prepass section checks the input specification for syntactic correctness and, in the event of the simulation option being selected, converts the MetaSyn specification into a set of LISP functions. These interact with the standard simulator package to provide an interactive, high-level, yet accu-

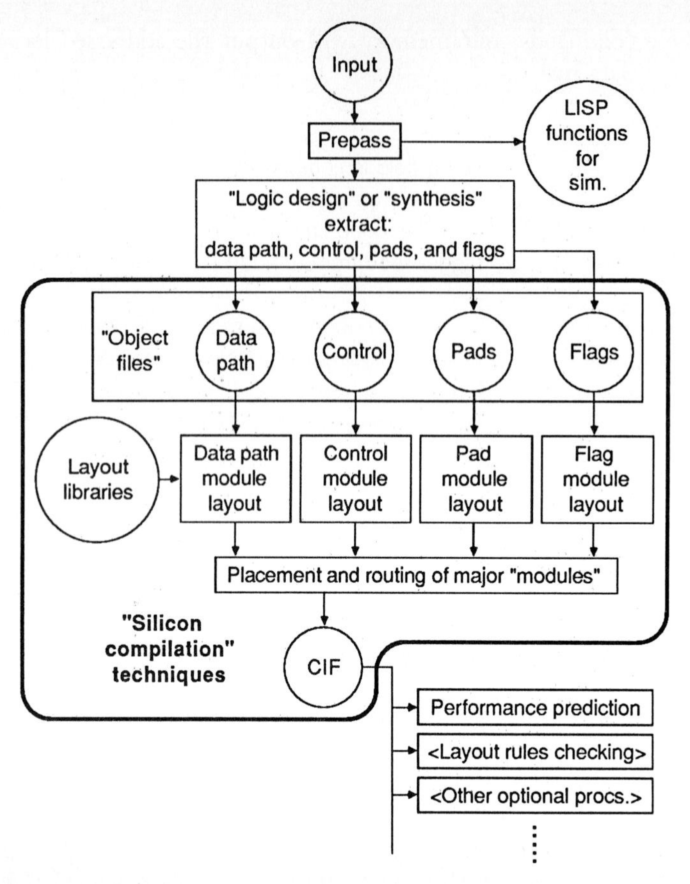

Figure 6.17 Components of the MetaSyn algorithmic system compiler.

rate, "hardware" simulation. Although not strictly part of "compilation," this simulation is very important. Finally, the prepass phase does "macro expansion." In fact, most of the "standard" MetaSyn operations are macro expansions of a relatively small number of primitives.

Simulation

Besides the requirement of accuracy, a good simulation should have three qualities. First, it should interact with the user at the "natural" level, which is the same as the input level. For example, the design specification "If reset then A: = A + 1 else A: = D" should never result in the error message "Node 327 Undefined." Second, a good simulation should have an interactive mode that operates with an acceptable response time. In practice this means that, starting from a state in which the simulated chip is "stable" (internally consistent), if the user changes the value of a variable, the simulation should respond with the change and its consequences completely propagated, and the chip again stable, within a few seconds at most. In addition, there should be a noninteractive ("batch") mode in which approximately 100,000 input-process-output cycles can be simulated in a few hours. Finally, the simulator should connect with an environment "cosimulation."

The last characteristic is one of several ways in which Algorithmic System Compiler requirements differ from those of true silicon compilation. For hardware design it is customary to provide a simulator with a means of reading a "test vector" file. The file consists of a set of test patterns with a code for each input and output pin. Each code tells a test interface program whether the particular pin is an input (and, if so, what its value is), an output (and, if so, what its value should be), or a "don't care." The interface program decodes one group of these pattern codes corresponding to the current simulated "time." This group is called a "test vector." The program then presents the values for input to the chip and tests that the outputs from the chip match the predictions in the file. This facility corresponds exactly to the way conventional hardware testers work on real chips.

On the other hand, it is not the system designer's goal to implement a logic design in hardware (or in an IC). Rather, it is to implement system functionality in an economical manner. The designer expects to try several architectural and algorithmic solutions, which may have little or no resemblance to each other on a logic or "per cycle" basis. Here only the "gross" effects are of interest. For example, in designing a real-time control chip for an automobile fuel combustion subsystem, the simulated chip will be tested against a simulation of the environment, consisting of sensors, actuators, and the physical dynamics. On a time basis consistent with the control problem (i.e., tenths of seconds), the chip is expected to achieve its goals. In Sections 6.5.5 and 6.5.7, we saw that the Algorithmic System Compiler user is able to easily modify the exact timing of operations while maintaining the underlying algorithm. It is also very likely that the algorithm will be modified in the search for good engineering tradeoffs. Each such modification would invalidate a test vector sequence, which would then have to be rewritten. In contrast, an environment cosimulation will always be valid.

For example, when designing a processor chip (several of which have been done with MetaSyn), the designer will often try out various degrees of parallelism, overlapping instruction fetch with execution, and so on. Creating an environment cosimulation, consisting of a simulated memory with program and data for the processor, involves work of less than three pages (depending on the program). Such a program may be good for several tens of thousands of clock cycles. To specify these cycles once for just one version of the processor would take even longer than the processor design. Of course, once the processor's environment has been proven to work for a simple (and slow) version of the processor, it will suffice, without change, to check out more complex, faster versions. Worrying about such changes and their testability is, of course, not the job of the IC or "post-logic" hardware designer, but of the system designer.

Macro Expansion

Both the programming world and the IC hardware world use the term *macro*. Of course, they each use it differently. In this chapter, the programming orientation of macro as "text substitution" will be maintained, rather than the idea of large cells per se.

One of the fundamental flaws of so-called standard-cell design is the constant proliferation of cells. When the time comes for a new technology to be supported, each cell must be redesigned. A good silicon compiler will keep the number of primitive cell types low so that it can be easily retargeted. On the other hand, users want to work with operations with which they are familiar. The standard macro library helps achieve a balance of these two desiderata.

For example, the word-permute operation is a basic primitive, from which all of the shift and rotate operations are constructed. This is "construction" by text substitution. The permute operation is rather complex in use, and operations like "arithmetic right shift variable x 3 places (a >> x 3)" are much more intuitive.

Another use for a macro would be where a new primitive is under consideration but not yet implemented, for example, a multiply-accumulate operation. Temporarily the multiply-accumulate operation could expand into a series of addition operations and a register. Later, if it were desirable, the custom operator could be substituted for the macro and the source specification recompiled (presumably with greater density—we will see why this might not be so in the next section, on logic synthesis). Finally, when the chip is to be recompiled into a new technology for which the custom operators have not yet been designed, the macro expansion can be used again. Note that we have referred only to macro expansion of operations into other operations, not operators into other operators. We will see why this is important in the next section.

6.6.2 Logic Design (Synthesis)

The logic design section is unique to algorithmic system compilation. Fundamentally, algorithms are temporally local, while structures are topologically local. In other words, in a program, statements ordered sequentially on the coding sheet are likely to be executed in sequence. On the other hand, components on the same schematic diagram sheet are likely to be interconnected.

Let us take a look at a simple algorithm:

```
cycle 1:   a:=a+b-c
cycle 2:   r:=r-a+d
```

The first step is to convert each of these steps to a structure. Basically every simple (i.e., unconditional) computation and assignment is equivalent to a simple structure (see Figure 6.18).

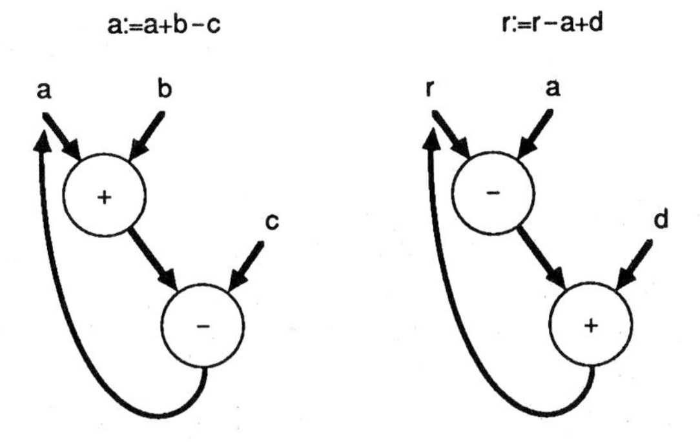

Figure 6.18 Two structures, each equivalent to its respective "algorithmic" statement.

The next task is to merge these two substructures efficiently. Certainly, we have no desire to construct each substructure independently, since only one is active at any one time and both use the same operations (and, hence, may share operators). One efficient way is shown in Figure 6.19.

This setup is indicative of the way the MacPitts/MetaSyn compiler generates structure from algorithm. For each statement a substructure is implied. The compiler checks to see if the operator units of the substructure already exist due to the compilation of previous statements. If they do, then they are free during this statement's time step. The compiler creates or extends the multiplexers (muxes) at the input of these operators to tap the appropriate buses. Finally, the control equation is updated to provide the additional control signals to the multiplexers based on the "state encoding" (or program counter value) of the current time step. Control signals go to the registers to allow or inhibit the storage of new data values.

Not only inter-time-step operations can share units. Logically, mutually exclusive operations within a single time step can share operator units also. The mechanism by which this is effected is, however, more complex.

These "logic design" or merging mechanisms depend on the initial algorithmic expression. With a few exceptions, they cannot be applied to the

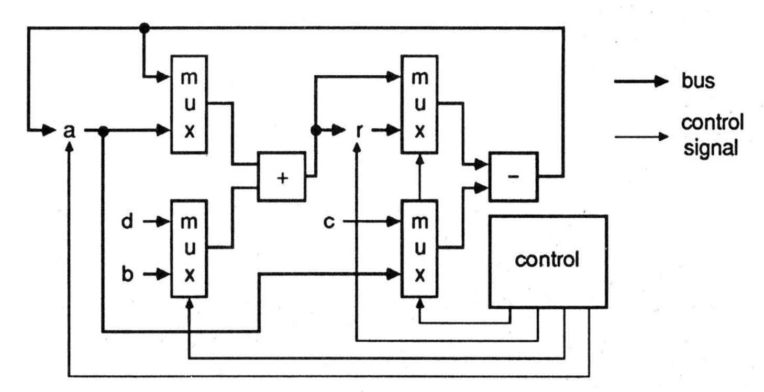

Figure 6.19 Efficient structure to execute the algorithmic statements of Figure 6.18.

structure once data flow is separated from control flow and the algorithm is hidden in the "control" logic. Such merging has proved highly effective. A PDP-8 design was compiled into 4200 transistors, mostly due to the logic synthesis efficiency of the compiler.

Let us return to the example of the multiply-accumulate operation. From the above discussion, it is clear that when this is designed in terms of a series of addition operations and a register, the addition operators can be used in those time steps of the algorithm in which multiply-accumulate is not operating. On the other hand, if the entire chip were recompiled with a custom multiply-accumulate operator, some number of addition operators would generally be required *as well as the multiply-accumulate operator*. Thus, the overall effect may be even *less* efficient than the macro expansion.

For a more concrete example, consider the "programmable modulus counter" discussed in Section 6.5.2. Suppose an operator existed that took in two operands, incrementing the first and checking to see if it is equal to the second:

```
(1+= count limit).
```

We can assume that this operator is marginally larger than either the incrementer or the equal-check operator, but smaller than both put together. If we had such an operator, the line

```
(if (= limit (1+ count))
```

would be replaced by

```
(if (1+= count limit).
```

The resulting hardware would be a more efficient implementation *of that line*. However, since

```
(1+ count)
```

is specified later, we would end up with both a 1+ = operator and a 1+ operator *for the whole chip*. Thus, the overall effect of a higher-level operator is to reduce the effective density of the chip. (As a word of warning, the change that we have analyzed would improve the standard measure of IC density, transistors/mm^2. For system designers, however, the appropriate measure of density here is counters/mm^2, which will be perceptibly diminished.)

The lesson here is that the more complex the operator, the more likely it is to be special-purpose and useful only part of the time. We discussed why larger operator libraries may be counterproductive over time in Section 6.6.1. In this section we discussed why higher-level primitive operators are not necessarily more efficient from a system point of view, even when they are pre-existing. Many advocates of the structural silicon compilation approach realize this as well. As further support for this position, recent computer science theory does not glorify the use of more instructions, but rather suggests better ways of combining a few simple ones.

In short, from the system designer's point of view, and given the inherent limitation of such a strategy, ever higher abstraction levels along the structural axis of the Y chart (see Chapter 1) are not satisfactory. Better ways of managing intermediate abstraction levels (somewhere around the level of a good computer instruction set) are definitely preferable.

6.6.3 Cell Library

As previously intimated, the cell library consists of bit slices of operators such as addition, increment, register, and so on. For maximal flexibility these cells should be parameterized by word length and bit position within the word. For example, the least significant bit of an adder may only need to consist of a half adder circuit, while every other bit must be a full adder. Other kinds of parameterization are useful for modifying drive capacity based on fan-out requirements, optional tristate outputs for maximal bus sharing, and so on. The tristate output option involves a tradeoff between area and speed, since buses with many devices hanging on them are slower than dedicated buses.

Because of the special nature of registers in the implementation of the all-important instruction or time step cycle, it may be instructive to observe their implementation. The default registers must juggle the competing requirements of size, speed, and testability. The time step orientation of the language necessitates the use of master-slave flip-flops. The standard master-slave NMOS flip-flop using a two-signal, nonoverlapping clock is shown in Figure 6.20.

This master-slave flip-flop relies on the dynamic (charge storage) aspects of an MOS transistor. When $\varnothing_a$ is low, the transistor is off, and the last voltage value at the input of inverter M is retained no matter what happens subsequently to the value of the input node. Eventually the charge holding up the voltage will leak away. The time for appreciable leakage to occur is orders of magnitude longer than any normal clock frequency. However, if the clock is stopped (for test purposes, for example), then all data in the chip will be corrupted.

To avoid this problem of the stopped clock, MetaSyn uses a three-signal clock mechanism (see Figure 6.21). The third clock signal is used to control a static feedback loop that will maintain the state of the flip-flop indefinitely. The clocks may thus be stopped (in the appropriate phase, of course) for testing. It can be seen that if $\varnothing_c$ were always off, this circuit

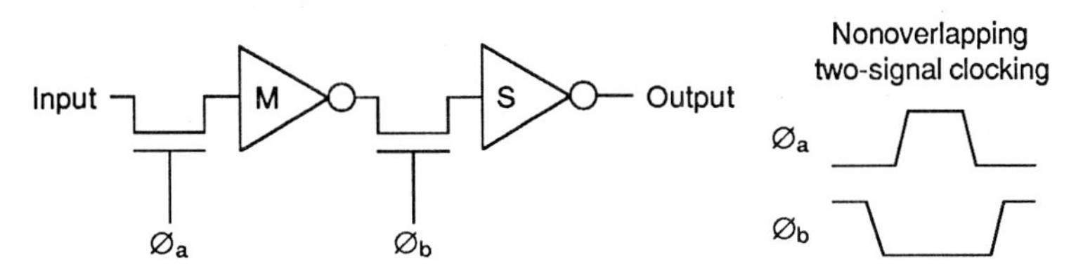

Figure 6.20 Simple NMOS master-slave flip-flop.

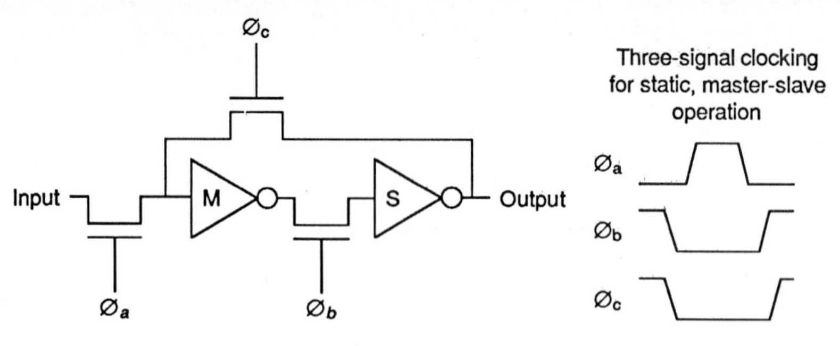

Figure 6.21 NMOS master-slave flip-flop with a "pseudostatic" phase.

would become equivalent to the simple master-slave flip-flop of Figure 6.20. As an interesting aside, for certain pipeline-type circuits, setting $\varnothing_a = \varnothing_b = 1$ and $\varnothing_c = 0$ converts the pipeline to a "cascade" circuit (see Section 6.5), which may be more easily testable.

There are several alternatives to the circuits presented here. For example, a two-signal, static, master-slave flip-flop could be constructed where the inverters are replaced by cross-coupled NOR or NAND gates. Another modification would be to generate the multiple clock signals with local logic based on a single broadcast clock signal (however, the note in the preceding paragraph would no longer hold true). Each of these alternatives trades off register cell logic for clock signal interconnection area. If, for some technology, it becomes apparent that one of these, or some other alternative register cell, is beneficial, such a cell may be substituted without change to the user level specifications.

6.6.4 Silicon Compilation

One could regard a "silicon compiler" as an efficient way of performing "placement and routing" (P&R) of component cells (such as one bit of a half adder, one bit of a register, or a logic gate). Since other chapters here cover silicon compilation techniques per se, this chapter will only give a brief perspective from the standpoint of algorithmic system compilation.

Silicon compilers attack the overall P&R problem in three parts:

1. Partition the system into blocks or "modules" for which locally efficient P&R techniques are known.
2. Efficiently lay out each module.
3. Place and route the major modules using compute-inefficient but layout-efficient methods.

One might compare this with the standard-cell or gate-array methodology, which performs homogeneous P&R on the small cells. One of the recurring themes of system design, of course, is that "local optima, optimally combined, do not yield a global optimum." Algorithmic system compiler implementers should always be aware that "silicon compiler" layout methods may not always be superior. Standard-cell and gate-array techniques (collectively called "cell net list"; see the knowledge tree in Figure 6.3) are theoretically, and may in time become practically, more efficient.

Data Path Module

As previously discussed, David Johannsen "invented" silicon compilation with his "Bristle Blocks" data path generator. It had been observed that cell net list automatic P&R routines (typically used in standard-cell or gate-array designs) were bungling the very "regular" or array-like structures.

As a simple example, two registers may feed an adder, which in turn feeds another register. There is a natural two-dimensional layout of these units that minimizes interconnection space (see Figure 6.21). The key observation here is that bit 0 of the adder needs to communicate with bit 0 of the registers and with bit 1 of the adder. However, bit 0 of the adder need never (almost never!) communicate with any other bit of the other units. In the properly placed array, communication area is (nearly) zero; the units communicate by "abutment." Now a minor perturbation of the array placement will cause a blow-up in the interconnections. If just two elements are swapped (see Figure 6.22), there will be a need for 4 or 5 extra interconnection channels. More misplaced elements cause the channel width to go up almost linearly. Observers have noticed factors of up to 3x increase in size when using cell net list P&R versus a smart manual layout of the same cells. Incidentally, note that each row of the data path could be considered a bit-slice machine, which could then be μ-coded. This is an important usage style for the algorithmic system compiler, as noted in Section 6.5.2, and one of the motivations of the original MacPitts work.

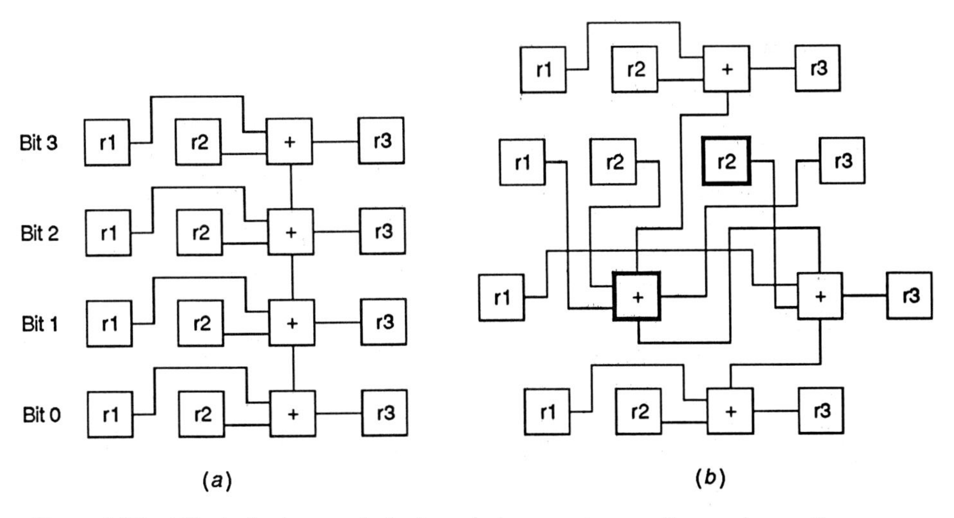

Figure 6.22 Effect of minor perturbation of placement on routing and overall area.

Johannsen realized that this two-dimensional regularity is not merely an accident requiring intelligence to take advantage of it; rather, it is a consequence of the fundamental nature of integer- or word-oriented operators. A P&R system that takes advantage of the implied regularity of word operators is thus indicated. "Implied" regularity means that the input specification must be composed of *only* word operators for this technique to work. By designing in terms of word operators (including storage), a user can simultaneously obtain a higher level of abstraction, and achieve superior performance over standard-cell or gate-array design.

Returning to the example of Figure 6.22, we note that the "+" unit must communicate with r1, r2, and r3. It is impossible to place the units so that all communication can take place by abutment. Thus, there must be some communications channels, even in the best case. If we observe various IC layouts of microprocessors, we see that there are often one or two buses within a data path (shared on a time cycle basis depending on control) and, possibly, a series of local "buses" which share the same physical channel (see Figure 6.23). Taking their cue from this observation, most data path layout programs are limited to a small, fixed number of global and partitioned channels. This is typical of IC-based thinking, which emphasizes density and electrical parasitic considerations, both of which are improved by limiting the number of buses.

On the other hand, to take full advantage of the parallelism inherent in an algorithm, it is not sufficient to have enough of the right kinds of operators. The operators must be connectable with enough parallelism also. What experienced programmer has not cursed a computer architecture with obscure restrictions as to which instructions can be used with which registers? Such restrictions are typically due to restricted bus interconnection and invariably require extra time steps for data to be passed from one kind of register to another. Although recent general-purpose computers have nearly eliminated such restrictions, they are still endemic in bit-slice and array processor machines. As a general rule, any restriction on the number of buses that results in improved IC performance criteria will also cause overall system degradation.

Thus, a MacPitts-style data path layout program will allow any number of interconnection channels between bit slices. It is this capacity that allows

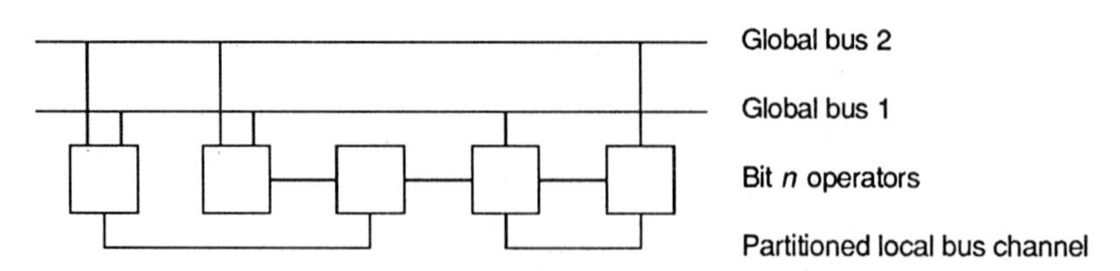

Figure 6.23 Data-path with 2 global buses and one partitioned channel.

the algorithmic system compiler user to twist the bit-slice μ-coding process as discussed in Section 6.5.2.

Incidentally, by analogy with Johannsen's observation (that the array layout is optimum for the large class of word-oriented operators), the initial impetus for algorithmic system compilation stemmed from the observation that word-oriented operators are typically an implementation of "program-like" activities.

Control

A data path module is not in itself sufficient to create any digital system. As well as integer and word operators, a digital system also requires control. However, implementation of control is available via the program logic array (note the "array" motif). The PLA is very analogous to the data path. It is a regular structure, which implies that it is easy to write an automatic layout program for it and it is performance-efficient. It is a "secret" rediscovered by thousands every few years that data flow (as in the data path) and control (as in the PLA) are sufficient for a large class of digital design.

Placement and Routing of Major Modules

Once the special array layouts have been taken care of in the creation of the major modules, a general-purpose P&R routine must be called upon to finish the chip layout. One might ask whether anything has been gained; although the array layouts have been performed efficiently, the partitioning into, and subsequent placement of and routing between, major modules must create its own inefficiencies. Moreover, some control functions would be much better performed by routing signals (and perhaps adding a logic gate or two) within the data path. However, the net result is almost always more effective than the current crop of cell net list P&R programs.

It is instructive to note that the cell net list P&R task is to place and route a large number of small components. Conversely, the silicon compiler's task is to place and route a small number of large components. Since there are only a few relative placements of the small number of blocks, exponentially exploding algorithms can be used. These are algorithms that execute quickly and thoroughly for five to ten blocks but would require longer than the "mean time between failure" of even the most reliable computer for 100 components.

6.6.5 Performance Prediction

The two fundamental axes of "value" are "How much does it cost?" and "What is it worth?" The worth of an integrated circuit depends on whether it meets its performance criteria. The three most generally useful measures of performance are:

- size
- power consumption
- speed

"Size" and "power consumption" are self-explanatory. However, "speed" has two components (for a synchronous digital system): the maximum clocking frequency and the number of clock cycles needed to perform the system function.

To illustrate the interplay of the two speed components, consider two CPUs with the same instruction set. Implementation 1 has a maximal clocking frequency of 10 MHz, and implementation 2 has a maximal clocking frequency of 8 MHz. Will implementation 1 execute a given program faster than implementation 2? Not necessarily! Implementation 2 may have a greater degree of overlap of instruction fetch with instruction execute than implementation 1. Thus, implementation 2 will execute a code sequence in fewer clock cycles than implementation 1. Such improvements can easily account for a factor of 4 in overall "speed" and have been known to account for a factor of 10 in complex designs, entirely overwhelming minor variations in clocking frequency.

In Figure 6.17 "performance prediction" is shown as an operation on the CIF (or other layout) data base. This is not completely accurate. Any engineer, in whatever medium, will have some heuristic model of the performance effects of various input specifications. This model is used during design to reject certain approaches and gauge the tradeoffs between others. For example, in Section 6.5.5 certain performance tradeoff implications of parallelism were examined without generating the final layout.

Prelayout Performance Prediction

An obvious advantage of the algorithmic approach to design is that the number of clock cycles needed to perform a system function is immediately observable and controllable. This is in contrast to a structural approach, in which such information is almost completely hidden at IC design time. On the other hand, at first glance it would appear that the clocking frequency aspect of speed performance is more reasonably observable and controllable by a structural approach. However, consider Figure 6.23.

To determine the maximum clocking frequency (for register y and flag z), one must first determine the critical path of the circuit. Unfortunately, the critical path is not determined only by the structure, but also by the control.

Here are two possible algorithms that might be equally well implemented by the structure of Figure 6.24.

```
Algorithm A:
  step 1:   (setq y (- (+a b) c)
  step 2:   (setq z (= (- e c) d))
Algorithm B:
  step 1:   (setq y (- e c))
  step 2:   (setq z (= (- (+a b) c) d))
```

Although the critical path is not observable from the structure, it is from the algorithm! In algorithm A, the critical path is due to step 1 and includes

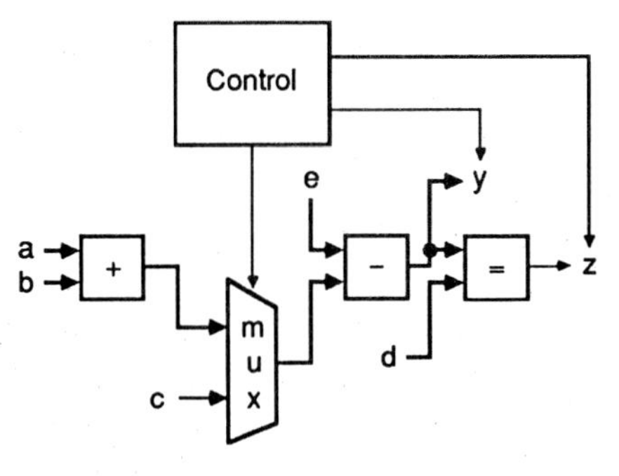

Figure 6.24 Performance prediction—structure is not enough! The "critical path" (and, hence, the maximum clock frequency) depends on control.

the addition time from inputs a and b, plus the multiplexer delay, plus subtraction time and the setup and hold time for register y. In algorithm B, the critical path is clearly due to step 2 and includes addition time, multiplexer delay, subtraction time, equality check time, and setup and hold time for flag z. The second critical path for algorithm B is also clearly longer than that for algorithm A.

The preceding analysis, of course, is not rigorous, but it does provide more constructive guidance during the design phase than does structure alone. One serious caveat is that the designer cannot easily and directly assess the effect of "fan-out" from the algorithm, whereas such effect is obvious from the structure. In short, both algorithm and structure have some advantages and some disadvantages in predicting and controlling speed performance during design.

Besides speed, the other two important performance criteria are "size" and "power." It would appear that a structural design approach would yield a greater degree of observability and controllability over these factors than the algorithmic approach. This is especially true in the first design "pass." Fortunately, the time to perform the first design pass is so much shorter for the algorithmic approach that the difference becomes meaningless.

During the initial design phase, the Algorithmic System Compiler user will generally be able to tell what resources (adders, incrementers, and so on) are being used in each time step. If the compiler is doing its job, the resulting structure will be the *union* of the resources required in each time step plus some multiplexers. After compilation (while the structural designer is still figuratively sharpening his pencils), the Algorithmic System Compiler designer knows the exact size and power consumption of the circuit. Some of the heuristics developed to trade off these parameters with speed have been indicated in Sections 6.5.5 and 6.5.7.

Postlayout Performance Prediction

Once layout is completed, the size of the chip is obvious. The power consumption can be calculated in several ways, and algorithms for that purpose are something of a specialty area. A simple and extremely conservative method (in NMOS) is to assume that all transistors are on and compute the power consumption based on the "on" resistance of the various-sized transistors as they are layed out on the chip. In any case, a power consumption estimate is mandatory during the silicon compilation phase since power buses should be automatically sized to prevent metal migration and debilitating IR voltage drops to circuitry on the chip. Thus, the power consumption statistic reported to the designer is exactly that required by the compiler when it is laying down metal to the chip's ground pad.

There are several ways to estimate speed performance from the layout. The two major categories are:

- "dynamic," or test-vector-dependent
- "static," or test-vector-independent

In this case, "dynamic" is not particularly better than "static," and theoretically it is inferior. However, in a practical sense, the dynamic predictor will share code with a "timing level" simulator, so it is often used.

In either case, there must be some model of time delays caused by active logic elements (transistors, gates, or cells, depending on preference), the fan-out from each element, and, for greater accuracy, interconnection parasitics. For example, MetaSyn extracts a transistor-level net list from the layout, which includes drive capacity and an *RC* (resistance-capacitance) model of parasitics and fan-out.

With a test-vector-dependent predictor, each test vector is run through a timing simulator. Such a test inputs the first test vector and then computes when the circuit will stabilize. A record is made of this time, which is the current estimate of minimal clock interval, and the next test vector is input. The time it takes this test vector to stabilize is compared to the current estimate, and if it is larger, the new time becomes the current estimate of minimal clock interval. This continues until the test vectors are exhausted. The drawback to this method is that it does not guarantee that the test vectors that were used actually exercised the critical path.

With a test-vector-independent predictor, an attempt is made to find the critical path from circuit considerations alone. There are some difficulties with respect to bidirectional switches (such as NMOS and CMOS "pass gates"). These are overcome through a combination of the compiler passing to the predictor information that determines the direction, wherever possible, and the compiler using cells that are not ambiguous in this way. Such a predictor is incorporated in MetaSyn.

6.7 Envoi

Once upon a time, there were three digital systems engineers sharing an office. Having time on their hands, they decided to build a digital system using "Mead-Conway" VLSI design techniques. The digital system was partitioned into a data path section, a control section, and a UART. The one responsible for the data path, having read David Johannsen's paper on data path generation, said he would write a data path generator. The one designing the control unit said he would write a FSM to PLA-with-register compiler. The engineer in charge of the UART made a block diagram showing a data path and a control section. As they looked over each other's shoulders, there was a simultaneous "Aha!" followed several breakthroughs later by a language and an early logic synthesizer. The commercial descendant is a tool for those who "peak" at applications and digital system engineering. Let us hope it will be embraced as "the way we always did digital design anyway."

References

M. R. Barbacci, *et. al.*, *The ISPS Computer Description Language*, Department of Computer Science, Carnegie-Mellon University, Pittsburgh, PA, August, 1979.

F. J. Hill and G. R. Peterson, *Introduction to Switching Theory and Logical Design*, Wiley and Sons, New York, NY, 1968.

T. J. Kowalski, *et. al.*, "VLSI Design Automation Assistant: From Algorithms to Silicon," *IEEE Design and Test of Computers*, August, 1985.

CHAPTER 7

The Yorktown Silicon Compiler System

R. K. Brayton
R. Camposano
G. De Micheli
R. H. J. M. Otten
J. van Eijndhoven

7.1 Introduction

7.1.1 History of the Yorktown Silicon Compiler

The Yorktown silicon compiler is a result of an ongoing project at the IBM T. J. Watson Research Center at Yorktown Heights, N.Y. The goal of this project is to study all aspects related to automating the synthesis of digital systems to be integrated on a semiconductor carrier. The compiler grew out of two initially independent clusters of activities in that project in the early 1980s.

One of these involved logic synthesis and minimization and began with research in automating the synthesis of logic to be implemented with cascode current switch circuits. In the summer of 1981, the scope of the logic synthesis activity was widened considerably:

- The table method of Ashenhurst and Curtis for decomposing switching functions was modified to incorporate knowledge about the arrival times of input signals.
- The observation that unate Boolean functions could be efficiently manipulated led to the idea of recursively expanding functions with respect to variables chosen such that the cofactors are as close as possible to unate functions. The efficiency of simplification, complementation, tautology, and other important operations in logic minimization was considerably improved by applying this idea to most functions encountered in practice.

- Algorithms used in the best heuristic minimization programs available at that time, MINI and PRESTO, were programmed. Better algorithms for several tasks were discovered and implemented. Many of the improvements in efficiency were obtained by applying the result of the previous task, the unate recursive paradigm. The activity resulted in a new package for two-level logic minimization called ESPRESSO.
- Activity in the decomposition of logic functions, aimed at different kinds of cascode circuits, resulted in the discovery of the kernel theorem. This allows one to use a subset of all common divisors for locating all nontrivial common subexpressions. Thus, the extraction of common subexpressions for use as an intermediate became a much more manageable task.

At the end of that summer, it was not difficult to see that a powerful package for multilevel logic synthesis, quite different from conventional local transformation methods, was within reach. This approach would combine many of the ideas discovered during the summer and would be basically independent of the final target technology chosen. By the end of 1982, a very powerful package of algorithms for logic synthesis and minimization existed and, for reasons of attracting other users, was called a *logic editor*.

Simultaneously, but on a smaller scale, a high-level layout design procedure was being developed. The net effect was a refinement and ordering of a given hierarchy. Special routines were invented that, based on eigensolutions of connection matrices, would try to discover a hidden structure and translate this into a slicing structure. The slicing constraint was imposed mainly because of its consistency with hierarchy concepts and its complexity-reducing effects on many optimization and decision procedures that were to follow. This constraint hardly could be disadvantageous if individual cells ("leafs" in the hierarchy) were being assembled according to shape and pin position constraints derived from a top-down floor plan design procedure. This required a characterization of the set of feasible shapes and an efficient algorithm that could determine the optimum geometry of a floor plan using only feasible module shapes. These algorithms were implemented and combined by the end of 1982. In addition, a version of the recently invented "annealing method" was implemented for handling a variety of objective functions in designing floor plans or for various placement problems.

Following development of the logic editor and a program for creating preliminary cell environments, the natural next step was the implementation of an assembler for logic macros. In selecting a technology for implementing the logic, we concluded that the domino circuit was the best option. In addition to the well-known advantages of complementary MOS circuits, the domino circuit added speed, potential to implement a complex single function in a simple discharging network, and the ability to adjust the driving power behind its output signal. An investigation into its properties with respect to testability turned out favorable as well. The problems with this family, reported in the literature, were overcome by IBM's advanced production techniques and special features that could be incorporated in the basic circuit, if necessary. Experience in handling such a large logic family did not exist, and the technique of local transformations does not lend itself to synthesizing logic in so large a family. Thus, only through a package such as the logic editor could the potential of domino logic be reasonably investigated.

Early exercises in implementing an arithmetic-logic unit (ALU) exclusively with domino circuits had demonstrated the advantages of using linear transistor arrays to realize the discharge network. Thus, algorithms for obtaining area-efficient linear transistor arrays formed the core procedures for generating domino gates. To put these gates together in a module that implemented a multilevel logic specification, we developed the pluricell layout style. At a first glance, pluricell modules seem to be polycell or standard-cell-type layouts. They use the columnar arrangement of the individual gates, with many of the interconnections made in the channels between the columns. Both realize the supply and clock lines by abutment of gates. A pluricell design procedure, however, does not copy the layout of its gates from a cell library. The gates are constructed algorithmically when called for in the result of the logic synthesis procedure. Also, the interconnections between gates bordering different channels are realized in second-level metal. The placement procedure has the task of optimally using the resources (wiring layers) in obtaining a time- and area-efficient layout. The availability of the second-level metal gives the macro better accessibility for global signals. The pluricell layout style requires a channel router that makes signals available in the first-level metal, where they must contact a second-level metal track.

The logic editor with a technology-dependent tail for domino logic, combined with the pluricell layout program, constituted what we called a *macroassembler*. Immediately after its completion, in September 1983, its results were compared with PLA realizations and gate array estimations for the same logic. All the comparisons turned out in favor of the macroassembler [Bray84c].

The next step was the development of a logic language. The first version of this language was available in the spring of 1984. In the meantime, due to experience from demonstrating the macroassembler at several sites, we decided to use the tools developed so far as a basis for a complete environment for designing digital systems. Operated in a fixed sequence, starting with a high-level description of the system and ending with production of data that uniquely determine the instructions for fabricating an integrated circuit, such a set of procedures is called a *silicon compiler*.

Once this decision was made, we needed to add the additional procedures necessary and to select an example as a test case. The 801 architecture was chosen because it was complex enough to exercise all the tools and, because it was the first RISC-type architecture, was of general and growing interest. The logic language was extended with a capability for interpreting tables, which made the specification of control modules easier, and extracting "don't cares" led to better implementations. To communicate the hierarchy of the system, a simple hierarchical specification language was devised, giving us, in effect, a register transfer level of input. Every combinational module was specified by using the logic language. All this was considered a preliminary solution for exercising the lower-level tools. An input format more acceptable for other users was planned as a future activity.

Simultaneously, the global wiring of the floor plan design tool was rewritten in a compiler language to accelerate the process, and at the same time it was provided with more powerful heuristsics. Also, additional work was being done to use the annealing algorithm in a fully automatic way. With improved understanding of the method, more improvements were added over the next two years.

By the end of 1985, we had specified additional interfaces and data structures in order for the various modules of the compiler to communicate with each other. Some verification tools were added for gate-level simulation at the system level, and some preliminary timing analysis was made available. Finally, during 1986, a fully automatic timing optimization strategy was implemented using logic resynthesis, resizing of output devices, and replacement of critical modules. In addition, a behavioral-type input specification was adopted and implemented as the main input method for the compiler. This used the previous register transfer level input as an intermediate format to communicate with the lower-level components of the compiler.

7.1.2 The Yorktown Silicon Compiler

We use the word *compiler* in a way similar to its use in a software context. A compiler processes a program in a language that is expressive in describing hardware. It is reasonably high level in the sense that the programmer need not know assembly code or the details of the final object code—in this case, the gates or the target technology, and the layout design rules. The program in the high-level language is automatically compiled into a valid chip image.

The Yorktown silicon compiler (YSC) project has three major goals. The first goal is to achieve automatic silicon compilation from a behavioral high-level description into a chip image. The compile time should be short and negligible in comparison to the time spent by the designer in describing the chip as a program. This allows the user to experiment with different chip architectures and evaluate in real time the circuit tradeoffs. The second goal is to be able to design chips that are competitive with custom manual design in performance and silicon area. This is made possible by using several optimization techniques that are coupled to the entire top-down synthesis process. As a final goal, the Yorktown silicon compiler provides a design environment whose backbone is the automatic synthesis operation. This environment includes verification techniques and alternative synthesis and optimization paths. Therefore, the Yorktown silicon compiler can be used as a workbench by the CAD researcher to develop new synthesis or optimization techniques and seek their validation in the final image of the chip.

As much as possible, the Yorktown silicon compiler has been kept independent from the target technology. Only the circuit layout phase depends on the implementation technology. This phase is the equivalent of the code generation step in the software context. At present, the target technology is a CMOS implementation. Different compiler back-ends have been implemented to explore different flavors of CMOS styles, such as dynamic and static mode of operation.

The Yorktown silicon compiler views a VLSI chip as a hierarchy of data sets, which we call *modules*. Each module represents a portion of the chip. The level of abstraction of the representation, as well as the hierarachy, are refined during the compilation process. Eventually, each module will correspond to a rectangular area of the chip and will represent the geometries of the masks needed to fabricate that part of the chip.

The hierarchy of modules is constructed by the compiler from the high-level system description, which is given by the circuit designer in the form of a program. The hierarchy may be suggested by the system designer by structuring the program appropriately. In the beginning, each module is a subroutine describing a finite automaton. Therefore, the entire circuit to be implemented can be seen as a hierarchy of finite automata. The first creative step of silicon compilation, called *structural synthesis*, is the processing and optimization of this hierarchy with the goal of devising an interconnection structure of data operators without memory, data storage elements, a set of input/output (I/O) ports, and, possibly, special function units. At this point, this structure determines another hierarchy: the leaf modules, called *cells*, are elementary building blocks of the chip and can be classified into the different types specified previously. Data operators without memory are implemented by combinational logic circuits, and they are referred to as *combinational macros*. Storage elements include various kinds of registers and arrays of read/write memory units (referred to as *register files*). The input/output ports are implemented by off-chip drivers and receivers. Finally, special function units may be described as calls to special procedures, provided by the user. The structural synthesis step optimizes the circuit timing performance by means of architectural-level decisions, such as the selection and the partitioning of data operators and the high-level determination of the control structure.

The Yorktown silicon compiler processes the representation of the combinational modules to obtain an optimized multilevel logic representation. This step is called *logic synthesis*, and it aims at determining a circuit implementation of the combinational module that minimizes the active area and the worst-case switching time. The logic synthesis step provides an estimate of the area needed to implement the combinational module. At this point, it is possible to estimate the timing performance of the chip and determine which interconnection nets are critical for the global circuit performance.

The chip floor plan is designed after the logic synthesis step using the information about the area estimates of the modules, the interconnections among modules, and the critical nature of these interconnections. Modules are characterized by constraints on their shape during the floor plan design phase. The functional hierarchy is then refined into a *slicing structure*—each module is associated with a rectangular area (see Figure 7.1). The estimated physical position of each module is then used to estimate the length of the interconnections, which will help to obtain a better estimate of the chip timing performance. If the design does not meet the required

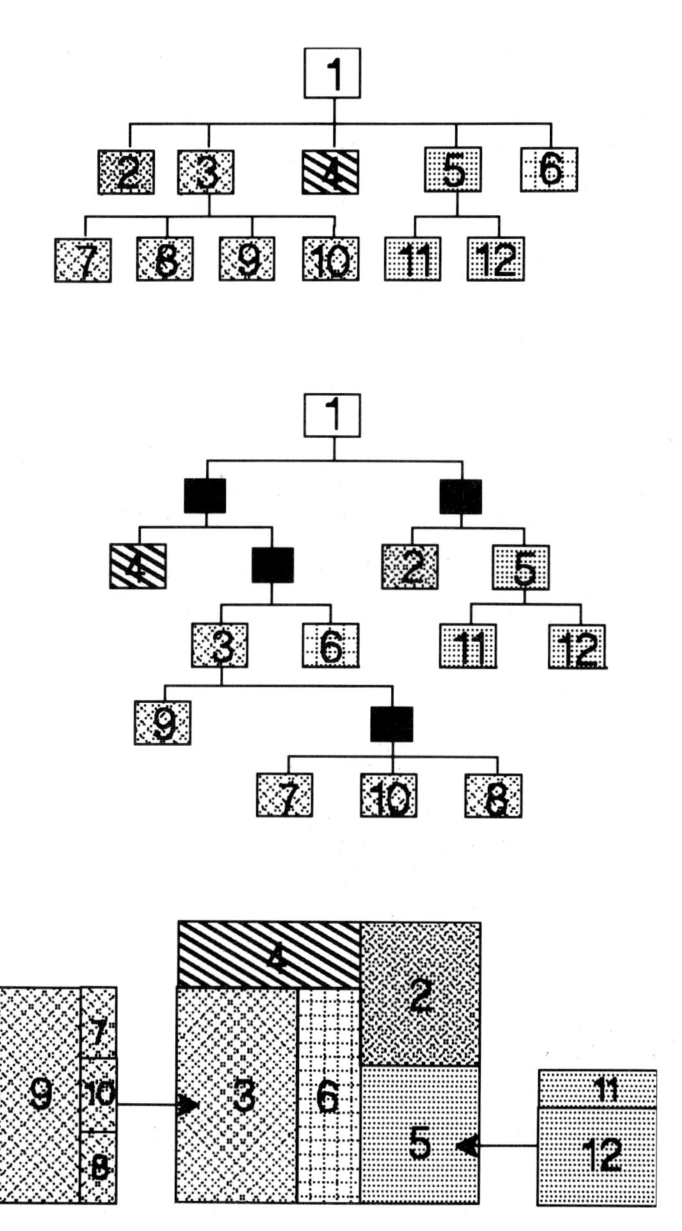

Figure 7.1 Slicing tree and corresponding floor plan.

specification, the circuit is "tuned" by adjusting appropriate parameters corresponding to the final device sizes, by iterating the logic synthesis step, or by repositioning some of the modules.

The circuit layout comprises two steps: (1) the generation of the images of the geometries of the cells and (2) the global chip assembly. The first step is accomplished by the use of *macroassemblers*, which have the ability of building each logic block or storage block in a rectangular area using an estimated length-to-width ratio and indications for the pin positions. Thus it can adjust the implementation of each module to the global floor plan requirements. The combinational logic modules are in pluricell layout style—that is, each multiple-input multiple-output combinational logic function is implemented as an interconnection of generalized gates, with each gate implementing a multiple-input single-output function. In turn, each gate is implemented by a linear transistor array. Gates are stacked in rows separated by wiring channels. The structure is determined by the shape constraints, and the microcell position is optimized by a placement procedure. The geometry of each gate is obtained and optimized by a procedure according to its logic functionality. The gates are interconnected by wiring the channels that separate them (see Plate 1, page 000). Other macroassemblers are used for generating the image of the registers and the register file as rectangular blocks, also according to their shape constraints. Eventually, in the global chip assembly step, the macrocells corresponding to each rectangular area are placed on the chip image and interconnected using the channels that separate them.

These steps of the top-down synthesis are represented in Figure 7.2. Circuit verification tools coexist in this environment and support functional-level and gate-level circuit simulation, propagation delay analysis, and connectivity verification.

The strength of the Yorktown silicon compiler, in comparison to other approaches to automated synthesis, relies on the following facts:

1. Synthesis starts from a behavioral description; therefore, it qualifies as a silicon compiler (as opposed to a module generator).
2. Logic synthesis uses powerful algorithms with global optimization routines (as opposed to local rule-based transformations).
3. The floor planning creates preliminary environments for the modules in a top-down sequence without fixing the geometry of the cells to be designed by special procedures.
4. The combinational modules are synthesized as multilevel logic macros laid out in pluricell style. This gives superior results in terms of area and switching time, as compared to two-level logic PLA and gate array implementations.
5. The timing performance of the circuit can be optimized by tuning the circuit in different ways: finding optimal device size, finding an appropriate logic structure of the combinational modules, and by influencing the positions of the macrocells by timing considerations.

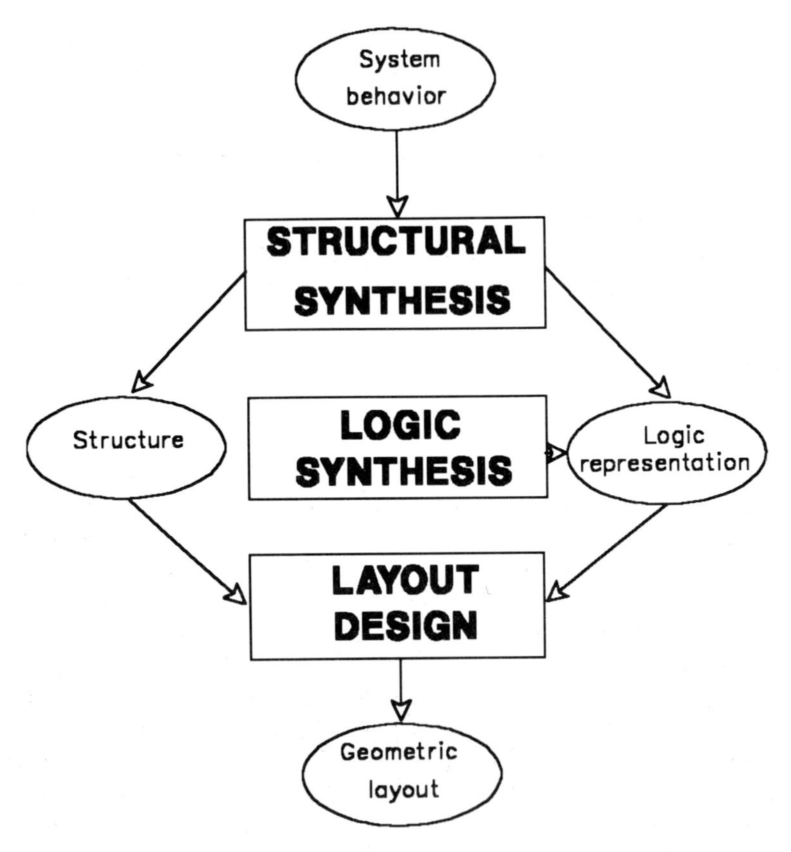

Figure 7.2 Synthesis in the YSC system.

The techniques that are implemented in the major components of the Yorktown silicon compiler will be discussed in Section 7.2. Section 7.3 will explain how the compiler achieves the automatic synthesis of a chip from high-level specifications. And, finally, Section 7.4 will present the results of our experience in using the compiler and the experimental evidence of the strength of the YSC approach in producing competitive chips.

7.2 The Functional Units of the Yorktown Silicon Compiler

7.2.1 Hardware Description

The starting point for the synthesis of a digital system is a system description. This representation captures the system behavior and provides the necessary information to design the hardware as well as the software system to be used in conjunction with the hardware. The description must be such that the hardware and software development can be performed independently. In principle, the hardware description should not imply any implementation choice and, therefore, should be purely behavorial. In practice, the semantics of most high-level representations used for hardware synthesis tend to suggest some architectural implications: there is a wide spectrum of high-level representations ranging from purely behavorial to merely structural ones [Gold85]. In the case of digital computers, the hardware specification is traditionally given in the *principles of operations*, which is a description in a natural language.

Since the high-level behavior of digital hardware systems can be described by algorithms, there are several advantages in describing them as programs in a programming language. First, the hardware description can be processed by silicon compilers in a manner similar to software programs processed by software compilers. Experience in software compiler technology has been key to the development of the front-end silicon compilers. Second, program descriptions can be simulated before proceeding to synthesis. Third, programs can be used as synthetic documents describing hardware systems and as replacements for the principles of operations texts, just as program descriptions of classical algorithms have replaced textual descriptions.

The choice of the programming language involves a tradeoff between generality and power of expressing hardware. In principle, programming languages such as C, Pascal, or APL could be used. The main advantage is the popularity of these languages: compilers and interpreters already exist for these languages. The main disadvantage in using software programming languages is the lack of constructs that are typical of hardware programming languages (e.g., instructions for handling I/O pins). A solution to this problem is to extend the capabilities of software programming languages. For example, Brenner has developed the Yorktown Logic Language (YLL) as an extension of the APL notation. The YLL will be fully described later in this section.

In the Yorktown silicon compiler, the systems to be designed are specified in the Yorktown Intermediate Format (YIF). The definition of the library elements to be used in synthesis is done in the YLL. Whereas YIF can be used to describe behavior involving both combinational and sequential aspects of a system, YLL is restricted to purely combinational, memoryless circuits.

The introduction of a format (YIF) rather than a language for the initial system specification separates the real silicon compilation task from the front-end processing of a programming language. The former task has all the essential challenges of hardware automatic synthesis; the latter task is analogous to the well-known problems of parsing, syntax checking, and semantic analysis. Like several programming languages, such as C, PLI, or Pascal, YIF has imperative semantics. Statements are evaluated sequentially according to the specified control flow. Consequently, programming languages with imperative semantics can be mapped into a YIF representation in a straightforward manner. For practical reasons, the YSC system supports the compilation of only one language into YIF. We have chosen the V language because of compatibility with other research activities within IBM and the possibility of having access to several examples [Bers85].

The YLL is used to describe the data operators in the module library as well as the control constructs that result after structural synthesis. The need for a module library derives from the frequent use in VLSI design of common modules, such as arithmetic-logic units (ALUs), shifters, and so forth. These modules are often very critical to the final hardware performance, and their behavioral description must have a predictable structure that can be highly optimized. However, instead of using just a small number of predefined library modules, an open book philosophy is followed. The library consists of parameterized YLL lines describing data operators (e.g., an addition). During synthesis, several of these lines are composed to form a YLL program that will yield, after logic optimization and layout, a hardware module that meets exactly the specific needs of the system being designed. Due to its powerful constructs for the description of combinational logic, YLL is very well suited for this purpose.

The Yorktown Intermediate Format (YIF)

YIF is a format to specify the behavior of a digital system at the algorithmic level. The main goal in designing YIF was to allow an easy compilation of programming languages with an imperative semantics into YIF while capturing all the information necessary to start a design. YIF is somehow inspired by the internal form defined in [Camp85]. Related work can be found in [Knap85], [McFa78], [Orai86]. In the YSC system, YIF is the input to structural synthesis.

YIF contains:

- *The definition of the variables necessary to describe the system behavior.* These include variables in the sense of a programming language. Also their observability in the final hardware can be described: architected registers might be specified as registers already at the beginning; variables that represent the interface to the outside world (pins) must be specified.
- *Operations that use the variables as inputs and outputs.* Operation sequences are specified giving the control flow. The basic control flow elements are sequence, conditional branching, and iteration. The operators can be chosen according to the application and must be described in YLL in the operator library.
- *A mechanism for hierarchical and modular design.* An operation can be specified by another YIF description. This corresponds to a procedure call in the behavioral program.

The primary representation of YIF is a textual format with a very simple syntax. This form is used to interface different programs, such as the V compiler with structural synthesis. Structural synthesis keeps an internal representation of YIF in matrices and lists.

A YIF description $Y = (V, S, R_s, R_i, R_o)$ consists of:

- *A set of operations*—V. Operations represent atomic computations (e.g., and, addition, shift-left, etc). One element, $v_f \epsilon V$, called the first operation, must be given.
- *A set of variables*—S. They are equivalent to variables in a program. A

variable s in YIF has a type(s) $\in$ { array, external, internal, register} . The type indicates whether a variable is an array, whether it represents a port to the outside (external), whether it is an internal variable, or whether it is an architected register. During synthesis, the internal variables might be converted automatically into registers. Thus, the initially specified architected registers are a subset of the registers of the final design. (An initial specification might contain no architected registers at all.)

- *A precedence relation*—$R_s \subseteq V \times V$. It defines the control flow for the operations:

  $$(v_1, v_2) \in R_s \text{ iff } v_1 \text{ before } v_2$$

  "before" here means "immediate predecessor," as implied by imperative semantics. The relation R_s is not reflexive.
- *An input relation*—$R_i \subseteq S \times V$. This indicates the variables used as inputs (are read) by each operation.
- *An output relation*—$R_o \subseteq V \times S$. This indicates the variables used as outputs (are written) by each operation.

The pair (V, R_s) can be seen as a digraph, the so-called *precedence graph*. Node v_1 is said to be the (immediate) predecessor of node v_2 and vice versa; v_2 is called the (immediate) successor of v_1. *Conditional branches*—that is, the selection of one among many possible successor operations—are represented by a node with more than one successor, leading to a fork. *Iterations* or *loops* are represented by circuits in the graph. The operational semantics of YIF is defined as follows:

> Operations are executed one at a time, starting with the first operation v_f. The next operation is a successor of the executed operation. If there is more than one successor, exactly one of them is chosen depending on the value in one of the variables. An operation v writes a new value to all variables s_o for which $(v, s_o) \in R_o$. It uses the values in the variables s_i such that $(s_i, v) \in R_i$. A variable retains its value until it is overwritten.

For synthesis it is necessary to eliminate the cycles in R_s. This is done by removing some pairs L (edges in the precedence graph) such that

1. The transitive closure A of $(R_s - L)$ is a partial order
2. The graph remains connected so that $\forall_{v' \neq v_f \in V}[(v_f, v') \in A]$ that is, all operations are still reachable from v_f.

Since a fork in the graph represents the alternative execution of one of the successors, all operations not ordered in A will be called *alternative*.

YIF also includes a mechanism for modular and hierarchical design. One YIF description is referred to as a module. *Modules* are synthesized separately and connected after structural synthesis ("module binding"). To relate the modules to each other a "module call" operation is used. It replaces an atomic computation. The semantics is as in imperative languages: control is passed to the called module until it is ready and returns. There are some restrictions however. Global variables that can be referenced in different modules are not allowed. All the information flow between the calling and the called module must be stated explicitly by listing the corresponding variables as inputs and/or outputs in the module call operation. In the called YIF description, these variables are of type external. These "parameters" are passed by name; they will be connected during module binding, also passing the required control information.

In addition to this information, YIF includes some attributes such as the width (number of bits) of a variable; its dimension, if it is an array; an index identifying an operation uniquely; the type of the operation (addition, shift, etc.); constraints on the operations; and so forth. A special constraint, the *map constraint*, specifies that one or more operations are to be mapped onto a specific hardware module. Thus, it is possible to specify the implementation of different operations using the same hardware. During synthesis, map constraints will be generated automatically.

Although the semantics of an imperative programming language and YIF are basically the same, some extensions (e.g., handling constraints) and restrictions (e.g., limiting the allowed data types) are necessary. During compilation, variables in the language are mapped on variables in YIF, expressions are decomposed into single operations, and the control constructs in the language are represented by the edges in the graph (V, R_s). The edges L are obtained directly from the iteration constructs such as DO UNTIL or DO WHILE. Procedure calls are replaced by module calls. The compiler must monitor the passing of all parameters as inputs and/or outputs in the module call; it may also need to monitor the scope of global variables according to the semantics of the language.

To illustrate these concepts, consider the self-explanatory V program in Figure 7.3, which describes a possible instruction fetching stage of the 801 processor.

The V compiler converts this program to the YIF code given partially in Figure 7.4 and plotted in Figure 7.5. The YIF code starts with a SYMBOLTABLE containing the used variables followed by their type and dimensions, the first pair of numbers being the width and the second pair being the array dimension limits. Variables T1 and T2 were generated by the compiler. The NODES represent the operations:

- INDEX is a unique number identifying them.
- TAG represents the kind of operation (e.g., CBR means conditional branch, SO means single operation, etc.).
- OPERATION gives the function of the operation (e.g., TR means transfer, CON means concatenation, etc.).
- LINE and COLUMN give the position in the original program, in this case token numbers indicated in the listing.
- INPUTS is the list of input variables for this operation and their used width and array dimension.
- OUTPUTS is the list of output variables.
- PREDECESSORS are the predecessors.
- SUCCESSORS represents the successors and their CONDITIONS (i.e.,

```
MODULE FETCH (BPC:IN, BRANCH:IN, IBUS:IN, IRE:IN,          1
              PCT:OUT, IRT:OUT, IA:OUT);                   20
EXTERNAL FETCH;                                            33
   DCL BPC BIT(32),      /* New PC                 */      36
       BRANCH BIT(1),    /* Use BPC                */      43
       IBUS   BIT(32),   /* Instruction Bus        */      49
       IRE    BIT(1);    /* Instruction ready      */      55
   DCL PCT    BIT(32),   /* PC                     */      61
       IRT    BIT(32),   /* Instruction register   */      68
       IA     BIT(32);   /* Instruction address    */      74
INTERNAL FETCH;                                            80
   DCL PC     BIT(32),   /* PC for prefetch        */      83
       IR     BIT(32),   /* Instruction register   */      90
       OLDPC  BIT(32);   /* PC for output          */      96
                                                          102
BODY   FETCH;                                             102
   PCT := OLDPC; IA := PC; IRT := IR;                     105
   WHEN (IRE || BRANCH)  /* catenation */                 117
     CASE 2;             /* 10 */                         123
     CASE 0;             /* 00 */                         126
       DO UNTIL IRE LOOP  ENDDO;                          129
     CASE 1,3;           /* X1 */                         135
       PC   := BPC;                                       140
       IA   := PC;                                        144
       DO UNTIL IRE LOOP  ENDDO;                          148
   ENDCASE;                                               154
   IR    := IBUS;                                         156
   OLDPC:= PC;                                            160
   PC    := PC+4;                                         164
END FETCH;                                                170
```

Figure 7.3 Basic instruction fetch in the 801 processor.

the value at the output of this operation to select the corresponding successor).

In the graphic representation, nodes represent an operation. Inside the nodes, the INDEX is given. To the left, OPERATION is found. INPUTS are listed above the nodes. OUTPUTS are below them and to the right. If an operation has more than one successor, the CONDITIONS that apply to each of the edges are also listed. The edges in $L = \{(v_8, v_7), (v_{10}, v_9)\}$ are not drawn. They represent the two UNTIL loops. The loop bodies are empty represented merely by the "dummy" operations v_7 and v_9, which also have no inputs and outputs (indicated by asterisks). Operation v_{11} is also a dummy operation that "closes" operation v_4, which represents the CASE statement

The main features of YIF not yet described are the specifications of constraints and of parallelism. *Constraints* include area, time, power, and mapping. They are given for sets of operations (e.g., a set of operations should be implemented in less than a given area or by some specific library cell). A similar approach can be found in [CaKu86].

Parallelism is introduced using an additional relation, $R_p \in V \times V$. But since we restricted synthesis initially to systems that can be represented by one finite automaton, it is not used yet. The parallelism achieved automat-

```
SYMBOLTABLE;
    IRE EXTERNAL 1 1 0 0;
    IBUS EXTERNAL 1 32 0 0;
    ...

    IR VARIABLE 1 32 0 0;
T1 VARIABLE 1 2 0 0;
T2 VARIABLE 1 32 0 0;
NODES;
    INDEX 15 TAG SO  OPERATION TR  LINE 164 COLUMN 168;
    INPUTS T2(1..32)[0..0];
    OUTPUTS PC(1..32)[0..0];
    PREDECESSORS  14;
    SUCCESSORS *;

    INDEX 14 TAG SO  OPERATION ADD LINE 166 COLUMN 168;
    INPUTS PC(1..32)[0..0] 4;
    OUTPUTS T2(1..32)[0..0];
    PREDECESSORS  13;
    SUCCESSORS 15 CONDITIONS *;
    ...

    INDEX 11 TAG CBE OPERATION * LINE 117 COLUMN 153;
    INPUTS *;
    OUTPUTS *;
    PREDECESSORS  8 10 4;
    SUCCESSORS 12 CONDITIONS *;

    INDEX 10 TAG UTS OPERATION TR  LINE 129 COLUMN 132;
    INPUTS IRE(1..1)[0..0];
    OUTPUTS *;
    PREDECESSORS  9;
    SUCCESSORS 9 CONDITIONS 0;
    SUCCESSORS 11 CONDITIONS 1;

    INDEX 9 TAG UB  OPERATION *   LINE 129 COLUMN 132;
    INPUTS *;
    OUTPUTS *;
    PREDECESSORS  10 4;
    SUCCESSORS 10 CONDITIONS *;
    ...

    INDEX 4 TAG CBR OPERATION TR  LINE 117 COLUMN 153;
    INPUTS T1(1..2)[0..0];
    OUTPUTS *;
    PREDECESSORS  3;
    SUCCESSORS 5 CONDITIONS  3 1;
    SUCCESSORS 9 CONDITIONS  0;
    SUCCESSORS 11 CONDITIONS 2;

    INDEX 3 TAG SO  OPERATION CON LINE 118 COLUMN 122;
    INPUTS IRE(1..1)[0..0] BRANCH(1..1)[0..0];
    OUTPUTS T1(1..2)[0..0];
    PREDECESSORS  2;
    SUCCESSORS 4 CONDITIONS *;
    ...
END;
```

Figure 7.4 YIF code for the example.

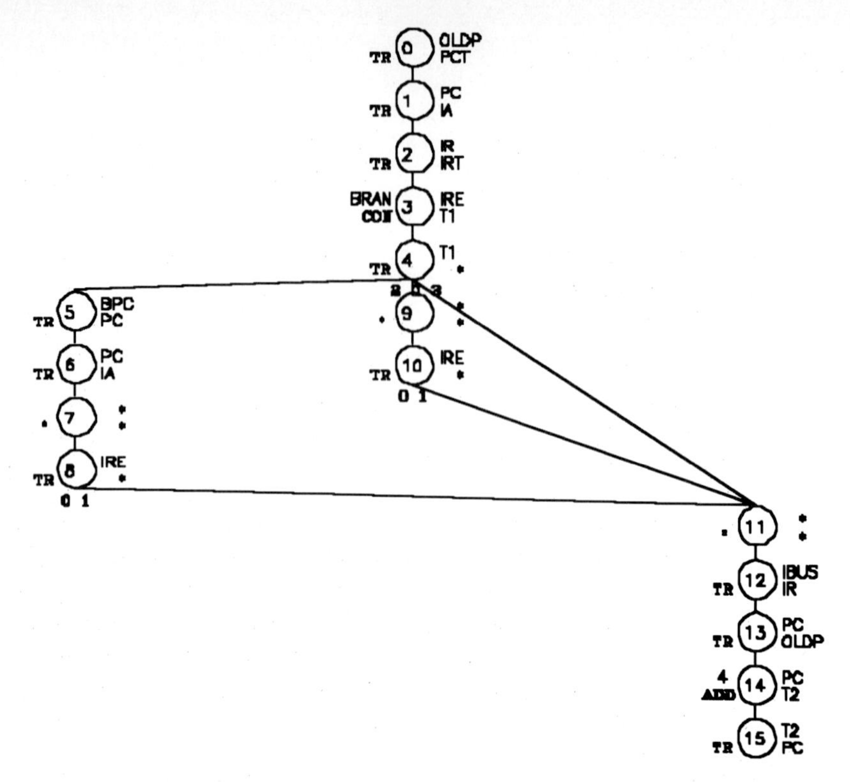

Figure 7.5 YIF represented as the graph $(V, R, -L)$.

ically until now has been obtained by scheduling different operations to be executed in the same cycle. Two or more state machines running truly concurrently must be described as two or more YIF specifications, including some synchronization mechanism.

The Yorktown Logic Language (YLL)

GUIDING PRINCIPLES YLL was developed to describe multilevel combinational logic. Its main guiding principle has been that a block of combinational logic is a sequence of computations on vectors of 0's and 1's, which produce similar such vectors. As such, any software language that can execute these computations would be a candidate for describing the logic. Of course, a language that easily and concisely expresses this computation is preferable. The software language when compiled and executed, or interpreted, processes these 0's and 1's numerically as numbers. A specification language for logic can be immediately derived if all "logic" operations are executed symbolically. In this case, the domain of the symbolic manipulation is the set of logic functions. Since the capability to create and manipulate logic functions already resides in the logic synthesis module, the language processing should be straightforward.

A second principle has been that since the YSC system was to be a tool for silicon compiler development, we needed the capability of extending the description language easily. Thus we adopted the strategy of embedding YLL in the language, APL, used for all our tool development. This decision brought with it several attractive features:

1. Interaction and easy debugging
2. Ease of manipulating vectors and arrays
3. Definition of powerful operators and their combinations, which could be extended naturally to the logic domain
4. Recursion
5. Easy syntax, simplifying the language parsing task

The idea of embedding YLL in an existing language has not only simplified the language processing task, but has also made language extension easy. For example, new operators can be created by writing an APL subroutine, named for the new operator. The subroutine is written in APL by the language developer, and uses the subroutines and conventions already established for manipulating logic functions for YLL processing.

We also decided, since the operators and combinations of operators of APL had been carefully chosen and constructed based on usefulness, that all those having a natural meaning in the logic domain should be defined. Thus any combination of the ten nontrivial binary logic operations, "AND," "OR," "XOR," and so forth, operating on arrays of arbitrary dimension are allowed. Perhaps this is more power than will ever be needed for specifying logic, but who knows?

Finally, we constrained the YLL processor so that the variables mentioned in the user's YLL source program and the set of logic functions produced by YLL are in exact one-to-one correspondence. This was done because of some initial experience with an input language that produced an equivalent and correct logic macro; however, only the identity of the module inputs and outputs was preserved. We felt that it was important that YLL could be used to describe the internal structure of the logic module in those cases where careful algorithmic design of the combinational logic was desired and where logic synthesis may not be capable of rediscovering this careful design.

YLL LANGUAGE The syntax of YLL is the same as APL, so a user of APL will find it easy to learn YLL. We will assume some knowledge of APL and discuss YLL in terms of its differences with APL. In principle, the YLL program written by the user to describe a module of combinational logic should closely resemble an APL program that performs the same computation numerically with 0's and 1's. Conversely, any APL program that operates in the Boolean domain should define, except for minor variations, a valid YLL program. The differences between APL and YLL are due largely to the fact that the APL program does numerical processing and the YLL program does symbolic processing. Thus, while the result of all statements in the APL computation is a 0 or a 1, the result in the YLL computation is a logic function. Other differences are due to choosing to extend YLL for further convenience.

Variables: There are two kinds of variables in YLL, logic variables and metavariables. *Metavariables* are those used for controlling the flow of the YLL execution and are typically used for parameterizing the subfunctions (e.g., to work with different bit widths). *Logic variables* correspond to those used to describe the hardware. Thus initially, each logic variable can be associated with a signal that will exist in the initial implementation description given to the logic synthesis task. To distinguish a metavariable computation from a logic computation, the user may choose to execute (⍎) a quoted string of characters, such as (⍎)$'N \leftarrow \rho V'$. The general rule is that if a string of characters in APL has an operator that has a logic meaning, like ←, ∧, ≠, ~ , but its APL meaning is intended, then the string is quoted and executed. Thus the full power of APL is available to YLL.

Constants: 0 and 1 are the only constants allowed for the logic variables. There is no restriction for metavariables. Reference to an integer that is to represent a binary vector can be provided by an APL function. For example, 32 DB 105 would produce a 32-bit binary representation of the decimal number 105. The function DB can be written trivially by the user.

```
    ∇Z←D DB N
[1] Z←(Dρ2)⊤N
    ∇
```

This is a good example of how easily YLL can be extended: simple or quite complicated new operators can be easily defined and used to extend YLL.

Assignment: Assignment (←) is a logic operation that takes the logic function defined by the right-hand side and "OR's" this with the current logic function held by the variable of the left-hand side. We assume that all logic functions start with logic function 0 and are built up by this "OR-ing" operation.

Operators: All operators keep their usual meaning, but only those that are Boolean operations will give rise to a hardware computation. For example, ∧, ∨, =, >, <, ⍲, ⍱, ≠, ≤, ≥, and ~ are the only true logic operations. The rest of the operators, such as ⍳, ↑, +, etc., are useful for manipulating metavariables or indexing into or selecting from an array. For example, $V[(\lceil N \div 2) + \iota K;]$ selects a set of rows of the logic array V. N and K are metavariables, and by convention because the operations are inside square brackets, the metavariable computation need not be quoted (in this case, since there is no ambiguity in what is meant, we would not have to quote the metavariable computation anyway).

Combinations: Combinations of logic operations operating on arbitrary arrays of logic functions are allowed along any dimension if such combinations are defined in APL. For example, (≠ \[3]*A*)∨.∧,(< ⌿ *B*) is allowed as long as the dimensions of A and B are compatible with this operation. The result is an array of logic variables with the same dimension as the corresponding APL result.

Index origin: We have adopted index origin of 0 for compatibility with hardware conventions. Thus a 32-bit vector is indexed 0,1,2,3,. . .,31.

Declarations: It is sometimes necessary to give a dimension to a logic variable before it is used. This is done by $\triangleq$ (e.g., $A \triangleq 3\ 5$ defines A as a 3-by-5 array of logic variables). The statement $A \leftarrow 3\ 5 \rho\ 0$ would also initialize A, but all bits of A would have a logic function, namely 0. With the other declaration only the dimension is given. By convention, the individual logic variables exist in the macro only if they are assigned to it. Thus if $A \triangleq\ 32$ is defined but only A[29] and A[30] are computed, then only these two will be logic functions of the logic macro.

Equivalence: For convenience, we allow equivalencing between names. For example, $B \triangleq 'A[\iota 3; \iota 4]'$ provides the upper left 3-by-4 array of A with another name B. No new variables B are created. $B \leftarrow 1$ is equivalent to the statement $A[\iota 3; \iota 4] \leftarrow 1$, but the first is easier to write.

Selection: We use the symbol $(A) \subset (Z \leftarrow B; Z \leftarrow C)$ is the same as the if-then-else statement; if $(A = 1)$ then $(Z = B)$ else $(Z = C)$. The convention is that each statement on the left, if true, selects the corresponding statement on the right. If there is one more statement on the right, then it is selected by the complement of the disjunction of all the statements on the left. The previous statement can be written also as $Z \leftarrow (A) \subset (B; C)$. Arbitrary nesting of selection is allowed.

Iteration: We borrow the "foreach" operator, $\ddot{}$, from APL2. Thus, $'A[\omega;] \leftarrow B[\omega;] \wedge C[\omega;]' \ddot{}\ \iota 32$ executes the quoted statement 32 times for $\omega = 0, 1, 2,. . .,31$.

Inputs/outputs: Inputs and outputs are passed to a YLL function, both through arguments, and through global variables. Local logic variables are intermediate variables of the hardware description that may disappear later in the logic synthesis process, but inputs and outputs will always be present in the final hardware implementation.

Consider the following example:

```
    ∇Z←A FCNY̲ B; C; N
[1] ⍉N←⌈(ρA)÷2'
[2] G≜N
[3] C←(2×N)↑A∧B
[4] G←C[2×ιρG]
[5] Z←(CNT)⊂(C;G,G)
```

This example illustrates the following:

1. Line 0 declares A, B as inputs, C as a local logic variable, and hence an intermediate, and Z as an output. Since CNT and G are not made local, they are global variables. CNT is not assigned to, so it is an input, while G is an output since it is assigned to in the body of the function definition. N is a local metavariable. The last letter of a YLL function definition must end in $\underline{Y}$. Reference to a function in another subroutine is made by dropping the $\underline{Y}$.
2. Line 1 was enclosed in quotes and executed (⍎) since it is a metavariable computation with a logic operation (←) in it. Once a statement is enclosed in quotes and executed, it is an APL, not a YLL, statement.
3. In line 2, $\underline{G}$ is allocated a dimension depending on the metavariable $\underline{N}$. This illustrates a convenient means of parameterizing subroutines.
4. In line 3, the computation of $\underline{C}$ is a mixture of a metavariable calculation $(2\times\underline{N})$ and a logic computation $(\underline{A}\wedge\underline{B})$, but there is no ambiguity in what it means, so no quoting is required.
5. In line 4, the selection of indices $(2\times\iota\rho G)$ is a metavariable calculation. It need not be quoted since it is unambiguous and inside square brackets anyway.
6. Line 5 illustrates selection, in this case equivalent to

   if (CNT = 1)**then** (Z = C) **else** (Z = G,G),

   which produces the logic equation $Z \leftarrow (CNT\wedge C)\vee(\sim CNT)\wedge(G,G)$. Another way to produce the same result is $Z \leftarrow (CNT, \sim CNT)$ $\vee.\wedge(2,\rho C)\rho C, G, G$.

We have also implemented in YLL the ability to handle tables that are useful in specifying finite state machines or large CASE statements. For example, the APL function GTABLE, described as follows, is one that was added later in the YLL development and is a good illustration of how the embedding of YLL in APL makes it easy to impart additional capabilities to YLL. Also, the example illustrates one way don't cares may be extracted and used in the logic specification. We use six tables, the first five of which are encoding tables, for the inputs and outputs.

LIGHTTAB:	*COLOR_NEW*		*TRANSTAB:*	*R_G*	
	YELLOW	*0*		*G_V*	*0*
	RED	*1*		*R_G*	*1*
	GREEN	*2*			
TIMETAB:	*SHORTY*	*0*	*CARTAB:*	*CAR*	*1*
	SHORT	*1*		*NOCAR*	*0*
	LONG	*2*			
	OUT	*3*			
STARTTAB:	*START*				
	YES	*1*			
	NO	*0*			

TRAFFIC Table

.input	*1*	*2*	*LIGHTTAB*				
.input	*2*	*1*	*TRANSTAB*				
.input	*3*	*2*	*TIMETAB*				
.input	*4*	*1*	*CARTAB*				
.input	*5*	*1*	*CARTAB*				
.output	*6*	*2*	*LIGHTTAB*				
.output	*7*	*1*	*TRANSTAB*				
.output	*8*	*1*	*STARTTAB*				
GREEN	*?*	*SHORT*	*CAR*	*?*	*GREEN*	*?*	*NO*
GREEN	*?*	*LONG*	*CAR*	*?*	*GREEN*	*?*	*NO*
GREEN	*?*	*SHORT*	*NOCAR*	*CAR*	*YELLOW*	*G_R*	*YES*
GREEN	*?*	*LONG*	*NOCAR*	*CAR*	*YELLOW*	*G_R*	*YES*
GREEN	*?*	*OUT*	*?*	*CAR*	*YELLOW*	*G_R*	*YES*
YELLOW	*G_R*	*SHORTY*	*?*	*?*	*YELLOW*	*G_R*	*NO*
YELLOW	*R_G*	*SHORTY*	*?*	*?*	*YELLOW*	*R_G*	*NO*
YELLOW	*R_G*	*SHORT*	*?*	*?*	*GREEN*	*?*	*NO*
YELLOW	*G_R*	*SHORT*	*?*	*?*	*RED*	*?*	*NO*
RED	*?*	*SHORT*	*?*	*CAR*	*RED*	*?*	*NO*
RED	*?*	*LONG*	*NOCAR*	*CAR*	*RED*	*?*	*NO*
RED	*?*	*OUT*	*NOCAR*	*CAR*	*RED*	*?*	*NO*
RED	*?*	*SHORT*	*?*	*NOCAR*	*YELLOW*	*R_G*	*YES*
RED	*?*	*LONG*	*CAR*	*?*	*YELLOW*	*R_G*	*YES*
RED	*?*	*OUT*	*CAR*	*?*	*YELLOW*	*R_G*	*YES*

The last table is the description of a finite state machine in symbolic tabular form. The first eight lines of this table state information about the columns of the table. For example, line 1 states that column 1 is an input encoded by the table LIGHTTAB using a two-bit encoding. The name of the encoding variable is, by convention, given at the top of any table; in this case, the name of the two-bit variable is COLOR_NEW. The rest of the table, TRAFFIC, gives state transitions. For example, line 9 states that if the light is GREEN and transition is anything (given by ?, an "input" don't care), and the time is SHORT, and there is a CAR on the highway, and we don't care if a car is on the side road, then set the light to GREEN, the transition to anything, and do not start the timer.

The following is the YLL program that uses this table.

```
 ∇ TRAFFIC_CONTROLLER
[1] COLOR_OLD ≜ 2
[2] R_G_OLD ≜ 1
[3] TIMER ≜ 2
[4] HIGHWAY ≜ 1
[5] SIDE_ROAD ≜ 1
[6] 'TRAFFIC 6 7 8' GTABLE COLOR_OLD,R_G_OLD,TIMER,HIGHWAY,SIDE_ROAD
```

Lines 1 through 5 are used to set the dimensions of the inputs. Line 6 calls the subroutine GTABLE, which is told to access the table TRAFFIC and process output columns 6, 7, and 8. This represents all the outputs in this case, but a subset could be specified. The inputs for this table are given in the right argument of GTABLE. They correspond to input columns 1 through 5 of the TRAFFIC table and should agree in dimension with the input dimensions for those columns given at the top of the TRAFFIC table.

Note that LIGHTTAB has only three colors; the code point 3 is never used, giving rise to a don't-care point. Also note that LIGHTTAB is used as the encoding table for the output column 6 where we see the three colors RED, YELLOW, and GREEN used as mnemonics in this column.

The APL function GTABLE does several things. First, it extracts all mnemonics for each output column and forms a don't-care function consisting of the complement of all code points appearing with a valid mnemonic. A "?" in an output column serves as an invalid mnemonic and hence the input conditions for that row are don't-care conditions. GTABLE accesses each input encoding table to determine the encoding for each input mnemonic. Thus, if a code point is not used in an encoding table, it also gives rise to a don't-care condition and will become part of the don't-care function derived. The encoding table also specifies the name of the output encoding vector for that column. One logic function for each output encoding bit is derived by GTABLE. Finally, all output functions for that output column are simplified using an ESPRESSO–based minimization procedure with a don't-care set given by the derived don't-care function.

Although it does not occur in the TRAFFIC example, the encoding table may be omitted for an output column. Then each valid mnemonic in that column is derived as a separate logic function, giving rise to a "one hot code" for that output column. Finally, if no output columns are listed in the left argument of the call to GTABLE, all output columns will be processed and a special signal, named TRAFFIC_VALID, will be created to indicate when a valid input was received. This can be used to detect an error condition and cause an interrupt, and indirectly, may be the mechanism used to allow the other signals to be don't care when there is an invalid input.

7.2.2 Structural Synthesis

Structural synthesis has become quite popular recently due to the need for more powerful design aids. Much thought has gone into the planning and the design of such systems, but comparatively little concerning the algorithms has appeared in the literature. Although structural synthesis is not yet a well-understood problem, several subproblems have been identified. Subproblems include scheduling operations given a certain number of resources, data path synthesis, control synthesis, pipeline synthesis, high-level optimizations, and so forth. Other efforts to develop design systems including structural synthesis have been and are being undertaken at the University of Karlsruhe [Wojt79] [Rose85], the University of Kiel [Zimm80] [Marw84], Carnegie-Mellon University [Dire81] [Kowa83] [Tsen86], University of Southern California [Knap83] [Park86], Bell Labs

[Kowa85] [McFa86], and GTE [Blac85] [Fox85] among others. More general overviews can be found in [Camp85] [Gajs86] [Thom86].

In addition to the usual restriction to synchronous systems described by finite automata, two principles were used in the YSC system in order to more precisely define the goal of structural synthesis:

1. Logic synthesis—that is, synthesis of the combinational parts of the design—will be used extensively at a later stage. This is justified by the powerful logic synthesis tools available in the Yorktown silicon compiler. Moreover, the combinational logic usually takes only a small part of the total chip area. At a high level, not too much effort should be spent to minimize this part of the design.
2. Structural synthesis will be restricted initially by a high-level architectural decision—that is, to minimize the number of control states needed for executing all operation sequences. The tradeoff, between the number of control states (the number of clock cycles) needed to perform a sequence of operations and the time period required to execute the operations within each control state (the cycle time), is always decided initially in favor of the minimum number of control states.

The principle advantages for structural synthesis in using those principles are:

- They yield to a natural separation between the *sequential* and the *combinational* aspects of a system. At the beginning, only sequential aspects of the circuit to be designed must be considered.
- The goal is clearly defined. Initially, unknown parameters such as absolute speed or area, which are often estimated very poorly in structural synthesis systems, are not taken into account.
- Using as few states as possible in the control part also minimizes the number of latches required for storing data used in different states. This can be advantageous if latches are expensive compared to combinational logic, for example if static latches with scan-path capabilities are used.
- The design obtained initially can be modified in subsequent steps using basically one transformation—control state splitting.

The procedure for obtaining a structure from the behavior is based on the following ideas. Assume initially the design contains just combinational logic. If this is correct, the circuit can be obtained just using logic synthesis. The same is true if there are registers present (and identified) but no sequential control is necessary. If more than one control state is necessary, then it must be decided what operations have to be executed in what state (control state assignment). This is done by subsequently splitting an initial state that holds all operations. The operations are distributed among two new states. In each state splitting, as many operations as possible are scheduled in each of the new states. This assures that two states never will be merged again, as far as the number of states for performing the operation sequences is concerned. Operations can be assigned to more than one state.

In the YSC system, the main steps in structural synthesis from YIF are:

1. *Memory assignment*. Variables in the behavioral YIF description that must be implemented by memory arrays are identified and replaced by memory references. The memories themselves are generated only during layout design.
2. *Control state assignment*. A control state machine is built. Operations are assigned to the states of the control automaton; that is, it is decided which operations must be performed in which control states. Also data path registers are allocated in this step. Thus all sequential aspects of the system being designed are defined.
3. *Control state splitting*. After control state assignment, states contain as many operations as possible. Splitting a control state means assigning part of its operations to one or more new control states. This is done for two reasons. First, if sequential operations within one control state are distributed among two or more states, the necessary time to execute all operations within one control state may be shortened (at the cost of a new control state). Second, if two operations that were executed in parallel within one control state are distributed between two control states, they are sequenced and may share hardware, thus trading increased execution time for less area.
4. *Variable unfolding*. Variables are replicated ("unfolded") until single assignment is achieved. Each unfolded variable has only one source and thus corresponds to a net in a network.
5. *Operation folding*. Operations performing the same function can be implemented by the same hardware module if they are alternative or executed in different control states. The mapping of two or more operations to the same hardware module is called "operation folding." Control state splitting heavily influences which operations can be folded.
6. *Register folding*. Registers in the data path are assigned during control state allocation and splitting, by examining each state independently. Just as with operations, values held by registers might be alternative, in which case they can be mapped onto the same register. Register folding (also called "register assignment") is guided by "lifetime" analysis of the variables.
7. *Control and network generation*. Multiplexers are created for nets that get different alternate values. The select signals for these multiplexers as well as enable signals for registers must be added. Operators are chosen from the operator library. This library contains YLL subroutines, which are parameterized logic specifications for each of the operators, not particular implementations. The final network for each module is generated.
8. *Partitioning and module binding*. Cells such as registers and memories are selected from the cell library. The cells in this library are generated during layout design. Unlike operators, they do not need to be logically synthesized. The combinational logic still described as a function in YLL is partitioned into modules of appropriate size for logic synthesis or for the subsequent cell construction during layout compilation.

Steps 3 to 6 can be iterated or redone to obtain different designs. The algorithms for structural synthesis are described here only briefly. More details can be found in [Camp87].

Memory Assignment

Variables in YIF in general will be mapped either on storage elements (e.g., registers or memories) or onto nets (wires) during structural synthesis, depending on whether their value must be kept for more than one control state or not. The type of a variable indicates how to implement it. Variables initially of type register are the so-called *architected registers*—that is, registers already specified as those in the initial specification. Variables of type external are ports (pins) to the outside of the specified system. Variables of type internal will be implemented as registers only if necessary, otherwise as nets (wires). Variables of type array possibly will be converted into memories. Only one-dimensional arrays are allowed. Multidimensional arrays must be converted into one-dimensional arrays during compilation. Notice that each element of an array has a given number of bits (width). Each of these bits can be referred independently.

Example: Let $S_a = \{s \in S | \text{type}(s) = \text{array}\}$. For $s \in S_a$ basically two situations can be distinguished:

- There is at least one reference to s that uses a variable index whose value cannot be determined during structural synthesis. Although there are special cases in which it would not be necessary, s is mapped onto a memory. For example, let $a \in S_a$ be of dimension m and width w (both constants). If in the YIF description there is a reference $a[j]$, j being another variable, then a will be implemented by a memory of m words of width w.
- All references to s use constant indices, that is, indices whose values can be determined at synthesis time. In this case, either s is mapped onto a memory or it is converted into as many variables as the dimension of s. The system decides which action is taken depending on the dimension of s and some other parameters. Large dimensions usually lead to memories, while small ones lead to single variables. For example, let $a \in S_a$ be of dimension m and width w (both constants). If in the YIF description all references use constant indices and m is small, such as 4, then a will be converted into a_1, a_2, a_3, and a_4. All references to $a[1]$ will be replaced by a new variable a_1, and so on. If m is large, such as 1,024, then a will be implemented by a memory of 1,024 words of width w.

Notice that the replication of variables might lead just to nets in the final implementation. Moreover, all variables may be referenced in parallel. Implementations as memories are usually smaller but only a small number of references in each control state (as many as the number of memory ports) is allowed.

Memory assignment gets more complicated if multiport memories are allowed. The decision of which reference to a variable is assigned to which port of a multiport memory cannot be made at this stage and must be delayed.

In YIF, variables are mapped on memories by changing their type from array to internal and introducing additional operations that represent read and write. The read and write operations refer to the variable using an explicit address (no index). A constraint maps each of them to a memory cell. These map constraints will be taken into account during module binding.

The design after memory assignment is still represented in YIF. Each variable, whose type was array, has now been converted to another type.

Control State Assignment

The aim of state assignment is to create a control automaton and to determine which operation is executed in which state. Since this can be done in many ways for a given YIF specification, additional constraints are necessary. In our case, the initial goal is to minimize the number of control states needed for the execution of each of the operation sequences, starting with the first operation and ending with the last operation. All operations scheduled to one control state are implemented by combinational logic and thus executed "in parallel." Operations scheduled into different control states are executed in sequence; thus they are alternative in time and can be folded.

Given a YIF description $Y = (V, S, R_s, R_i, R_o)$ and the type of each variable, type(s) $\in$ {external, internal, register}, the control state assignment algorithm works in four steps, constructing a set of states Z and assigning to each state $z_i \in Z$ a set of operations ops(z_i) $\subseteq V$ to be executed during that state.

1. *Control states due to loops.* For each loop (iteration, cycle in a graph) at least one control state must be introduced to allow the repetition of the operations in it. Loops are eliminated, and the iteration is encoded into the control by providing appropriate state transitions. The algorithm proceeds as follows. It first finds the feedback edges $L \subset R_s$ such that, in forming the transitive closure, A, of $(R_s - L)$, a partial order relation results. In our present approach, L is derived syntactically from operation tags. These tags are generated during compilation by considering the syntax of the iteration constructs of the high-level language (e.g., WHILE and UNTIL iterations).
   The operations of the resulting acyclic graph $(V, R_s - L)$ are now assigned to states. Let

   $$V_L = \{v_f\} \cup \{v \in V | \exists_{v'}[(v', v) \in L]\}$$

   V_L contains all those operations that must form the starting point of a state. A state starting point is the first operation scheduled in that state, according to A. Operation v_f (the first operation of the whole YIF) clearly is such an operation. The other operations in V_L are the first operations in each loop; since according to the semantics of YIF loops must possibly be repeated, also these operations must be the starting point of a state.

A first automaton is built with as many states Z as $|V_L|$. The operations executed in each control state $z_i \in Z$ are

$$\text{ops}(z_i) = \{v_i \in V_L\} \cup \{v \in V | (v_i, v) \in A\}$$

Since our goal is to minimize the number of required states to execute all operations sequences, all successors in A of the first operations V_L are initially included in each state. Particularly, all successors from v_f, that is, all operations, are scheduled in one state. An example of the states generated by loops is given in Figure 7.6.

2. *Control states due to module calls.* An operation can be either an atomic computation, such as an addition, or a reference to another YIF description. The latter is called a *module call*. If the called YIF description generates a combinational circuit, it can be treated as an atomic computation. If the resulting circuit is sequential, then it cannot be treated as an atomic computation any more and an additional control state in the calling module must be introduced, splitting the original state. In the most general case, the number of control states (cycles) that a module call will need is unknown. For each of these module calls we introduce a new "waiting" control state. This module call waiting state does not perform any operation besides enabling the called module, until a *ready signal* is generated. Ready signals can be generated automatically by identifying control states that contain final operations in a module.
3. *Control states due to constraints.* Timing constraints, such as specifying that an operation should be performed in one cycle, might produce new control states. These new control states are added as specified.
4. *Control states due to data flow restrictions.* The first three steps yield a state assignment that is the starting point for data flow analysis. Data

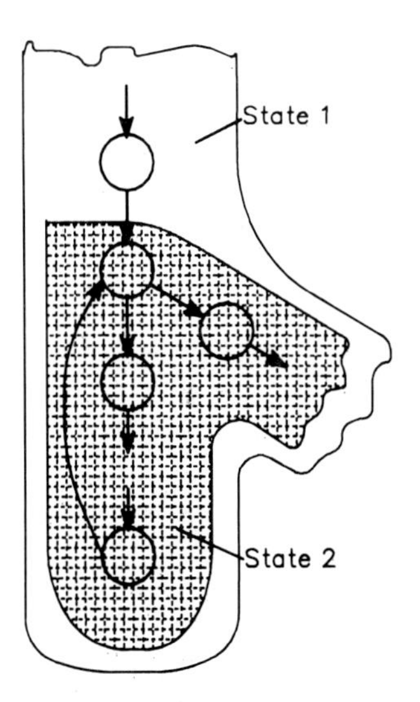

Figure 7.6 Control state assignment for a loop.

flow analysis detects which variables of type internal must be converted into registers and detects if new control states are necessary due to data flow restrictions. Data flow restrictions arise from the fact that during one cycle, a register can be written only once and, similarly, external variables (pins) may also hold only one value. Since new registers might introduce new data flow restrictions, and new control states due to data flow restrictions might introduce new registers, an iterative algorithm of two steps is necessary:

a. Internal variables that are used as inputs without having been used as outputs before in the same state must hold a value that was written in another state. They must be converted into registers. These variables are defined by

$$S_{ir} = \{ s \in S \mid \text{type}(s) = \text{internal} \wedge$$

$$\exists_{z \in Z, v \in V}[v \in \text{ops}(z) \wedge (s, v) \in R_i \wedge \forall_{v' \in \text{ops}(z)}[(v, v') \in A \vee (v', s) \notin R_o]]\}$$

They are used as inputs by the operation v in state z without being used as output by any operation v' before v in that state. Thus, the value of these variables must have been written in another state. A storage element is necessary to keep a value from one control state to another in hardware. Hence, the type of all variables $s \in S_{ir}$ is changed to type(s) = register. Unless architected, registers are only introduced where it is necessary.

b. Variables with type(s) $\in$ {register,external} create *data flow restrictions* if they are used as inputs and/or outputs more than a given number of times within one control state (cycle). A register can be written only once in a cycle, an external port (or memory port) might have only one value associated with it in each cycle. Control states in which this limitation is violated must be split into two or more states. For data flow constraints due to multiple use of a variable as output, these control states are obtained as follows. Given a state $z \in Z$ then if for some variable $s \in S$ with type(s) $\in$ {register, external} the set

$$V_{ZS} = \{ v \in \text{ops}(z) \mid (v, s) \in R_o \}$$

has a subset $V_c \subseteq V_{ZS}$ such that

$$\forall_{v_1 \neq v_2 \in V_c}[(v_1, v_2) \in A \vee (v_2, v_1) \in A] \wedge |V_c| > c_s$$

then z must be split. c_s is the maximum number of times the variable s can be written (used as output) in one state (during one cycle). Usually c_s = 1. The control state z must be split between both references (in this case output references) to the variable s. Input or mixed (input and output) data flow restrictions are treated in an analogous way.

Steps a and b are iterated until no changes occur.

After control state assignment the sequential control is explicit. A register holding the control state is introduced, and the state transitions are given by operations that load new values into this register. State assignment (i.e., encoding the states) may be done here if desired. The design at this stage is a finite state machine. All paths from a unique first operation to one last operation are performed in one cycle. This first design has the "extreme architecture" that minimizes the number of control states required for all operation sequences.

Since the registers in the data path were also determined during control state assignment, the data path also is designed to some extent. Hence, control and data parts are synthesized together in a single step.

The result is again represented in YIF. Consider the example of Figure 7.4. It produces the YIF description given in Figure 7.7. Here the variable *PC* was converted into a register (and unfolded, see the variable unfolding description that follows); also the state register STAT was created, encoding the states by simply numbering them.

Control State Splitting

Further modifications to the basic structure obtained so far are:

- Splitting states that are too long, that is, the combinational part in the data path that is too slow. Possibly new registers are introduced as explained in control state assignment.

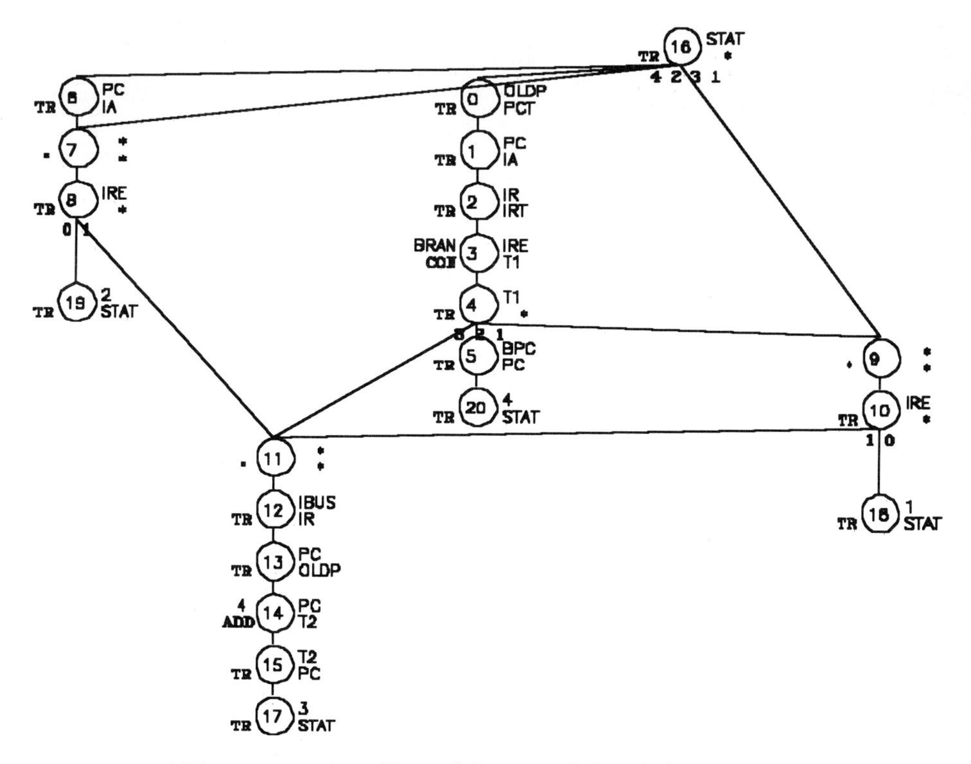

Figure 7.7 YIF representation with explicit sequential control.

- Modifying the design so that the critical paths in the data path assigned to each control state have a similar estimated delay or meet timing constraints.
- Splitting states that contain operations requiring hardware that uses a large area. Since control states are alternative in time, this allows for the mapping of such operations onto the same hardware gaining area.

To identify the states to be split, a goal must be defined, such as a given maximal cycle time (specified as a timing constraint). State splitting already requires an estimation of the delays and areas not yet known at this stage. Successful state splitting can only be achieved with good relative estimates. In the YSC system, area, power, and delay are estimated by using functions of arbitrary complexity stored in the module library or by exercising logic synthesis using only fast algorithms.

Variable Unfolding

In an imperative description, variables hold their values until they are overwritten; therefore, they can be used many times. Variable unfolding introduces additional variables so that single assignment results, and changes the references accordingly. Variable unfolding occurs both for registers and for internal variables implemented by nets.

Storage elements already meet the single assignment criterion due to the data flow restrictions imposed earlier. Nevertheless, it must be identified, if an input or an output is being referenced: outputs are only loaded at the end of a cycle and are available only during the next cycle. The unfolding consists of introducing new variables for the inputs whenever this is necessary. An example of register unfolding is given in Figure 7.8.

Internal variables that will be implemented by nets do not necessarily meet single assignment. For sequential assignment to these variables, new variables must be introduced. In doing this, the necessity of introducing additional multiplexers might arise (see Figure 7.9).

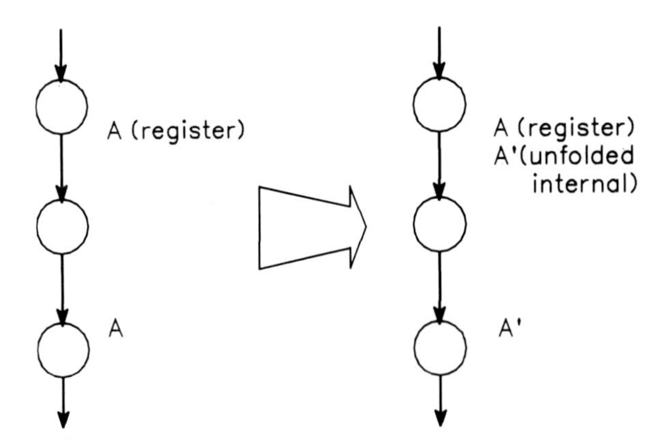

Figure 7.8 Register unfolding.

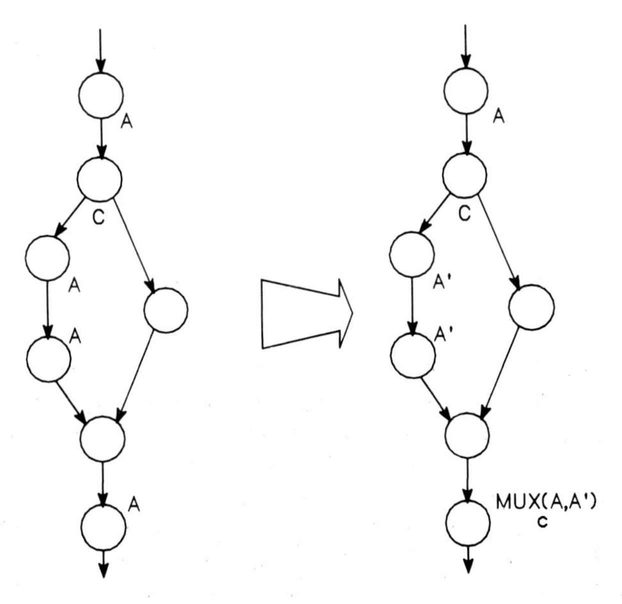

Figure 7.9 Internal variable unfolding.

Operation Folding

Although we relieved structural synthesis from performing so-called high-level optimizations of combinational parts that can be done during logic synthesis, not all optimizations can be done by logic synthesis. Some optimizations arise from the sequential nature of the circuits. As an example, consider the circuits in Figure 7.10.

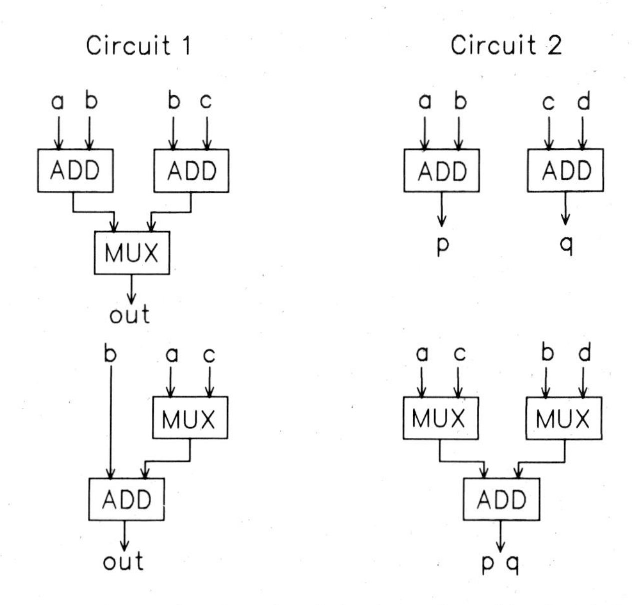

Figure 7.10 Two examples of optimizations of combinational circuits.

The optimization of circuit 1 can be performed by logic synthesis; it will yield an optimal circuit (optimal according to the goals defined in logic synthesis) for any width of the operators, for some of the operators being constants, and so forth. Doing this at a higher level would be extremely difficult since the properties of each functional block (an addition in the example) vary (e.g., they might be commutative or not or they might have a neutral element 0 or 1).

The optimization in circuit 2 cannot be performed by logic synthesis; the operands used in the original circuit are different for both adders. But using the fact that both additions always occur in different control states, that is, they are alternative in time, both adders can be collapsed by adding some multiplexers as indicated. This kind of optimization should be done by structural synthesis. In the YSC system, this is called *operation folding*.

Two operations v and v' can be folded, if

$$(v, v') \notin A \wedge (v', v) \notin A$$

that is, if they are alternative. It can be shown that folding more than two operations results in the clique partitioning problem for the complement of the symmetric closure of A[Golu80].

Folding should result in some advantage. Our criteria are:

- The operations should be complex compared to a multiplexer. This excludes, for instance, simple gates from being folded.
- The operations should be similar so that folding results in a smaller circuit. Similarity in the simplest sense means requiring the same function (e.g., an addition), only allowing different input and output widths. More refined similarity criteria can be table driven. The number of different functions that an operation can perform is limited to the order of 10^2, so this approach is practicable.

Operation folding is done simply by appending a *map constraint* to the operations being folded. The design is still represented in YIF. During control and network generation, the multiplexers for the inputs will be created and the outputs will be merged. The combinational circuit for the operations themselves will be obtained and optimized during logic synthesis.

Register Folding

Registers are generated according to the initial specification and during control state assignment and splitting. Besides the architected registers, registers are introduced if the value of a variable must be passed between two or more states. Clearly some of these variables might be "alternative," that is, there is no situation in which they are both "alive." These variables can be mapped onto a single register. Merging two or more of the registers initially generated is called *register folding*. To detect which registers can be folded, techniques similar to "lifetime" analysis in compiler construction are used.

Control and Network Generation

At this point, the design is considered ready with respect to its sequential aspects. The YIF data format now satisfies the following properties:

- The predecessor-successor graph, which orders the operations (nodes in the YIF graph) is acyclic.
- The data flow graph defined by R_i and R_o, giving the connection between variables and operations, is acyclic. In this context, the input and output sides of registers are considered separate.
- Indices selecting parts (bits) of variables are constants. Variable indices have been replaced by operations or memory calls.
- All operations can be implemented by combinational logic. The only YIF nodes that specify an operation requiring a sequential implementation are module calls.

The goal at this stage is to produce:

- The description of a network connecting module inputs, module outputs, registers, blocks of combinational logic, and calls to other modules
- The logic specification of the combinational blocks

Descriptions of the different steps to reach this goal follow.

OPERATION-TO-OPERATOR MAPPING For each operation specified in the YIF graph, a combinational function ("operator") must be chosen. A library of operators is supplied, given by their logic specification, as well as connection and control information. For instance, an ADD operator in the library might not be used only for the "+" operation in the YIF graph, but also for an increment operation "++". The library specifies that ADD can be used for incrementing, by connecting the single operand to the first input of the ADD and a constant "1" to its second input. Simultaneously an ALU could be present in the library, capable of performing among others the "+" and "++" operation. In the same way, constants will be specified for their control inputs, instructing the ALU to perform the required function. Besides this so-called mapping information, expressions are provided in the library for every feasible operation/operator combination to predict roughly the area, delay, and power requirements. The expressions yield these estimates as a function of the width and the number of operands, and they are used to make a default selection when several mappings are feasible. By default, one combinational function is instanciated for each operation. The choice of the operator might be dictated by previous decisions, such as the folding of more operations onto a single operator.

It is worth noting here that, for example, the implementation of an increment by an ADD unit is *not* a waste of area or power. The library fixes only the logic behavior of the ADD unit. The actual hardware (gates) is only generated after a logic synthesis step, thus reducing ADD with a constant input to a "real" incrementer.

WIDTH ADJUSTING The widths of variables, constants, and operators need to be matched in order to write a real network definition. Normally a (sign-) extension of the variables and constants (operands) is sufficient to match the inputs of the operator. Sometimes the operator itself must be resized, for instance when operations involving different widths are folded onto each other. Further, the output of the operator must be matched with the specified output variable. This can be done by removing the most significant (carry-) output bit of the ADD unit, for instance.

MULTIPLEXING The insertion of multiplexers is needed to write to nets, which are assigned to more than one operator, and to write to the inputs of operators, which are assigned more than one operation due to folding. The "select" inputs of the multiplexers are decoded, in order to match the control (enable) signals discussed in the next section. The multiplexers are stored in the YIF description in the same way as any other operator, and they are fetched from the same library.

A more complex control, such as for an ALU, is automatically generated by this scheme. The control input of the ALU was assigned a constant for every operation that was folded on it. By multiplexing these constants, and subsequent logic synthesis of this multiplexer with constant (data-) inputs, the proper control logic is obtained.

CONTROL SIGNAL GENERATION Control signals are needed for the select inputs of the multiplexers, for the enable inputs of the registers, and for the synchronization and communication with the "outside world" and other module calls.

Note that the entire YIF graph handled at this stage is only one module out of a hierarchy of modules specifying the intended circuit. For each module, one finite automaton is created. In general, the synchronization between these automatons is arranged in such a way that automata above in the hierarchy provide a start (enable) signal for those in the level immediately below (their children), and await their completion, which is signaled by an outgoing ready (enable) signal.

Inside each module, control signals are generated, enabling the paths through the graph that should be executed according to the conditions in the forking nodes. In general, each module gets one enable input, assigned by the outside world, controlling the activity of the module. All starting nodes of the (acyclic) YIF graph are controlled by this incoming enable signal. Proceeding down through the graph, new enable signals are assigned. At the joining of paths, a new enable is created by "or-ing" the enables corresponding to the incoming paths. At a fork, new enable signals are created for the outgoing paths according to the branch decision and the incoming enable. The different logic functions required for the creation of the control signals are obtained from the same YLL library as the data path operations.

A module output enable signal is created by "or-ing" the enables of a subset of the terminating nodes of the graph, indicating to the outside world the completion of the activity of the automaton. Further, outgoing enables are provided for the control of outgoing data, declared as registers in the outside (parent) module. Of course, module calls have an enable input, assigned by the enable in the current path. The output enable of the module call is used as the enable for the next node(s) downward in the same path.

The enable signals created this way are easily attached to the multiplexer inputs. The register enables might need a final "or-ing" of these signals if they are written by more than one operator. If the writing node was a module call, its corresponding outgoing register enable is used.

Basically, this scheme generates a so-called one hot encoding. For example, the node that implements the branch on the value of the state register will cause one enable signal for every (used) state register value. However, if the logic blocks that generate or use these enable signals are passed as one unit to the logic optimizer, these enable signals remain internal to this unit. They might be removed by minimizing the logic. Thus the large number of enable signals are just part of the logic specification and may not be present in the final realization.

NETNAME ASSIGNMENT For writing the network, a unique name should be given to each net. Normally, the name of the YIF variable can be used, but in practice a large number of operations (nodes) in the YIF graph specify an assignment or a catenate operation. For these operations, no logic is generated; they are resolved instead by a suitable naming, thus specifying the required interconnection. By this mechanism, many YIF variables are removed and replaced by constants.

Partitioning and Module Binding

The previous steps generated a network, connecting I/O variables, registers, and calls to logic functions from a library. As a last step, these logic function calls are grouped into combinational units that will be passed to logic synthesis. Of course, for the best minimization, all the function calls should be grouped into one large block. However, two reasons might exist for a partitioning into several smaller groups of functions. The first is the time and space complexity of logic synthesis, which limits the size of its input. The second reason is that the layout programs might be able to find a better floor plan if several smaller units are created. Thus a partitioning of the logic may be desired, which can be based on connectivity between the resulting groups and on the use of inputs common to the groups trying to construct blocks of a given size. Notice that control and data path are not separated.

The resulting network is written in a simple language, HND, suitable for machine and human reading. It allows a hierarchical definition of the network with separately generated modules. The interconnection through the different levels of the hierarachy is checked and established later. The logic specification of each group of logic functions is written in YLL.

After all different modules of the chip hierarchy are generated this way, a linking phase is done. Recursively, all HND files are processed, checking the consistency of the interfaces. For the calls to combinational logic, the results of logic synthesis are read, because the interface might have been affected by minimization, inputs might have been removed, and outputs might have been set to constants. Further, the hierarchy is expanded, generating one output module for each different instantiation, unless mapping constraints specify mapping of several module calls onto the same instance. The network information is written in a compact binary form suitable for interfacing with the layout compilation step.

The HND result of structural synthesis for the example of Figure 7.3 (also in Figures 7.4 and 7.7) partitioning the required logic into two groups is given in Figure 7.11.

HND consists of modules. Each module has a header indicating its identification (Module_id), its type, its technology, and the date and time it was synthesized. Module types can be:

- *Module*—for modules that are not leaves in the hierarchy and instantiate other modules.
- *Decomp*—for modules that are leaves in the hierarachy and contain combinational logic. For each of these modules a YLL program containing its logical specification is generated.
- *Latche*—for modules that represent latches with an "enable" input. Other types of latches (e.g., without an "enable" input, with "set" and "reset" inputs, etc.) are also available.

For each module, the input and output nets as well as the internal nets (Locals) are indicated. The names are generated, keeping the name in the original YIF wherever possible. A module that is not a leave may instantiate other modules giving them an instance name (first parameter after the keyword Instance). Each instantiated module is connected by giving a list of nets, which is connected in that order to the inputs and outputs as listed in the module definition.

Several things should be explained for the example shown in Figure 7.11. Module FETCH is the root of the hierarchy. It instantiates the decomp modules FETCH1 and FETCH2, which contain the combinational logic partitioned into two groups. A two-bit latch instantiated as STATE_REGISTER.1 corresponding to the state register and three 32-bit latches instantiated as OLDPC1 (holding OLDPC), PC1 (holding PC), and IR1 (holding IR) were generated. To the inputs of each module a CLOCK was added (SCVS implementation). The inputs and outputs of the

```
Module_id FETCH module scvs 198612102253;
Inputs CLOCK IRE IBUS[0:31] BRANCH BPC[0:31];
Inouts;
Outputs IA[0:31] IRT[0:31] PCT[0:31];
Locals STATE_REGISTER.[0:1] in.OLDPC[0:31] in.PC[0:31]
in.STATE_REGISTER.[0:1] enable_OUT_17[2:3] enable_OUT_12[0]
enable_IN_5 latch_PC1 latch_STATE_REGISTER.1;

Instance OLDPC1 LATCHE32 *;
Connects CLOCK enable_IN_5 in.OLDPC[0:31]
         PCT[0:31];

Instance PC1 LATCHE32 *;
Connects CLOCK latch_PC1 in.PC[0:31]
         in.OLDPC[0:31];

Instance IR1 LATCHE32 *;
Connects CLOCK enable_IN_5 IBUS[0:31]
         IRT[0:31];

Instance STATE_REGISTER.1 LATCHE2 *;
Connects CLOCK latch_STATE_REGISTER.1
in.STATE_REGISTER.[0:1]
         STATE_REGISTER.[0:1];

Instance EQL_OR1 FETCH1 *;
Connects CLOCK IRE BRANCH STATE_REGISTER.[0:1]
         in.STATE_REGISTER.[0:1] enable_OUT_17[2:3]
enable_OUT_12[0] enable_IN_5 latch_PC1
latch_STATE_REGISTER.1;

Instance ADD1 FETCH2 *;
Connects CLOCK BPC[0:31] in.OLDPC[0:31] enable_OUT_17[2:3]
enable_OUT_12[0] enable_IN_5
         IA[0:31] in.PC[0:31];
End_of_module
```

Figure 7.11 HND Example.

root module FETCH correspond to the interface specified initially with a CLOCK added.

Module FETCH1 results in just 22 transistors, and module FETCH2 results in 406 transistors after logic synthesis. Without partitioning the logic and using the same logic synthesis procedure, just one module with 417 transistors would result. Besides the latches, this is all the hardware required. Several other alternatives using different state splitting criteria, other partitioning of the logic, different logic synthesis alternatives, and trading time for area can be exercised very fast.

7.2.3 Logic Synthesis

Introduction

The YSC system is an optimizing compiler, and logic synthesis has exploited different optimization techniques. We have also emphasized certain styles, in implementing the combinational logic, that fit well with the philosophy of the other modules of the compiler. For example, we use multilevel logic, and we have emphasized the use of cells based on linear transistor arrays for the gates. The logic synthesis also has the ability to trade off the speed of an algorithm with the quality of results so that early estimates of area and delay may be obtained quickly. In addition, we are continuing to experiment with new algorithms for minimization, factorization, and decomposition of multilevel logic.

Complicated logic functions or sets of functions are often implemented in practice by using multilevel logic (multistage circuitry). This is because it is easier to control a number of properties of the final implementation, such as its area, delay, input/output pin positions, and flexibility of its layout. Although a set of logic functions is simply a mapping from its inputs to its outputs, if the mapping is defined as a sequence of smaller mappings whose intermediate results are useful, then a more efficient implementation may be obtained. Data flow mappings such as ALUs, rotaters, and decoders are such examples. Although "control" logic is generally considered suitable for a two-level implementation, our experience is that control logic also maps well into multilevel logic. Thus by adopting multilevel logic as the design style, nothing is lost, and we gain the advantage that control and data flow can be treated equivalently, allowing for merging the two where appropriate. We felt that in describing logic the user should be allowed the full power of a general programming language. The logic synthesis program must have the ability to deal with the output of such a description, which may intermix control and data flow logic.

Just as with software, there are different algorithms that compute the same thing, and some are more efficient than others. A software compiler generally preserves the overall sequence and flow of the algorithm. With a hardware compiler, a similar option should be available; but hardware and software compilers are different. With hardware, there is more opportunity to optimize since the compile time allowed can be longer and optimization has a bigger payoff. Thus the hardware compiler has the opportunity to change the sequencing and flow in order to optimize further, if the appropriate algorithms can be developed.

The task of logic synthesis in an optimizing silicon compiler is to map the logic equations into the specified target circuitry and to optimize the area while meeting the timing constraints. In the Yorktown silicon compiler, we use automatic synthesis and minimization for multilevel logic based on algebraic factorization and Boolean simplification. The synthesis process is carried out by applying a sequence of operators to a fixed ambient data structure. The operators are such that they can be applied in any order and repeated until no change occurs. They can both remove intermediates and synthesize new ones, guided by measuring their "value" based on area savings or delay reduction. If the incoming logic is already multilevel, the degree to which the initial decomposition is preserved during the synthesis process can be controlled, thus allowing the designer to specify more and more of the implementation details if desired.

The logic synthesis procedure transforms a more or less arbitrary logic description into a set of circuits that can be implemented in a given target technology. It also carries out the optimization of the circuits themselves. In many circuit families, such as dynamic domino CMOS logic [Erde84][Kram82][Pret84] differential cascode [Hell84], and conventional static CMOS, a complicated logic function can be implemented in a single circuit. Automatic optimization is essential to the success of these complex circuit families, since design by hand is impractical. For a general discussion of the complex CMOS circuit families, see [Chen84].

Although each function in the final result must be realizable, much of the synthesis procedure is technology independent. Even those operations that are guided by hardware considerations often require only a simple subroutine to check whether or not a given Boolean expression is simple enough to be realized as a single circuit. Due to this modularity, it is easy to change from one circuit family to another. One can also carry out synthesis of a single piece of logic several times, varying the target technology; in this way, a comparison between circuit families can be obtained without difficulty. The three types of complex CMOS circuits mentioned previously are among those supported by the logic synthesis programs.

An outline of the logic synthesis procedure follows. We begin with the process of constructing the logic from the YLL description. This creates an initial description of a "Boolean network," which is described in the next section along with transformations to be performed on it. We then detail the algorithms that carry out these operations and explain how they fit together to form a logic synthesis procedure. Finally, we demonstrate the effectiveness of this procedure with a practical example.

Logic Construction: YLL Processing

One way to view the YLL processing is to understand that its purpose is to build a structure and that the YLL program is a description of that structure. The structure is a set of nodes, one for each variable used in the YLL code. Associated with each node is a logic function. The structure is thus an interconnected set of logic functions. The variables of the YLL program are either inputs, intermediates, or outputs of the logic module being described. The inputs and outputs are variables passed into or out of the YLL subfunction, either as arguments or global variables. The intermediates are the local variables of the YLL subfunctions.

Unlike simulation languages, the order in which the lines of the YLL code appear, except for a few cases, is fairly arbitrary. The reason is that each line describes a piece of this structure and the sequence in which these pieces are described is not relevant. Thus a variable can be "used" before its value is computed.

Each variable is assigned a unique positive integer starting with 2. 0 and 1 are reserved for the logic constants. Associated with the integers

2 3 4 ... are the APL variables Δ2 Δ3 Δ4..., which will hold the logic functions of those variables, stored in sum-of-products form. Each time a new value (logic function) of a variable is computed, it is "or-ed" with the existing value, which is assumed to be 0 if there is currently no value. Since "or-ing" is commutative, the sequence in which these values are computed is of no relevance. Also if a variable *X* is used in another logic function, say *Y*, it does not matter that the value of the logic function for *X* has not been constructed yet. For example, in the statement $Y \leftarrow X \neq A$, the logic function $\underline{X}A + X\underline{A}$ is constructed, so the actual logic function for *X* is not relevant. Of course, in the final structure for the logic module, the logic function for *X* will affect the variable *Y*. For each of the positive integers, the user name for that variable is kept on a namelist, so these names may be passed to the rest of the compiler for other purposes.

We also have to process temporary variables, which arise as intermediate results. For example, in a line of YLL such as $Z \leftarrow (\vee/A)\vee.\wedge(B\vee C)$, the results $(B\vee C)$ and $(\vee/A)$ must be stored in temporary variables. Then the inner product is calculated, using these temporary logic functions, to obtain the logic functions for *Z*. After this line is completed, the temporary results may be erased. In this way, we ultimately create a structure that has exactly one node for each user-defined variable, and no other nodes. We use the convention of assigning temporary variables negative integers, starting with ¯1,¯2,... and storing the associated logic function in the APL variables Δ1, Δ2,....

Since YLL is embedded in APL, the processing is made easier. Processing a YLL function consists of two steps: translation and execution. Translation is a fairly straightforward process that alters the given YLL program and produces a similar-looking APL program. The translation is guided by the following table.

<u>L</u>XL1TAB

∨	*<u>L</u>OR*	⍲	*<u>L</u>NAND*
∧	*<u>L</u>AND*	⍱	*<u>L</u>NOR*
~	*<u>L</u>NOT*	⍎	*<u>L</u>EVALA*
>	*<u>L</u>GT*	∘	*<u>L</u>OUTER*
≥	*<u>L</u>GE*	⌹	*<u>L</u>PLA*
<	*<u>L</u>LT*	⊤	*<u>L</u>ENCODE*
≤	*<u>L</u>LE*	⊥	*<u>L</u>DECODE*
=	*<u>L</u>EQ*	⍟	*<u>L</u>PREV*
≠	*<u>L</u>NEQ*	⌶	*<u>L</u>RENAME*

Line 1 of the table gives the APL function <u>L</u>OR as a replacement for the APL operator ∨. <u>L</u>OR was written by the language developer and is a function that manipulates logic functions using the data structures defined. Thus any occurrences of ∨ not in quoted expressions are simply replaced by <u>L</u>OR in the translation process. Other symbols (e.g., ←, /, \, ., ∘, ⌿, ⍀, etc.) that have logic meaning are similarly translated, but since their syntax is more complicated, the translation process is embedded in a program. The YLL program FCNY that follows is an example of this translation process.

```
  ∇Z ← A FCNY B;C;N
[1]  ⍎'N ←⌈(ρA) ÷ 2'
[2]  G≜N
[3]  C← (2 × N) ↑ A ∧ B
[4]  G← C[2 × ιρG]
[5]  Z← (CNT) ⊂(C;G,G)
```

This YLL program is translated into the two APL programs that follow:

```
  ∇FCNA
[1]  ⍎' N←⌈ (ρ A)÷ 2'
[2]  SINK 'G' LALO N
[3]  SINK 'C' LSET (2× N) ↑ A LAND
[4]  SINK 'G' LSET C[2t×ιρ G]
[5]  SINK ' Z' LSET (CNT) LIMP ' (C;G,G)'
  ∇

    ∇FCNΔZ ← A FCN B;C;N;Z
[1]  ε Function names: FCNY
[2]  ε Global names set: G
[3]  ε Global names ref: CNT
[4]  ε Global names alloc:
[5]  LOCK
[6]  FCNA
[7]  UNLOCK
[8]  FCNΔ Z← LEVALA Z
  ∇
```

The first program, FCNA, is the direct translation of FCNY. Note that the lines of FCNY and FCNA are in a one-to-one correspondence. The differences are basically that certain logic symbols have been changed to the corresponding subfunction call. Thus

Symbol	*Replaced by*
∧	*LAND*
⊂	*LIMP*
←	*LSET*
≜	*LALO*

The subfunction SINK is placed at the beginning of each line, containing an assignment, to clean up the temporary variables that may have been created by that line.

The function FCN is the main call to FCNA, and its purpose is to take care of the input and output arguments, the local and global variables, and to do various cleanup chores. The call in FCN to FCNA is surrounded by LOCK and UNLOCK, which lock the current temporary variables required after FCN<u>A</u> is completed. Then they are unlocked so that they may be erased later. The final operation is to take the temporary result *Z* returned from FCN<u>A</u> and put it into the correct place, in this case into the dummy logic variable FCNΔZ. At the time FCN<u>A</u> returns its values, the variables *Z* have been assigned a permanent value, denoted by positive integers, but because *Z* is a dummy variable, these variables are now made temporary by the function <u>L</u>EVALA. Negative integers are returned by FCN, and the corresponding logic functions are stored in the APL variables, e.g., ≜ 1, ≜2, ...

After the translation process has been completed for all appropriate functions, the next phase is to execute the resulting APL program; in the previous example this is done by a call to FCN. If any arguments are to be passed, they must be allocated first; then the function call can be made. For example, in FCN, we would have to allocate the variables for the inputs *A* and *B*, and the global variable CNT. This can be done, for example, by another YLL program with lines

```
AA≜BB≜10
CNT≜1
ZZ← AA FCN BB
```

This causes the APL program FCN to execute, and the result is that the block of appropriate combinational logic is created.

When a variable is allocated in YLL, it is assigned an integer and its name is stored on the namelist. A variable may be allocated either the first time it is assigned a value (logic function) or by using an allocation (≜) statement. When the APL programs execute, the variables in the YLL programs become APL variables. These variables are arrays of integers, positive or negative, with the integers being the numbers allocated to the variable name or a temporary result. Thus the YLL variables may be reshaped, selected from, and generally manipulated like any APL array, since they are simply just arrays of integers.

It is also possible in a YLL program to refer to the value of a variable *X* using the symbol ⊗<u>X</u>, but the value obtained is the current logic function associated with *X*, so care must be taken here to describe what is meant.

If the variable *A* is allocated the numbers 33 34 35, then if during the execution of the APL programs we requested the APL value given to *A*, we would get 33 34 35. But if we request its logic value using an auxiliary function called SHOW, we would see the logic functions Δ33, Δ34, Δ35 displayed in symbolic form. For example, suppose that we define the program

```
∇MAINY
[1] D≜5
[2] E≜5
[3] CNT≜1
[4] Z← D FCN E
```

and then translate and run MAIN, stopping execution before the fourth line of *FCNA*:

```
MAIN
FCNA[1]  ⍎' N←⌈ (ρ A)÷ 2'
FCNA[2]  SINK ' G' LALO N
FCNA[3]  SINK ' C' LSET (2 × N) ↑ A L AND B
FCNA[4]  SINK ' G' LSET C[2×ιρ G]
```

Now we print both the APL value of *C* and its associated logic function using SHOW.

```
          C
16 17 18 19 20 21
    SHOW C
C[0]=D[0]∘E[0]
C[1]=D[1]∘E[1]
C[2]=D[2]∘E[2]
C[3]=D[3]∘E[3]
C[4]=D[4]∘E[4]
C[5]=Φ
```

The process is Similar for *G*.

```
         G
13 14 15
    SHOW G
G[0]=C[0]
G[1]=C[2]
G[2]=C[4]
```

Continuing execution, after MAIN has completed, we can display the result *Z* and its logic functions:

```
         Z
28 29 30 31 32 33
             SHOW Z
ZΔΔ 1[0]=CNT ∘ C[0]+CNT ∘ G[0]
ZΔΔ 1[1]=CNT ∘ C[1]+CNT ∘ G[1]
ZΔΔ 1[2]=CNT ∘ C[2]+CNT ∘ G[2]
```

Note that the name *Z* in MAIN has been renamed temporarily to *Z*ΔΔ1 because the dummy name *Z* used in FCNY conflicts with the name *Z* used it the subfunction MAINY, which will be an output of the combinational logic module created. The variable *Z* in FCNY is, in effect, a temporary variable and like any temporary variable will not appear.

Conditional clauses associated with ⊂ are processed using the notion of an environment. The environment is a logic function that is always "and-ed" with the logic function computed on the right side of an assignment (←) statement. The environment function is held in a special variable ENV.

As the environment changes, for example when another conditional clause is encountered (if conditional clauses are nested), the new condition is computed and "and-ed" with the current environment, the old environment is pushed onto a stack, and the new one is stored in ENV. When a condition ends, the stack is popped reestablishing the previous environment. Environment logic functions are stored in temporary variables ($\triangleq n$), and the associated negative integers are the ones actually pushed and popped from the stack. A special "locking" and "unlocking" mechanism is used to keep these temporary variables from being swept away when temporary variables are cleaned up in SINK.

The final phase is to write the results into a file, using the function WRITEL. We use logic description format, logic interchange format (LIF), but could also store the result in an internal data structure used in the logic synthesis programs. For LIF, two steps must be carried out. The first is to gather up each logic function and write it out in a PLA-type format. The second is to determine which variables are the inputs, outputs, and intermediate variables of this logic module. This is determined by which APL variables still exist in the workspace (global variables) and which have associated logic values. The global variables with associated logic functions are the outputs. The variables that exist, that are used by at least one logic function, and that have no logic value are the inputs. The logic functions that exist, that are used, and for which there are no APL variables with those integers as values are the intermediate variables. This information along with the users' names in the namelist is written to the output file.

Boolean Networks

Some logic synthesis programs are based on data structures that consist of a netlist of qualified gates. Local transformations are performed, mapping one set of such gates into another. The simple notion of a *Boolean network* provides more freedom in developing algorithms that are largely independent of the target technology. A Boolean network is a representation of a combinational module, that is, a block of logic implementing a set of logic functions. Eventually, a Boolean network will be implemented by a macro cell, that is, by an interconnection of cells each implementing a single logic function. Each logic function produces either a primary output of the module, an intermediate result, or both. These functions may depend upon the primary inputs to the module, upon the intermediates just mentioned, or upon both. In order to preserve the combinational nature and avoid race conditions, no cyclic dependencies are allowed. A primary input or an intermediate will be referred to simply as a *variable*. A Boolean network can be represented by a directed graph with a logic function associated with each node and an arc from node i to node j if function j depends on the variable associated with function i or input i.

The logic function at each node is represented as a sum-of-products Boolean expression. Internally, this expression is given as a matrix, the columns of which correspond to variables and the rows of which represent individual product terms. For example,

$$D = AB + B\underline{C}$$

$$F = D\underline{E} + \underline{D}E$$

A	B	C	D	E
1	1	2	2	2
2	1	0	2	2
2	2	2	1	0
2	2	2	0	1

Here we have used + to denote logical "or," juxtaposition to denote "and," and underscore to denote negation. Thus F is defined to be the exclusive or of D and E; note that D is actually an intermediate defined in terms of A, B, and C. In the matrix representation, variables occurring in their positive form are marked with a 1, those occurring in their negative form are marked with a 0, and a 2 indicates that the variable does not occur. A row defines one product term.

The function at each node is stored in the data base. The data structure used in the Yorktown silicon compiler is a single, large, bit matrix with a set of pointers to the rows associated with each function. Each variable is associated with two columns of the bit matrix, with 00 representing a 2; 01 representing a 1; and 10 representing a 0. As the synthesis proceeds, functions are retrieved form the data base, modified or eliminated, and rewritten back into the data base. Additional information about the function (e.g, its delay, suitability for implementation in a single gate, or if the function has changed recently) is sometimes stored in other arrays.

Input to the synthesis system is typically a two-level Boolean cover, an interconnection of two-level Boolean covers (LIF), or a Boolean network of functions produced by a logic description language such as YLL and stored in the data structure described previously. Thus the output of the logic synthesis can be an input as well.

A two-level Boolean cover can be interpreted as a Boolean network in at least two ways: (1) as a two-stage network where the product terms in the input plane produce intermediates, sums of which are then used to express the outputs; or (2) as a one-stage network in which each output is expressed as an independent sum of products. In the former case, the Boolean cover can be minimized by a PLA minimizer such as ESPRESSPO-II, which creates intermediates (in this case, product terms), that may be useful even for multilevel minimization. Thus for some logic, it may be suitable to use ESPRESSPO-II and direct its output into the multilevel logic synthesis program. In the latter case of the one-stage network, the creation of intermediate expressions to simplify the network is entirely the responsibility of the synthesis program.

Basic Logic Synthesis Operations

In the course of synthesis, a sequence of operators is applied to the Boolean network. There are five kinds of operations performed. They are of two different types—algebraic and Boolean—and different algorithmic variations are provided so that the synthesis process can be tailored to be faster (for estimation purposes) or slower (for more optimal results). The algebraic operations represent methods for increasing the speed of the operations by treating the logic functions as polynomials. Generally, a sum-of-products representation of a logic function can be viewed as a special type of polynomial—one where no power greater than 1 occurs and where the combination $X\underline{X}$ never occurs. Also, we normally manipulate only those sum-of-products expressions that are prime and are irredundant. In such cases, very fast manipulation techniques (e.g., for finding common subexpressions) can be found. The Boolean operations (e.g., logic minimization, complementation, etc.) allow the full power of changing a logic expression into a logically equivalent one, but require more time. A typical synthesis procedure is composed of a sequence of these operations, and the procedure is tuned to give a desired tradeoff between speed of the synthesis process and quality of its results. The five kinds of operations are:

1. *Extraction (algebraic)*. A subexpression common to two or more functions is extracted and used to create a new intermediate variable, which then replaces the subexpression wherever it occurs. The new intermediate can be used in both its positive and negative forms.
2. *Elimination (algebraic)*. An intermediate of low "value" is eliminated, and the subexpression it represents is "pushed back" into the functions that use it. This is the inverse of extraction.
3. *Simplification (Boolean)*. An unminimized logic function is replaced by a logically equivalent (for the module) but "simpler" expression. This transformation is essentially "static." Unlike the other two, it does not usually change the structure of the Boolean network, although arcs in the network may disappear or new arcs may appear if dependencies change.
4. *Resubstitution (algebraic and Boolean)*. An existing function is substituted or divided into another to see if the second can be usefully simplified. Substitution can be either of the algebraic (weak division) or Boolean type (strong division). Typically, every pair of functions is tried, but appropriate filters are used to quickly rule out some pairs.
5. *Decomposition (algebraic)*. After common subexpressions have been identified by extraction, some of the functions may still be too complex to be implemented in a single circuit of the target technology. Such functions must be further factored and decomposed. This decomposition is carried out locally because it is assumed that global commonality has already been identified by the extraction process.

After the decomposition step, an actual gate of the target technology is selected for each function in the network. For the complex cells used in the domino and static CMOS technologies, this gate is automatically designed with the aid of logic factorization algorithms.

EXTRACTION (ALGEBRAIC). The discovery of common subexpressions appearing in the Boolean network is the "creative" step in the automatic synthesis process. It is achieved by forming partial factorizations of each function and generating a list of *kernels*, which are essential subexpressions from which algebraic commonality between two or more functions can always be discovered.

To explain the notion of kernels, we must first introduce the idea of an *algebraic quotient* for a pair of Boolean expressions. A *logic expression* $f = f_1 + f_2 + \ldots + f_n$ is a particular sum-of-products representation of a Boolean function. Each f_i is a single product term or *cube*. The *algebraic product* of two expressions f and g is only defined when f and g depend upon disjoint sets of variables. It is given simply as the sum of all possible crossterms $f_i g_j$. Since f and g have disjoint variable sets, no zero products can occur (a variable is never multiplied by its complement). In this sense, the product is purely "algebraic."

Similarly, the algebraic quotient (f/g) of two expressions is required to depend on variables other than those on which g depends. It is defined to be the largest (most terms) expression such that

$$f = (f/g)g + r$$

where r is yet another expression (the remainder). Here the product between (f/g) and g is the algebraic product (which is defined), and the right and left sides of the equation are required to agree as expressions, not just logical functions. That is, the same set of cubes must appear on both sides of the equation. The quotient (f/g) is unique. For example,

$$f = AB + AC + AD + BC + BD$$

$$g = A + B$$

$$f/g = C + D$$

$$f = (f/g)g + r = (A+B)(C+D) + AB$$

By using sorting techniques, the computation of the quotient f/g can be carried out very efficiently. In fact, the division requires only $O(n \log n)$ steps, where n is the total number of product terms in f and g. Further, by keeping the product terms of all expressions sorted, which is possible to do, then algebraic division is linear in n.

Clearly the formation of quotients is closely related to the problem of factoring an expression. Similarly, the task of identifying subexpressions common to two or more functions is essentially the problem of finding common divisors. Here g is a divisor of f if f/g is nontrivial. Notice that if g is a divisor of f, then so is $(f/g.)$ The motivation for defining the kernels of an expression is to specify a manageable set of divisors that is still rich enough to allow all common subexpressions to be located.

An expression is said to be *cube-free* if no cube can be factored out of it. Thus $ABC + ABD$ is not cube-free (it equals $AB(C + D)$), but $C + D$ alone is cube-free. The set of *kernels* $K(f)$ of an expression f is defined to be the set of all quotients f/c such that c is a cube and f/c is cube-free. Note that a kernel must have at least two product terms, since a single cube is not cube-free.

The number of kernels in an expression is typically much smaller than the number of possible cubes c that can appear in the definition. Indeed, f/c is the empty expression for many choices of c, and for many additional choices it is not cube-free.

Since we represent our Boolean expressions in sum-of-products form, it is fairly easy to identify common divisors that are single cubes. It is much harder to locate cube-free common divisors, and this is the purpose of kernels, as illustrated by the following result [Bray84a].

> If two expressions f and g have a cube-free common divisor, then they have a cube-free divisor which is the sum of cubes common to two expressions h and k, where $h \in K(f)$ and $k \in K(g)$.

Thus to find a subexpression common to two or more functions in a Boolean network, we compute the set of all kernels and then build expressions out of cubes common to pairs of kernels. If no common subexpressions are found in this way, then the preceeding result guarantees that the functions in the network are "relatively prime"—that is, they have no cube-free common divisors. Consequently, the only remaining common subexpressions are single cubes; these can be located in a straightforward way.

For additional efficiency, we can limit the calculation to "depth zero" kernels (see [Bray84] for more details), a somewhat smaller set that still suffices to find useful new intermediates.

Once the candidate intermediates have been identified, they are given ratings, roughly measuring the number of literals they would save in a factored representation of the Boolean network. If this rating exceeds a critical value, the intermediate is created and the subexpression it represents is replaced by a single new variable wherever it occurs. More precisely, to create an intermediate variable v to represent the expression g, we first add the equation $v = g$ to our Boolean network. Then we replace each expression f already in the network by the new expression

$$f = v(f/g) + \bar{v}(f/\bar{g}) + r$$

(Of course, if both g and $\bar{g}$ do not divide f, then f is left unchanged.) We refer to this process as *Algebraic substitution* of v for g in f. (True Boolean substitution will be discussed in the section on simplification that follows.)

The new intermediate consolidates logic that would otherwise be duplicated in many functions. The rating just mentioned usually reflects a savings in transistors in a physical realization of the network, independent of the circuit technology used. For example, consider the three Boolean expressions f, g, and h given by

$$\begin{aligned} f &= AE + BE + CDE \\ g &= AD + AE + BD + BE + BF \\ h &= ABC \end{aligned} \tag{1}$$

We compute the set of kernels for each function:

$$K(f) = \{A + B + CD\}$$
$$K(g) = \{A + B, D + E, D + E + F\}$$
$$K(h) = \phi$$

Observing that the expression $A + B$ consists of two cubes common to the kernel $A + B + CD$ of f and the kernel $A + B$ of g, we create a new intermediate variable X to represent this expression, and replace $A + B$ with X wherever it occurs. The resulting Boolean network is

$$\begin{aligned} X &= A + B \\ f &= EX + CDE \\ g &= DX + EX + BF \\ h &= C\underline{X} \end{aligned} \tag{2}$$

Notice that the negation of X has been used to simplify the expression for h.

ELIMINATION (ALGEBRAIC) Elimination is the inverse of extraction. If we eliminate (push back) the intermediate X in the Boolean network (2) above, we obtain the original network (1).

This process is not carried out indiscriminately. Rather, we begin by assigning a value to each intermediate in the Boolean network. This value, like the rating used during extraction, measures the degree to which an intermediate helps to simplify the Boolean network. We then eliminate only those intermediates whose values fall below a certain critical level. This critical level mediates between the otherwise antagonistic extraction and elimination processes.

Elimination tends to reduce the total number of circuits and delay stages in the design. It is also useful in converting a large, sparse network of simple functions into a small, denser network of somewhat larger functions. A sparse network is sometimes produced by the extraction operation; it also might be given as input to the synthesis procedure—certain logic description languages produce such networks. Elimination also enhances the potential for simplification. Thus it is not unusual for elimination to be invoked several times in the course of synthesis.

SIMPLIFICATION (BOOLEAN) A single Boolean function corresponds to many equivalent Boolean expressions. It is obviously desirable to keep all expressions occurring in a Boolean network in as simplified a form as possible. The incoming logical data must be minimized by the synthesis program, and further simplification may be possible after an elimination operation. It is slightly less obvious that an expression involving intermediate variables contains implicit relations that also can be used in its simplification. For example, if it is given that

$$X = A\underline{B} + \underline{BC}$$

$$Y = \underline{A}D$$

then

$$Z = X + \underline{C}Y$$

simplifies to

$$Z = X + Y$$

This simplification results because the condition $\underline{X}CY$ can never occur; it is part of the don't-care set for this network; thus $Z = X + \underline{XC}Y + \underline{XC}Y = X + Y$.

In general, to any Boolean network we can associate a list of variable combinations that are logically impossible; the totality of all such conditions is the don't-care set for the network. The don't-care set includes, for each intermediate variable v, the condition

$$\bar{v}f + \bar{f}v$$

where f is the expression represented by v. In the preceding example the don't-care set is given by the expression

$$\underline{X}(A\underline{B} + BC) + X(\underline{AB} + B\underline{C}) + \underline{Y}(AD) + Y(\underline{A} + \underline{D})$$

The reader can verify that the cube $\underline{X}CY$ is contained in this don't-care set as claimed. These conditions describe implicit relations between the variables in the network, and the exploitation of these relations is essential to the optimization of multilevel logic.

The simplification routines we employ were abstracted from the ESPRESSO-II minimizer, described in [Bray84b]. These algorithms are based upon the "unate recursive paradigm," a divide-and-conquer strategy that has proven quite efficient. Simplification without a don't-care set can be carried out by a very efficient procedure (called SIMPLIFY in [Bray84b]), which does not guarantee that an irredundant cover of prime implicants will be produced, although this is often the case. If a don't-care set is available, there are two methods for producing an irredundant cover. One is based on tautology, and one is based on complementation of the function and its don't-care set. We have implemented the latter method since it is the general paradigm of the ESPRESSO algorithms, but if the don't-care set is extremely large, a tautology-based method is probably preferable. We use the ESPRESSO-based simplification operations to obtain the simplified result if don't cares are available.

In practice, the don't-care set for the entire network is too large to process at once, so when minimizing a particular function, we only compute the relations between the variables it directly depends upon. Note that elimination tends to bring distant dependencies closer together, so if done before simplification, there is increased potential for simplification.

The idea of what constitutes a "simple" function for multilevel logic must be discussed. It is not measured simply by the minimum number of product terms in the expression. In general, we would like to obtain an expression that factors simply. Since this is difficult to control with the two-level simplification techniques, we have employed a "minimum variable" or "minimum literal" heuristic to obtain "simpler" expressions.

For example, suppose we want to simplify

$$g = AB + CX$$

where

$$X = \underline{A} + B$$

We merely simplify f relative to the don't-care set

$$d = \underline{X}(\underline{A} + B) + X(A\underline{B})$$

associated to the intermediate variable X.

As a heuristic, we wish to reexpress f in terms of the least number of variables A, B, C, and X required. (A similar heuristic is defined as follows to determine the minimum number of "literals" A, $\underline{A}$, B, $\underline{B}$, C, . . ., required to express f.) This is done by computing the onset $f = g - d$ and offset $r = \sim g - d$; thus defining an incompletely specified function (f, d, r). In this example

$$\textit{on set} = f = ABX + \underline{A}CX$$
$$\textit{don't-care set} = d = \underline{X}(\underline{A} + B) + X(A\underline{B})$$
$$\textit{off set} = r = \underline{AC}X + A\underline{BX}$$

Each of the cubes of f do not intersect with the cubes of r. This means that for each pair of cubes, there must be at least one variable in "conflict." Thus ABX is orthogonal to $\underline{AC}X$, because A conflicts with $\underline{A}$. . We then build a "variable-blocking matrix" M, where each row represents a pair of cubes, one from f and one from r and where we put a 1 in the columns that represent the conflicting variables. In the example,

Variables

	A	B	C	X	*f-cube*	*r-cube*
	1	0	0	0	ABX	$\underline{AC}X$
$M=$	0	1	0	1	ABX	$A\underline{BX}$
	0	0	1	0	$\underline{A}CX$	$\underline{AC}X$
	1	0	0	1	$\underline{A}CX$	$A\underline{BX}$

It is easily shown that a literal may be removed from a cube of f if there remains at least one 1 in each row of the blocking matrix. For example, B can be removed from ABX. If an entire column of the variable-blocking matrix can be removed while maintaining at least one 1 in each row, then each occurrence of that variable may be deleted from the cubes of f. A minimum variable representation can be found by determining the maximum number of columns that can be removed. This is solved by a minimum "row covering" algorithm—that is, by selecting a subset of columns so that there is at least one 1 in each row of M. In the example, one determines by inspection that two minimum covers exist, corresponding to the variables A, C, and X or to variables A, B, and C.

If we now restrict f to the last subset of variables, we obtain $g = AB + \underline{A}C$ (the superfluous variable X has simply been discarded). At this point, normal simplification proceeds, but each cube is prime and irredundant, so the final result is

$$f = AB + \underline{A}C$$

Thus the minimum variable heuristic selects that X should be eliminated before any of the cubes of f are expanded to prime.

As an alternative, a minimum literal heuristic can be obtained by building, in a similar way, a "literal-blocking matrix," with a column for each literal appearing in the cubes of f. Thus

	Literals						
	A	B	C	X	$\underline{A}$	*f-cube*	*r-cube*
	1	0	0	0	0	ABX	$\underline{A}\underline{C}X$
$M=$	0	1	0	1	0	ABX	$A\underline{B}\underline{X}$
	0	0	1	0	0	$\underline{A}CX$	$A\underline{C}X$
	0	0	0	1	1	$\underline{A}CX$	$A\underline{B}\underline{X}$

We could eliminate the sets of literals $\{B, \underline{A}\}$ or $\{X\}$, obtaining either $f = AX + CX$ or $f = AB + \underline{A}C$, respectively. However, the first solution is the minimum literal solution, and it is the one found by the minimum covering algorithm.

This example illustrates how the minimum literal heuristic can give a preferred answer in terms of multilevel synthesis, since $f = X(A + C)$ is preferred over $f = AB + \underline{A}C$ (less literals in the factored form of f).

RESUBSTITUTION (ALGEBRAIC AND BOOLEAN) Resubstitution is the operation of attempting to use a function that already exists in order to simplify another existing function. There are two versions of resubstitution, algebraic and Boolean. In algebraic resubstitution, we select any pair of existing functions f and $v = g$, and then form $(f/g)v + (f/\overline{g})\overline{v} + r$ and unless $f = r$, we change f. Since there may be many pairs that have to be tested, it is expedient to eliminate some pairs by "filters"; for example, if the number of terms in g is greater than the number of terms in f, or if g contains a variable not in f, then the pair (f, g) is not tested. The set of algebraic filters eliminates most pairs, so algebraic resubstitution is extremely efficient. In theory, extraction will find any potential algebraic resubstitutions. However, it is less efficient since it is more general than resubstitution; therefore, we employ algebraic resubstitution quite liberally in the synthesis process.

Boolean resubstitution can be more expensive than algebraic and must be used with care. Don't-care simplification, already discussed, can be used to define true Boolean substitution (also called strong division). In general, if we wish to substitute a new variable X with associated logic function g into a given logic function f, we write down a don't-care set,

$$d=\underline{X}g + X(\sim g)$$

For example, if we wish to determine if the new intermediate

$$X=\underline{A}+B$$

can be usefully substituted into the existing expression

$$f=AB + \underline{A}C + BC$$

we merely simplify f relative to the don't-care set

$$d=\underline{X}(\underline{A}+B) + X(A\underline{B})$$

associated to the intermediate X.

The problem is the same as don't-care simplification—that is, to simplify f using the don't-care set d. We employ the extended notion of "simple" (see the previous section on simplification) using the minimum literal heuristic. Also, since we wish to substitute the intermediate X into the expression f, we require that X be included in any minimum literal subset so computed. In this example, we obtain $f = X(A + C)$.

Boolean substitution can be a powerful tool for multilevel logic synthesis. The following examples of its effectiveness occur in part of a 4-bit ALU description.

$$L=(B\underline{F}+\underline{B}F)(A+E)+\underline{AE}(\underline{BF}+BF)$$

substituted into

$$R=(B\underline{F} + \underline{B}F)(\underline{A} + \underline{E} + AE(\underline{BF} + BF)$$

yields

$$R=A(\underline{E}L + E\underline{L}) + \underline{A}(\underline{EL} + EL)$$

Here both the positive and negative polarities of L have proven useful in the simplification of the expression for R. On the other hand,

$$R=(BF + \underline{BF})(\underline{A}+\underline{E}) + AE(\underline{B}F + B\underline{F})$$

substituted into

$$L=(BF + \underline{BF})(A+E) + \underline{AE}(\underline{B}F + B\underline{F})$$

yields

$$L=A\ (E\underline{R} + \underline{E}R) + \underline{A}(\underline{ER} + ER)$$

This shows that R can be useful in expressing L. Remarkably,

$$T=(DH+\underline{DH})\ ((\underline{G}+\underline{C})\ ((\underline{E}+\underline{A})\ (\underline{B}+\underline{F}) + \underline{BF}) + \underline{CG})$$
$$+(D\underline{H} + \underline{D}H)\ ((G+C)\ (AE\ (B+F) + BF) + CG)$$

substituted into

$$S=(H+D)\ ((G+C)\ (AE(B + F) + BF) + CG) + DH$$

yields

$$S=D(H + T) + HT$$

This demonstrates that strong division can sometimes discover unexpected and dramatic simplifications.

Boolean substitution is more powerful that algebraic substitution, but it is also more expensive. In practice, similar results are often obtained more efficiently by algebraic substitution followed by simplification, and this is the technique we normally use. The new intermediate is first incorporated into those existing expressions for which it is an algebraic divisor, and then the more expensive Boolean simplification is carried out only for these expressions. In the model example discussed earlier, the algebraic substitution of X into f yields the expression $AB + XC$. When this is simplified relative to the don't-care set for X, the expression $X(A + C)$ is again obtained.

DECOMPOSITION (ALGEBRAIC) After extraction, it may still be necessary to decompose large functions into smaller pieces. Different technologies demand different decomposition styles; the common problem is to locate within a complicated expression, a large subexpression that can be implemented as a single circuit. Since the task of identifying global commonality has already been carried out, this additional decomposition can be applied to each function independently, without loss of efficiency in the final design.

Decomposition and factoring are basically the same thing. In factoring an algebraic expression, we select a divisor, g, and write the expression as $f = (f/g)g + r$ (where (f/g) is the algebraic quotient) and then recursively factor the three expressions (f/g), g, and r. Decomposition can be done in the same way, except that after selecting g we change f to $(f/g)v + (f/\bar{g})\bar{v} + r$ and store $v = g$ in the data base. The f and g are recursively decomposed further. This process stops if (1) f or g cannot be factored or (2) f or g meet some stopping criteria.

At this point, there may still be some functions in the data base that do not meet the technology specifications. Such functions are broken down further by (1) splitting a sum of products into two smaller sums or (2) breaking down single product terms, which do not pass the criteria, into the product of two smaller product terms.

After this, all functions will pass the technology criteria that assure that each can be built in a single gate of the target technology. Then, usually, a switch is set indicating that no expression should be subsequently changed into one that does not meet the criteria. The test for passing the technology criteria is coded in a single subroutine. This subroutine depends on the target circuitry used and could be simply a function that receives a logic expression as input and uses table lookup to determine if the expression is acceptable. For complex cell families, such as domino logic, there would be too many entries in the table, so we have used parameterized criteria (e.g., the maximum number of literals in any cube being less than a given value, typically 5). For complex static CMOS, the criteria might be no more than 5 literals in any cube of the function, and no more than 3 literals in any cube of the complement function.

All the operators previously described have been implemented such that when the switch is set, a test is made to see if any function has changed from one that was acceptable to one that is not. If so, the change is rejected and the operator attempts either to find a change that is acceptable or gives up and goes on to the next function. Thus in effect, each operator has two versions, one that knows nothing about technology and one that knows about technology but only through the simple device of using the single criteria subroutine to test if a change in an expression is acceptable. This allows for the creation of technology-independent algorithms for logic synthesis and a set of programs that can be easily altered to switch from one set of target circuits to another. We have implemented this with a set of overlays to the logic synthesis programs so that the user may select from a menu of different circuit technologies.

A Typical Set of Steps in Synthesis

The general philosophy used in the synthesis sequence is to collapse the logic into a small but highly useful set of logic functions, apply simplification using as much of a local don't-care set as can be allowed, and use extraction to decompose the network further. Thus, when simplification is applied, a more global result can be obtained than is possible by a set of local transformations. Further, the generality of logic simplification, in effect, defines all possible local transformations and is applicable to any type of target technology. After each function is simplified, its resulting sum-of-products form is most always in a canonical form; hence algebraic extraction will identify most of the common subexpressions.

The operations described previously are linked together in a sequence to form a complete synthesis procedure. A typical sequence of operations is as follows:

1. Apply fast simplification (SIMPLIFY) to the incoming data.
2. Apply algebraic resubstitution for maximum sharing of the intermediate or output functions.
3. Apply elimination. As usual, those intermediates whose values exceed a given critical amount are not pushed back. By setting this cutoff higher or lower, we can control the degree to which the original decomposition is preserved.
4. Perform Boolean simplification, using the implicit don't-care sets.
5. Extract common subexpressions. Even subexpressions with fairly low value should be extracted, since they help to disclose other subexpressions.
6. Eliminate again. Any intermediates of little value are now removed.
7. Simplify using don't-care sets again.
8. Decompose the logic into functions simple enough to be implemented in single circuits in the target technology. Set the technology switch.
9. Eliminate one final time. At this point, because the technology switch is set, we do not push back any expression if its removal creates a function that can no longer be realized as a single circuit. At the same time, attempt to reduce the number of circuits and delay stages by trying to push back all but the most valuable intermediates.
10. Select or design the circuits for each function in the network (using factorization for domino or complex static CMOS technologies).

Practical Examples

An APL implementation of the logic synthesis system just described has been in use at IBM T. J. Watson Research Center since 1982. Boolean networks with as many as 1,000 functions and PLAs with 30 inputs and outputs and over 2,000 product terms can be treated with reasonable expenditure of resources. A Boolean network that has not been minimized can be synthesized into a circuit family such as domino logic in roughly one to three times the amount of CPU time required to produce a minimized PLA for the same logic. When the input to the synthesis procedure is known to be a minimized PLA, the processing is faster because some Boolean operations can be suppressed.

Various stages in the synthesis of a small piece of logic extracted from an IBM System 38 CPU will be illustrated here. The incoming data represent six output functions in the form of a single PLA; there are no intermediates initially. The PLA depends upon 27 inputs, which are labeled A through Z, and A'. Intermediates introduced during synthesis will be labeled B', C', and so forth; the output signals are labeled 1 through 6.

This example illustrates the functioning of extraction, elimination, and a form of decomposition. The original PLA has already been minimized (as a PLA), and subsequent simplification has no effect.

The original expressions for the six outputs are shown below. For brevity and ease of comprehension, we show a factored form of each of these expressions, although internally each is represented as a sum of products. The factorization algorithm is itself based on kernel generation, so it is not surprising that the factored forms already reveal a good deal of the structure present in the network.

Input Data

$$
\begin{aligned}
1 = {} & (\underline{HI} + \underline{G}HI)(F(\underline{B}(C\underline{D} + E) + A) + N(B(\underline{C}D + \underline{E}) + \underline{A})) \\
& + (\underline{E} + \underline{C}D)B(GP(H + I) + \underline{GKL}O)) + (E + C\underline{D})\underline{B}(GM(H + I) + \underline{GJKL}) \\
& + (I + H)G(\underline{A}P + AM) + \underline{GKL}(\underline{A}O + AJ) \\
2 = {} & (\underline{HI} + \underline{G}HI)(F(\underline{S}(\underline{U}(T + V\underline{W}) + R) + Q) + N(S(U(\underline{T} + \underline{V}W) \\
& + \underline{R}) + \underline{Q})) + (I + H)G(M(\underline{S}(\underline{U}(T + V\underline{W}) + R) + Q) \\
& + P(S(U(\underline{T} + \underline{V}W) + \underline{R}) + \underline{Q})) + \underline{GKL}(J(\underline{S}(\underline{U}(T + V\underline{W}) + R) \\
& + Q) + O(S(U(\underline{T} + \underline{V}W) + \underline{R}) + \underline{Q})) \\
3 = {} & (ABCDENQX + \underline{ABCDE}F\underline{QX})(\underline{G}HI + \underline{HI}) + (ABCDEPQX \\
& + \underline{ABCDE}M\underline{QX})G(H + I) + \underline{GKL}(\underline{ABCDE}J\underline{QX} + ABCDEOQX) \\
4 = {} & ABCDEX(GP(H + I) + \underline{G}(\underline{KL}O + HIN) + \underline{HI}N) \\
5 = {} & \underline{ABCDEX}(F(\underline{HI} + \underline{G}HI) + GM(H + I) + \underline{GJKL}) \\
6 = {} & (\underline{Z} + \underline{Y})((ABCDENQRSTUVWXA' + \underline{ABCDE}F\underline{QRSTUVWXA'}) \\
& (\underline{G}HI + \underline{HI}) + (ABCDEPQRSTUVWXA' \\
& + \underline{ABCDE}M\underline{QRSTUVWXA'})G(H + I) + \underline{GKL}(\underline{ABCDE}J\underline{QRSTUVWXA'} \\
& + ABCDEOQRSTUVWXA'))
\end{aligned}
$$

Next we illustrate the results of extraction. Five new intermediate variables have been introduced. In addition, the program has discovered that primary outputs 4 and 5 can also be used to simplify some of the other functions. This could have been discovered by resubstitution as well.

Result of Extraction

$$1 = D'(\underline{B}(C\underline{D} + E) + A) + E'(B(\underline{C}D + \underline{E}) + \underline{A})$$

$$2 = D'(\underline{S}(\underline{U}(T + V\underline{W}) + R) + Q) + E'(S(U(\underline{T} + \underline{V}W) + \underline{R}) + \underline{Q})$$

$$3 = QF' + \underline{Q}G'$$

$$6 = (QRSTUVWA'F' + \underline{QRSTUVWA}'G')(\underline{Y} + \underline{Z})$$

$$B' = I + H$$

$$C' = \underline{B}' + \underline{G}HI$$

$$D' = J\underline{H}' + FC' + GMB'$$

$$E' = O\underline{H}' + NC' + GPB'$$

$$4 = F' = ABCDEXE'$$

$$5 = G' = \underline{ABCDEX}D'$$

$$\underline{H}' = G + L + K$$

The elimination process reduces the seven intermediates above to four somewhat larger ones, shown below. The degree to which the complexity of the original network has been reduced is evident when compared to the input data.

Result of Elimination

$$1 = D'(\underline{B}(C\underline{D} + E) + A) + E'(B(\underline{C}D + E) + \underline{A})$$

$$2 = D'(\underline{S}(\underline{U}(T + V\underline{W}) + R) + Q) + E'(S(U(\underline{T} + \underline{V}W) + \underline{R}) + \underline{Q})$$

$$3 = QF' + \underline{Q}G'$$

$$6 = (QRSTUVWA'F' + \underline{QRSTUVWA}'G')(\underline{Y} + \underline{Z})$$

$$D' = F(\underline{HI} + \underline{G}HI) + GM(I + H) + \underline{G}J\underline{KL}$$

$$E' = GP(I + H) + \underline{G}(HIN + \underline{KL}O) + \underline{HI}N$$

$$4 = F' = ABCDEXE'$$

$$5 = G' = \underline{ABCDEX}D'$$

This network is now decomposed, with domino logic playing the role of the target technology. In this circuit family, the main constraint on feasibility of a given logic expression is the length of its longest product term. The decomposition operation for this family discovers that if the primary outputs of the network are all complemented, then every expression becomes feasible. (Complementation is allowed in this case because the outputs are feeding a latch.) The new network, with the intermediates relabeled, follows.

Result of Decomposition

$$^{-}1 = B'(A(E(C + \underline{D}) + \underline{B}) + C') + \underline{A}C'(\underline{E}(\underline{C} + D) + B)$$

$$^{-}2 = B'(Q(R(T(V + \underline{W}) + \underline{U}) + \underline{S}) + C') + \underline{Q}C'(\underline{R}(\underline{T}(\underline{V} + W) + U) + S)$$

$$^{-}3 = QD' + \underline{Q}E'$$

$$^{-}6 = Q(\underline{R} + \underline{S} + \underline{T} + \underline{U} + \underline{V} + \underline{W} + \underline{A}') + \underline{Q}(R + S + T + U$$

$$
\begin{aligned}
&+ V + W + A') + Q(R + S + T + U + V + W + A') + F' + YZ \\
B' &= (H\underline{I} + \underline{H}I + \underline{F})\underline{G}(\underline{J} + K + L) + G(\underline{M}(H + I) + \underline{FHI}) \\
C' &= (\underline{N} + H\underline{I} + \underline{H}I)\underline{G}(K + L + \underline{O}) + G(\underline{P}(H + I) + \underline{HIN}) \\
{}^{-}4 = D' &= \underline{A} + \underline{B} + \underline{C} + \underline{D} + \underline{E} + \underline{X} + C' \\
{}^{-}5 = E' &= A + B + C + D + E + X + D'
\end{aligned}
$$

Finally, series-parallel circuits are constructed for each of the preceding functions. This operation, which is essentially the same as factorization, is also performed by the synthesis system. The series-parallel graph for the intermediate C' follows.

Circuit for $\underline{C}'$

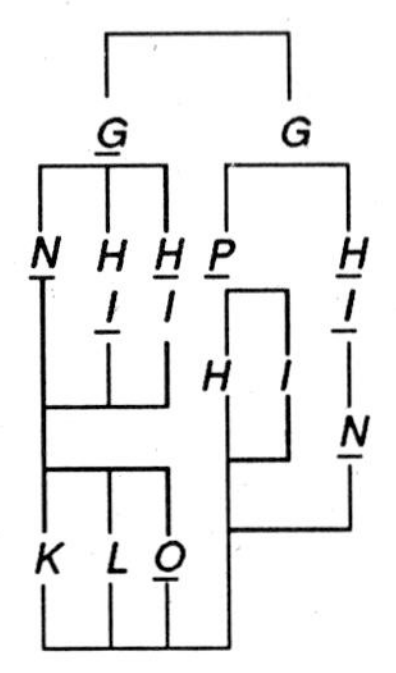

The Boolean network, together with the topological circuit descriptions, is now available for further processing by the layout program.

7.2.4 Layout Design

Introduction

The dominating guideline in the layout design part of the YSC system is to postpone implementation decisions in order to avoid committing the program prematurely to a specific implementation. Each decision should leave enough freedom for the remaining stages to satisfy the constraints it created and at the same time should rearrange the available data such that further meaningful decisions are possible in the next step. So, concurrent with the gradual stiffening of the design, the information is progressively organized so that more detailed decisions can be derived. Postponing implementation decisions is the characteristic of what has been called *stepwise refinement*. The usual perception of the principle is as a progressive decomposition of a task into subtasks in a sequence of refinement steps. This sequence of refinement steps terminates when all undecomposed tasks are straightforward translation routines. This method thus entails a hierarchical structure.

The principles of stepwise refinement obviously apply to any complex design task based on a top-down strategy rather than to a process of combining independently developed subdesigns. In general, completely specified subdesigns are difficult to handle because the flexibility and the information for adapting them to their environment are often not available when they are designed. Therefore, the concepts of stepwise refinement and a hierarchical structure are particularly compatible with the environment of automatic layout design in a silicon compiler.

The Environment of Layout Design

THE INITIAL HIERARCHY The input to the Yorktown silicon compiler and the structural synthesis in the early stages have imparted hierarchical structure to the design data. Each module in this hierarchy contains a possibly empty collection of submodules with an incidence structure (a net list). If the collection of submodules is empty, the module is a cell. In all other cases, the module is a compound. The hierarchy can be represented as a rooted tree. The modules are represented by the nodes. The root node represents the whole system. The leaves represent the cells. The internal nodes represent the compounds.

The hierarchy probably reflects some high-level viewpoints such as the functional interdependence of the modules. Functional interdependence and connectivity are often highly correlated, and the latter is an important basis for decisions in layout design. The functional hierarchy, therefore, represents useful data for the layout design part. In general, however, the considerations that lead to a functional hierarchy ignore important aspects of layout design. Logic decomposition and data path definition are just two examples of decisions that have a significant influence on the final layout. Many decisions that are functionally almost equivalent may have completely different consequences for the layout and its design process. Therefore, the layout design part may choose to ignore parts of the decomposition, initially or throughout. Nevertheless, it is assumed that after some clustering around seeds and some pruning the design is completely hierarchically structured. That structure is considered part of the initial data for layout design.

No layout design system is complete without algorithms to generate the geometrical specification of each cell (these algorithms are called *cell assemblers*), and algorithms for combining geometrically specified modules into supermodules. However, isolating the cell assembly tasks from the combination task is dangerous because of their mutual dependence. Taking this dependence into account by iterations over several design tasks is undesirable because of the time complexities involved and the convergence problems. In the YSC system, we have chosen an approach in which cell assembly is preceded by algorithms that in a top-down sequence create preliminary environments for submodules, so that part of the input for the cell assembler is the preliminary environment of the cell to be generated.

Cells can be of various types. The type determines the cell's flexibility (used in the creation of preliminary environments) and the cell assembler to be used for obtaining the complete layout. The most rigid cell type is an *inset* cell. Its layout and pin positions are fixed and stored in a library or are completely implied a priori by the algorithm generating the cell. The layout design system can only assign a location and an orientation to such a cell. Other cells have pointers to algorithms that can produce their layouts. These cells may have much more flexibility and therefore can be more easily adapted to a given environment. The most flexible kind of this type of cell is a *macro*. These cells have a decomposition of their own into circuits of a particular family. They are the ones used primarily in the Yorktown silicon compiler for implementing combinational logic, not only because high degrees of flexibility are desirable in the context of layout design in a silicon compiler, but also because with the appropriate choice of family they usually outperform other implementation styles. In addition, there are cells generated by algorithms, depending on a few parameters, that construct special-purpose subcircuits such as ALUs, rotators, and adders. These cells have limited flexibility (e.g., they stretch in one direction).

THE RECTANGLE CONSTRAINT The ultimate task of a layout design system is to produce a *layout*–that is, a set of masks that completely specifies the geometry of the circuit. For present-day technologies, the geometrical specification of eight to fifteen masks suffices. Lithography techniques quite often require orthogonal artwork. This leads to masks that are unions of perpendicular rectangles. Rarely is a restriction to rectangles and combinations thereof detrimental, whereas the cell design algorithms and layout data bases profit from such a restraint. The rectangle is therefore the basic construct in the YSC layout system.

The rectangle constraint is also imposed for the compounds and cells of the hierarchy. Consequently, each hierarchy will be a rectangle dissection in the final layout, that is, a rectangle subdivided into nonoverlapping rectangles. The restriction to rectangles might seem rather arbitrary; however, a constraint on the shape of a region is of great value in determining the shape and positions of its parts. If these parts have a high degree of flexibility, then the restriction to rectangles is of little consequence since the parts can be fit into the environment by using their flexibility. On the other hand, choosing rectangles as the only constructs simplifies the formulation and derivation of the design decisions.

It is also expedient that the results of the layout design process are stored in a way that is compatible with the data representation delivered by previous design procedures. The initial data available in the layout design stage is hierarchically structured. We use a layout design procedure that preserves that structure, although refined and ordered. *Refinement* here means that a module with all its submodules can be replaced by a tree with the same root and the same leaves, but with a number of additional internal nodes.

DESIGN RULES To improve chances for successful integration of the circuit and to increase yield when the circuit goes into production, patterns are required to satisfy certain rules, the so-called *design rules*. A first classification distinguishes roughly two classes: (1) *numeric rules* that quantify spacings between patterns in a plane or in combinations of planes and (2) *structural rules* that enforce and prohibit certain combinations.

There usually are a large number of numeric rules. Very few, however, are critical to the layout algorithms. The rules are formulated as minimum rules only, because it is assumed that the layout design techniques will try to keep the total chip small. Good algorithms working with these rules produce valid layouts in a wide range of values. Of course, the algorithms do not produce optimal layouts for all combinations of values in these rules, but they should produce acceptable solutions for all practical sets of values. In the YSC system, the numerical rules are stored as a vector. Specifying the technology causes values to be assigned to the components of this vector. The vector is accessible by all procedures in the layout design stage.

The structural rules are much more difficult because changes in these often require completely different decisions. These rules usually increase the dependence between different masks. This is particularly so if the metal layers are involved. Rules that forbid or enforce certain overlaps between patterns in the metal layer masks and other masks affect the wiring routines, because these routines are often based on generic algorithms solving some cleverly isolated interconnection problem. Introducing structural constraints often invalidates the assumptions made in the design of the algorithm. In the YSC system, the generic algorithm is usually followed by a mask generator in which the implications of the structural rules are incorporated. We hope that changes in structural rules will only affect the mask generator, which must be adapted in that case.

Floor Plan Design

SHAPE CONSTRAINTS For every cell in the hierarchy there is an algorithm that tries to adapt the cell to its estimated environment while generating its detailed layout. This preliminary environment must be created on the basis of estimates of the area needed by each cell, feasible (rectangular) shapes for it, and the external interconnections. The size and the shape of a cell are constrained by the amount and type of circuitry that must be accommodated in that cell. It is reasonable to expect one dimension of the enclosing rectangle not to increase if the other dimension is allowed to increase. Constraints satisfying that requirement are called *shape constraints*. Examples of shape constraints are shown in Figure 7.12.

Inset cells have piecewise linear shape constraints. Such constraints can be conveniently represented by a sequential list of their breakpoints. This

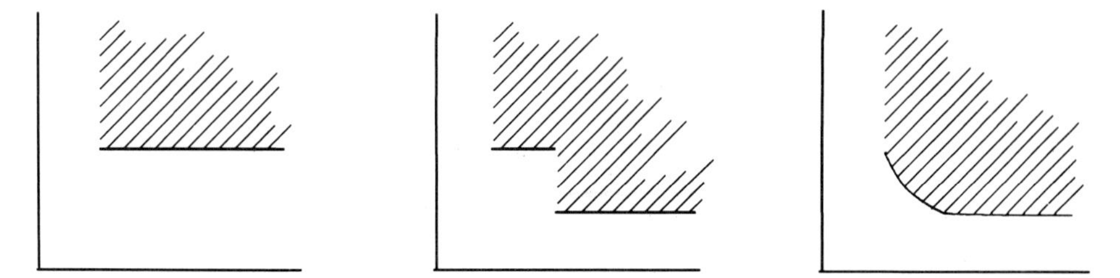

Figure 7.12 Examples of shape constraints. The set of feasible dimension pairs is indicated by shading. The examples are (a) an inset cell with a fixed orientation, (b) an inset cell with free orientation, and (c) a cell with a minimum area and minimum dimensions.

is not the case for flexible cells, and possibly other cell types occurring in practice. Of course, any shape constraint can be approximated by a piecewise linear bounding function with arbitrary accuracy. Usually, a piecewise linear approximation with three breakpoints will suffice, considering the limited accuracy of most area estimations.

Useful shape constraints of compound modules in the functional hierarchy can be derived in an obvious way from the shape constraints of its cells.

PLACEMENT VERSUS FLOOR PLAN DESIGN The estimation of the rectangle, in which the module is to be realized, is controlled by the shape constraints. Guidelines for the position of such a rectangle among all the other rectangles are contained in the functional hierarchy and in the incidence structures associated with the modules. In the context of layout, these incidence structures are called *net lists*.

Utilizing the data (shape constraints, net lists, and functional hierarchy), the cells must be arranged in a rectangle. This enclosing rectangle is often desired to be as small as possible; sometimes it is constrained in the aspect ratio or completely specified. If the cells were fixed objects, this would be a placement problem. However, in this context, the cells are allowed to take any shape allowed by its shape constraint. This generalization of placement is called *floor plan design*.

Both floor plan design and placement are guided by a number of objectives, which are not easy to formulate in a single function. This can be illustrated by the following typical combination of objectives. The first objective is primarily concerned with the realization of the interconnections. A common figure of merit is total wire length, often estimated by summing the perimeters of the rectangles that enclose all module centers connected to the same net. At the same time, the second objective is to give the cells optimal rectangular regions, minimizing such objectives as deformation, dead area, wiring space, and so forth. The first objective is topological in nature, while the latter is geometrical. To relate the two objectives, an additional refinement step (using an intermediate structure capturing much of the data affecting the topological objectives) is used in the Yorktown silicon compiler.

FLOOR PLAN TOPOLOGIES It has already been observed that in the final floor plan, the modules will be rectangle dissections in which each submodule is either a rectangle dissection itself, or, as in the case of cells, a rectangle. Creating a preliminary environment for the cells amounts to generating certain aspects of the rectangle dissection in which each cell is an undivided rectangle. Since the shape of the cells is not yet known at this stage, the geometrical details of the rectangle dissection cannot be determined. However, less restrictive aspects of a rectangle dissection are its neighbor relations—that is, which cells share a particular line segment in the dissection. The set of neighbor relations is called the *floor plan* of the rectangle dissection. A floor plan is useful information that can be generated at an intermediate stage of the refinement process. Usually, enough freedom is left for the cell-assembling procedures after fixing the floor plan, and further decisions concerning the environment of the cells can be derived from it. Therefore, the first task will be designing a floor plan. A reasonable decomposition of that task, certainly in the light of the discussion in the previous section on the layout design environment, is to take one module at a time, starting with the root of the hierarchy and progressing downward such that no module is treated before its supermodule. This translates the hierarchies into nested rectangle dissections [Otte82a].

It is desirable to design a floor plan that minimizes the chip area while meeting the shape constraints of its cells. A concise and common method of representing the floor plan of a rectangle dissection is by a *polar graph.* Many floor plans designed in practice and all the floor plans of the successful, more specialized layout styles have polar graphs that are two-terminal series-parallel digraphs. The optimization problem described previously can be efficiently solved for this class of floor plans.

A first observation is that a floor plan that can be represented by a two-terminal series-parallel graph can also be represented by a more manageable data structure, namely an ordered tree. By restricting floor plans in this way, the floor plan design procedure essentially orders and refines the initial hierarchy, and the hierarchy and the layout structure can be consistently represented and stored in a natural way during the entire procedure. A rectangle dissection, with a two-terminal series-parallel polar graph, is a rectangle dissected by a number of parallel lines into smaller rectangles that might be dissected in the perpendicular direction. Such structures are called *slicing structures* (see Figure 7.1), and the associated tree is called a *slicing tree.* Each vertex of the tree represents a *slice.* Each slice either contains only one cell or is a juxtaposition of its *child slices.* In the latter case, that slice is said to be the *parent slice* of its child slices, and these child slices are the *sibling slices* of each other. The sibling slices are ordered according to their position in the parent slice (e.g., left to right and top to bottom).

When a (preliminary) geometry is associated with the slicing tree, the *longitudinal dimension* of a slice is the dimension it inherits form its parent slice. The other dimension is its *latitudinal dimension.* The latitudinal dimension of the common ancestor slice is the length of the side it transmits to its child slices. The other dimension is its longitudinal dimension.

IMPLEMENTATIONS There are several ways of obtaining slicing structures. A well-known method is the min-cut algorithm. If applied in its pure form, it leads to binary slicing trees. However, it can be extended to produce general slicing trees. The min-cut method does not use an intermediate structure that captures globally a large part of the topological aspects of the input. Each dissection divides the problem into smaller problems, but it is difficult to take into account decisions in one part when handling the other parts.

There are also methods that use an intermediate structure [Otte82b]. One such structure is a point configuration in which the topological properties of the input are somehow translated into a closely related geometrical concept, namely distance, and since the configuration will be embedded in a plane, the geometrical concept is usually distance in the two-dimensional euclidean space. High connectivity is reflected in relatively short distances. A useful metric in this context is the so-called *dutch metric*. To obtain the distance between two modules under the dutch metric, first calculate a proximity by taking the quotient of the sum of weights of the net that connect the two modules together and the sum of the weights of the net that connect each of the modules with another module. Obviously, the proximity can never exceed 1. The distance is the square root of 1 minus that proximity. The weights of the nets can be used to grade the relative importance of nets. The distance space can always be embedded in a euclidean space, although seldom in the euclidean plane. However, it is straightforward to construct a two-dimensional point configuration in which the distances d_p are such that $\Sigma(d_d^2 - d_p^2)$ is minimum. This is done by determining the eigenvectors with the two largest eigenvalues of an associated matrix, the *schoenberg matrix*. Also by calculating a partial eigensolution, the weighted sum of distances can be minimized under the constraint that the configuration must have a certain spread. However, these are euclidean distances, and wiring is generally orthogonal, requiring a different metric. It is also difficult to take into account more detailed information (e.g., pin positions on the perimeter) in assessing the wiring objective.

Recently, a probabilistic algorithm has become very popular for handling combinatorial optimization problems [Kirk83]. This algorithm, called (simulated) annealing, can also be used for generating point configurations in the present context [Otte84]. The following formulation is illustrated in Figure 7.13. Each configuration allowed in the algorithm is represented by a pair of permutations over the set of modules. To obtain the score of that configuration, the two permutations are used to obtain the horizontal and vertical coordinate for each module. The horizontal coordinates, for example, are obtained by placing an interval between two neighboring elements of the horizontal permutation, proportional to the sum of the two areas of the associated modules. Thus the horizontal and vertical coordinates provide a point configuration in the plane for all the modules. The perimeter of a box that contains all the points representing modules connected by a particular net is used as an estimate for the size of that net. Several refinements (e.g., correction for nets connected to many modules, different weights for different directions, weights for nets, etc.) are possible. The sum of these size estimates is the score of the configuration. To generate a new configuration, a pair of modules is transposed in one of the permutations. The score function is then evaluated for this new state and is either accepted or rejected according to acceptation rules of the annealing algorithm. In this way, an optimal or near-optimal point configuration is obtained.

Properties of the final rectangle dissection are derived from this point configuration and the shape constraints. The topological considerations are taken into account by preserving relative positions in the point configuration and keeping modules close together if they are represented by points

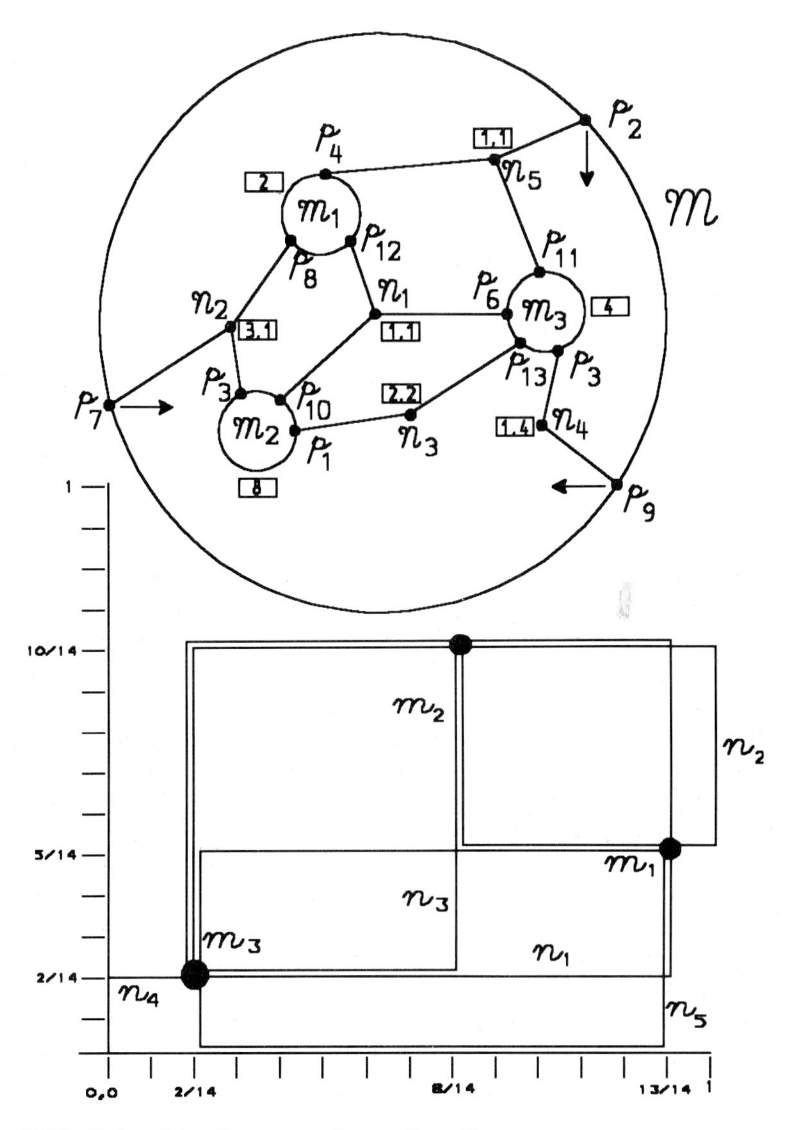

Figure 7.13 Determining the score of a configuration.

at a small distance from each other. The geometrical aspects are taken care of by keeping track of deformations implied by the dissections, for example. Shrinking is a method that efficiently achieves these goals for slicing structures. The discussion of shrinking will be limited to designs with exclusively flexible modules. For the most part, nearly square shapes are preferred for flexible modules because square shapes generally require less wiring space. Now, assume that a given slice with two or more submodules is to be sliced into a number of child slices. This slice is separated from its sibling slices by lines parallel to one axis, and the slicing must now be tried parallel to the other axis. Figure 7.14 is a visualization of a process

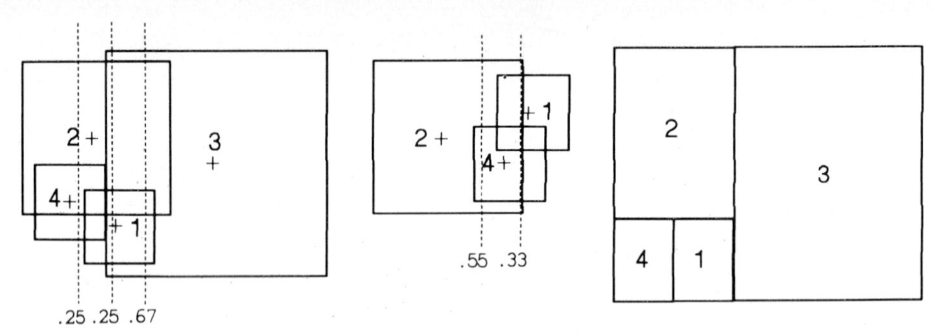

Figure 7.14 The determination of the shrink factors. In this example, the squares around the module centers are drawn in their position before the shrinking process. The amounts of shrinking necessary to keep the vertical dashed lines from intersecting are given for each line. After accepting the line with the highest shrink factor, the three modules at the left are treated in the same way.

of determining the shrink factor for possible slicing lines with the given orientation.

Think of a module as a square around the point representing it and with its sides parallel to the axes. The sizes of these squares are such that each pair of squares has overlaps and the ratio of their areas is equal to the ratio of the area estimates given to the corresponding modules. The squares are shrunk simultaneously, leaving their centers fixed and preserving the area ratios. At some point during this shrinking process, there will be a line with the given orientation that divides the modules into two blocks without intersecting any of the squares. The amount of shrinking necessary to reach that point is the *shrink factor* to be assigned to the corresponding line. Shrinking continues, and other lines with the same orientation and separating other blocks are determined along with their shrinking factors. The lines with relatively high shrink factors are candidates for slicing lines. The accepted lines partition the modules into groups. Each of these will be treated recursively in the same way. The selection of the lines usually considers other factors. One of these, *deformation*, will be discussed next.

The preferred shape for flexible cells is a square. However, in general it is not possible to accommodate cells as squares in a slice with the area being the area estimate for the supermodule. The deviation from the square can be measured by a simple function. The calculation of deformation on the basis of such a function starts with deriving the outer dimensions from the total area and the aspect ratio. One of these dimensions is inherited by the child slices from the parent. The other dimensions of these child slices are obtained by dividing the sum of the areas of the modules in each of them by the inherited dimension. Proceeding in this way, each slice inheriting the dimension calculated for its parent slice will yield the dimensions of all slices, which enables one to calculate the minimum amount of deformation (under the given function) incurred by each cell. Since the computation is performed top-down, this is applied during the slicing procedure. For every decision, proximity information (shrink factors) and shape information (deformation) is available. For floor plans with a large number of modules, of similar size, the first slicing lines are mostly determined on the basis of proximity information. Later, when few modules are allocated in a slice, deformation usually dominates the selection. In the final stage, one may even consider determining the deformation incurred by all possible slicing trees.

FLOOR PLAN OPTIMIZATION Given the slicing tree and the shape constraints of the leaf modules, the best rectangle dissection corresponding with that tree must be determined. By *best*, we usually mean area or perimeter of the overall rectangle; however, a more general objective function, with the outer dimensions as arguments, can be used as well. Thus we can construct the smallest rectangle dissection with (1) a given slicing structure and given shape constraints for its undivided rectangles and (2) with or without constraints on the aspect ratio (e.g., a given aspect ratio, or a lower and upper bound on the aspect ratio). The time complexity of the algorithm is polynomial.

The shape constraint of a compound slice can be derived from the shape constraints of its child slices. In the final configuration, these child slices must have the same longitudinal dimension, which is the latitudinal dimension of their parent. The inverse of a compound's shape constraint is only defined on an interval in which the shape constraints of all its children are defined. Its smallest possible longitudinal dimension for a given feasible latitudinal dimension x is the sum of the values of the shape constraints of the children at x. So, the shape constraint of a compound is obtained by the addition of the shape constraints of its children in the interval in which they are all defined and by the determination of the inverse of this resulting function. These operations are easy for piecewise linear shape constraints, represented by a list of their breakpoints ordered according to the respective longitudinal dimensions (see Figure 7.15).

The ability to obtain the shape constraints of a slice by adding the shape constraints of its child slices and inverting the result enables us to obtain the shape constraint of the entire enveloping rectangle. Appropriate objective functions (quasi-concave functions that are monotonically increasing in their arguments) always assume their minimum value at the boundary of the associated shape constraints. This is a consequence of the monotonicity of shape constraints and the objective function. For piecewise linear shape constraints, that minimum will be assumed at one of its breakpoints. To find an optimum pair of dimensions for the common ancestor slice, the objective function only has to be evaluated at the breakpoints of its shape constraint in the convex set of permissible pairs.

Given the longitudinal dimension of a slice and its shape constraint, the latitudinal dimension can be found by evaluating the shape constraint at the given longitudinal dimension. After deriving the shape constraint for the common ancestor and determining a dimension pair for which the objective function assumes a minimum, the longitudinal dimensions

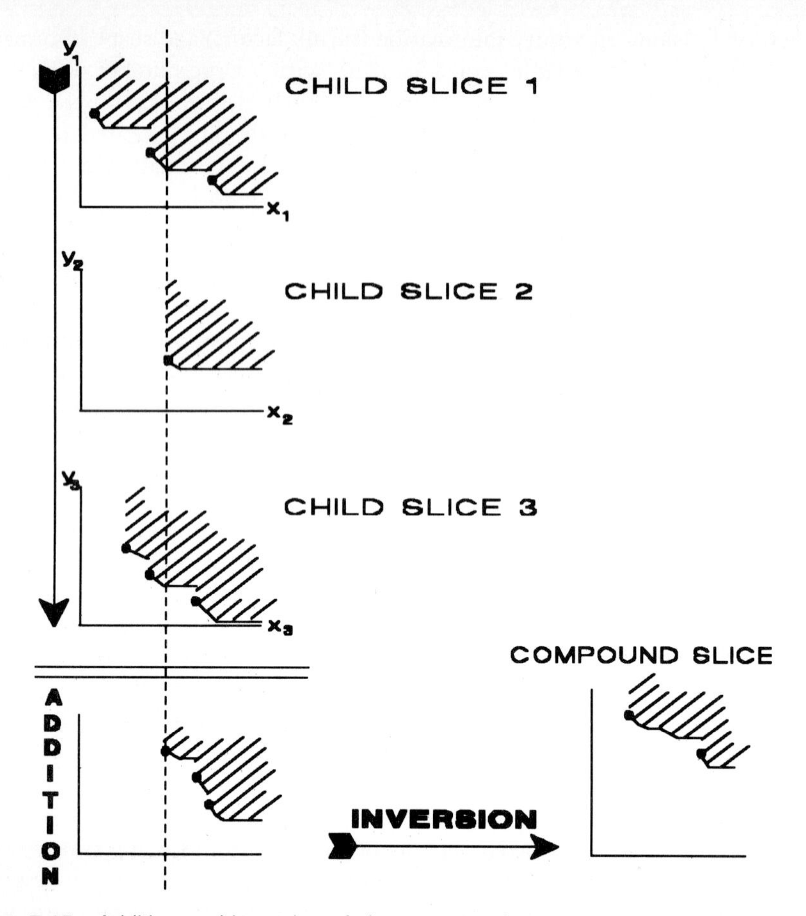

Figure 7.15 Addition and inversion of shape constraints.

of its children are known. For each of them, the latitudinal dimension, which in turn is the longitudinal dimension of its children, can be calculated. Continuing in this way will yield the dimensions of all slices in the configuration.

The algorithm for finding the optimum dimensions for all slices consists of three parts:

1. Visit the nodes of the slicing tree in depth-first order; and, just before returning to the parent, determine the shape constraint by adding the shape constraints of its children and inverting the result.
2. Evaluate the objective function for each breakpoint of the derived shape constraint of the common ancestor, and select a dimension pair that corresponds to the smallest value.
3. Visit the nodes of the slicing tree in depth-first order; and, before going to any of its children, determine the latitudinal dimension by evaluating its shape constraint at the inherited longitudinal dimension.

Completing all three steps will yield the dimensions of all slices in an optimum configuration for the given floor plan and cell shape constraints. To determine the position coordinates of the slices from these dimensions and the floor plan is straightforward. It is also easy to determine what orientation the inset cells can have in this optimum configuration.

NET ASSIGNMENT The more complex the circuit is, the more dominant the wiring is in the final layout. Although part of the wiring can be realized on top of active devices, particularly when there is more than one metal layer, a considerable portion of the chip is used exclusively for wiring (wiring space). This part of the chip will be implemented using "wiring cells." If these wiring cells are realized in rectangular regions, the wiring space can be seen as the union of nonoverlapping rectangles. The selection of the rectangles for the wiring space affects the efficiency of the wiring procedures. It constrains the sequence in which the wiring can be generated, determines the algorithms that can be used for the wiring, and determines the number of different algorithms needed to perform that task.

Again, slicing structures have considerable advantages over general rectangle dissections. First, because they imply a decomposition of the wiring space into the minimum number of rectangles: the slicing lines are, in effect, the wiring rectangles. To distinguish these undivided rectangles from the ones that correspond to the cells in the functional hierarchy (in the slicing tree both kinds are represented by leaves), they are called *junction cells*. Second, a feasible sequence for generating the wiring can be easily derived from the slicing tree. Third, all these rectangles can be wired by using the same kind of algorithm, usually called a *channel router*.

Since the wiring consumes a high percentage of the total area, it is useful to have early estimates for this space so that the sequence of floor plan calls can take these estimates into account when designing the nested floor plans. Several objectives may be important in realizing the interconnections, and many of these are directly related to the length of these nets. This immediately presents a problem since a floor plan is a topology rather than a geometrical configuration. Yet, in order to measure the length of a net, a metric is necessary.

A commonly used structure for approaching these problems is the plane graph determined by the rectangle boundaries in the rectangle dissection. Each rectangle corner is a vertex, and the line segments between them are the edges. This graph depends on the geometry of the rectangle dissection, and this geometry is not known in the floor plan design stage. A closely related graph can be defined for slicing structures. For there is a one-to-one correspondence between the junction cells of rectangle dissections with the same slicing structure. Also, the relation generated by the T-intersections between junction cells is invariant over the rectangle dissections with the same slicing tree. Based on these observations, an environment can be created in which any steiner tree heuristic can be used for routing each individual net.

Cell Assemblers

The refinement steps described in the previous section determine a topology for every compound of the functional hierarchy. These topologies are restricted to slicing structures, so floor plan design replaces each subtree, of which the vertices represent a certain module and all its submodules, with another tree, of which the root represents the selected module and the leaves its submodules. Another type of refinement is the replacement of a leaf in the functional hierarchy (a *function cell*) by a tree decomposition. The reasons for not having this decomposition in the initial tree can be quite diverse. For example, the decomposition suitable for the functional design may be far from optimal for layout design. In that case, such a hierarchy is pruned, and a data base problem must be resolved when this happens. However, most often there is no need for further decomposition from the functional design point of view, but flexibility is increased if the layout design procedures use some inherent decomposition.

Algorithms for designing cells, possibly using such a decomposition, are called *cell assemblers*. The task of a cell assembler is to determine the internal layout of its cell (with respect to a reference point in that cell's region) on the basis of a suitable specification and data about its environment. There may be quite a diversity of cell assemblers in a silicon compiler system. The applicability of the silicon compiler is highly dependent on the set of implemented cell assemblers. While most of the decisions during floor plan design are largely technology independent, cell design is dominated by the possibilities and limitations of the target technology. The numeric design rules are stored as numbers, the value of which is assigned to certain variables in the cell assembler. The structural rules are mostly incorporated in mask generators.

The layout of a slice is obtained by first determining the layout of all its child slices, except junction cells, and then by calling the appropriate assemblers for the junction cells. Visiting the slicing tree in depth-first order and performing the above operations when returning to the parent slice enables the program to determine the *chip coordinates* (coordinates with respect to a unique point on the chip) of all layout elements in the parent slice before leaving the corresponding vertex. The translation of that result into the rectangles of the various masks is also performed at this point. This translation is straightforward.

FUNCTION CELLS The task of a cell assembler is simple for inset cells since the internal layout is stored in a master or user library. From the topology (determined in the floor plan design process) and the shape constraints of all cells (including the junction cells), an estimate of the rectangle of that module has been derived. Reasonably accurate data about the position of the nets to be connected to that cell have been generated by the net assignment process. On the basis of that data, the assembler must decide which orientation is given to the inset cell and how it is to be aligned with its sibling slices.

For function cells, where the layout is not stored in a library, the layout may be still largely implied by the specification. For example, if a cell is a programmed logic array, the specification is either a personality matrix or a set of Boolean expressions. In the former case, the assembler does nothing but translate into a layout, so the result is obtained in the same way as the stored inset cells. If the array is specified by a set of Boolean expressions, the assembler uses a so-called pla-generator. The shape of the resulting array is still difficult to control; but the pin positions can be adapted to the results of the net assignment, which are performed during the floor plan design stage. Some sophisticated pla-generators use techniques such as row and/or column folding to make the area of the array smaller. This constrains the choice in pin positions considerably and might lead to a higher area consumption because of the complex wiring around that array. A pla-generator in a cell assembler should be able to take the results of the net assignment into account [Demi85].

Regular arrays (e.g., programmed logic arrays and memories) are heavily constrained, and it is not easy to manipulate the shape of the array. Other cells, called macros, have a natural or given decomposition that can be used for that purpose. They are decomposed into circuits that either are selected from a prespecified catalogue or can be designed with a simple algorithm from a function specification. The reason for having such a macro as a cell assembler, rather than part of the floor plan design, is that the circuits have certain properties that make special layouts very efficient. For example, the catalogue or the simple algorithm may have a constraint that gives all cells in the macro the same width and the same positions for power supply and clock pins. In that case, a pluricell layout style is suitable for the macro. It forces the cells to be distributed over columns; but the number of columns can be chosen freely. Therefore, the aspect ratio of the macro can be influenced. Also, the pin distribution around the periphery can be prescribed on the basis of environment data. In the YSC system, we use predominantly a cell generator that implements combinational logic in pluricell style [Bray 84c].

Pluricells are similar to polycell or standard cell layouts as can be seen in plate I. Both have their circuits placed in columns, either with two sides accessible for signal nets or with one side accessible and the other abutting cells in a neighboring column. The latter is chosen in the YSC system. Supply and clock lines are realized by abutting subsequent cells in each column. The signal nets are realized mainly in channels between the (double) columns. Two wiring layers, polysilicon and first-level metal, are used to realize the nets in the channels.

One important difference is in the way that the individual cell layouts are obtained. Polycell layouts find their cells in a master library that is built and maintained independently from the individual applications. Pluricell cells are constructed ad hoc. In case of a domino circuit, the switching network is realized as a one-dimensional transistor array along the edge of the channel. Such an array is formed by crossing a strip of *n*-diffusion with

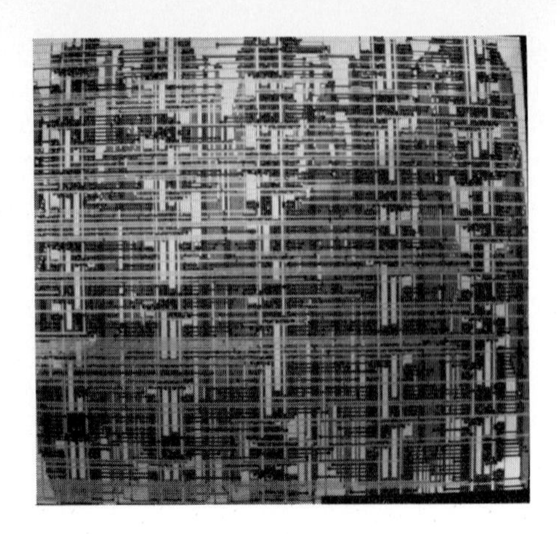

Plate I

polysilicon tracks carrying the associated logic signal. This is not always possible without modifying the network, that is, duplicating transistors, introducing transistors with grounded gates, changing sequences in series connections, and so forth. As long as the logic function and the electric properties of the result are still acceptable, these modifications may be used for making the diffusion strip short and the number of tracks to realize the network topology small. The rest of the circuit is realized between the power supply and ground lines that run parallel to the diffusion strip on the side opposite from the channel. The sizes of those devices are adapted on the basis of critical path information generated by the algorithms in the timing optimization of the Yorktown silicon compiler. Between those two parts of the circuit runs the clock line. Any number of additional first-level metal tracks can be allocated between that line and the ground line of the inverter. However, one is always there without increasing the circuit area. This is used primarily for sense node. Another track, which is free because of the spacing rules, is between the clock line and the transistor array and is used for making the output available via first-level metal. If this cannot be used, the output will enter the channel via polysilicon (see plate II).

Of course, other types of circuits can be incorporated in pluricell layouts as long as their periphery is compatible (i.e., pitches of clock and supply lines are matched, signal pins are on the correct side). Carrying this to the extreme with completely specified circuits from a library leads to polycell layouts. Therefore, pluricells can be seen as a generalization of polycell styles.

To connect parts of nets in different channels, polycell layouts use either so-called feedthrough cells or the net may leave the channel at one end and make the required connection outside the macro. In pluricell layouts, the feedthrough connections are made by second-level metal tracks. This also allows signals to enter the macro from all four sides. Each signal requiring a second-level metal connection has exactly one track assigned to it.

Channels in a pluricell layout are different from channels resulting from

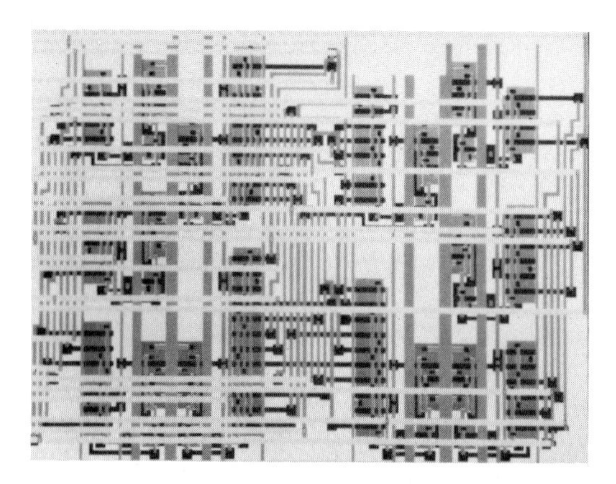

Plate II

a *classical* channel router, because signals can be dropped into the channel by contacts between first-level and second-level metal. Thus there are three ways of bringing signals into the channel, whereas the formulation has only two—the sides and the ends of the channel. The channels shown in plate II differ in one more aspect from classic channels. Design rule sets have developed smaller and smaller widths for metal and polysilicon tracks. However, the area taken by contacts between polysilicon and first-level metal has not decreased as fast. Currently, with minimum pitch wiring under the presence of these contacts, there is enough space for a first-level metal track in both directions. One router implemented in the YSC makes use of this fact by allowing twice as many tracks in the channel as would be allowed by a classic channel router, but contacts in neighboring tracks are not allowed. Moreover, in latitudinal direction, the router uses the extra space for adjusting the position of a net by first-level metal jogs. Therefore, the wiring space is reduced by a factor of about two, as compared to what a classic channel routing would produce.

The results of the logical synthesis and timing are available to the layout program. Also, information about desirable device sizes is given to the layout part. For example, functions in a critical path may be marked so that the associated domino circuit gets a more powerful inverter. All this information is stored in a specific format readable by the layout procedures. After the design of the whole chip has entered the layout stage and after floor planning and the global net assignment, this information is read to obtain information about the environment of the macro. The construction of a pluricell style layout is performed in the following steps.

1. **The topology of the switching network**. The construction of the domino circuit in a pluricell layout is executed by first performing a topological analysis of the logic function and encoding the result in a compact vector from which periphery data, such as dimensions and pin positions, can be derived. Domino circuits are specified as factored logic functions. These have an obvious translation into a two-terminal series-

parallel network of switches. The translation of a switching network in a linear transistor array is illustrated in Figure 7.16. This translation process is possible for the network in Figure 7.16, because the graph of the switching network has a special property; only two vertices have an odd degree. Switching networks that do not have this property cannot be translated into a linear transistor array by the process illustrated in Figure 7.16. However, different graphs may be used to realize the same logic function. In the case of switching functions with a series-parallel graph, the commutativity of the AND operation makes the sequence of the series connections irrelevant. Also, duplicating a transistor and connecting the two in parallel does not change the switching function. Thus any series-parallel logic function can be transformed to one that has at most two odd nodes. An interesting problem now is to find one that realizes the given logic function with the minimum number of transistors, using the permutation of the series combinations and duplication of transistors. This corresponds with a transistor array of minimum length under the given constraints. In practice, however, we also want to restrict the number of metal tracks required to connect the various diffusion regions to realize the required graph. Typically, we try to keep the number of metal tracks to three in order to keep all the circuits in a column of the pluricell at the same width. Using an APL

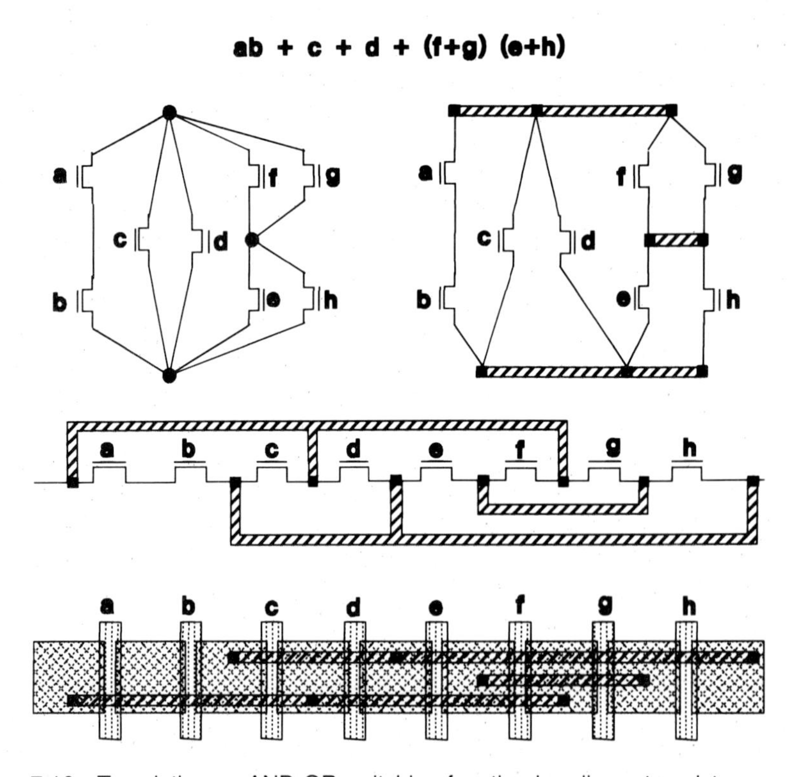

Figure 7.16 Translating an AND-OR switching function in a linear transistor array.

implementation, all the circuits of a controller represented in plate I were created in less than one second of CPU time on an IBM3081. The controller had 81 single output switching functions, which were realized by almost 800 transistors in the linear arrays.

2. **Placement in (double) columns**. Once the circuits are designed, a placement procedure consistent with the structure rules of the slicing structure is used. First, a configuration of points, each representing a domino circuit, is created in the plane on the basis of proximity data such as net length and delay. Slicing will preserve the relative position within each slice. The determination of the final slicing structure is based on approximately equal-length double columns.
3. **Second-level metal assignment**. Once the distribution of the circuits over the columns is known, their position in the column is fixed. This enables the program to determine the intervals at which the nets must occur for each pair of double columns in order to connect the corresponding pins along the sides of the channel. If the same net occurs in several channels or enters the macro from one of the directions perpendicular to the columns, a second-level metal track is reserved for that net. The position of these second-level metal tracks is optimized with respect to net resistance and net size as permitted by the environment of the macro. This problem has been formulated as a bipartite matching problem.
4. **Mask generation**. After finishing the second-level metal track assignment, the data for the channel-routing problems are complete. Pluricell floor plans are, as a slicing structure, always represented by a five-level slicing tree. The leaves of that tree represent the circuits and the channels. By generating the mask data of these leaves in the same sequence as they are visited by a depth-first search over that tree, the coordinates of the elementary mask elements are immediately assigned. Consequently, circuit geometry and wire specification are interleaved, and in one pass the mask data are complete.

Decompositions in macros occur very often; however, for special cells that are frequently used, it sometimes is worthwhile to implement a special algorithm producing a highly optimized layout. In word-organized digital systems, these special cells often process a number of bit vectors. The layout as a whole may benefit from aligning these cells so that the wires carrying these bit vectors do not have to be matched to the pitch of each individual cell. Also, wires that pass over such a cell without making any contact must be accommodated. These requirements imply a certain kind of flexibility, such as stretchability and variable pitch. If possible, such an algorithm should be able to produce these highly optimized cells for several bus dimensions and a range of performance requirements.

JUNCTION CELLS. Junction cell assemblers are closely related to channel routers [Burs85], because of the way they are isolated and called in the layout process. The junction cell is a rectangular area of which the latitudinal sides are part of the longitudinal sides of junction cells that are represented in the slicing tree by vertices closer to the root. When the assembler is called in a bottom-up sequence for a certain junction cell, the longitudinal coordinates of the entering nets are known. There are several ways a net may enter the junction cell: from the longitudinal sides of that cell, from a higher metal layer, from the latitudinal sides, and perhaps in still other ways. The task of the assembler is to realize all the required interconnections in a rectangle with the smallest possible latitudinal dimension. The longitudinal dimension must be commensurate with the latitudinal dimension of the parent slice. Therefore, increasing the longitudinal dimension of the channel should be avoided if possible.

7.2.5 Verification Tools

The design synthesis should be a strictly automatic process. However, its effectiveness depends on a compatible debugging environment where the "program" can be run and errors corrected. The YSC system is equipped with a number of facilities for accomplishing this. Two gate-level verification tools are described here: a logic simulator and a timing verifier. With the logic simulator designs can be verified by inspecting the contents of the registers for a given input stream. Information about the arrival times of the signals can be obtained by means of the timing simulator. This information can be used for verifying whether the circuit operates within the required cycle time as well as for providing timing information to the timing-performance optimization tools.

The Gate-Level Logic Simulator

A gate-level logic simulation provides a means of verifying the logic behavior of the circuit being designed. The simulator allows the monitoring of the logic values (1 or 0) of some of the variables in the original description. The values of the variables that are always observable are those that are latched into registers (architected or not), those corresponding to I/O pins, and the primary inputs and/or outputs of the blocks of the hierarchical circuit structure (e.g., I/Os of combinational modules).

Gate-level logic simulation uses the information of the structure of the circuit provided by the structural synthesis step and the logic representation of the combinational modules generated by the logic synthesis step. The behavior of these modules is characterized by the Boolean function that they implement. This information is easily accessible after logic synthesis. The logic simulator assumes zero-delay in each circuit block. The behavior of special-purpose modules, or circuits, external but interacting with the hardware being designed (e.g., a memory unit) may be modeled by an APL program.

Simulation consists of loading the registers with a set of logic vectors representing the "initial conditions" and then feeding the circuit primary inputs with a stream of input data. The stream of values of some (or all) the observable variables may be monitored or stored. In particular, if the circuit being designed has a synchronous operation, the simulation may be halted at each cycle.

Simulation may be used to verify the correctness of the input to the YSC system. In the case of processor design, it provides a mechanism for verifying the architecture. In particular, architectural verification may be achieved by "running" a stream of instructions and by monitoring the circuit primary outputs and the values of the architected registers.

The Gate-Level Timing Simulator

The gate-level timing simulator provides information about the propagation delay through the circuit. It serves two purposes. First, it checks a circuit against its specifications. For sequential synchronous circuits in particular, it is important to know whether the circuit meets the cycle time specification. Second, it evaluates the signal delays needed by the time-performance optimization programs.

The delay computation requires several assumptions on the circuit. An accurate timing evaluation requires not only the knowledge of the circuit topology, but also layout data such as the length of the interconnections. However, it is important to obtain timing information before the final layout is available. Such information may be used for early verification or for driving the layout procedures to produce images that satisfy timing requirements. For example, if timing information is available before the floor plan is designed, critical nets can be identified and incorporated into the objective function of the floor plan.

The timing simulator of the YSC system has been designed for different stages of the synthesis process, using all the information that is available at that stage. Just after the logic synthesis and before the layout stage, the wiring load is either estimated from the fan-out information or neglected. After the floor plan has been determined, the wiring length is computed on the basis of the position of the modules. Of course, when the layout has been completed and the wiring lengths are known exactly, the simulator can provide more accurate timing data.

The timing simulator is capable of simulating combinational circuits or the combinational portion of sequential circuits, including multiple-clock circuits. Combinational logic is described to the simulator as an interconnection of *logic gates*, by means of *nets* carrying *logic signals*. The hierarchy in the circuit structure is not required to explain the timing model. Therefore, we can assume that there is no hierarchy for the sake of this description. However, the hierarchical information is retained in the actual implementation of the simulator for computational efficiency.

The circuit being described has a set of primary inputs and outputs. The primary inputs to this network are also interconnected to the logic gates by means of nets. The primary outputs are identified by the gates generating the corresponding signal. We consider each signal stored in a register as both a primary input and output to the combinational subsystem. Each logic gate is assumed to be unidirectional because of the implementation technology. Therefore, we can associate a source and one (or more) sinks to each net. The interconnection can be modeled by a direct graph $G(V, A)$, whose node set $V = \{v\} = V^g \cup V^i = \{v^g\} \cup \{v^i\}$ is in one-to-one correspondence with the set of logic gates (V^g) and primary inputs (V^i) and whose edge set A is in one-to-one correspondence with the source-sink pairs of the nets. The circuits synthesized by the logic synthesis stage are unidirectional to avoid race conditions, that is, $G(V, A)$ is acyclic. A node v_i is said to be a *predecessor* (successor) of gate v_j if there is a directed path from v_i to v_j (from v_j to v_i) in $G(V, A)$. A predecessor (successor) is said to be direct if the path has a length of 1.

The signal propagation is modeled by associating a *propagation delay*, $d(v^g)$, to each logic gate or, equivalently, a weight to each node in the set V^g. The propagation delay through a physical gate is modeled by an empirical equation. Although the procedure for computing the delays is fairly general, the propagation delay model for a gate depends on the target implementation technology. Delay equations for different logic families and technologies can be obtained by: (1) determining a certain number of circuit parameters that characterize the gates (e.g., gate sizes, capacitive loading, etc.); (2) simulating a wide set of gates with different parameters using a circuit simulator such as ASTAP to compute the propagation delay; and (3) using regression analysis to tabulate the delay as a function of the characteristic parameters. Therefore, the propagation delay $d(v^g)$ can be computed quickly from the parameters describing the gate and its interconnection.

For each node, we compute a *data ready time*: $t(v_i^g), i = 1, 2, \ldots, |V^g|$. The data ready time of a gate is the time at which the signal generated by that gate is ready. Similarly, we associate to each primary input a data ready time: $t(v_i^i), i = 1, 2, \ldots, |V^i|$. For our purposes, we synchronize the computation of the data ready times to the system clock. We assume the data ready time to be zero for each primary input corresponding to a register. The data ready times of the remaining inputs are set to the delay of the corresponding input signal with regard to the system clock. (The extension to multiple-clock systems is straightforward.) The data ready times at the logic gates can be computed by tracing forward the signal propagation—that is, by computing:

$$t(v_i) = d(v_i) + \max_{k \epsilon K} t(v_k) \quad K = \{k | (v_k, v_i) \epsilon A\}$$

for each node v_i corresponding to a gate in a sequence consistent with the partial order represented by the graph.

The gate-level timing simulator computes the set of data ready times for all gates in the circuit. It allows comparison of the computed arrival times of the primary outputs with the expected arrival time. With this information, the user can determine whether the circuit being synthesized satisfies the circuit specifications or not. It also provides the basic information for the timing optimization procedures.

7.2.6 Timing Performance Optimization

We present here some algorithms to improve the timing performance of the circuit being designed by the YSC system. Circuit performance is related to the worst-case propagation delay of signals between two register boundaries, because the system clock must be adjusted to allow the arrival of each signal at the destination registers within the clock cycle. In this context, the optimization of circuit performance is equivalent to the minimization of the critical path delay. The circuit timing depends on the structure of the circuit (e.g., register boundaries). Therefore, timing optimization can already be considered at the structural synthesis level (see section 7.2.2.). We assume now that the global interconnections among combinational logic, registers, and I/Os are frozen, and we survey timing optimization techniques that are compatible with this assumption. We refer the interested reader to [DeMi86] for the details as well as for a set of references to other work in the field.

The logic synthesis stage in the YSC system can be programmed to optimize the silicon area taken by each combinational module, the timing performance of a module, or both. Logic synthesis of each combinational module is performed one module at a time, and timing information during synthesis is limited to the domain of the module being synthesized. A gate-level timing simulation of an entire circuit can be done only after all its combinational logic blocks have been synthesized as an interconnection of gates. Since the measure of the timing performance is not an additive function (e.g., silicon area), the circuit being synthesized may not be optimal in terms of timing performance.

Similarly, the layout design stage (in particular, the floor plan design and the module and gate placement) can be programmed to optimize some figure of merit of the design, such as total wiring length or wiring density. Timing considerations may be included in the layout, by considering, for example, the length of the wires carrying signals that are critical to the circuit performance. However, these critical wires can be identified only after a timing simulation, the accuracy of which improves as more information about the actual layout is available. Similar considerations apply to the selection of the sizes of some devices in the macrocells. The timing performance can be optimized by choosing the gate sizes appropriately; however, the optimal choice depends on the structure of the logic network.

For these reasons, we consider a global approach to timing optimization which involves operations at the logic, topological, and layout levels of circuit description. This approach exploits the hierarchical circuit representation used by the YSC system as a collection of modules. In particular, at the logic level, timing optimization is achieved by modifying the internal structure of the logic gates and their interconnection inside each combinational module. At the topological level, we position (or reposition) the modules and, as a consequence, the corresponding gates; this reduces the delay on the wires along the critical paths. At the layout level, we select the gate sizes to improve the switching speed. These operations are interlaced with the synthesis steps of the YSC system and can be seen as the "code optimizer" part of the compiler that may be invoked when compiling circuits with critical timing performances.

In principle, the strategy for timing optimization can be viewed as an iteration of two steps: (1) evaluate the critical path delay; and (2) modify the circuit appropriately. These steps are repeated until a satisfactory performance is obtained. Because of the relations between the synthesis steps and the accuracy in the delay evaluation, the timing performance optimization is done as a stepwise refinement. The data ready times are estimated first, after the logic design phase, by neglecting the capacitive load due to wiring through the use of the gate-level timing simulator described previously. The corresponding logic representation of the circuit optimizes the number of gates and their internal complexity, which correlate to the optimization of the silicon area. Similarly, the devices are assumed to have minimum size, because this choice corresponds to a minimal area implementation. The results of the timing estimate can then be used to identify critical nets that can be assigned an appropriate weight in the objective function of the floor plan design. After the floor plan is designed and the modules are placed, a more accurate estimate of the delays can be obtained, because the positions of the modules are known. If the performance is not satisfactory, the circuit may be redesigned for optimal performance by:

1. Resizing the active devices
2. Resynthesis of the combinational modules
3. Repositioning the modules

These operations are iterated until a satisfactory performance is achieved or no improvement is detected (see Figure 7.17). Other figures of merit in the design, such as silicon area and wiring length, are also taken into account.

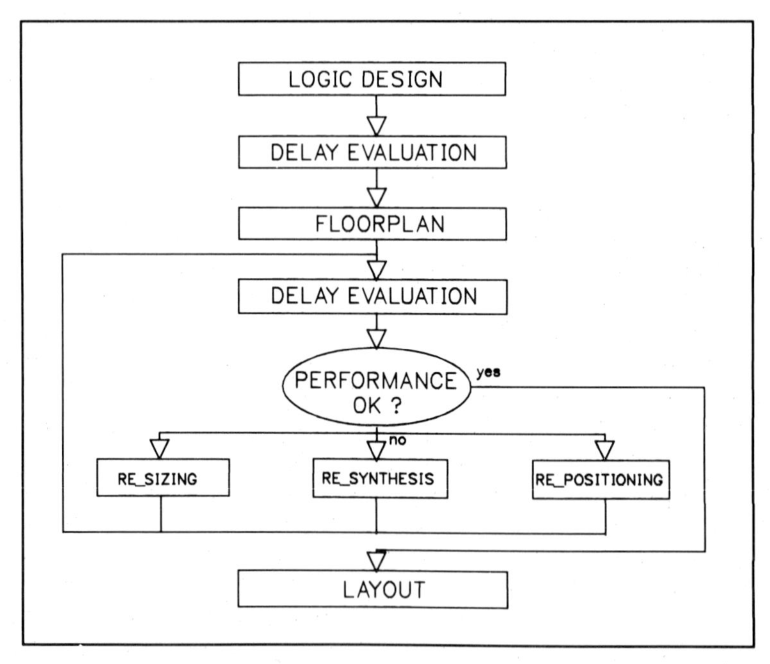

Figure 7.17 Timing optimization in the YSC system.

The performance-oriented redesign algorithms trade off these figures of merit for a faster timing performance. Due to the complexity of this approach and the interrelations among the effects of the changes at the logic, topological, and layout levels, it is not possible to guarantee an optimal procedure in rigorous terms. However, the heuristic procedures used for resizing, resynthesis, and repositioning have shown to be very effective.

Critical Paths

For a given circuit description, the gate-level timing simulator is used to compute the signal delays at each gate of the circuit. For this reason, we will use the terminology and notation introduced in the section on the gate-level timing simulator throughout this section. Important information about signal timing includes the slacks of the signals generated by the gates (also called slacks at the gates). The *slack* of each primary output (or at the gates generating a primary output) is defined as the difference between a chosen time $\bar{t}(v_i)$ (e.g., the minimum system cycle time) and the computed data ready time $t(v_i)$. For the other gates, the slack $s(v_i)$ at node v_i is defined to be:

$$s(v_i) = \min_{j \in J}\{s(v_j) + \max_{k \in K}\{t(v_k)\} - t(v_i)\}$$

$$J = \{j \mid (v_i, v_j) \in A\} \quad K = \{k \mid (v_k, v_j) \in A\}$$

The slack at each gate measures how much additional delay each signal can tolerate, while satisfying the relation $t(v_i) < \bar{t}(v_i)$ for each node v_i corresponding to a primary output signal. The slacks can be computed by tracing the signal propagation backward in the circuit—that is, by computing the slack at each gate in a reverse order consistent with the partial order represented by graph $G(V, A)$.

Let $\varepsilon \geq 0$ be an arbitrary constant. The set of *critical nodes* is the subset of nodes $C_\varepsilon = \{ v \in V | s(v) \leq \varepsilon \}$. The *critical graph* $H(C_\varepsilon, B)$ is the subraph of $G(V, A)$ induced by C_ε. A *critical section* (of the critical graph) is a maximal subset of nodes $\mathcal{C} \subseteq C_\varepsilon$ with no mutual predecessor/successor relations. A *critical path* is a maximal directed path in $H(C_0, B)$. Any critical path intersects any critical section.

The meaning of critical node, critical graph, and critical path depends on the choice of $\bar{t}(v_i)$ at each node v_i corresponding to an output signal and on the parameter ε. Let $\bar{t}(v_i)$ be the required circuit cycle time (or a required arrival time for circuit output signals), and let ε be a number taking into account safety margins and tolerances. Then, the circuit meets the timing specification if and only if the set of critical nodes is empty. If this set is not empty, its elements indicate those gates for which data ready times are to be reduced to meet the timing specifications. Reducing the data ready time at the noncritical nodes does not help in meeting the circuit timing specification. Therefore, a timing optimization procedure may concentrate its efforts on the set of critical gates (nodes).

The concept of critical nodes and a critical graph is also useful in detecting those gates that limit the timing performance of a circuit, regardless of a timing requirement. Let us consider the case in which the circuit output nodes of interest are latched into registers. A design strategy may try to achieve the best timing performance of that circuit and then adjust the system clock accordingly. In this case, let t^* be the largest data ready time among the circuit primary outputs of interest. The system clock period is bounded from below by t^*. Timing performance optimization aims at reducing t^*. Let $\bar{t}(v_i) = t^*$ for each node v_i corresponding to an output signal. The critical graph $H(C_0, B)$ is the set of nodes having the property that any independent variation in the data ready times of any critical section implies a variation in t^*. Therefore, the nodes (gates) along the critical graph are "critical" because the timing performance of the circuit (limited by t^*) can be improved by decreasing the data ready time at critical nodes. This can be done, for example, by reducing the propagation delays at those nodes. Note also that a variation of the data ready time of a critical node v_i may influence t^* by very little if the direct successor of that node has a noncritical direct predecessor v_j, such that $|t(v_i) - t(v_j)|$ is small. In other words, node v_j is "almost critical." Considering "almost critical" nodes is important in a timing optimization procedure, as far as computational efficiency is concerned. For this reason, "almost critical" nodes are made "critical" by choosing $\varepsilon > 0$. The choice of positive values for parameter ε widens the set of critical nodes.

Although the definition of the critical nodes and critical graph is fairly general, we consider in the following sections the problem of minimizing the maximum data ready time t^*; therefore, we choose $\bar{t}(v_i) = t^*$ for each node v_i corresponding to an output signal, and we set ε to a proper fraction of t^*. We then use critical sections to determine those gates for which data ready times must be decreased to reduce t^*.

Delay Modeling and Electrical Considerations

We consider here only one family of circuits, namely dynamic CMOS circuits, operating in the domino mode. We consider only the *evaluation phase* of the domino cycle—that is, we assume that the timing of the *precharge phase* is correct. The gates consists of (1) a series-parallel connection of transistors that provides a discharging path to ground from a precharged node and (2) one (or more) *driver*, implemented by a static CMOS inverter. The size of the device implementing the driver can be adjusted by changing its width. The width is a linear function of the parameter w; the device length is kept constant. The sizes of the devices of the discharging path are kept constant, because the capacitance of the sense (precharged) node is negligible with regard to the gate and wire capacitance connected to the driver output.

The propagation delay through a physical gate is modeled by a delay equation that depends on a set of characteristic parameters, namely: (1) the size w of the drivers; (2) the capacitive loading c at the output, which in turn depends on the fan-out of the gate and the wiring capacitance to ground; and (3) the structure of the discharging path and in particular the maximum number of devices in a discharging path l. It is convenient to express the propagation delay as $d = \alpha(w) + \beta(w)c + \gamma(w)l$. The coefficients α, β, and γ are tabulated for a finite number of values of w. These coefficients are computed by simulating a wide set of gates with different parameters w, c, and l with a circuit simulator, such as ASTAP. Note that l is known for each gate after having completed the logic design phase of the silicon compilation process. The parameter w is set to its minimal value during logic synthesis; it is then adjusted appropriately by the resizing procedure. The evaluation of the capacitive load c on a gate requires the knowledge of the gate fan-out and the wiring load. The fan-out information is also known after the logic design phase. The wiring information is known exactly only after the layout design is completed; an estimate is used before.

From the model of the circuit and its propagation delay, it is clear that the timing performance can be improved by changing the sets of parameters $\{w\}$ (driver size), $\{c\}$ (load capacitance), and $\{l\}$ (gate structure) and/or the structure of the graph $G(V, A)$. Device resizing aims at optimizing the set of parameters $\{w\}$. Circuit resynthesis improves the timing performance by changing the structure of the graph $G(V, A)$ and, as a by-product, the parameters $\{l\}$ and $\{c\}$. The module replacement aims at speeding up the circuit by modifying the parameters $\{c\}$ through changing the wire lengths.

Device Resizing

Before introducing the device sizing strategy, it is important to consider the particular delay model for domino CMOS circuits and how the device sizing affects the propagation delay. As pointed out in the previous section, the only design parameter of interest is the width of the driver. We limit the choice of the driver device widths to a finite number for practical reasons. Therefore, we assume that the parameter w can take a finite set of values $\{1,2,\ldots,p\}$. In our delay model, the functions $\beta(w)$ and $\gamma(w)$ are monotonically decreasing with w, while $\alpha(w)$ is monotonically increasing with w. This is consistent with the facts that the current flowing through the MOS drivers is directly proportional to the gate width (the higher the current, the lower the propagation delay) and that the driver gate capacitance increases as the size increases (the higher the gate capacitance, the higher the propagation delay). As a net result, for typical values of c and l, the propagation delay is monotonically decreasing with w—that is, $\Delta^n d(v) \equiv d(w + n, c, l) - d(w, c, l)$ is always negative for each node $v \in V^g$ and $n > 0$. Note also that the gate input capacitance does not change with w, because only the driver size changes. Therefore, an increase in the parameter w is always beneficial in reducing the propagation delay through the corresponding gate and does not increase the data ready time at any other gate—that is, the mapping $w \rightarrow t^*$ is monotonically decreasing.

With this model, it is clear that a minimum value of t^* can be achieved by selecting the maximal size w for each gate. However, the drawback of this trivial size assignment is the cost in silicon area ($\mathcal{A} = f(w)$) and power consumption, which are a monotonic increasing function of the sum of all driver widths, that is, Σw_i. Conversely, assigning a minimal value to the size w for each gate would optimize these last two figures of merit but would also correspond to a maximum value of t^*. Therefore, the problem of device resizing can be seen as finding an assignment of gate sizes that yields a good area/speed tradeoff. For example, the problem can be cast as finding the gate size assignment corresponding to a minimal value of t^* with minimal area consumption.

We consider two heuristic approaches to the solution of the problem. A first algorithm starts by setting all device sizes to their minimal size and then raises those that limit the circuit performance. The algorithm iterates the three steps: (1) compute the set of critical nodes; (2) determine a particular critical section $\mathscr{C}$; and (3) raise the size of the drivers in $\mathscr{C}$ by 1. The detection of the critical nodes is achieved by computing the data ready times and the slacks, as described previously. The algorithm then constructs a critical section $\mathscr{C}$. The critical section is required to include a critical section of $H(C_0, B)$ whose corresponding drivers have $w \lessdot p$. If no such set exists, the algorithm terminates. The critical section $\mathscr{C}$ is constructed according to a greedy strategy by looking at the values $\Delta^1 d(v)$ of the critical nodes. The rationale is to maximize the minimum value of $\Delta^1 d(v), v \in \mathscr{C} \cap C_0$, which in turn is an upper bound on Δt^*. Then all the sizes corresponding to the gates in $\mathscr{C}$ are incremented. At this point, all the data ready times and slacks need to be recomputed; this is done at the next iteration of the algorithm.

The algorithm terminates when there is not a critical section of $H(C_0, B)$ whose corresponding drivers have $w \lessdot p$. Equivalently, the algorithm terminates when the maximum data ready time t^* is determined by a critical path with maximum-sized drivers. At this point, the maximum data ready time cannot be improved by resizing other devices. Note, however, that the algorithm does not guarantee the minimality of the silicon area—that is, $\mathscr{A} = f(w)$.

A second algorithm uses an opposite heuristic strategy. All device sizes are set initially to their maximal size, and then those device sizes that do not limit the circuit performance are decreased. These gates, which we can informally call "oversized," correspond to the nodes such that $s(v) \geq \Delta^{-1} d(v)$. The algorithm iterates the three steps: (1) compute the data ready times and slacks; (2) determine a subset $\mathscr{V} \subseteq V$; and (3) lower the size of the drivers in $\mathscr{V}$ by 1. The algorithm selects $\mathscr{V}$ as a maximal set of nodes that have the following properties: (1) $s(v) \geq \Delta^{-1} d(v)\ \forall v \in \mathscr{V}$; (2) no minimal-sized driver (i.e., with $w = 1$) is included; and (3) no mutual predecessor/successor relations exist among the elements of $\mathscr{V}$. If no such set exists, the algorithm terminates. The size of the drivers corresponding to the nodes in $\mathscr{V}$ are reduced by one.

The algorithm terminates when there are no more "oversized" gates—that is, when the gate size assignment corresponds to a local minimum of silicon area. Note that the value of t^* is not affected by the algorithm, because the data ready times at the successors of $\mathscr{V}$ cannot increase.

In the program implementing the sizing procedure, the first algorithm is used to determine a gate assignment corresponding to a minimal value of t^*. Then, this assignment is used as the starting point for the second algorithm, which ensures a local minimum of silicon area. In the actual implementation, for practical reasons, there are two other criteria that can terminate both algorithms. The algorithms terminate if the number of outer iterations reaches a predefined quantity. The first algorithm terminates also if the area estimate $\mathcal{A} = f(w)$ reaches a predefined bound $\mathcal{A}_{max}$. Note that by adding these two other criteria for termination, the optimality of the solution cannot be claimed. However, a "good" solution can be found with limited computing time and/or silicon area and power requirements. In addition, the quality of the solution can be traded off for computing time by choosing the parameter ε when running the first algorithm. Note that by choosing $\varepsilon = 0$, only the sizes of the gates that affect t^* are incremented. This corresponds to using a local strategy. Local optimization might be very inefficient, as far as speed of computation is concerned, when it aims at decreasing one output data ready time and when another output gate has a close data ready time. This other gate would probably generate the signal with the largest data ready time at the next iteration of the algorithm, and the algorithm would take many iterations to improve the circuit timing performance. A global strategy is achieved by choosing $\varepsilon > 0$. There is a trade off in the choice of ε: by increasing ε, fewer iterations are needed to achieve a given timing performance, but possibly more devices are resized.

Note that both algorithms terminate at a point when the maximum data ready time is constrained by the technological limitation on the size of the drivers on a critical path. To improve further the timing performance of the circuit with the given technological constraints on the maximum size of the drivers, it is necessary then to change the structure of the circuit by resynthesis or to change the position of the gates, as described in the next two sections.

Circuit Resynthesis

We consider in this section the problem of improving the circuit performance by changing the gate interconnection. We attempt to optimize the circuit by modifying the graph $G(V, A)$ by adding and/or deleting nodes and/or edges. Note that a change in the graph structure affects both the gate fan-out (and therefore the set of parameters $\{c\}$) and the gate fan-ins (and therefore the gate structure and the set of parameters $\{l\}$).

We consider different strategies for resynthesis. They can be classified as global or local. *Global strategies* apply a set of transformations to the entire circuit simultaneously. *Local strategies* assume a partition of the circuit into blocks and are applied to each circuit block in a sequence. Clearly, global strategies can achieve results that are at least as good as those obtained by applying local strategies. However, their computational cost increases as the size of the circuit, and it may be impractical to apply a global strategy to a large circuit. Local strategies can be made very effective by using an appropriate circuit partition into blocks. Resynthesis strategies may preserve the technological constraints on the gate structure or not. For example, in the case of CMOS domino circuits, gates are subject to a constraint on the longest discharging path, that is, $l \leq l_{max}$. A strategy that preserves the technological limitation would use transformations such that no gate at any step has $l > l_{max}$. A strategy that does not preserve the technological constraints would allow more general transformation that would at some point permit gates with $l > l_{max}$, for example. These gates would be eventually decomposed to satisfy the technological limitation. Note that the set of transformations is richer in the case of a strategy that does not preserve technological constraints. However, in this case, it is hard to determine a sequence of transformations corresponding to a descent in the maximum data ready time t^*. In fact, gate decompositions, which may be required at the end of the resynthesis procedure, may affect the maximum data ready time.

We consider two basic circuit transformations: gate elimination and gate factoring (extraction and decomposition), as introduced earlier. Note that by using a sequence of these two transformations, the structure of the interconnection of gates represented by $G(V, A)$ can be modified arbitrarily to any other structures implementing the same combinational function. Note also that by changing the circuit structure, the optimality of the gate size assignment no longer holds. Therefore, gates are assumed to have maximal size during circuit resynthesis— that is, the circuit speed is assumed to be limited by the structure of $G(V, A)$. A gate resizing is then applied after resynthesis.

As far as circuit timing is concerned, gate elimination and gate factoring can be represented as follows. Let gate j be a direct predecessor of gate i. The elimination of gate j into i is the replacement of gate i by gate i'. The inputs to gate i' are all the signals that are input to gate i and to gate j, except for the output of gate j. Gates i and i' are equivalent in the sense that their outputs coincide for any valid combination of the circuit primary inputs. Note that gate j need not be implemented if it has no successor and if its output is not a primary output of the circuit. In this case, gate j is deleted from the network—that is, node j is deleted from the graph $G(V,A)$. A gate elimination may violate a technological constraint, $l' > l_{max}$. If technological constraints are required to be preserved, such an elimination cannot be considered.

Gate factoring is the transformation opposite to elimination. Gate i is replaced by gate i'. A new gate, j, is added to the network—that is, a new node, j, is added to graph $G(V, A)$—and it is a direct predecessor of gate i'. The inputs to gate i' are some of the signals that are input to gate i and the output of gate j; the inputs to gate j are some other inputs to i. Also in this case, gates i and i' are equivalent in the sense that their outputs coincide for any valid combination of the circuit primary inputs. Note that factoring cannot introduce a violation of the technological constraint $l \leq l_{max}$.

Gate transformations change the number of stages, or logic levels, needed to implement one (or more) primary output of the circuit. In general, for a gate elimination (extraction), gate i' is more complex (simpler) than gate i. Therefore, its propagation delay may get larger (smaller). The data ready times of all the gates that are successors of gate i are affected by the transformation. Moreover, the fan-out of the direct predecessor of gates i' and j may change, and, as a result, so may the data ready times at some other circuit outputs. Let $t'(v)$ denote the data ready time after the transformation, and let $\Delta u(v) \equiv t'(v) - t(v) - s(v)$. A transformation of gate i is said to be "locally favorable" if in the modified circuit: (1) $\Delta u(v_i) + s(v_i) < 0$, that is, the data ready time decreases; and (2) $\Delta u(v_j) < 0 \ \forall j \neq i$, that is, any increase in the data ready times is less than the slack. Since a gate transformation affects the entire circuit, it is important to measure its global impact on the timing of the circuit. Let $\Delta \underline{u} \equiv [\Delta u(v_1), \Delta u(v_2), \ldots, \Delta u(v_{|V|})]$. Then a figure of merit of the global impact of the transformation is measured by $\underline{\eta}^T \bullet \Delta \underline{u}$, where $\underline{\eta}$ is a vector of coefficients. Gate transformations affect the silicon area (estimated by $\mathcal{A}$) of the circuit implementation, because of the change in the number of gates and in their internal structure. Therefore, a comprehensive figure of merit is $\Delta e \equiv \underline{\eta}^T \bullet \Delta + U\underline{\theta}\Delta\mathcal{A}$, where θ is a scalar coefficient.

We consider first an example of an algorithm that uses a global strategy. This strategy guarantees a descent of the maximum data ready time at each iteration. If, in addition, we require the strategy to preserve the technological constraints, the transformed circuit can be implemented without additional gate decompositions; therefore, its performance is guaranteed to be better than the original one.

Circuit resynthesis with a global strategy that preserves the technological limitations can be done by applying a sequence of gate eliminations and/or factoring to a circuit. An algorithm for circuit resynthesis can be implemented by an iteration among the steps: (1) compute the set of critical nodes; (2) determine a particular critical section $\mathcal{C}$; and (3) apply the "best" transformation to the gates represented by $\mathcal{C}$. The detection of the critical nodes is achieved by computing the data ready times and the slacks, as described before. The algorithm then constructs a critical section $\mathcal{C}$ as in the case of the first resizing algorithm. The critical section is required to include a critical section of $H(C_0, B)$ the corresponding gates of which have a locally favorable transformation (according to the technological limitations). If no section exists, the algorithm terminates. The critical section $\mathcal{C}$ and the type of transformations are constructed according to a greedy strategy by examining the values $\Delta e(v)$ of the critical nodes. Then all the gates corresponding to the gates in $\mathcal{C}$ are transformed. At this point, all the data ready times and slacks need to be recomputed; this is done at the next iteration of the algorithm.

As mentioned previously, this algorithm guarantees a descent of the maximum data ready time at each iteration. However, it is inefficient to apply it to large circuits because for each node there are several possible transformations to be evaluated. Because of the computational complexity, only few transformations can be considered in practice. On the contrary, it would be useful to use other circuit transformations that may be beneficial, such as applying simplification after elimination or factoring the gates with the goal of determining common subexpressions. For this reason, it is interesting to develop algorithms with local strategies. In this case, the resynthesis of the circuit is done on circuit blocks, and a richer set of transformations may be applied. Unfortunately, while resynthesizing a circuit block, it is not possible to compute the effect of a transformation on the circuit data ready times and on the slacks, if we are monitoring only the structure of a block. Therefore, some heuristic rules must be used to determine which transformation to apply and where.

The strategy for circuit resynthesis, which has been implemented, is a local one. It exploits the hierarchical structure of the circuit. The circuit is partitioned into blocks, and this partition corresponds to the one used by the YSC system for synthesis purposes. The blocks of the partition are the combinational modules specified in the hierarchical circuit description. Transformations are applied only inside the combinational modules. Each combinational module is described by a subgraph $G^m(V^m, A^m)$ of $G(V, A)$. A *critical module* is a module whose subgraph has critical nodes.

At first, the critical graph for the entire circuit is computed. Then, the critical modules are redesigned one at a time. The data ready times and slacks are computed for the whole circuit after the resynthesis of each block. Circuit blocks are sorted by first choosing those subcircuits with nodes that have no predecessor in any other critical module that has not yet been resynthesized. The reason is not to change the data ready times at the input gates of a module after its resynthesis, since the circuit structure has been tuned to this particular input signal arrival-time distribution. For the resynthesis of each subcircuit, each subgraph corresponding to a critical module is considered along with its boundary conditions, which are the data ready times at the module inputs and the slacks at its outputs. While doing resynthesis of a module, the goal is to reduce the data ready times at the critical nodes. Note that since the module is "extracted" from the whole circuit, the slacks cannot be recomputed after each transformation. Therefore, the critical nodes are computed only once, and the transformations are applied to reduce the data ready times at these nodes.

The overall goal of the algorithm is an area effective and time effective synthesis. An elimination leads to a more compact implementation, if the node being eliminated is not implemented. Therefore, if the eliminated node had more than one direct successor, it would be convenient to have it eliminated into its direct successors, even if these eliminations would not improve the timing performances, since they would improve the silicon area. The factoring of a gate requires the implementation of an additional gate; the cost of the additional area can be (partially) offset by the internal simplification of the successor gate. For this reason, it is convenient to consider the extraction of common subexpressions of the logic functions implemented by two (or more) gates. In this case, any node extraction would lead to the simplification of the two (or more) direct successors of the generated node and eventually to a reduced loss (or possibly a saving) of silicon area.

The algorithm for the resynthesis of a subcircuit can be summarized as follows: (1) label the critical nodes; (2) eliminate labeled nodes; (3) simplify the gates where an elimination took place; (4) extract common subexpressions; and (5) perform area/delay tradeoff. The critical nodes are detected at the beginning. All subsequent operations attempt to decrease the data ready time at these nodes. Direct predecessors of labeled nodes are eliminated into them only if the transformation is locally favorable and the estimated area $\mathcal{A}$ is less than a bound $\mathcal{A}_{max}$. Simplification is applied to those gates where an elimination took place. Then, for each subexpression common to at least one labeled node, a gate is added to the circuit to implement the subexpression if the extraction is locally favorable and $\mathcal{A} < \mathcal{A}_{max}$. Finally, area/delay tradeoffs are exploited by moving subexpressions across gate boundaries to reduce gate and device count.

The algorithm may be constrained to preserve the technological constraints or not. In the latter case, some gate decomposition may be required as a last step. Even though the algorithm does not guarantee a descent of the data ready time of t^*, experimental results have shown that performing resynthesis with local strategy (with or without preserving the technological constraints) yields a considerable improvement in the circuit timing performance.

Circuit Repositioning

The repositioning algorithm aims at improving the timing performances by reducing the capacitive load $\{c\}$ of the critical gates. This is done by shortening the length of the nets carrying critical signals. In the YSC system, each module is designed as a rectangular macrocell consisting of an interconnection of smaller cells each implementing a logic gate. Gates are interconnected by wires that run inside and outside the macrocell. We assume that the capacitive loading on a gate depends primarily on the length of the wires that go across the module boundary. Therefore, we restrict our attention to the intermodule wiring, and we estimate the wire lengths by the macrocell positions.

The goal of the algorithm is to reduce the length of the critical nets by changing the mutual positions of the modules. The rationale of the repositioning algorithm is to bring the critical nodes (gates) closer to their direct successors in other (not necessarily combinational) modules. Since any change in a module position affects the length of all the nets connected to it, we require that any change in the module position correspond to an improvement in timing performance. For this reason, we introduce the *geometrical slack* $g(v_i)$ for each net connected to the output of the gate denoted by v_i. The geometrical slack $g(v_i)$ is derived from the slack $s(v_i)$ and represents the additional length that each net can tolerate while satisfying the relation $\bar{t}(v_i) \geq t(v_i)$ at each output node v_i.

The repositioning algorithm is based on pairwise module interchange. The moves that are allowed are position interchanges between modules with compatible shape, such that the increase in wire length for each net connected to each gate v_i of the modules under consideration is less than

$g(v_i)$. With this choice, each move cannot increase t^*. The objective function is a weighted sum of the net lengths connected to the critical nodes. The algorithm uses a greedy strategy in determining a sequence of allowed moves corresponding to a maximal decrease of the objective function.

The algorithm computes first the critical graph $H(C_\varepsilon, B)$, the set of critical modules M, the slacks $s(v)$, and the geometrical slacks $g(v)$. Then it iterates the following steps: (1) select a critical module; (2) determine the set of allowed position interchanges for the selected module; and (3) select and perform a position interchange. The algorithm first determines a candidate module for interchange by using a weighted sum of the parameters of the critical nets connecting it to other modules of compatible shape. Then the set of allowable moves is determined by considering the geometrical slacks. If no move is possible, the candidate is rejected. Otherwise, the local best interchange is performed and another interchange for that module is searched for. When all the critical modules have been examined as candidates, another pass is done if at least one module interchange has been performed at the present pass. Otherwise, the algorithm terminates. If the original placement was obtained by minimizing the total wire length, this number is monitored during the interchange. The algorithm trades off the total wire length for the wire length of the critical nets. An additional termination criterion that may be used is reaching a bound on the total wire length.

7.3 Using the YSC System

7.3.1 The YSC System Implementation

The YSC system implements the ideas and the synthesis algorithms presented in section 7.2. The system consists of a coherent set of programs that perform the synthesis of a digital system from its description in the hardware language to the specifications of the geometries of the masks needed to fabricate the chip. We describe here some implementation details and how the Yorktown silicon compiler is used to synthesize a VLSI chip.

Since the YSC system is experimental, we have chosen to implement most of its tasks in the APL language. APL is an interpreted language, and the APL notation allows for complex operations specified by simple statements. Thus APL programs are generally compact, easy to write, and easy to debug. The various synthesis tasks of the YSC system are each organized around a set of functions that operate in an environment called an *APL workspace*. The APL workspaces provide facilities for interaction and easy debugging of code. As a result, APL permits fast code development and easy experimentation with new ideas. On the other hand, a few synthesis tasks are based on well-established, computationally expensive algorithms (e.g., floor planning using simulated annealing). These tasks are still implemented by APL functions in APL workspaces, but the computationally expensive parts are executed through function calls to compiled programs, written in Pascal, for faster execution.

Each major task of the YSC system is implemented as an APL workspace. The workspaces communicate with each other by writing intermediate files to the operating system file mechanism (in this case, CMS operating system). These files may contain binary or textual information and are linked by name through a hierarchical set of references. This hierarchy corresponds to the initial hierarchy of the hardware description or its refinement during the synthesis process. The file names are those given by the user or those created by the structural synthesis process. Different file types are used to specify the type of information contained in the file. For each intermediate stage of the synthesis process and for each module, there is a set of files containing the corresponding intermediate data. The set of files at a given stage of the synthesis represents a view of the system being designed, which corresponds to a particular level of abstraction. For example, a hardware system can be described by a set of files containing either a YIF or a YLL description. This data will be transformed into another set of files describing the logic equations of each module, then into files describing the topology of the transistor implementation, and eventually into a set of files containing the layout geometries.

The compilation of a digital system consists of a sequence of calls to the appropriate APL workspaces. This sequence will be described in section 7.3.4. The result is the automatic transformation of the chip description from the high level into a set of geometries representing the mask data.

7.3.2 Describing Hardware to the YSC System

The hardware description principles were described in section 7.2.1. We consider here the task of the hardware designer in describing an architecture to the YSC system. The Yorktown silicon compiler supports several approaches to hardware descriptions, ranging from pure behavioral to pure structural and including combinations of both.

A behavioral description can be given by a program in the V language or, equivalently, in a YIF file. In this case, the circuit structure is determined during the structural synthesis phase of the YSC system. On the other hand, purely structural descriptions can be given by a hierarchical set of HND and YLL files. In this case, the HND files specify directly a system partition into combinational logic modules, registers, drivers, receivers, and so forth. Structural synthesis is thus bypassed, being performed by the human designer directly. Of course, a behavioral description frees the designer from having to describe the details of the circuit implementation. However, it is sometimes desirable to specify the structure of the circuit more directly. For example, computer architects sometimes have in mind an implementation when they specify an architecture at the behavioral level. Of course, it would be desirable that the structural synthesis stage discover an equivalent or better implementation. In general, this is not yet possible. For example, structural synthesis does not support the detection and efficient use of concurrency required for an efficient pipelined implementation. Therefore, to be able to synthesize chips with specific structures, it is necessary to specify the structure directly. This is a limitation of the present knowledge and algorithmic development of the structural synthesis stage, which can be expected to improve in the near future.

In the YSC system, it is also possible to use a high-level hardware description that is neither purely behavioral nor purely structural. Structural implications in the hardware specification may be achieved by arranging the description as a set of module calls (in either V or YIF). Each module would correspond to either a combinational block or a finite automaton. The distinction between a module call and a call to an operator in the YLL library is that the operator call is similar to a macroexpansion mechanism, while a module call is similar to a procedure call. Thus a module call produces a separate piece of hardware, and an operator call is expanded and merged with the logic of the present module. Our mechanism for describing concurrent automata is to describe each as separate modules, using a user-written HND module to connect them through their I/O names. Note that such a partition is forced by the design specification and is preserved through synthesis. It is also possible to specify systems with mixed behavioral and structural representations by using two (or more) descriptions, one structural and one behavioral, which communicate through their I/O ports and are connected with an appropriate HND module description. One description may be purely in V or YIF, and the other purely HND or YLL.

7.3.3 Application-Specific Cell Descriptions

Everything that can be described in the input languages of the YSC system can be synthesized by the programs. This does not always lead to acceptable implementations. The libraries are available for cells that are quite common and in which special features must be realized. The user can extend the library with cells that he or she wants to be treated in a special way. This can be done at several levels.

If, for example, the user wants a special logic implementation of an arithmetic unit, a YLL description can be added to the library. In this description the user must make sure that the intermediates wanted for this implementation are referenced as global variables. The logic synthesis system will view these as outputs of the logic module and, therefore, will preserve them in the gate implementation of that module. During the linking their intermediate character will become apparent again, and proper connection to the environment is guaranteed.

If the requirements are more specific, even down to the layout level, more work must be done. If the layout is completely fixed, there is the possibility of adding an inset cell and its periphery description to the system. This will suffice for the incorporation of the cell in the final layout. If this cell is to take part in simulations that are carried out for verification purposes, a consistent functional description is needed as well.

If there is still some flexibility needed in the cell, a special cell assembler must be written. In the YSC system experience described in section 7.4 such a cell occurs; it is called the general-purpose register. We have supplied a cell assembler for that module. Part of that module, the address decoder, is described in YLL but is passed through the logic editor, without doing any optimization effort, in order to obtain the appropriate data structures describing that gate implementation. The layout program treats each decoder as a normal logic macro with the constraint that all circuits must be placed in one column. The other part of the module contains the latches, the bit line prechargers, the sense amplifiers, the write drivers, and the shut-off feedback circuitry. The layout of all these circuits, except the feedback circuitry, is completely implied by the procedure as soon as the basic cell dimensions and the number of read and write ports are given. The feedback chain, however, can be completely tuned—that is, every transistor in that chain can be sized individually on the basis of simulation, experience, or other criteria. The two parts, the address decoding part and the latch array with its supporting circuitry, are connected together by a generic river routing routine followed by the right mask generator.

In the high-level input language, all these application-specific cells must be treated in the same way as the more permanent library cells.

7.3.4 The Synthesis Stream

Silicon compilation must be capable of being a completely automatic process. The YSC system provides both an automatic and a hand-driven synthesis mode of operation. The latter is useful for stepping the compiler through its various stages and for verifying whether the circuit synthesis is correct and optimal. Errors may be introduced in the hardware description that may be detected after the first stage of the compilation. The compiler developers may wish to study new algorithms for synthesis and optimization and verify their correctness. Also, decisions about which procedures to apply may be made by the user as certain intermediate reslts become available.

Each major task of the YSC system is implemented by a set of functions in an APL workspace and can be executed by a single command. This command loads the appropriate APL workspace and executes a main execution program. The exact mechanism used in the CMS operating system is the Rexx interpreter executive command language. The Rexx program initiates the APL processor, loads the appropriate APL workspace, and starts the main program. The workspace also contains all the parameter and option settings for executing the corresponding silicon compilation task. The main program is stored along with the workspace and need not be known or understood by the user of the YSC system. However, it is of extreme importance to its developers, since the option settings affect the quality of the results and should be set properly. For example, in the case of logic synthesis, the execution program specifies the sequence of operations (e.g., extract, eliminate, simplify, etc.) used to optimize and synthesize the multilevel logic. Certain default settings for the synthesis procedure can be made to depend on the name of the macro being synthesized. In this way, an ALU can be treated differently from a piece of control logic or a multiplexer for example.

The entire silicon compilation process can also be executed by a single command, which invokes a main Rexx program that is a sequence of commands each related to a synthesis step. As an example, a main execution program could be as follows:

- Execute structural synthesis
- Execute logic synthesis
- Estimate delays
- Execute floor planning
- Estimate delays
- Execute timing optimization
- Execute macrogeneration
- Execute layout assembly

A "standard" main execution program can be invoked by a user of the YSC system who does not need to know its content. However, other main execution programs may be executed either in the development and experimental phase or to compile chips with special features or requirements. For example, if the design is a microprocessor and the designer wants to keep all the data flow portion in a stack organized with a particular structure, an appropriate command must be introduced (execute data flow stack construction) in the main execution program.

This programmability feature of the compilation process allows the compiler to be extended easily to incorporate new ideas. It is conceivable that different users of the YSC system will have different interests and capabilities. The novice may need to know only how to describe the hardware and how to invoke the main execution program. A more sophisticated user may want to understand the mechanism of the compiler and explore the use of the different execution sequences. An expert in hardware and software design may want to add a special feature to the compiler for his or her own needs (e.g., an on-chip self-test pattern generation module) by writing a program for a specific task and invoking it from the main execution program.

7.4 Experience with Using the YSC System

As an exercise for the compiler and as an opportunity to learn more about system-level issues, we have developed a design for the IBM 801 processing unit [Radi83] and described it in the languages used as input to the YSC system. We chose the 801 architecture for two reasons. First, since the 801 was developed at IBM Watson Research Center, we had access to a detailed description of its operation as well as to the developers themselves. Second, the 801 is sufficiently complex to provide a good test for the compiler and, at the same time, sufficiently simple to be understood by novice system designers.

This section describes the process of spanning the gap between the architectural specifications and the input to the silicon compiler. We present two descriptions of the 801 processor as inputs to the SC system. The first is a behavioral description in the V language and implies only a few features of the circuit structure. The second is a description in the HND/YLL notation, which allows the user to specify the structure in detail. We also report the results obtained by the Yorktown silicon compiler in synthesizing the chip.

7.4.1 The 801 Architecture

The architecture of a system defines its attributes as seen by the programmer—that is, the conceptual structure and sequential behavior of the machine as distinct from the organization, the logical design, the layout, and the performance of any particular implementation. Typical parts of an architecture specification are the instruction set, the addressing capabilities, the registers visible to the assembly language programmer, and the interrupt handling. In spite of the clear separation provided by this definition, it certainly would be unwise to isolate architecture design from its target operating system and compiler on one side and from its hardware implementation on the other. Therefore, it was not surprising that during this exercise we were confronted repeatedly with the fact that the 801 architects had the application and the implementation constantly in mind. Their decisions were mainly guided by three basic design principles:

1. A compiler should be able to produce object code with an efficiency comparable to the best hand code so that assembly language programming is never needed for performance.
2. Each instruction should be able to be executed in one short machine cycle, and yet the path lengths of the computations should not be commensurately larger than those required by machines with more complex instructions.
3. The idle time of the processing unit due to storage access should be kept short.

What follows is meant only as a brief introduction to the architecture of the 801 processor. For more details the interested reader is referred to [Radi83].

The 801 is a 32-bit minicomputer architecture. The arithmetic and logic operations deal with 32-bit words. Shifts and rotates can have lengths of up to 32 bits. Addresses are 32 bits long. The 801 architects foresaw a compiler capable of a very effective utilization of a large number of registers, and therefore decided to prescribe a register file of 32 full-word general-purpose registers (GPR).

Instruction length and format greatly influence the complexity of the hardware for decoding and the efficiency of a pipeline if implemented [Mati84]. Therefore, accepting some alignment constraints can be very beneficial with respect to hardware simplification and pipeline implementation. This was recognized in the architecture design of the 801. Each instruction is exactly one word long and is aligned on word boundaries. Only a few instruction formats are used. All operands are aligned according to their size. These constraints, the emphasis on register utilization, and the single-cycle instruction principle, move the 801 into the class of streamlined architectures [Henn84]. Although many of its properties are shared with the so-called reduced instruction set computers (risc), the size of the instruction set alone would make such a qualification a misnomer in the case of the 801.

Typical instructions are provided for storage access, address computation, branching, and comparing. Consistent with the third basic design principle, the processor is allowed to continue with instruction execution after the initiation of a storage access. The processor will only stall if the memory system has not yet provided the requested data when the content of the affected register is needed. To make better use of this facility, the optimizing compiler can reschedule the storage access instructions to reduce the number of stalls. Also, a branch with execute instruction (similar to the delayed branch in reduced instruction sets) allows the compiler to move inevitable instructions into the normally idle period following a taken branch instruction.

Arithmetic operations are done using 32-bit two's complement, with the usual add and subtract instructions as well as some special instructions for computing the minimum and the maximum of two values. There are also several instructions not normally found in a reduced instruction set, such as "the multiply step" and "the divide step," which allow complex operations to be easily decomposed into sequences of simple instructions. Also, a rich set of rotate and shift operations, controlled by a register or an immediate field are provided.

The third basic principle immediately led to a store-in-cache strategy. But instead of a single conventional cache that delivers a word every cycle, an instruction cache and a data cache were both included separately, thus effectively doubling the cache bandwidth and allowing asynchronous fetching of instructions and data. Explicit instructions for cache management are introduced to make a reduction in unnecessary loads and stores of cache lines possible at the software level.

Finally, the number of interrupts defined for the processing unit is reduced by imposing software protocols for interrupt handling and prescribing an external interrupt controller. Protection is achieved by making some instructions privileged in the sense that they can be executed only in a special state that is not accessible to application programs.

7.4.2 System Description

An Initial Behavioral System Description

The *Principles of Operation*, a document describing in precise and verbose detail the 801 architecture, was our starting point. In order to condense this bulk of information (as well as to provide us with an understanding of the operation of the machine), we first translated these principles of operation into a high-level behavioral description in V, specifying the effect of each instruction on the architected registers, the details of when interrupts may occur and how they are processed, and so forth. This description is also very useful as the system documentation; for example, from within an editor, we could easily locate all of the instructions affecting a particular special-purpose register.

The main routine is nothing more than an infinite loop. For each cycle this loop fetches (P8RI) an instruction, executes (P8EXE) it if no instruction storage interrupt has occurred, and tests the various interrupt conditions (P8CHI) as shown in Figure 7.18.

Architected registers were modeled by variables in the program. The general-purpose registers (GPR) were modeled by an array with 32 elements. Other variables not mentioned in the principles of operations were necessary to hold intermediate results or to "remember" certain special situations.

The V module P8RI roughly corresponds to the example presented in sections 7.2.1 and 7.2.2. In the execution module P8EXE, each instruction was coded as a CASE. The decoding of the instruction was done by first separating the two main instruction formats with an IF statement and then decoding the appropriate bits with a CASE statement (see Figure 7.19).

Interrupt checking (P8CHI) is done by examining various interrupt flags and external pins in the order established by interrupt priorities. The flags are set during instruction fetch and execution according to the allowed interrupt conditions. In the case of an interrupt, P8CHI will load a new machine state and save the old one in some special-purpose registers.

Besides the interrupts, the most difficult aspects to model using only an imperative language were:

- *The GPR management.* All data memory access goes from and to GPRs. To minimize the idle time due to data memory requests, these operations

```
MODULE P801                 /* This is the main 801 module */
  ...
BODY P801
  ...
 DO INFINITE LOOP           /* Fetch-Execute-Check-interrupt loop */

    P8RI;                                         /* Fetch */
    IF ¬IN_STOR_FLAG THEN P8EXE; ENDIF;           /* Execute */
    P8CHI;                                        /* Check interrupts */

 ENDDO;

END P801;
```

Figure 7.18 Main V program for the 801 processor.

```
MODULE P8EXE;
  ...
BODY P8EXE;

 IF ¬((IR::0,6)=63)
   THEN                         /* Simple opcodes    */
     WHEN (IR::0,6)
       CASE 16;                 /* LC RT,D(RA)       */
         ...
       CASE 18;
         ...
     ENDCASE;

    ELSE
      IF ¬((IR::21,4)=0)
        THEN PGM_FAULT:=0;     /* not defined       */
        ELSE                    /* Extended opcodes */
          WHEN (IR::25,7)
            CASE 16;            /* LCX RT,RA,RB      */
              ...
            CASE 18;
              ...
          ENDCASE;
      ENDIF;
 ENDIF;

END P8EXE;
```

Figure 7.19 Program structure for the execution module.

can be initiated during an instruction and finished when the requested data arrive. Hence, for memory pending read operations, it is necessary to implement a GPR management that controls which registers contain valid data and which still must be actualized.

- *The instruction storage access*. Instructions are prefetched, and so it is necessary to keep track of whether a prefetched instruction can be used or must be discarded because of a branch or an interrupt. Also, an instruction fetch might result into a processor stall if the instruction cannot be (pre)fetched fast enough.
- *The data memory access*. Data memory access is done in parallel to instruction execution. Each access might imply a processor stall if a previously initiated data access is still occupying the data bus.

The technique used to model these three aspects consisted of writing one module for the management of the GPR, one for the instruction storage bus, and one for the data storage bus. To use these resources the respective module call was issued.

This description could be used as the starting point for structural synthesis. However, this raises the following problems:

- the design is not pipelined. We have no means of automatically constructing a pipeline out of a nonpipelined description and including all the pipeline control (e.g., flushing the pipe in the case of an interruption or a branch, feeding back results to previous stages of the pipe without storing them first in registers to avoid processor stall, etc.) To our knowledge, no such methods have been exercised in practice with realistic designs.

- The GPR originally was modeled as a single port memory. In practice, this is not a viable solution because it results in very bad performance. Multiport memories are necessary. Although we can handle multiport memories, the problem of assigning ports to accesses of a single array representing the memory is not done automatically in our system yet. To use multiport memories, each port must be modeled as a separate variable, thus deciding in each access (by specifying the particular variable) which port will be used.
- The execution module coded as presented results in a huge number of control states. The reason for this is that during each instruction that uses registers or the data storage bus (i.e., almost all instructions), a processor stall is possible. Since waiting states are created automatically in this case, a different waiting state for each instruction (basically "remembering" the particular instruction) results. Since many instructions access more than one register, the total number of these control states is huge. This is not a problem in principle—that is, the state transitions are quite simple and result in a small amount of logic, the memory required to hold these states is small, and everything is generated automatically. Only the running time of structural synthesis will increase.

As a consequence, the performance of a processor synthesized in this way would not be acceptable. As an exercise and a test for the Yorktown silicon compiler, we synthesized the structure of the 801 in this version. The results are given later in this section. The real design required some modifications, such as the introduction of a pipeline and the separate representation of each port of the GPR. As will be seen, this also solved the problem of the excessive number of control states.

System Decomposition

To introduce a pipeline, we decomposed the system into pipeline stages, specifying each stage as a separate design. An additional module implemented the pipeline control. The pipeline design was done manually. Since some of the 801 instructions seem naturally tailored to a pipelined implementation, the first behavioral description could be easily decomposed into what seemed to be a natural four-stage pipeline. This choice was different from the design discussed in [Radi83].

The stages, as illustrated in Figure 7.20, are:

1. *Fetch (F-stage)*—Load into the instruction register the instruction pointed to by the program counter, a branch address, or an interrupt address. Load the program counter with an appropriate new value.
2. *Decode (D-stage)*—Load the appropriate registers (RA,RB,RS) with the contents of the general-purpose registers pointed to by the instruction.

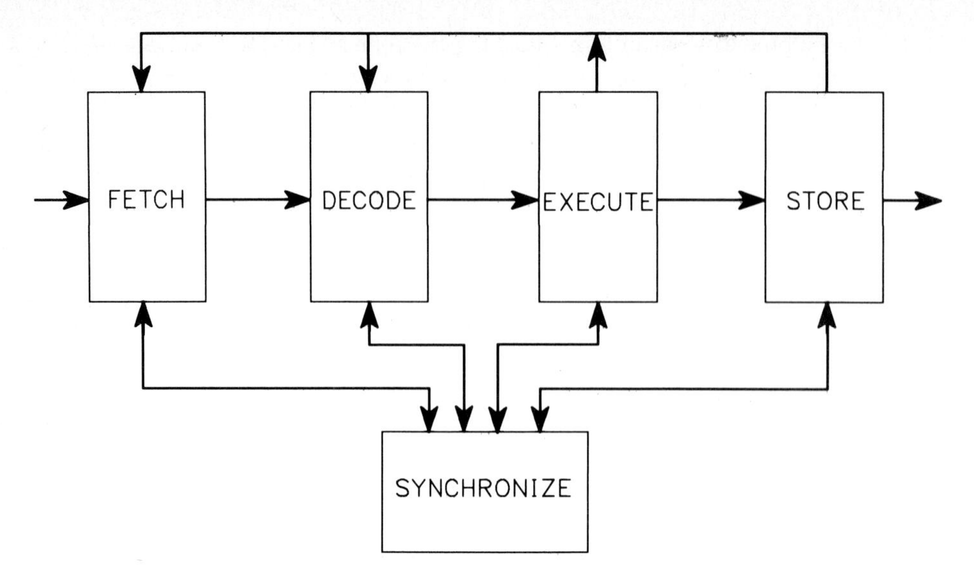

Figure 7.20 Pipeline structure for the 801 example.

3. *Execute (E-stage)*—Generate control signals from the operation code, make the arguments available to the execution unit, and latch the results. Generate a branching address if necessary.
4. *Store (S-stage)*—Place the result in a register. Initiate requests to main memory.

The behavioral description of the pipeline consists of five V programs, one for each stage plus one for pipeline control. Each of these programs is synthesized independently, obtaining five finite automata that run concurrently. Each of them has only a small number of states, because the part chosen to be performed by one stage naturally corresponds to parts that had fit in one state in the initial description! Instructions are decoded in each of the last three stages, so it is not necessary to encode the instruction in the control states. Additionally, each stage decodes the instruction differently, allowing optimization to a large extent (e.g., the logic needed by the decode stage uses only a few transistors).

The pipeline works as follows. In the absence of branches and interrupts and if data are delivered fast enough from the data and the instruction cache, each stage of the pipeline performs its operation in one cycle. In the case of a branch instruction, if the branch is taken, the program counter is loaded with the target address at the end of the E-stage. The partially executed instructions in the other stages of the pipeline are now incorrect; in an unpipelined design, they never would have been fetched. To handle this situation, each stage of the pipeline has a "kill flag," which is propagated from one stage to the next as the instruction flows down the pipe. A branch taken sets the kill flag for all the other instructions currently in the pipe.

When such an instruction reaches the S-stage, its effect on architected storage is inhibited because the kill flag is set. In this way, a branch instruction effectively flushes the contents of the pipe.

Interrupts are very similar to branches. In fact, the action of the processing unit under an interrupt is to store the current status in special-purpose registers and then to branch to a fixed memory location. One additional consideration is required because interrupts can be generated at any stage of the pipe. To handle this, each stage has interrupt bits that flow along with the instruction in the same way as the kill flag described previously. As interrupts are generated, these bits are set; when the instruction reaches the S-stage, the interrupt is serviced.

An instruction can generate an interrupt in the D-stage, for instance, and then be killed by a branch instruction ahead of it. In this case, the instruction in the D-stage never should have been executed and the interrupt it generated is ignored.

Another consideration is "waiting for data." For example, when loading a register from external memory, the load instruction generates a memory request that is sent off-chip. Execution continues normally until a reference is made to the register that was supposed to be loaded. If the memory request has not yet been serviced, we must wait for it before proceeding. The waiting function is accomplished by inhibiting the transfer of data from one stage of the pipeline to the next. In this way, we effectively halt the pipeline until the external request is serviced.

The flow through all four stages of the pipeline is altered somewhat by certain bypasses in the pipeline. These speed the flow of the instructions through the pipe. For instance, if an instruction in the D-stage requires the contents of a register that is the target of a load instruction or the target of the result just computed in the E-stage, the data will be routed into the appropriate latch on the boundary between the D- and E-stages. On the next cycle these data are stored in the correct register but are also already in the E-stage. This allows, for example, a sequence of additions to be executed without interruption. Another complication is that there are instructions that refer to writing data directly into some of the architected latches. This could be done at the end of the S-stage, but this might cause some delay in the instruction flow. If we know that no interrupt will be processed on the next cycle, then this store can be performed at the end of the E-stage. In general, these decisions corrupt the clean design of the pipeline, complicate the control logic, and may affect the cycle time of the machine.

A Structural System Description

Although the behavioral description is sufficient input for the YSC system, we also described the complete structure of the 801 in HND and the combinational logic in YLL. In addition to the historic reasons for this further description [Bray85], it was very useful to validate the results of structural synthesis by comparing them with this "hand design."

The major steps in defining the structure were:

- *Identification of all registers.* In addition to the specified architected registers, many others were necessary.
- *Definition of all the necessary control signals.* The control within each pipeline stage was treated locally. Local control is convenient because it reduces the wiring. Moreover, it can be efficiently designed by exploiting the unused or inappropriate operation codes as don't-care conditions for the local control blocks.
- *Partitioning of the combinational logic.* While writing the behavioral description of the 801, the frequent operations were identified and coded as a separate module. This already represents a first structure, which guided the partition of the logic. Sometimes further partitioning was advantageous, such as when the execution unit was divided into an arithmetic-logic unit and a rotating and merging logic.

After the structure was obtained, it was directly coded into HND. Writing YLL programs for each logic macro in the 801 turned out to be a fairly straightforward and modular programming task. For the logic involved in the control part, YLL's ability to handle tables proved to be very convenient. Our structural design led to the diagram in Figure 7.21.

The major difference between our behavioral description and our structural description is that the latter does not include an explicit module to

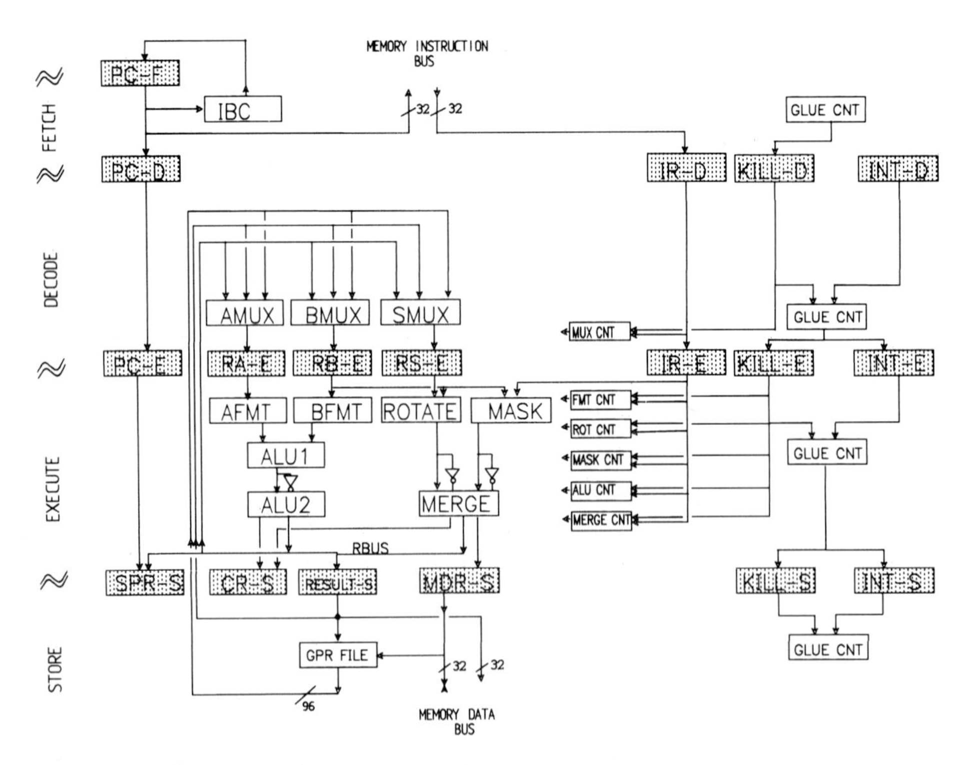

Figure 7.21 Schematic of our structure of the 801 processor.

synchronize the pipeline. This module only controls the synchronization aspects of the pipeline stages, and consequently it is small.

7.4.3 Design Statistics

In this section, we summarize very briefly some results obtained by using the YSC with our 801 processor design.

Input Code

Our two behavioral descriptions and the structural description resulted in the input code given in Table 7.1. The operator library included only 16 operators coded in YLL, enough to implement any of the YIF operators. The cell library contained only a register file and bit images for single latches, drivers, and receivers. The data in Table 7.1 contain the entire chip description.

Structural Synthesis

The synthesis of the nonpipelined version led to the results shown in Table 7.2. "Operations" refers to the number of YIF nodes before control and network generation, "states" refers to the number of generated control states, and "latches" refers to the number of bits requiring a latch. The module EXE contains the complete data path, but the control was generated for only 40 percent of the instructions.

The results obtained for the pipelined version are given in Table 7.3. They do not include some purely combinational modules that are also necessary.

The pipeline stages automatically synthesized led to the combinational logic shown in Table 7.4. The transistor count given includes only the transistors required for the gate logic (pull down tree). Those were the best results achieved concerning transistor count. Trading 15 percent transistor count for 25 percent of speed improvement was possible, for example, for FETCH. The delay is estimated for a 1μ CMOS domino (single voltage cascode switch or SCVS) implementation. A straightforward implementation of the logic specification generated by structural synthesis would have

Table 7.1 Input Code Statistics

Version	*Modules*	*Code (lines)*	*Code (statements)*	*Tables (lines)*
Behavioral, unpipelined	13	V: 1,487 YIF: 10,212	V: 1,370	
Behavioral, pipelined	10	V: 1,820 YIF: 12,481	V: 1,190	
Structural, pipelined	YLL: 64	YLL: 1,877 HND: 132	YLL: 771	YLL: 421

Table 7.2 Results of Structural Synthesis for the Nonpipelined Version

Module	*Operations*	*States*	*Latches*
CHD	34	4	5
MAIN	29	5	289
CC	70	0	0
EXE	706	113	114
CHI	55	10	4
RD	54	5	3
EFFA	72	9	91
GPR	20	2	1
INTE	82	2	1
SELBIT	66	0	0
RI	34	4	2
MASK	36	0	0
DATA	44	1	67

contained up to nearly four times more transistors! The pipeline stages DECODE and EXECUTE were partitioned automatically because their combinational logic resulted too large.

Our manual decomposition led to a total of 118 different modules. The combinational logic modules have been synthesized to the gate level. Also a global net list was created for the chip. These data have been used for gate-level simulation and timing optimization at the chip level.

Some statistics of the synthesis results have been entered in Table 7.5. The totals in this table do not include the internal powering of signals (e.g., by clock repeaters), but they do include all the off-chip drivers and receivers. In compiling the statistics, all output buffers are assumed to be of minimum size. (The driver-sizing factor is implemented by inserting parallel transistors of minimum width.) As a result, the transistor totals for the domino circuits include only the transistors required for the gate logic and five transistors for the buffer and clock. The decomposition of the logic was done with the possibility of distributing the control to or near to the modules being controlled. Thus, of the 58 combinational logic modules, 25 generate control signals.

Table 7.3 Results of Structural Synthesis for the Pipelined Version

Module	*Operations*	*States*	*Latches*
FETCH	16	4	98
DECODE	329	2	181
EXECUTE	806	1	180
STORE	50	4	35
SYNCHRONIZE	312	1	99

Table 7.4 Logic Generated for the Pipelined V Description

Module	Inputs	Outputs	Levels	Trees	Transistors	Delay(ns)
FETCH	68	69	11	85	417	23
DECODE	219	131	–	320	2,098	–
EXECUTE	255	152	–	1,607	10,414	–
STORE	73	44	4	46	191	8
SYNCHRONIZE	202	167	13	212	1,107	28

Timing Optimization

The programs for timing optimization have been tested on the microprocessor synthesized from the structural description (Table 7.5). The maximum data ready time in the circuit, t^*, was 72.7 ns before optimization. A pass of a resynthesis and a resizing step could reduce the maximum data ready time to 53.2 ns, or 26.9 percent. The critical path of the processor was related to the generation of condition codes, as a result of a trap condition based on the outputs of the processor ALU. Therefore, the path affected 9 critical modules, or 15 percent of the total.

The repositioning algorithm was applied only to the part of the chip consisting of the glue logic circuits implementing the interrupt and global processor control. One pass of repositioning and resizing reduced the maxi-

Table 7.5 Design Statistics for the Pipelined Structural Description

	Total blocks	Total gates	Min/Max Gates in block	Min/Max Transistors in block	Total transistors
Combinational logic blocks	58	1,415	1/110	10/1,344	17,660
Inverter blocks	8	150	1/97	3/291	450
Subtotals	66	1,565			18,110
Regular latch blocks (LSSD)	5	38	1/32	24/768	912
Latch/Enable blocks (LSSD)	31	514	1/32	32/1,024	16,448
Register file	1	1,184	1,184	19,000	19,000
Subtotals	37	1,736			36,360
Driver blocks	4	69	1/32	4/128	276
Reciever blocks	10	41	1/32	0/0	0
Tristate driver/receiver	1	32	32	320	320
Subtotals	15	142			596
Totals	118	3,443			55,066

mum data ready time by an additional 2.1 ns, so that the total improvement in one pass was about 30 percent. This corresponded to a decrease of the total estimated length of the critical nets of about 4.9 percent.

After repositioning (and the corresponding resizing), the set of critical modules included only two modules that were not critical before. Resynthesis could reduce their delay by about 1 ns, which corresponded to an equivalent reduction of t^*. For this circuit example, no other timing improvement could be achieved; the maximum data ready time was 49.9 ns and the total reduction of t^* was 31.3 percent. At this point, the performance of the circuit was limited by the structural decomposition on which these timing techniques have no control.

Layout Design

The chip image, except for the bonding pads which form the perimeter of the chip, is shown in Figure 7.22. The total area with bonding pads is 72 mm^2 and has been constructed using a 1.3 micron production standard CMOS process with two levels of metal. The major modules seen in the figure are a stack, a general-purpose register (GPR) file and the remainder, which comprises mainly the control logic and related latches. One of the

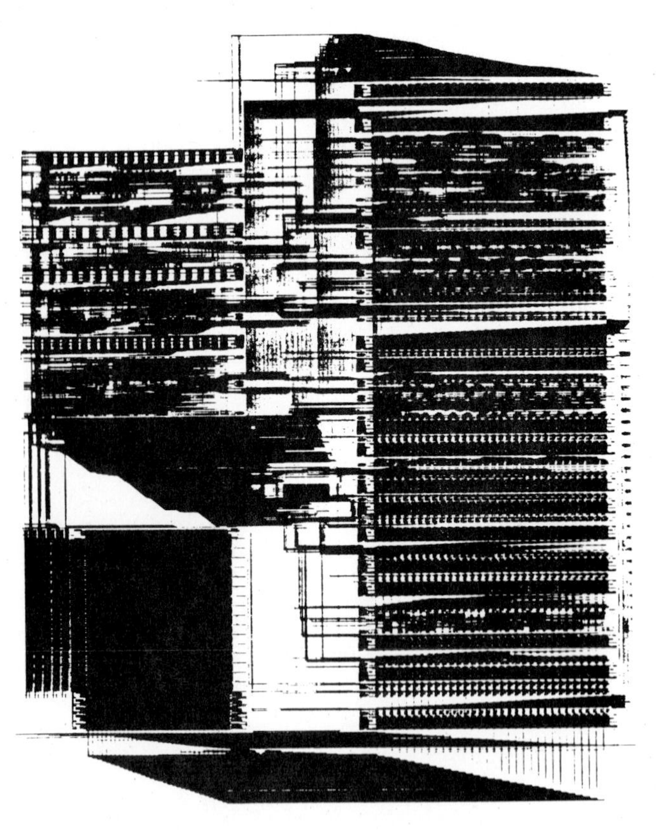

Figure 7.22 Chip image (excluding bonding pads) of the 801 processor.

major constraints dictating the floor plan at this level was that the stack, which ideally would contain the GPR (note 160 nets connecting the top of the stack with the top of the GPR), had to be folded or else the chip would be too long in one dimension. The remainder was originally decomposed so that it could be distributed with the macros being controlled. However, once the stack is folded, the space under the GPR is free, and hence, in this case, is used for the remainder for this chip floor plan. Finally, the remainder was constructed as a stack because a large clock driver was required for each macro to ensure that clock signals were well distributed.

Clearly, more area can be made free by distributing the control as part of the stack along one side, thus essentially eliminating the two major global wiring channels. However, for better area efficiency more function should be added to this chip.

Nevertheless, the chip as is, in area and delay, is slightly smaller and somewhat faster than another chip containing the same functionality and technology ground rules, but designed by hand. We attribute these results to:

a. The use of domino logic, a two-phase domino clock, and linear transistors for the gates
b. The extraction and use of don't cares for logic minimization
c. The automatic design of a normal stack which includes much more than a normal data flow stack

Even with the floor plan shown, some improvements should be tried, something that is quite easy with a silicon compiler. For example, better area can be obtained by turning the GPR so that its 160 data nets enter the side of the stack directly. Similarly, the remainder should be rotated 90° clockwise so that its nets enter the middle vertical channel directly. Both of these changes require reconstructing the stack. This is a relatively easy task for the YSC system, but would be quite difficult for other design methods, including other silicon compilers.

7.5 Conclusions

The YSC system has been designed to satisfy its primary goals: automatic compilation, competitive chip design, and creation of a design environment for experimenting with microsystem architecture and CAD algorithms. The synthesis algorithms have been chosen because they are guaranteed to complete without conflict or the need for intervention. In other words, they support automatic silicon compilation. The synthesis and optimization algorithms are the strength of the YSC system, because they allow the user to explore different system solutions and they offer leverage over manual design procedures. A key part of the system is the ease with which it can be enhanced, thus offering the possibility of future extensions as algorithms are improved.

The compiler structure and hardware description languages are sufficiently general to allow a wide range of digital systems to be designed with relative ease. The compiler input languages allow the user to accurately control the implementation if desired. Otherwise, the user need only specify the system behavior and let the compiler make the major decisions about the system implementation.

Our original goals have been fulfilled by the results of the compilation of a medium-scale, 32-bit processor. We could achieve different system solutions, explore the benefit of pipelined architectures, and eventually achieve a chip image that compares favorably with manual design of the same architecture.

7.6 Acknowledgments

Several people contributed to the Yorktown silicon compiler system. We take this opportunity to express our appreciation to all those who helped us in building the system. In particular we would like to acknowledge the contribution of Curtis McMullen to the logic editor (YLE) and the linker, of Norman Brenner to the YLL language and processor, and of Lukas van Ginneken to the annealing algorithm and the global routing. Within IBM Research, Marian Mack supported the layout design programs, C. L. Chen provided us with insight into the design and technology problems, and Yiannis Yamour helped in understanding logic and architectural issues. Several visitors to IBM Research helped in writing experimental tools and in preparing the description of the 801 processor: Willem van Bokhoven, Jochen Jess, Jacob Katzenelson, Richard Rudell, and Jan-Edzard Talsma. Within IBM Corporation, we also benefited from interactions with other IBM divisions, especially Jim Davis at Boca Raton, Bill Griffin and Ralph Kilmoyer at Burlington, and Gunther Machol at Los Gatos. We would also like to thank Carla Otter for helping us in preparing this manuscript.

CHAPTER 8

CATHEDRAL-II: A Synthesis System for Multiprocessor DSP Systems

J. Rabaey
H. De Man
J. Vanhoof
G. Goossens
F. Catthoor

8.1 Introduction: Goal, Motivation, and Summary

This text reports on the status of the work on a silicon compiler for the automatic synthesis of synchronous multiprocessor system chips starting from a high-level behavioral description. The work on this compiler is the subject of the ESPRIT project 97 sponsored by the EC and undertaken by IMEC, Philips, Siemens, Bell Telephone Manufacturing Company, Silvar-Lisco, and Ruhr University Bochum. This report is on the work done at IMEC.

A silicon compiler is defined here as a software system supporting chip layout synthesis starting from a behavioral system description. This work is a continuation of the work on an operational silicon compiler CATHEDRAL-I for bit-serial digital filters, which has been reported in [Jai86]. While doing CATHEDRAL-I we have experienced that, in order to be successful in silicon compilation, it is necessary to start with a careful definition of the target architecture and its associated design strategy.

We believe that *the* silicon compiler simply does not and probably never will exist just as *the* software compiler does not exist since many source and target languages exist, each of which is optimized for a given task. Therefore we believe that efficient silicon compilers will necessarily be strongly tied to a particular target application area for which it is first necessary to define a target architecture. As a result a good many silicon compilers will evolve in the future.

A target architecture is defined as the hardware design style into which the behavioral description is to be mapped. Again, if this is to be successful, we think that this design style must be clearly defined and constrained in order to allow for a formal approach. A good compromise must be found between the constrained design style necessary for the compiler to be feasible and the efficiency to realize the target application area. This requires a careful study of the architecture and the design methodology in the context of realistic problems in the application area before any tool is developed.

Based on the results of such study on handcrafted designs, the synthesis formalism can be defined and the implementation of CAD tools and silicon libraries can begin. In doing so, we can identify target-dependent and target-independent tools. For the latter, is is useful to derive a software development environment, which is reused when the next target is to be implemented. We call that the *diversification environment.*

In CATHEDRAL-II, the target application is a subset of digital signal processing (DSP) algorithms to be architecturally realized by a set of concurrent dedicated bit-parallel processors on a single chip. Such an architecture is especially suited for so-called third generation DSP algorithms involving large blocks of sampled data subject to complex decision-making algorithms. It will be shown in this report that such architectures lead very often to much more efficient implementations than general-purpose DSP processors. As shown in Figure 8.1, we typically address very complex algorithms in the audio and telecommunications intermediate frequency spectrum. We will show, however, that the diversification environment also allows applications in the video domain.

Typical applications are: speech synthesis and analysis, modems, digital audio (compact disc signal processing), matrix-based processing, ISDN, and so forth.

In Section 8.2 we will discuss the general design methodology, which is called "the meet-in-the-middle design strategy." It refers to the fact that we opt for a strict separation between system and silicon design levels. The system designer performs a top-down design, starting from a description in a high-level language. This design is mapped into an architecture consisting of an interconnection of instances of a minimal set of primitive silicon modules at the level of large arithmetic blocks. These modules are, in fact, software procedures called by the high-level compiler. These procedures have been composed by silicon specialists using a module generation programming environment. The system designer then can use an interactive floor planner to compose the chip. It is as if the system designer

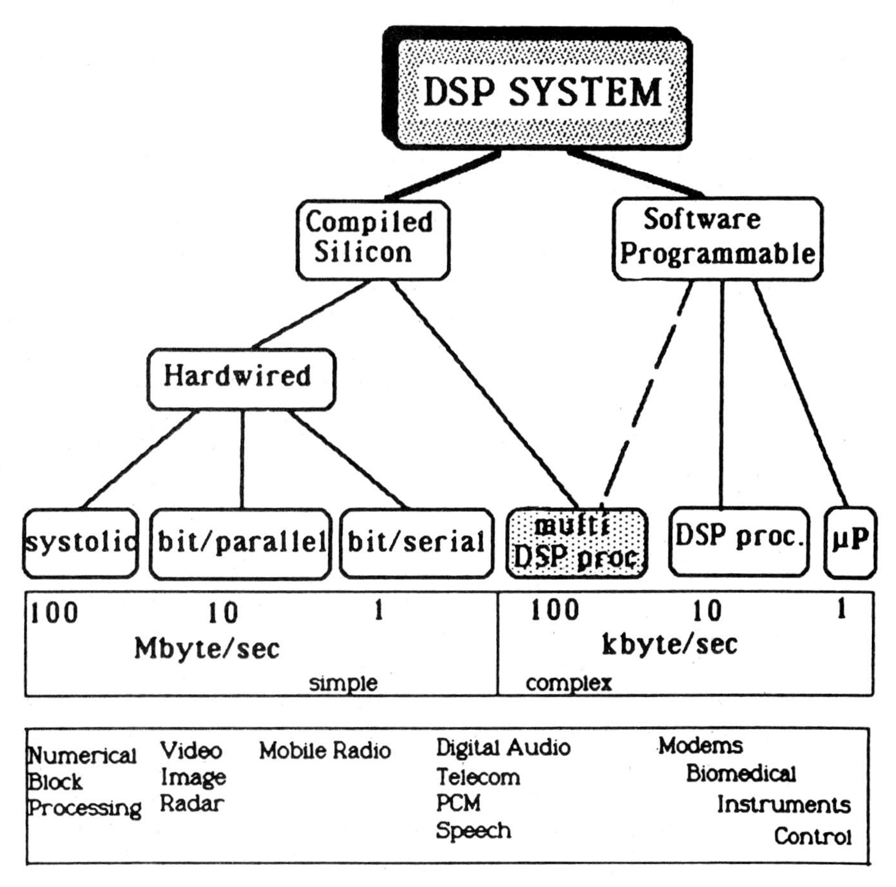

Figure 8.1 Different implementation styles for DSP algorithms. CATHEDRAL-II aims at the multiprocessor implementation.

meets the silicon designer in the middle of the design abstraction levels. This avoids the problem of the "long thin man."

This separation is clearly reflected in the structure of CATHEDRAL-II, in which we can identify two main parts: the architectural synthesis part (system designer) and the module generation part (silicon designer). Both parts will have a need for their own verification parts, which are based both on simulation and on knowledge-based verification that is made easier by the imposed design constraints. The module generation concept will be introduced in Section 8.2; however, only the architectural synthesis part will be discussed in detail. More information about our vision on the module generation environment and its tools can be found in [Six86a] and [Six86b].

As stated previously, no compilation is possible without a careful definition of the target architecture. This definition is given in Section 8.3.

We will show, based on a speech processing application, that a general multiprocessor architecture can be defined in a hierarchical fashion.

The basic components of the architecture are six so-called execution units (EXUs), which are prototype data paths and can be constructed from an elementary set of functional building blocks (FBBs) (e.g., shifters, adders, etc.). Other fundamental units are memories (RAM, ROM), I/O units, and controller modules.

Based on that architecture, in Section 8.4 we discuss the architectural synthesis part. As a high-level language we have chosen the SILAGE applicative language developed at the University of California-Berkeley [Hil85]. It will be shown that the translation from behavior to structure for this particular architecture consists of four steps:

1. Partitioning into processors communicating using dedicated protocols and hardware
2. Mapping partitioned code into structure defined as EXUs interconnected by a minimal set of dedicated buses
3. Scheduling of the register transfer operations in order to generate the microcode that forms the basis for step 4
4. Controller generation

In this chapter, prototype CAD tools for tasks 2, 3, and 4 will be presented. In fact, manual design exercises have shown that these tasks (especially 3 and 4) are the most labor-intensive and error-prone components of the synthesis task. It turns out that the partitioning of the algorithm into processors is relatively easy for a designer, based on his or her knowledge of the system hierarchy and the complexity of the different system functions. However, after completion of the synthesis process, it may turn out that the initial partitioning was not optimal (in terms of the usage of the hardware resources and the time frame). Therefore, it is important that the partitioning can be adapted with minor effort by changing only a few statements in the higher-level description. An automatic approach to the partitioning problem will be tackled in the future.

We will start from the hypothesis that the partitioning is done, and we will concentrate on the results obtained in mapping, scheduling, and controller generation per single processor executing its share of the SILAGE code. During the mapping operation, an optimal time-and-area mapping of the SILAGE code into the architecture within the throughput requirements of the algorithm is to be found. Since this is strongly dependent on the target architecture, we have chosen to do it in a *rule-based environment.* This process requires a considerable amount of unification between the preprocessed code and the hardware operators and registers, so we have selected PROLOG as an implementation language. The present implementation of the mapping operation can be considered as an *architecture knowledge data base,* which contains all knowledge of how higher-level constructs can be implemented in the defined target architecture.

The outcome of this program is the data path of each processor whereby all data communication is done over dedicated buses. The data path operators correspond to the ones available from the module generators supporting the architecture.

Another output of the mapping process is a register transfer language (RTL) description of the primitive operations to be performed as transfers from register to register (or memory).

The problem of scheduling can now be described as the technique to find a timing order to fire the operations in such a way that data precedence is respected, that no resource allocation conflicts occur, and that pipeline constraints in operators and controllers are respected. It will be shown that such a problem can be formulated as an integer linear programming problem (ILP). Since ILP is NP complete we have developed heuristic solutions to this problem. We will demonstrate the feasibility of the approach with some examples.

Finally, when scheduling is done, the number of cycles per processor is known. At this stage the bottleneck processor can be identified and, using pragmas (or hints) in the SILAGE description, one can try to optimize the bottleneck processor if the required throughput is not reached. However, more often, one will have the opposite case in which one should exchange speed for area. The first mechanism present in the system is a bus merging strategy. In the future we will add more automated optimization to redistribute time for area automatically.

Finally, in Section 8.5 we will draw some conclusions regarding the actual status of the work and discuss future extensions.

8.2 The Meet-in-the-Middle Design Strategy

It is well known [Tuc85] that by 1990 nearly 50% of all systems implemented in ICs will be ASICs and nearly 40% are expected to be designed by the end users. As shown in Figure 8.2 we can classify VLSI design in a performance complexity plane. Performance is defined as data rate/maximum clock speed. Complexity ranges from simple control logic or pure data flow to implementations of complex decision-making algorithms. Today system-designer-oriented design, truly supported by commercial CAD in the low-performance/low-complexity range, is supported by standard cell and gate array type of design. The challenge posed by advanced 1.2 micron CMOS is the design of chips containing nearly 300,000 devices. This amounts to potentially more than 15 processors of 16 bits each with 256 bytes local memory or nearly 0.5 billion multiplications/sec (0.5 Mega MIPS!) in a data path high-performance realization. This kind of complexity, hard to realize with off-the-shelf standard DSP processors, opens challenging system applications in the fields of telecommunications, digital audio, video, robotics, speech analysis, image processing, dedicated matrix processing, and so on.

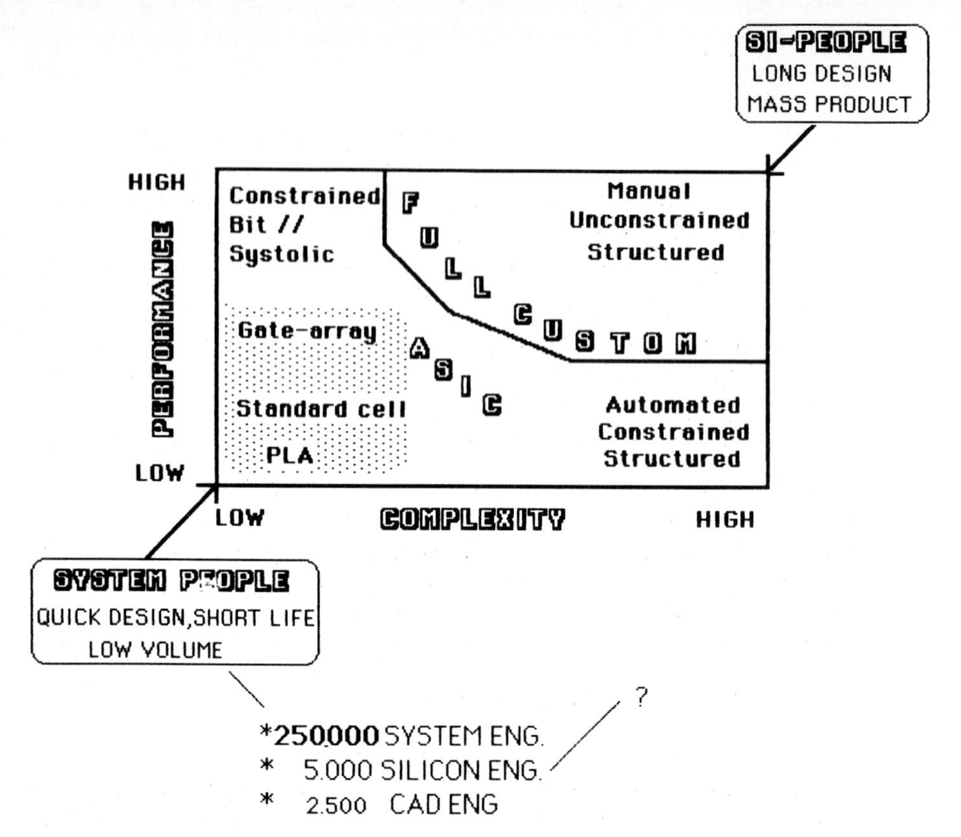

Figure 8.2 Evolution of design will be toward ASIC by the shortage of silicon designers and the required small design time to product life ratio.

8.2.1 Meet-in-the-Middle Design Strategy—Separation of System Design from Silicon Design

It is clear that the level of such designs exceeds by far the circuit or even the logic design level. Furthermore, as shown in Figure 8.2, silicon and CAD engineers are so scarce that the latter cannot cope with the demand for such system designs, even if they were capable of being so "long and thin" that they would span the whole design spectrum in order to make a full custom chip. In addition it is very likely that the production volume as well as the product lifetime of such complex chips is rather low, such that full custom design is totally excluded.

As a result we believe that such promising systems must be designed by the system designer without any need for detailed silicon knowledge. It is therefore necessary that silicon design knowledge (at the 1 micron level) be localized in reusable silicon modules at the usual LSI/MSI level familiar to the system designer. This leads to a design scheme such as that shown in Figure 8.3 that we have called the *meet-in-the-middle design methodology*.

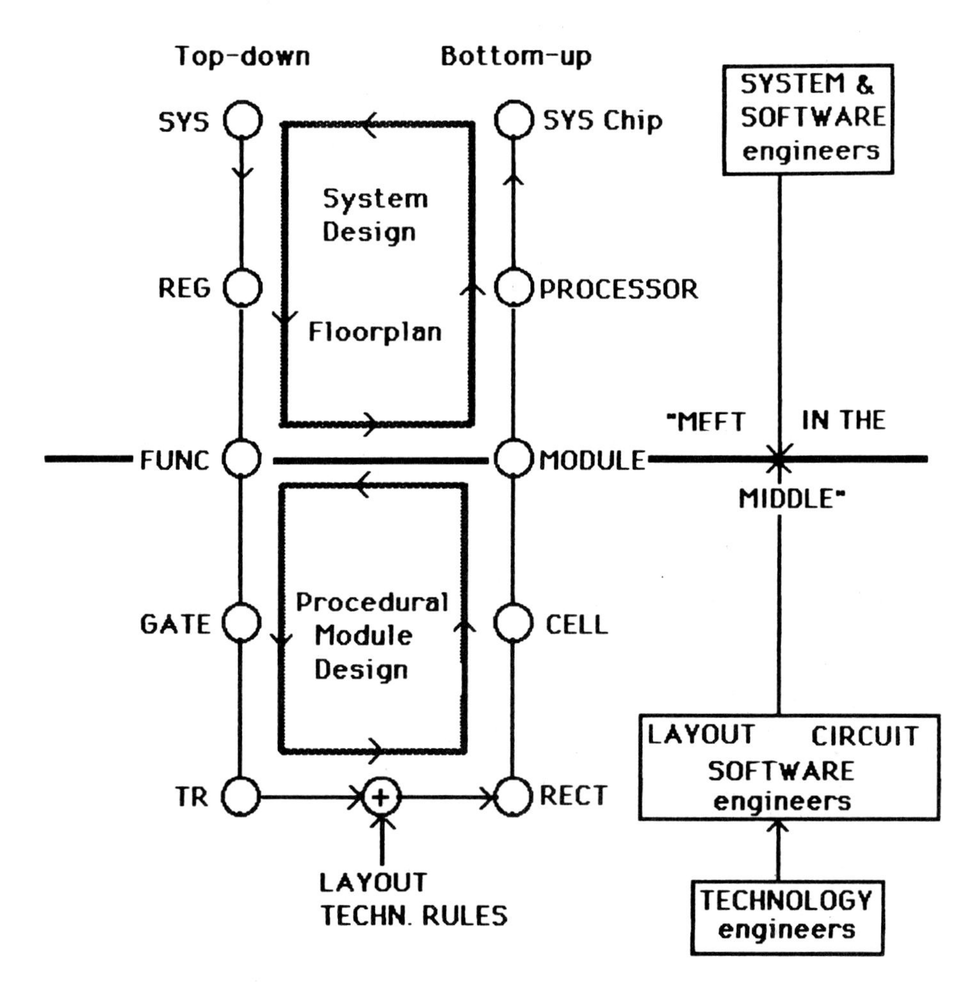

Figure 8.3 Meet-in-the-middle design strategy.

This technique is characterized by the following characteristics:

1. System design is strictly separated from silicon design. The interface is located at the level of arithmetic/logic operator blocks, data storage, controllers, and I/O units. We will call that the *functional level.* The silicon primitives used at that level are called *modules.* Design at the system level consists in translating a system specification into a structure that is a net list of module instances. Placement and routing of layout instances of the modules completes the chip design.
2. Silicon modules are reusable just as standard cells are. In this way the costly investment in high-performance, advanced silicon technology design is limited and the cost is written off over as many designs as possible.

3. Silicon modules are much more complex than standard cells and, in order to save even more in terms of silicon design cost, these modules must also be capable of surviving a number of technology updates. Therefore modules must be *technology adaptable.*
4. Silicon modules, however, are not the privilege of a particular foundry or CAD vendor. Since the competitive edge between system houses will not be in the technology, but in the architectural technique and in the implementation of it, we expect that module design will require a powerful design environment itself for a local team of silicon designers. This may not be the case today, but we expect this to happen in the future.

Notice that in this design style system designers design as usual in a top-down fashion down to their usual intermediate level. The silicon people design in their usual bottom-up fashion, composing the LSI level modules from functional building blocks that, in turn, are composed of logic leaf cells at transistor level.

It is as if both parties meet each other in the middle of the design abstraction levels. In this way scarce talent is optimally used and the design process corresponds to usual patterns.

However, due to characteristics 2, 3, and 4 defined previously, some fundamental deviations from classical design at silicon and system level do occur. These differences will become clearer as we give an outline of the CAD toolbox that we are developing for CATHEDRAL-II.

8.2.2 The CATHEDRAL-II CAD Toolbox for a Meet-in-the-Middle Design Strategy

Figure 8.4 shows a CAD toolbox as we are developing it for the design of multiprocessor implementations of high-complexity/low-performance DSP algorithms. It is clear from the figure that there is, indeed, in the middle of the design abstraction a separation between the silicon and the system designers. The link between them is a "call" to a (hopefully) limited set of silicon modules that must be carefully defined in the target architecture.

The system designer defines his or her system at the behavioral level in a high-level language. In our system, as we will explain in Section 8.4, we use the SILAGE [Hil85] language which is especially suited to describe complex DSP algorithms in an applicative way, that is, as a set of simultaneous equations rather than in a procedural way. Coupled to this language is a high-level simulator to verify the behavioral correctness of the algorithm. Based on the throughput requirements and a set of expert design rules, the SILAGE code is first optimized in function of the target architecture. It is subsequently compiled directly into structure, that is, a net list in terms of the limited set of predefined silicon modules. Clearly most of this synthesis trajectory is dictated by the target architecture and therefore quite a large part of it is rule based and in our case implemented in a PROLOG program as we will discuss in Section 8.4.

Notice in Figure 8.4 that we also allow, in principle, the designer to specify the design at intermediate levels, all the way down to structure. However, the price to be paid is an increasing amount of lower-level simulation, a higher redesign risk, and a longer time to market. Perhaps a smaller chip area or a better layout may result. Here the economics of the end product will dictate the choice.

When all calls to the modules are successful then the chip can be composed with the aid of a floor planner. This can be done either interactively or by automatic place-and-route techniques.

Notice that in CATHEDRAL-II we really aim at a *true silicon compiler* since we include the direct synthesis of the algorithm into data paths and

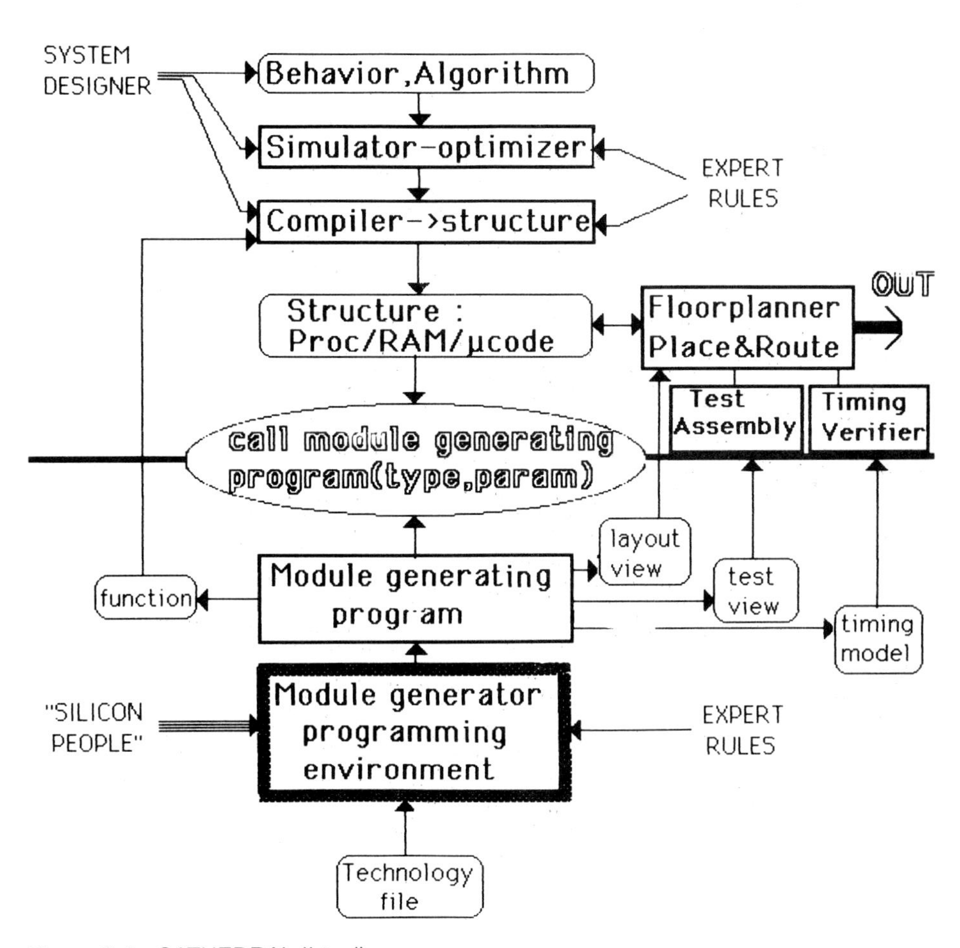

Figure 8.4 CATHEDRAL-II toolbox.

control logic. This is in contrast to most commercial systems today that are limited to floor planning and to module generation.

In order for this scheme to work, a clear definition of the module library is essential. Clearly, the library cannot have the same variability as is offered by a TTL or CMOS LSI part catalogue: it would take too long to design it and no compiler would be powerful enough to map into it. Neither can we design the modules in a fixed technology, this in view of the vast design efforts and the evolving technology. We should be able to adapt it to technology changes.

On the other hand, just as is necessary in traditional PCB design using LSI components, we need from a module generator a lot more information than just the *layout view,* which in itself should consist of a bounding box view (BBV) and a full layout view (LV). These views can be used by the synthesis tools (for cost function evaluation) and the floor planner.

In Figure 8.4 we show that we need a *functional view,* that is, an RTL level parameterizable function. This view is useful in case design is done at the structural level, since then simulation is necessary. Even if full synthesis is used, designers still will want to simulate the actual chip. Furthermore, a *timing view* is necessary since, although the synthesis compiler can take first order throughput requirements into account, it is only after placement and routing that a full performance check can be done. In CATHEDRAL-II we are following a bottom-up hierarchical generation of timing models in a knowledge-based program, SLOCOP, that follows closely the composition procedure of a module [Van86]. At the floor plan level the interconnection parasitics are taken into account to check global timing. If this is unsatisfactory, first buffer sizes are adjusted. If this is not satisfactory, one might consider different placement or, finally, a pragma could be formulated at the SILAGE level to call for higher-performance modules or ultimately to increase parallelism in the algorithm.

Finally, although not present in the actual CATHEDRAL-II version, there must be a *test view* with each module. We envision for the future that most modules with a high degree of structure will be C-testable, that is, dependent on their particular structure, a small set of wordlength independent patterns exists that guarantees the module testability. We then want to implement a test assembly program at the level of the floor planner that, for the particular architecture, will generate total testability for the whole chip. This is a topic for future research, but we believe that a silicon compiler without ATPG for the chip is like a hospital building without doctors!

8.2.3 From Fixed Standard Cells to Flexible Module Generation

The problem with the above is that the variability of modules is much larger than the variability at the logic standard-cell level. In order to have maximum reusability as well as technology independence it is not possible to design modules in the standard CALMA type way. Modules must be written as parameterizable procedures or, to be more in vogue with actual object-oriented thinking, as classes from which modules are objects. Parameterizability is to be understood in a broad sense ranging from simple word length to conditional composition in terms of functional building blocks or size of output buffers, and so forth. Moreover these procedures should also generate the other views needed by the synthesis programs.

In order to reduce the variability, one has to constrain the design space. This can be achieved through the careful selection of a well defined target architecture as dictated by the application area. As an example, the area of the high complexity/low performance DSP applications was selected as the target application area for the CATHEDRAL-II system. As we will explain in Section 8.3, a multi-processor structure was selected as the most appropriate architecture for this class of applications. A detailed study of a variety of applications showed that a limited set of so-called execution units is sufficient to support the intended application range. These EXUs perform the basic arithmetic and logic operations comparison, scaling and address computation. Other units needed are RAMs, ROMs and PLAs for realization of data memory and control structures. Each of the above-mentioned modules has a well defined parameter domain.

We believe that the feasibility of our compilation approach is founded on the fact that we are restricting ourselves to this carefully chosen set of EXUs. In the software world of compilation one can compare this choice to the choice of the target microcode for the compiler to be built.

The task of the module generator environment therefore consists of composing the EXUs from so-called functional building blocks (e.g., adder, register, shifter), which themselves are composed of logic cells. We will examine the composition of such a module generator next.

8.2.4 Anatomy of the Module Generator

Figure 8.5 shows the anatomy of the module generator as used in CATHEDRAL-II. The different arrow types indicate the "create" (silicon designer), "generate" (call from silicon compiler), and "adapt" (to technology rules) functions. Notice that the CATHEDRAL-II module generation environment not only provides the layout environment but also will be equipped with an expert system that is necessary for the following tasks:

1. Generation of the functional/timing and test models during "generate" phase.
2. Verification of the modules during the "create" and "adapt" design phase.
3. In the future we would also like to store the design history based on the accumulated experience.

The design of a module is based on a parameterizable, functional description at the register transfer level. This description (and its simulation) can be done by the HILARICS-LOGMOS system developed at IMEC [Mari86]. This functional description is the documentation link between the system and the silicon designer. In the case of controller synthesis it will be shown

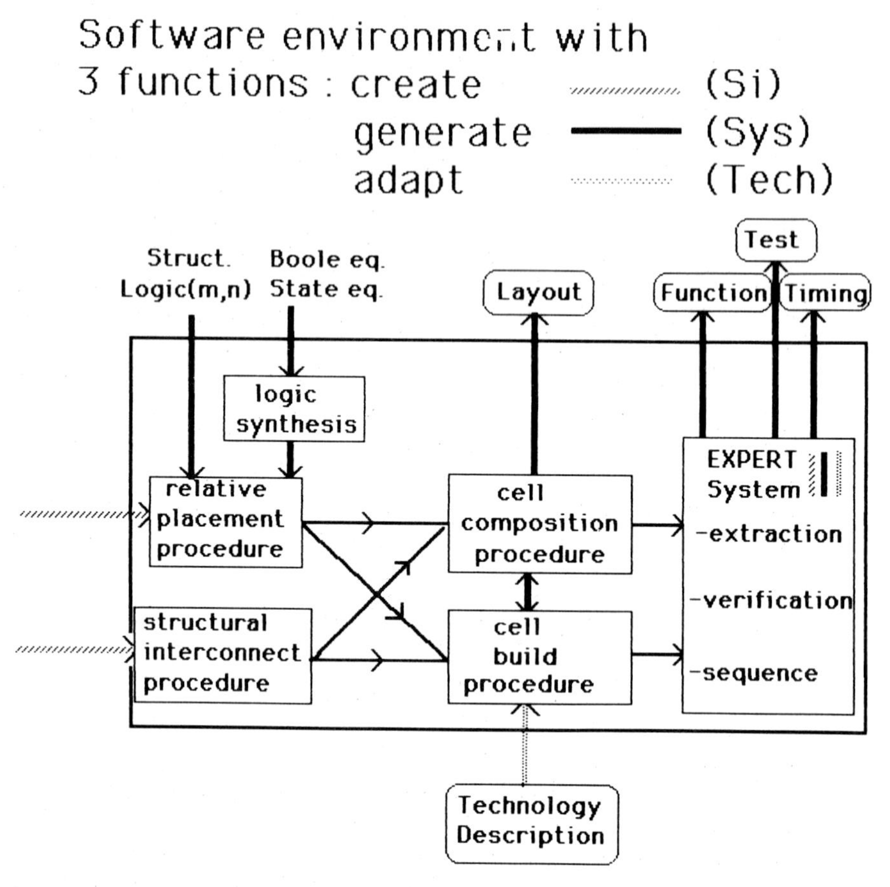

Figure 8.5 Anatomy of a module generator.

in Section 8.4 how also a finite state machine description can be an input to the module generator in order to generate PLA-based controller structures.

The big difference between a module generator and a fixed layout is that, because of the parameterizable nature of the description, the layout also must be parameterizable and this requires that the relative placement of cells as well as the connectivity needs to be procedural. In CATHEDRAL-II this is done by using a LISP interpretative programming environment whereby, as much as possible, the code for procedural composition of a module out of cells is generated from a graphics definition. Only complex mathematical relationships are directly programmed in LISP.

The primitive cells themselves are generated mainly with the aid of a symbolic layout tool called CAMELEON. CAMELEON supports symbolic cell editing and a critical path driven layout compaction. It also allows for direct circuit extraction from the symbolic layout. The extracted model serves as an input to the verification and modeling tools, which include a timing verifier and a logic and electrical verification and simulation part. Major advantages of the symbolic technique are the technology updatability and the support for automatic cell abutment.

The module generation environment described here forms the foundation of the architectural synthesis: it delivers the information (area, connections, delay, power, etc.) necessary for a reliable evaluation of the optimization cost functions. The CATHEDRAL-II synthesis will be discussed in Section 8.4, after a clear definition of the target architecture.

8.3 The Target Architecture

It has already been stated that efficient and acceptable design synthesis is only feasible with a restricted target architecture in mind. The definition of such an architecture, even for a (restricted) application field such as digital signal processing (DSP), is, however, not straightforward. In fact, the architectural composition needed for the implementation of a certain algorithm is heavily determined by factors such as the ratio between maximum hardware clocking frequency and sample frequency, the arithmetic complexity per sample, and the regularity or the streamlined nature of the algorithm. Therefore, a large set of totally different architectures have been envisioned as roughly sketched in Figure 8.1.

At the same time, a substantial increase in the complexity of the typical signal-processing functions an algorithmic designer wants to put on a circuit can be noticed [All85]. Vector and matrix operations are becoming commonplace, algorithms are more and more block or frame oriented, and decision making becomes almost as important as pure number crunching, all of this paired with an increase in sampling frequency.

Based on these considerations and the study of a number of typical applications, we have selected a flexible, customizable multiprocessor architecture as the target for our design synthesis system. The defined architecture is flexible enough to evolve into the single, general processor at the lower end of the frequency spectrum and into a connected set of parallel, hardwired data paths at the upper end and thus spans a large application set in the fields of speech, audio, robotics, telecommunications, and image processing. We have attempted to add enough flexibility to the architectural composition to attack the three basic bottlenecks that normally limit the data throughput in general-purpose signal processors: the limited arithmetic throughput, the congestion in data transfer, and the controller delays (when performing decision-oriented operations).

8.3.1 The Overall Multiprocessor Architecture

A typical signal processing system can easily be subdivided into a set of distinct subtasks, each of which requires fairly different signal processing

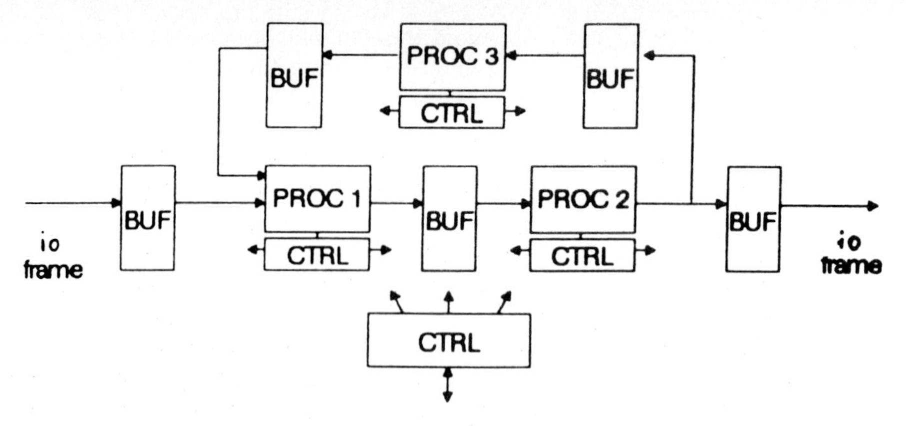

Figure 8.6 Composition of multiprocessor architecture.

operations. This decomposition is reflected in the higher-level block diagram of the system. The proposed architecture is an attempt to map this system decomposition directly into the hardware, by providing a set of parallel operating processors, each of which executes one particular subtask and is optimally tuned to just that one task (Figure 8.6). Each of those processors operates relatively independently from its neighbors and communicates with them via a set of transparent communication protocols, exchanging only data that are global between them. Communication with the outside world proceeds over an I/O frame. This frame can support a large range of I/O protocols, ranging from parallel to serial, from synchronous to asynchronous, and from word oriented to buffered block oriented.

As stated above, each of the processors is optimized to perform one particular part of the algorithm and consists of a dedicated data path and controller. The data path is composed of a cluster of strongly connected execution units, communicating with each other over a restricted number of dedicated buses. In this way, the bus contention that occurs in the case of a single or dual bus architecture is largely avoided. The number of selectable EXU-types has been deliberately restricted to six, which are stored in a parameterized format in the module library.

A multibranch, microcode-based controller structure has been selected to control the data flow through the data path. This structure is flexible and powerful enough to handle a large span of algorithms in an elegant and efficient way. Some of the EXUs (or their subblocks) can also have a local controller. This helps to reduce the complexity and the number of instruction bits in the central processor controller. Examples of local controllers are the decoder of the register file and the slave controller of the divider.

It is interesting to see how the hierarchy of the overall architecture is also reflected in the controller hierarchy; at the system level, we have a central controller that governs the data flow between the processors and to the outside world. Each of the processors has its own controller that determines the functional behavior of the processor, and, at the lowest level, a number

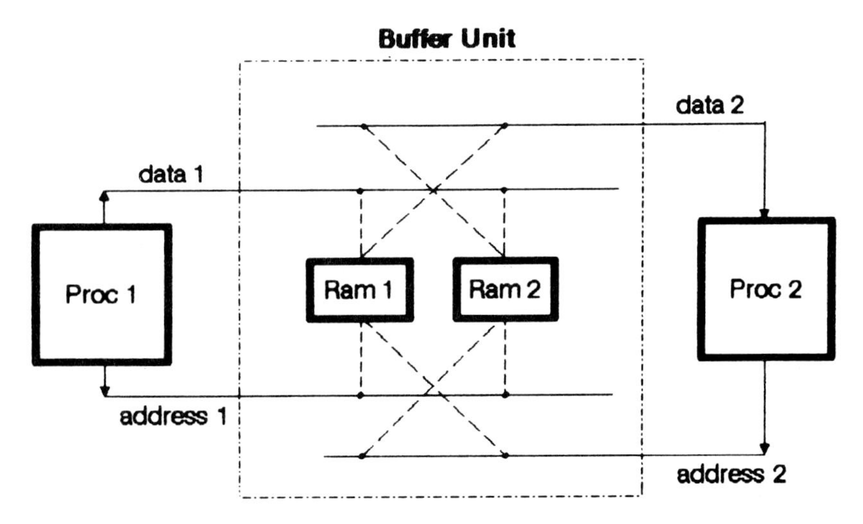

Figure 8.7 Interprocessor communication protocol based on switched RAMs.

of local controllers can be attached to the EXUs. This strict hierarchical division allows for a modular approach to the synthesis problem.

A number of the topics raised above will be discussed in more detail in the sections below.

8.3.2 The Interprocessor Communication

During the partitioning of the algorithm over the processors, one of the main goals is to keep the mapping and scheduling of the identified subtasks as independent as possible. This requires that the communication between the processors is completely transparent to the internal operation and the timing of those processors. The type of protocol used to achieve this goal is heavily dependent upon the properties of the required transfers: the required transfer capacity, the ordering of the read and write operations, and the synchronicity. Therefore, a set of different communication protocols (and supporting hardware) has been provided, so that for each particular case an optimal solution can be selected.

The most general, synchronous protocol is presented in Figure 8.7. The communication hardware consists of two identical RAM units that are alternatively switched between processors 1 and 2 (and their address computation units). During time frame 1, processor 1 writes its data into RAM1, while processor 2 reads the data from RAM2. This situation is reversed in the next frame. This technique allows for the transmission of large, unordered data blocks at a high rate without any restriction, but at a considerable hardware cost. When the sequence of the generated data is identical to the order in which it should be received, a substantial hardware reduction can be obtained by replacing the switched RAMs by a single FIFO unit, as shown in Figure 8.8.

Other simplifications become feasible with decreasing communication

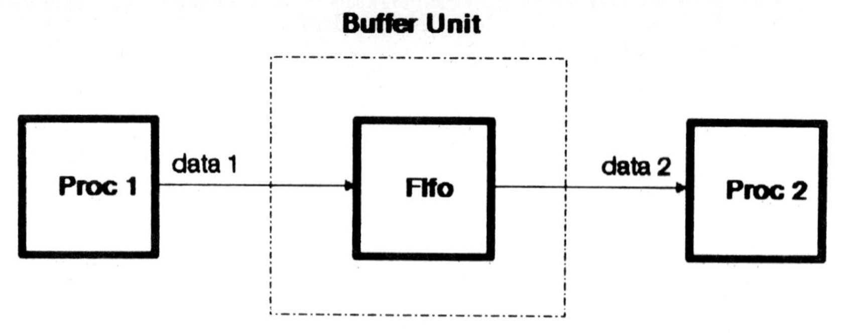

Figure 8.8 FIFO-based interprocessor communications.

rates; in those cases, it is often possible to skew the programs of source and destination processors in such a way that no bus conflicts will occur (Figure 8.9). A single bus structure with connected RAM buffer will then be sufficient. The last solution, however, assumes that the absolute timing of the transfers is known at compile time. When this is not the case (e.g., when conditional operations are allowed), a two-way handshake between the controllers of the communicating processors can be used to synchronize the data transfer.

Other hardware reductions are feasible by using buffered bit-serial transfers (as in [Pop84]). The communication hardware can also be multiplexed over multiple sources and/or destinations by providing multiplexer and demultiplexer units at the processor sides.

8.3.3 The Processor Data Path

According to our architectural approach, each of the processors consists of a dedicated data path and a controller. The data path is optimized for only the particular tasks it has to perform and is assembled from a set of selected

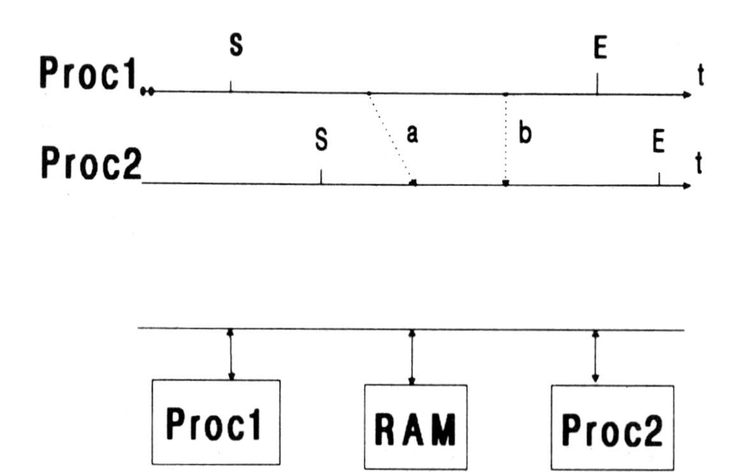

Figure 8.9 The application of skew on the programs of two processors makes it possible to use a single bus + buffer protocol.

EXUs, interconnected by a restricted number of customized buses. Each of the EXUs contains a register file of variable size at its input side. This local storage helps to resolve the memory bottleneck that is often a problem in general purpose signal processors.

Studies have shown that a restricted set of EXUs is sufficient to span most of the application area. EXUs can be divided into general purpose units such as an ALU/Shift and an address computation unit (ACU) (address arithmetic, modulo counting, loop counters) that are very general in terms of functionality, but are rather inefficient for critical operations such as multiplication, and accelerators such as a parallel multiplier/accumulator for high-speed number crunching, a parallel/serial divider, a comparator (max-min operations), and a normalizer-scaler (e.g., for fixed to floating point conversions). All these blocks have been designed in a parameterizable fashion and have been implemented in the module generation environment. Typical parameters are the word length, the size of the register files, the size of the multiplier array, the type of adders used (in function of the speed requirements), the maximum depth of the shift-operations, and so forth. Unused operators can be removed from the data path, resulting in a further increase in area-efficiency; for example, the saturate unit of the ALU/Shift EXU can be removed when no saturate operations are required in the task to be performed. In this way, the units can be tailored to the particular requirements of an algorithm.

As an example of such an EXU, we will study the ALU/Shift unit, which is shown in Figure 8.10. The EXU is constructed of a set of so-called functional building blocks that are in this case two register files,

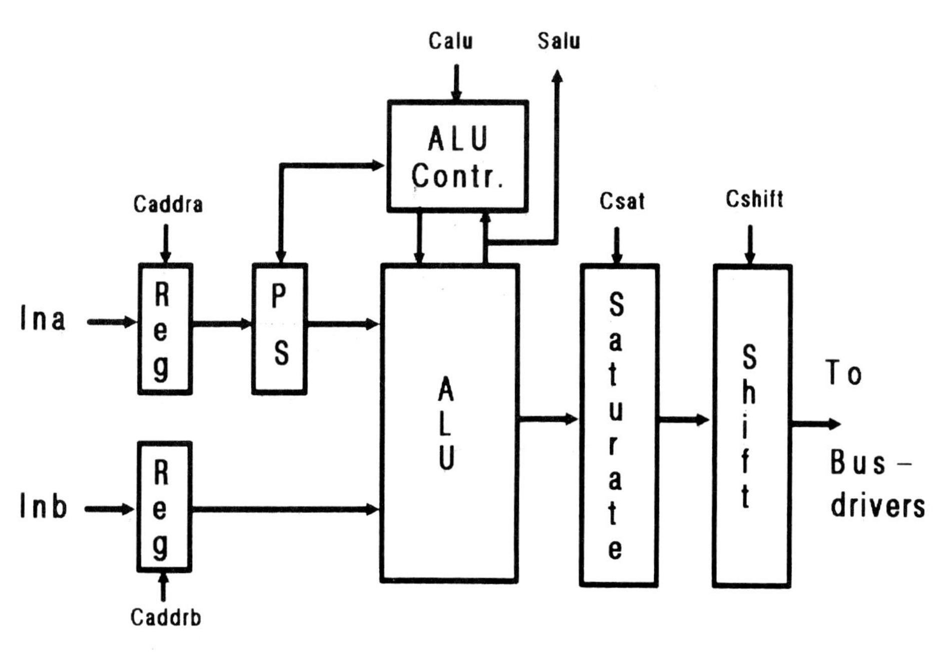

Figure 8.10 ALU/Shift Unit.

a parallel-serial converter (for parallel-serial multiplications), an ALU (or adder/subtractor), a saturator, a logarithmic shifter, and some bus drivers. The parameters are the word length, the sizes of the register files, the depth of the shifter, the presence of a saturator, the type of adder unit, the need for multiplications, and the number of bus drivers. The shifter, P/S, adder, or saturator can be removed if not needed. The unit is controlled by a number of control signals (names starting with a "C") and a set of status signals (marked with an "S"), which is routed to the control unit in order to evaluate decisions.

An example of a processor data path, constructed using the proposed strategy, is shown in Figure 8.11. This processor is used to compute the amplitude spectrum of a signal, given the complex frequency domain spectrum, and to determine at the same time the maximum amplitude. The real and imaginary values are obtained sequentially from the ADFT processor and are transferred through the I/O buffer and bus1 to the multiplier/accumulator (8 by 8) where they are squared and added (Ampl = Re*Re + Im*Im). After this, the amplitude values (16-bit) are transferred to the next processor over bus2. At the same time, the comparator unit compares the computed amplitude value with the old maximum, which is updated over bus3. The final maximum is also transmitted to the next processor. Note that the amplitude processor is buffered from the next processor by a switched RAM structure, as defined in Section 8. The addresses for the data in this RAM are computed concurrently by the address computation unit (7-bit), which also implements the loop counter. This processor with three concurrent units can perform an amplitude computation and a maximum update in two cycles (average)!

8.3.4 The Processor Controller

The main task in the actual implementation of an algorithm after a suited data path has been selected consists of the design of the associated control section. A single control task can be realized by a large variety of alternative control architectures, ranging from random logic over sequencers to microcoded controllers. Which structure is preferable depends upon the

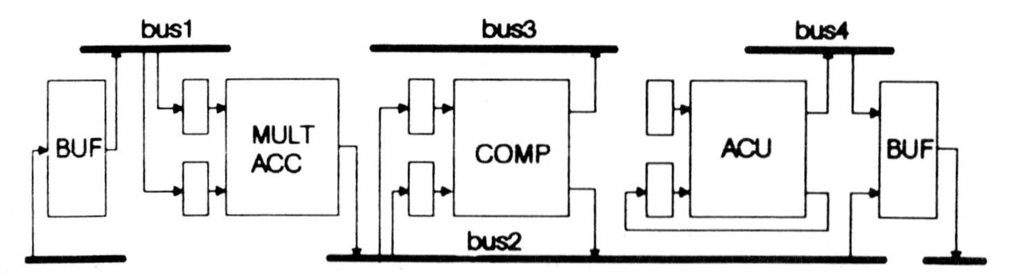

Figure 8.11 Example of customized processor data path (processor 1 of the pitch extractor example.)

nature of the algorithm to be implemented (e.g., repetitiveness, number of branches). This can, for instance, be demonstrated by the case of the general-purpose signal processors. The controller of these processors is optimized for repetitive operations (such as filtering, FFT). This results in a rather poor behavior for decision-oriented algorithms. The selection of a particular controller architecture is also influenced by the availability of optimization and simulation tools (such as logic optimization and state assignment).

Based on these considerations, we have selected a single, flexible microcode-based multibranch controller as the standard controller structure for our processors. This controller has the flexibility to support heavily decision-making-oriented algorithms as well as regular, repetitive algorithms in an efficient way.

The selected architecture is presented in Figure 8.12. The core of the controller is identical to the conventional microcode-based controller. It consists of a program ROM that stores the horizontal microcode words and a program counter register (PC) that is incremented in normal operation mode, but can be loaded with a jump address through the multiplexer MUX in jump mode. In a conventional architecture however, multidirection

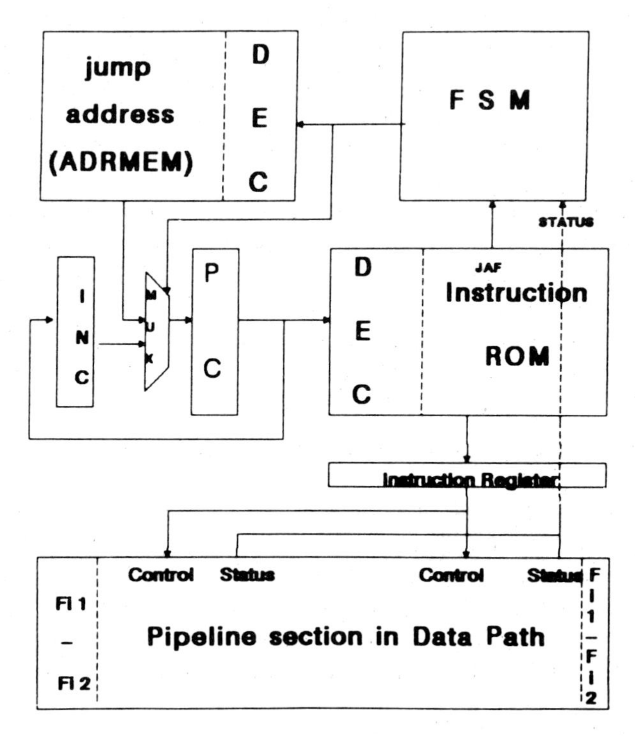

Figure 8.12 Multibranch controller architecture.

branches must be split into a number of consecutive conditional jumps with the jump address stored in the program ROM. In this way, instruction cycles as well as ROM area are sacrificed. The controller architecture of Figure 8.12 avoids this by storing the jump addresses (or states) in a separate memory called ADRMEM. When some type of jump is required, the output of ADRMEM is loaded into the PC. In the case of a data-dependent branch, the status signals generated in the data paths have to be interpreted first in a finite state machine (FSM) in order to decide on the appropriate next state address. The updating of the state and the computation of the jump address can proceed simultaneously with the normal arithmetic operations and do not ask for extra cycles.

It should be noted, however, that three pipeline stages are present in the critical control path (setting of the status bits, evaluating of the ADRMEM, and fetching of the next instruction). In the case of a conditional branch, this results in a necessary delay of two cycles between the test and the actual branch. This could result in a number of NOPs (no operation), similar to the situation occurring in the RISC architecture [Pat81]. Most of the time however, the microcode scheduler can introduce independent instructions in between and avoid the loss of those cycles.

The power of the controller can best be illustrated by the following example: a decision feedback equalizer, together with the timing recovery circuit [Tze85], could be executed by this controller in 25 cycles (on a single Add/Shift data path). The decision feedback equalizer by itself took 80 cycles on a TMS32010 [Mag84]. The algorithm in question relies heavily on decision making.

Other alternative controller architectures are also under consideration at present. These include the implementation of subroutines, sequencers (instead of the combination of PC plus ADRMEM), the addition of cache, and so forth. The most important task here is to identify the application domain of each of those structures in order to help the synthesis process in its selection.

8.3.5 Floor Plan Considerations

A number of general chip and processor floor planning considerations are appropriate at this time, since they will help to determine the area efficiency of the architecture.

As stated above, a processor data path is composed of a number of EXUs connected by customized buses. Each of those EXUs is composed out of a set of so-called FBBs (e.g., adders, registers, shifters). The following layout strategy has been applied within the FBBs, that have been developed in a double metal, 3 micron nwell CMOS technology: buses and supply lines are running in metal 2, parallel to the bit slices, whereas control signals and clocks are routed orthogonally on the metal 1 level. The FBBs normally connect to each other by abutment. When this fails, a number of the metal 2 tracks (normally five of them are available per slice) can be used, as shown in Figure 8.14. The remaining tracks can be used to route the inter-EXU

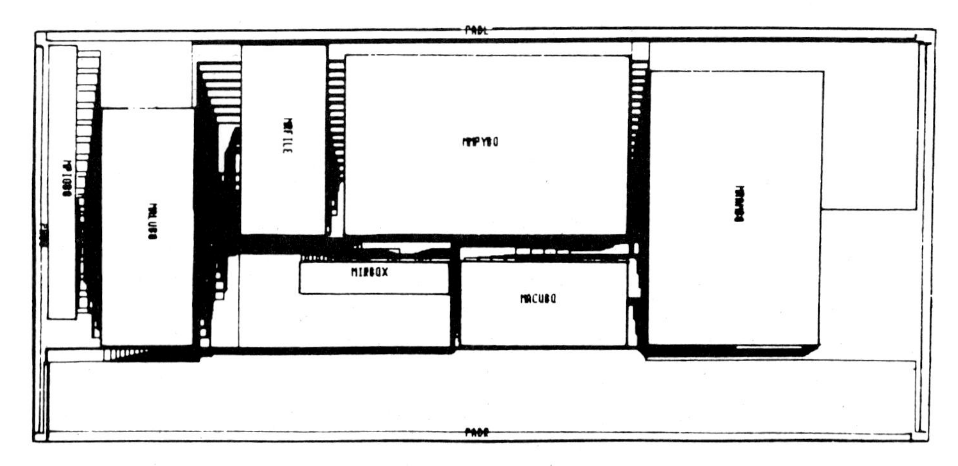

Figure 8.13 Processor floor plan using over-the-EXU routing.

bus connections, thus avoiding large routing channels (that are only needed for the routing of the control signals or when not enough spare tracks are available). An example of a floor plan realized in this way is shown in Figure 8.13.

A generally applicable floor plan strategy for the controller structures is difficult to determine, as it depends upon the selected controller arch-

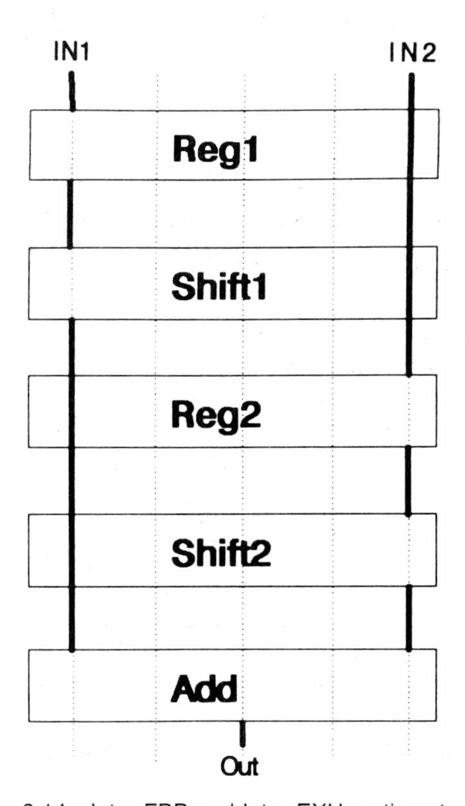

Figure 8.14 Inter-FBB and Inter-EXU routing strategy using metal two tracks.

itecture, the complexity of the controller, and the constraints imposed by the adjoining data path(s). The floor plan generation is therefore an interactive procedure, in which the aspect ratios of the controller modules are gradually and iteratively refined in order to obtain an optimal match between data path and controller.

The placement and routing task at the processor level has to be performed by a floor planner. The connections will most probably run in channels along the function blocks, as normally too many obstructions are present already on the internal metal 2 tracks. Due to the sometimes very long communication tracks between processors, the capacitive loads can grow reasonably large. Therefore, these transfers have to be latched and consume (at least) one cycle.

8.3.6 Example

The effectiveness and the efficiency of the multiprocessor architecture presented have been validated with a number of practical test cases. Applications in the fields of digital audio, telecommunications, speech coding, and speech recognition have been studied. The implementation of a high-quality pitch extractor for speech [Slu80] will be discussed in more detail.

In this algorithm, the pitch of the speech signal is estimated by studying the matching between the maxima of the frequency domain spectrum of the signal (obtained from a DFT on the windowed signal) and a set of predefined patterns (40 in total). The system can be divided into four subtasks (or processors) as illustrated in Figure 8.15. In a first step, the amplitude spectrum and the absolute maximum (as threshold) are computed. The first 8 maxima (above the threshold) are derived in the second processor and compared with the 40 predefined patterns in processor three. In a last step, the precise value of the pitch period is computed.

The algorithm has been manually and partially automatically (using the emerging synthesis tools described in Section 8.4) mapped into the defined target architecture. The results of the mapping process are collected in Table 8.1. The active area needed for the total system equals 37 mm^2 in a 3-micron CMOS process. About 20 additional mm^2 will be needed for the interconnect. The figures also show that the architecture is flexible enough to handle computation-intensive functions (processors 1 and 2, large ratio

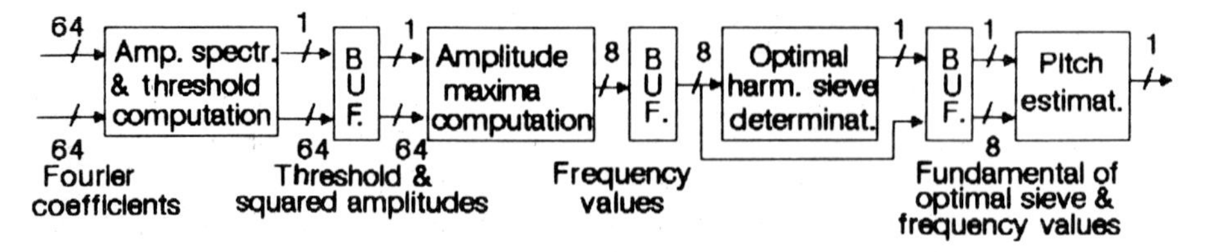

Figure 8.15 Pitch extraction block diagram and processor definition.

Table 8.1 Pitch Extractor—Results of Mapping Exercise (for 3 micron nwell CMOS process)

	Proc1	Proc2	Proc3	Proc4	Total
Data Path (mm^2) Control (mm^2) Buffer (mm^2)	3.0 0.6 3.3	4.3 1.6 3.7	6.4 2.6 1.0	6.1 4.1 0.5	19.8 8.9 8.5
Total (mm^2)	6.9	9.6	10.0	10.7	37.2
Cycles	1,225	661	2,854	240	2,854

between data path and control area) as well as decision-making-oriented tasks (processors 3 and 4).

The exercise gave us important feedback concerning the nature and the quality of the synthesis tools needed and it strongly determined the underlying synthesis philosophy of our synthesis tools, as will be described in Section 8.4. Synthesis is an *iterative* and *interactive* procedure, in which the synthesis tools generate a suboptimal solution in a first step (e.g.,with important timing mismatches between the different processors), that can be gradually refined in the next iteration steps. It turns out that a system designer is very good at making compositional and architectural decisions, as long as these are at a high level and as long as he or she does not have to bother about details such as microcode scheduling, register assignment, bus connections, and so forth.

8.4 The Architectural Synthesis

8.4.1 The Synthesis Process: General Principles

The automatic synthesis of digital systems from behavioral descriptions has drawn a lot of attention recently. However, most of the techniques described in literature ([Tho83], [Marw84], [Cam85]) adopted a purely top-down methodology: the behavioral model of the system is translated into structure through a sequence of transformations and optimizations. This top-down methodology has the advantage of being very general and flexible, but lacks the ability to incorporate the information about the silicon implementation of the low-level primitives into the synthesis process. Besides, the generality of the approach (and thus the enormous dimensions of the design space to be explored) explains why these techniques have failed to produce acceptable designs until now. We attempt to overcome this deficiency by restricting the synthesis process in the following ways:

- The synthesis is targeted towards one single (but flexible) architecture, that limits the application range of the compiler but results in a far more optimal design. The architectural limitations result in a considerable pruning of the design search space and make it possible to use dedicated optimization techniques (exploiting the particular properties of the architecture in the form of heuristics). The developed synthesis tools should be flexible enough to accommodate changes or additions in the target architecture. This concept formed the base of the MacPitts [Sis82] and LAGER [Rab85] compilers, which produce acceptable designs but whose architecture is too restricted to span a large application field.

- The architectural primitives are a set of parameterizable modules, that are defined, described, and specified in a module generation environment. This environment acts as a *module knowledge data base* that can be queried by the synthesis tools. In this way, design decisions can be based on physical implementation details (e.g., speed, area, power). This methodology, a mixture of top-down and bottom-up strategies, is the meet-in-the-middle strategy described in Section 8.2.

These ideas are reflected in the overall graph of the CATHEDRAL synthesis system (Figure 8.16), which consists of three major parts: the architectural synthesis, the module generation, and the floor planning. In this section, we will show how this design philosophy allows for an efficient mapping of a high-level behavioral specification of a DSP algorithm into a customized multiprocessor architecture.

An overview of the architectural synthesis process is given in Figure 8.17. The algorithm to be implemented is described in SILAGE [Hil85], a language that is optimized for the description of DSP algorithms. SILAGE

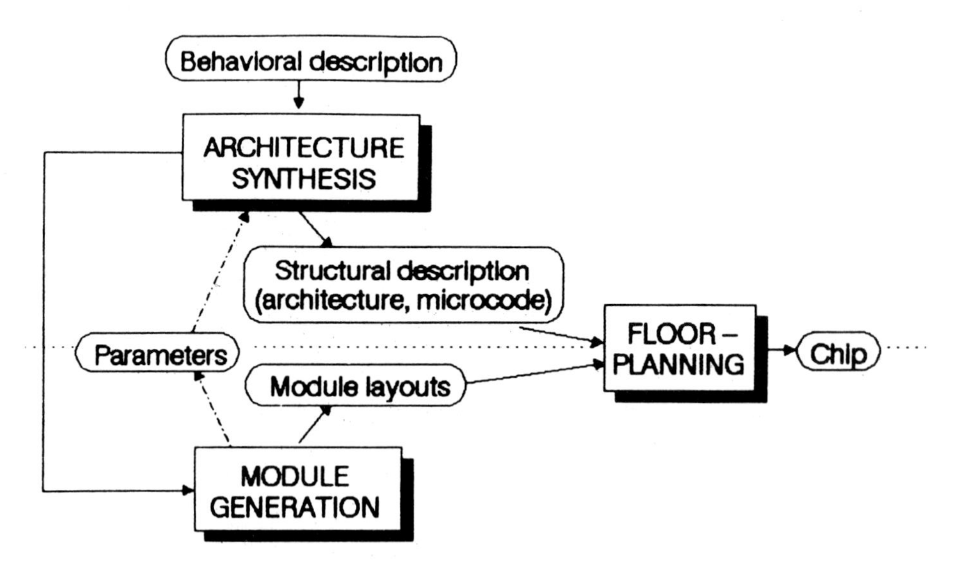

Figure 8.16 The basic functions in CATHEDRAL-II: Architecture Synthesis, Module Generation, and Floorplanning

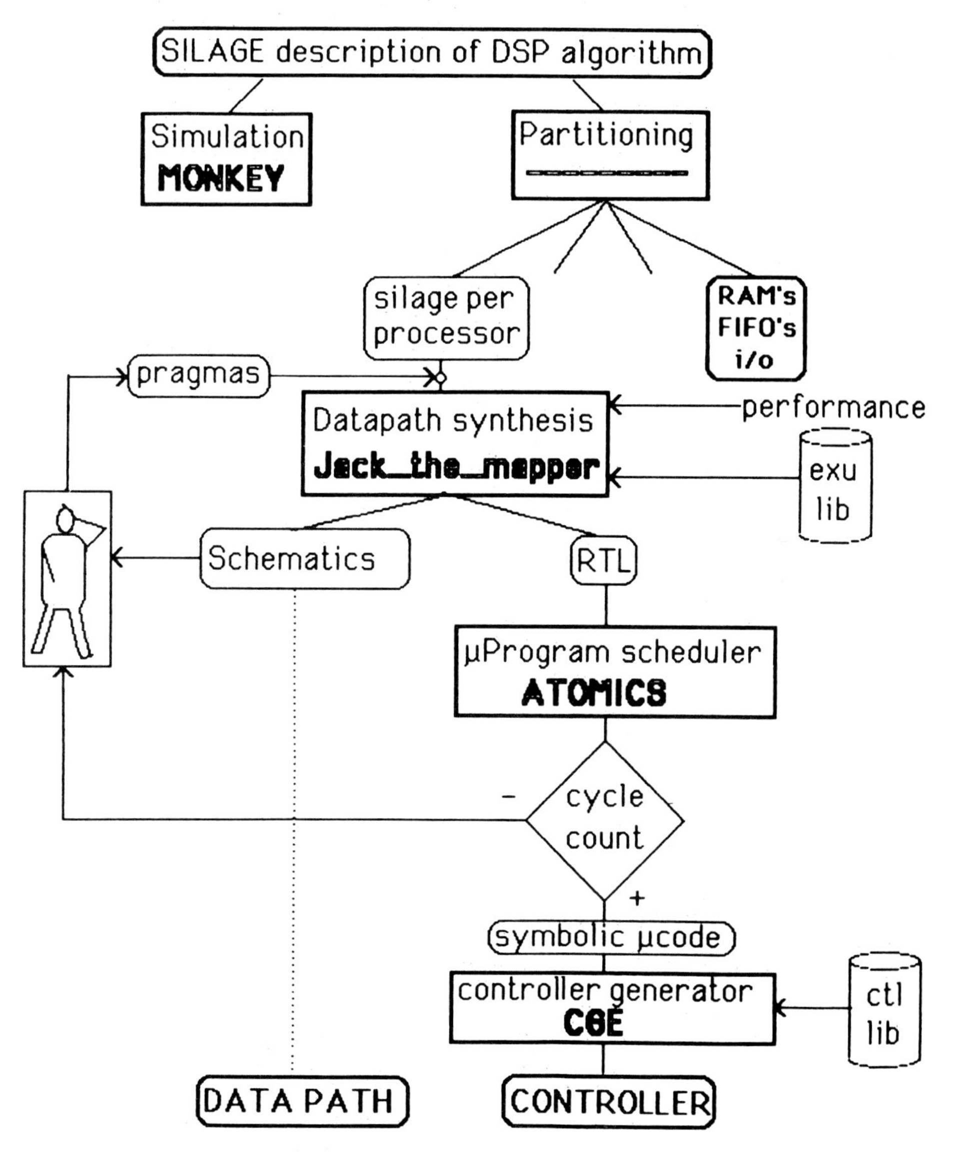

Figure 8.17 General overview of functions, languages, and simulators in the CATHEDRAL-II synthesis process.

has been designed as an applicative language (a language having no assignment or side-effect producing operations) in an attempt to describe the signal flow nature of signal-processing algorithms. In a first step, the algorithm has to be partitioned over a set of processors. In order to optimize each of those processors for the particular subtask it has to perform, a customized data path has to be generated (as a composition of the primitive modules or EXUs). This task is performed in the data path synthesis. The second step in the synthesis consists of the generation of a dedicated controller, which implements the algorithm on the synthesized data path within the constraints of the allotted time frame (microcode scheduling and controller synthesis). It is interesting to note that the synthesis process can be considered as a stepwise refinement from an applicative (pure behavior, no notion of control) into a procedural (complete definition of the control sequence) representation of the algorithm.

From the design exercises, we have found that the synthesis process has to be considered as an *iterative* and *interactive* optimization procedure with the cycle count and the area as main cost functions. Therefore, it is important that the experienced system designer can interfere with the synthesis process at the different levels. This implies that he or she should be able to check at all those levels if these interferences did not cause any disturbances from the requested behavior. Based on those considerations, we decided that the synthesis system should be an open system with multiple entry points (well defined and readable description languages) and verification (simulation) tools at all those levels. Therefore, we have defined a register transfer language that can be used to describe the behavior of a processor once the structure is known, a structural specification language, and a controller definition language. However, low-level user interference (and probably disturbance) with its accompanying verification steps should be avoided as much as possible. A more appropriate way for the designer to steer the synthesis process without jeopardizing the correctness is by using the SILAGE pragma constructs. A pragma is a high-level structural user directive, which allows for user interference with the synthesis process at the highest level.

The different tasks defined above will be discussed in detail in the following subsections. The example of the pitch extractor for speech will be used as a test vehicle to illustrate the different stages of the compilation process.

8.4.2 The Input Language (SILAGE)

There is no doubt that different application fields require different description or specification languages. The data objects handled by a microprocessor designer (e.g., micro instruction sets) are definitely different from the design entities used by a modem designer (e.g.,filters, delays, decimators). Therefore, we have selected SILAGE [Hil85], a language targeted toward digital signal processing, as the input specification language for the CATHEDRAL-II system.

SILAGE is aimed at the description of signal-processing algorithms at a high behavioral level. The basic object in SILAGE is a *signal,* which is an infinite vector in time (comparable to a stream), and the basic operation is function application on those signals. In this way, a SILAGE description of an algorithm is equivalent to a signal flow graph, in which nodes represent instances of functions and arcs represent the paths followed by the signals. SILAGE allows for the description of equations, delays, decisions, repetition, hierarchy, finite word length specifications, and quantization characteristics.

The basic idea of SILAGE, as stated previously, is to describe the behavior of an algorithm, such as the mathematical relationship between input and output. A SILAGE description does not include structural or timing information (in contrast to HDLs as VHDL [Lip86]). In some cases however, the designer may wish to pass some structural or timing constraints (hints) to the compiler. In SILAGE, a construct named *pragma* is used to pass this kind of information.

The SILAGE description of the amplitude and threshold computation part of the pitch extractor is presented in Figure 8.18. The function AmplitudeSpectrum computes the amplitude of the frequency domain spectrum, given the real and imaginary parts of the spectrum as obtained from a DFT. The computation has to be performed over 64 discrete points (as given by the constant NrOfPoints). At the same time, the maximum of the amplitude spectrum is computed (Thresh). The following important aspects of the SILAGE language can be pinpointed in this example:

```
# define NrOfPoints 64
# define SQUARE(x)        num<16,0> (x * x)
# define WORD8 num<8,0>
# define WORD16 num<16,0>

/* Computation of the amplitude spectrum and the threshold function. */

func main(Re, Im : WORD8[NrOfPoints]) : WORD16 =
begin
    Spectrum = AmplitudeSpectrum(Re, Im);
    return = Spectrum.Thresh;
end;

func AmplitudeSpectrum(freal, fimag : WORD8[])
                        Ampl  : WORD16[]; Thresh : WORD16 =
begin
    Max[0] = WORD16(0);
    Thresh = Max[NrOfPoints];
    (i : 1 .. NrOfPoints)::
    begin
        Ampl[i] = SQUARE(freal[i-1]) + SQUARE(fimag[i-1]);
        Max[i] = if (Ampl[i] > Max[i-1]) -> Ampl[i] || Max[i-1] fi;
    end;
end;

pragma "(AmplitudeSpectrum, proc, 1)";
```

Figure 8.18 SILAGE: description of amplitude and threshold computation in the pitch extractor.

- In an applicative language, every signal can be defined only once. The subsequent values of the max-signal are therefore indexed.
- The order of the equations is irrelevant, as shown by the position of the Thresh equation.
- SILAGE supports fixed-point typing; that is, num<8,0>denotes a fixed-point data type with a total word length of eight and zero bits behind the binary point (or an integer of word length eight). The typing of signals is done in an implicit way, so that no explicit declarations (e.g., as in Pascal) are needed. A coercion operation has been defined to change between data representations (as shown in the SQUARE definition).
- The iteration construct, used in the definition of the AmplitudeSpectrum function is only used to express repetitivity and does not imply any control or sequencing.
- In the same way, the function construct is only used to structure the description in a hierarchical fashion. This description will normally be expanded for hardware implementation.
- Note also the pragma construct, that forces the compiler to implement the AmplitudeSpectrum function on processor one in the hardware realization.

A SILAGE simulator (called MONKEY) has been developed. Simulation at this level can be used for algorithm development and for code debugging. It was found that a demand-driven simulation technique gives the best speed performance.

8.4.3 The Algorithm Partitioning

A partitioning of the algorithm into a set of parallel operating processors is the first task in the synthesis process. A processor is defined here as an entity that consists of one central controller and a set of connected EXUs. Each processor can be synthesized independently from the other processors due to the transparent interprocessor communication protocols.

The automation of the partitioning task is a nontrivial job and is influenced by a set of often contradictory considerations. From a number of manual exercises, however, we have found that a human designer is capable of making a good first guess based on simple inspection of the behavioral description. This is the philosophy we use in the first version of CATHEDRAL-II, in which the processor partitioning is left to the user by means of pragma statements in the SILAGE description. A reiteration on this initial partitioning can be performed after the synthesis of the processes themselves. At that time, precise data concerning the load balancing and the sizes of the interprocessor communication buffers are available.

We envision a partial automation of the partitioning task in the near future. This partitioning tool will probably be rule based. It will be able to suggest a number of feasible solutions and will help the designer to evaluate the drawbacks of partitioning at a high level. The rule base will contain a precise knowledge of the interprocessor communication protocols and a set of selection rules that will often be contradictory. A number of those selection rules for partitioning have already been identified in the manual exercises:

- The identification of a set of algorithmic subtasks with diverging computational requirements (as can be observed from the systems block diagram).
- The minimization of the data flow between processors. This can be based on the selection of a set of cutsets (in the data flow graph) that can only pass through delay operations and that are selected such that the number of connections crossing the cutsets is minimal. Function loops with only one delay operator (e.g., an adaptive predictor) must be implemented on one processor as a result of this observation.
- The minimization of the buffering and storage requirements. Changes in data rates can cause data transfer bottlenecks. This occurs fairly often in block-oriented algorithms, in which an algorithmic subtask can only start its execution after a complete block of data (from a previous computation) is available. This requires complete data buffering and allows for the introduction of interprocessor pipelining. The inherent capabilities to pipeline the algorithm for blocks of data can be detected from the data dependency graph, as illustrated in Figure 8.19. In this directed graph, the vertices represent computations and the edges indicate the required data. Parallel paths with identical operations on the vertices correspond to samples to be processed in the same way. These samples can be processed sequentially on the same processor. A point of convergence (or divergence) on the graph indicates a bottleneck at which data buffering is needed. This technique was used to partition the pitch extraction algorithm in four processors, as indicated in Figure 8.15.
- One of the most important factors in the algorithm partitioning is the load balancing, which makes sure that all the hardware is used in an efficient way. At the partitioning time, the load of an algorithmic subtask can only be guessed, based on a rough evaluation of the critical path in the

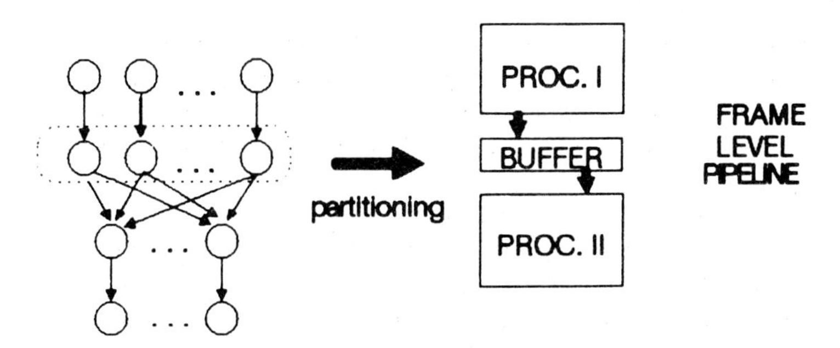

Figure 8.19 Bottlenecks in the data flow graph of block-oriented algorithms.

data flow graph of the algorithm. Therefore, it is clear that the partitioning (as well as the whole synthesis process) will be an iterative task with stepwise refinement.

8.4.4 The Data Path Synthesis

The goal of the data path synthesis operation is to map the SILAGE description of a single processor into a minimal structure (in area), that will execute the algorithm within the allotted time frame. The result of the operation is a definition of the structure of the processor (in terms of parameterized modules and their interconnections) and a black box description of the controller that consists of an unordered list of the operations that have to be performed on this hardware structure.

The data path synthesis process consists of three basic functions:

1. *The hardware allocation* [Tho83]. In this step, it is estimated how much arithmetic power is needed in order to execute the algorithm within the timing constraints. The type and the number of EXUs in the data path are selected in this step.
2. *The translation.* High-level behavioral code is translated into primitive operations that can be executed directly on the target architecture. Since this translation is dependent upon the data path structure, it has to be performed *after* the allocation step.
3. *The hardware assignment.* The translation step produces a set of primitive operations that must be executed on the selected data path. The assignment step will select the EXU instances on which each primitive operation will be executed. It will assign the variables to exact register or memory positions and will determine the bus interconnection structure. The operations previously mentioned can be redefined as a set of optimization tasks with a common cost function; given the maximum allotted time frame, they find the solution with the minimal area (and/or power). The assignment task is closely interwoven with the controller (or microcode) scheduling as will be discussed later.

Manual design exercises and discussions with designers have shown that the major and most important part of the synthesis job for our target architecture is located in the translation and assignment phases, which are also the most error-prone parts of the synthesis task (register allocation, bus assignment, loop implementation, function expansion, delay implementation, etc.). Therefore, we decided to tackle these problems first. The selection of the EXU types and of the amount of parallelism needed (allocation) can be enforced by the designer using the SILAGE pragma concept. In this way, the designer only has to make large grain decisions, and all the nitty-gritty and error-prone details are filled in automatically.

The translation and assignment processes will be heavily influenced by the target architecture. Therefore, we have selected a rule-based system for most of the process, in which the rules are used to describe the basic properties of the architectural components. This technique allows for a flexible definition of the architecture and eases the introduction of architectural changes or additions. In fact, the synthesis program acts as an architecture knowledge data base. A number of procedural (algorithm-based) tasks can also be identified.

The program flow graph of the processor synthesis is pictured in Figure 8.20. A quick glance at this graph already reveals the basic synthesis guidelines: synthesis is a sequence of relatively simple design steps. This is the only way to have sufficiently simple rule bases. Design iterations should be done locally as much as possible, although overall iterations cannot be avoided.

The following sequence of design steps that are currently in the prototyping phase can be identified:

- Preprocessing of the SILAGE source code.
- Generation of the data path structure of a single processor, while also generating the register transfer description of the controller (translation).
- Scheduling of the controller register transfer statements (mapping the algorithm on the time axis). This task will be described in Section 8.4.5. This step also fills in the final assignments (if not yet done in the translation step).

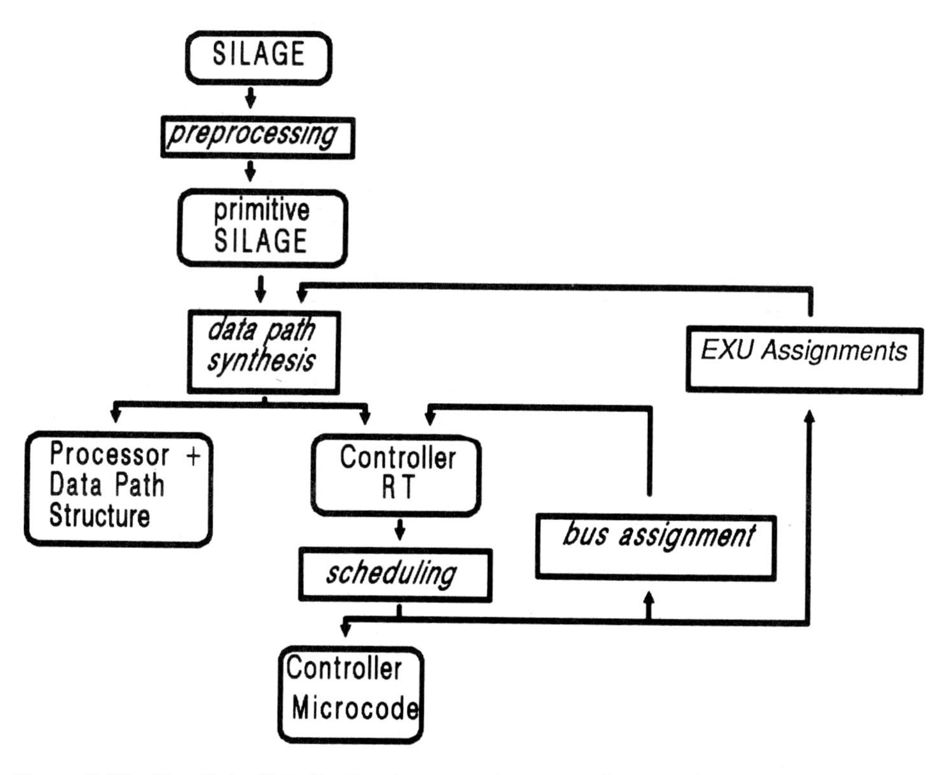

Figure 8.20 The Data Path Synthesis: general program flow graph.

- Register binding.
- Bus allocation and binding.

The implementation of these tasks will now be described in detail.

The Preprocessor

In order to understand the functions of the preprocessor and the subsequent paragraphs, a few definitions must be given first.

Basic equations are identities or equations that can be solved by an EXU in one single machine cycle. For example, {reg1, mult, c} = {reg2, alu, a} + {reg2, alu, b} which takes the variables a and b from register files reg1 and reg2 of the alu respectively, and stores the result c in the register file reg1 of the multiplier.

A *singular variable* is a variable that actually appears in the SILAGE description of an algorithm. A *compound variable* is a variable that can be considered as an intermediate result that has no explicit name. The generation of these variables depends on the kind of mapping rules used. For example,

$$s = a - b + c$$

In this example, the variables s, a, b, and c are singular variables, whereas the variables '$a - b$' and '$b + c$' (depending on the precedence rules of the operators + and −) are compound variables.

The tasks of the preprocessor are:

- Parse the SILAGE description
- Perform syntax and semantic checks
- Determine the datatypes of all signals (based on the production rules as defined in SILAGE)
- Perform a number of local transformations common to most general-purpose software compilers (these include the elimination of common subexpressions, the removal of constant or manifest expressions, and the flattening of the conditional constructs)

The output of the preprocessor is an expanded and reordered list of SILAGE equations, in a format suited for the data path synthesizer (which is a PROLOG program). Also forwarded to the synthesizer are a set of allocations and partial assignments, as obtained from the SILAGE pragmas or from a potential allocation tool. These constraints, which will steer the data path synthesis process, are passed in this format: {_*_, mult, _} states that all multiplications should be implemented on a parallel multiplier (without specifying the particular instance name), while {x*_, alu,2} forces the data path synthesis tool to use alu 2 to execute all multiplications involving the variable x. {read(Coef), rom,_} forces all the variables Coef to be stored in ROM memory. Note that the underscore must be interpreted as a wild card.

The preprocessor is implemented in the C programming language. It is currently limited to the transformations mentioned previously.

The Translation Step

In order to transform the preprocessed applicative high-level language description of the algorithm into a customized processor structure, the mapping tool has to assign primitive SILAGE operations to EXUs and both singular and compound variables to registers. It has to expand higher-level constructs such as repetitions, functions, conditional equations, delays, and decimation/interpolation. It must also figure out how the execution units should be interconnected by parallel buses, in other words, it should assign transfers of SILAGE variables to buses. Finally, it must decide which parameter sets (e.g., word lengths, shift ranges, etc.) are to be assigned to the functional building blocks inside the execution units.

The assignment tasks (binding the operation to EXU, register and bus binding) are procedural optimization steps and are implemented in Pascal. The translation part from primitive equation to basic equation must be flexible with respect to architectural changes and has therefore been implemented in a rule-based fashion. PROLOG was selected as the implementation environment for its pattern matching capabilities.

The PROLOG environment acts as master and calls the assignment procedures when appropriate in the design iteration. In this section, we will describe the translation mechanism. It consists of two main entities: an expert system shell (called JACK) and a knowledge base.

The rule base of the translation step captures the knowledge of the architecture designer and forms the real creative step in the synthesis process: it tells how to translate behavioral primitives into architecture primitives. This might be straightforward for an addition, but is far more complicated for constructs such as multiplication (parallel, parallel-serial, constant multiplication), algorithmic delays, matrix operations, and repetition. All these translation rules are very much dependent on the target architecture.

A second set of rules implements the interconnection strategy. It generates the necessary buses, input multiplexers, and tristate output buffers.

The expert system shell is tuned towards synthesis problems: it offers data structures and manipulation routines for data flow graphs and data path structures, an inference engine, specialized graphics, and a rule editor to update the rule base.

The inference engine uses the information in the data flow graph to decide which equation is to be mapped next. The code is traversed in a demand-driven way. The initial demand-list consists of the set of output signals. For each triggered equation, the following actions are performed:

- select an appropriate rule in the knowledge base, using an ADVICE-RULE cycle.
- put demands for the input operands of the equation in the demand list.

The translation process finishes when the demand list is empty. The rule selection is based on an ADVICE-RULE cycle. A high level equation is passed to the ADVICE function. ADVICE decides, based on the format of the particular equation, which rules should be tried in which order.

The ADVICE function has the following general (PROLOG) format:

adviceEq (Equation Format, Unit, Selected Rule): −conditions.

The conditions add some extra constraints to the matching of an equation with this particular advice function, for example, one of the arguments has to be a power of two.

Before a particular equation is passed to the ADVICE function, the supervisor first tries to match the equation with a pragma statement that may enforce the selection of an EXU.

A typical set of advice equations may look similar to the set below, defined for the multiplication operation (note that the underscore once again has to be interpreted as a wild card. C has to be interpreted as a variable):

adviceEq (_=_*C, div, ruleMult1): −PowerOfTwo (C), !.

adviceEq (_= _*C, alu, ruleMult2) :− constant (C), !.

adviceEq (_=_*_, mult, ruleMult3): −!.

adviceEq (_=_*_, alu, ruleMult4): −!.

In plain English, these advice functions can be interpreted as: If one of the arguments of the equation is a power of two and a divider unit has been allocated or assigned, implement the operation on this divider. The expansion of the primitive equation into basic equations is performed in ruleMult1. If one of the arguments is a constant, try an add/shift–based multiplication on an ALU. Variable-variable multiplications can be performed on a multiplier (parallel) or an ALU (parallel-serial).

Note that the ordering of the advice functions is important, for example, if a multiplier as well as an ALU have been allocated, then the multiplication will be performed on a multiplier (using ruleMult3) if no overruling pragma assignment exists.

The RULES unify or expand the high-level equation into a set of low-level or basic equations that can be dealt with immediately by the target architecture. Such low-level equations are, for instance, the addition, subtraction, division, and multiplication of variables. At the same time, some low-level register transfer operations have to be generated that transfer the input arguments to the appropriate input register files of the selected unit. Finally some new high-level equations, producing the input arguments, have to be triggered. These equations are added to the demand list. The ADVICE-RULE cycle continues (in a demand driven fashion) until no high-level equations are left. At that moment, the equation set contains only basic register transfer operations that can be passed to the microcode scheduler and the assignment tools.

As an example of a simple rule, consider ruleMult3: the equation VAR = OP1 * OP2 is expanded by ruleMult3 in the following equivalent set of low-level equations, identities, and new high-level equations:

```
[
VAR =[reg1, mult, OP1]*[reg2, mult, OP2],
[reg1, mult, OP1]= OP1,
[reg2, mult, OP2]= OP2,
OP1 = OP1,
OP2 = OP2
]
```

and the message "Multiplication on parallel multiplier" is displayed.

Note that OP1 and OP2 may be compound variables as $'a + b'$, which are stored as intermediate results. The construction rule for these variables ($'a + b' = a + b$ or OP1 = OP1 in the rule) is then passed to the ADVICE cycle for further mapping.

For example, the equation $d = a * (b + c)$ is expanded in following equations:

```
d=[reg1, mult, a ]*[reg2, mult, 'b+c']
[reg1, mult, a]=a,
/*bring variable a to register file of the multiplier*/
[reg2, mult, 'b+c']='b+c',
/*bring variable 'b+c' to the register file of the multiplier*/
a = a, /*identity, will be ignored */
'b+c' =b+c
/* new high-level equation; will be passed to the ADVICE-RULE mechanism*/
```

Note that the produced lower-level equations are in a so-called half register transfer format (VAR = [REG,_,_] . . . and [REG,_,_] = VAR). These half equations will be merged into full register transfers ([REG,_,_] = [REG,_,_] . . .) by the interconnection rules that also allocate buses and input multiplexers.

At present, the translation rule base for the multiprocessor architecture consists of 105 rules. They can be divided into the following knowledge fields: logic operations, addition, subtraction, multiplication/accumulation, division, shifting, accelerated nonprimitive SILAGE functions, algorithmic delays, address computation, constants, conditional equations, loops, function application, and decimation/interpolation.

Although these rules express different types of knowledge, they all have the same format. This makes it possible to develop a general knowledge acquisition system that allows for a simple introduction of new rules by a system designer. In fact, we have found that this is a crucial point: efficient implementations of certain algorithms will ask for the addition by the user of specific translation mechanisms (e.g., for delay handling) or for special accelerators (e.g., a square root function). The knowledge acquisition system translates the user definition into PROLOG clauses, inserts them at the right place in the knowledge base, and checks for consistency.

The result of the mapping operation is a description of the structure of the processor (in terms of basic components and interconnections) and a description of the controller as a set of unordered register transfer operations that have to be executed on this structure. Note that these register transfers are only allocated to a type of EXU, but not yet assigned to a particular instance (except when there is no choice or when the assignment has been dictated by a pragma statement). At this point of the synthesis process, we assume that the transfer of a variable (singular or compound) proceeds over a bus dedicated to that variable (leading from the output of the EXU to the destination register). This results in a large number of buses. After the scheduling of the transfers and the simultaneous definition of the remaining assignments, it will be investigated which buses can be merged.

For the example of the AmplitudeSpectrum processor, the mapping tool produced a processor data path containing a multiplier/accumulator and a comparator, as pictured in Figure 8.21. The data path also contains an address computation unit (ACU) for address computations and loop counting and an input and output buffer RAM. A connection has also been provided between the data path and the controller (called rom_ctrl). This

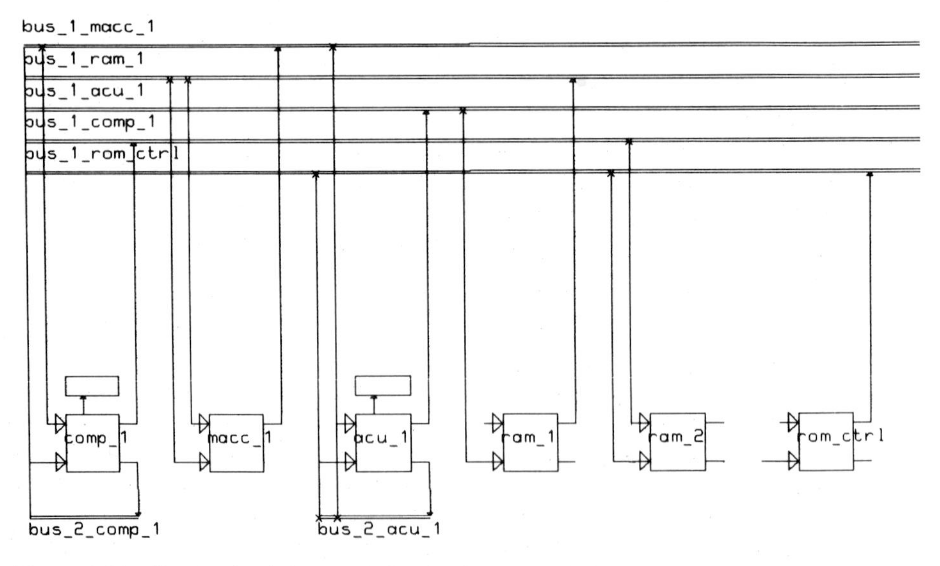

Figure 8.21 Synthesized data path for the AmplitudeSpectrum function.

bus will be used for the fetching of immediate memory addresses. A total of seven customized buses has been allocated. These will be reduced later using the bus merging process.

Figure 8.22 contains the generated set of register transfers to be executed on the data path. This description (called ATOMICS) serves as the input to the microcode scheduler (and the EXU assignment).

The register transfer language ATOMICS differs fundamentally from other existing RT languages in the sense that the language itself does not contain any keywords to indicate functions or operators. A typical ATOMICS construct contains a transfer and an operation part, separated by a vertical bar, | , as shown in the example that follows:

x: reg_1_alu_*a* ← *y*: reg_1_mult_1, *z*: reg_2_mult_1 |

MULT = multiply, bus_1_mult_1 = *x*;

means that the variable *x* stored in register file 1 of the alu *a* is generated by the multiplication of variables *y* and *z*, stored respectively in register files 1 and 2 of multiplier 1. The result of the multiplication is transferred to the alu using bus_1_mult_1.

The description in Figure 8.22 contains twenty-one basic register transfers and one iteration statement. A close inspection of the description shows that the synthesizer added index counters and address arithmetic to the original SILAGE operations.

In order to evaluate the conditional expression, a number of condition or status bits have been generated (called ampl[*i*]>=[*i* − 1] and ampl[*i*] = = max[*i* − 2]) as a result of a comparison operation. These bits will be used by the controller to select the right piece of code. Besides RT operations, the description also contains one renaming operation (max[*i*]:reg_2_comp_1 = max[*i* − 1]), which allocates the same register field for both variables max[*i*] and max[*i* − 1]. From a controller point of view, this is an equivalent to a NOP (NO Operation).

Register Binding

During the assignment process, the variables have been bound to specific register files attached to EXUs. However, the dimensioning of these register files as well as the assignment of the variables to precise fields in those files can only be performed after the completion of the scheduling operation. At that time, a procedure can be called to compute an optimal storage allocation and to minimize the total register count.

This optimization procedure, which tries to share as many registers as possible between variables, is based on a foregoing lifetime analysis of the variables. The lifetime of a variable is equivalent to the continuous interval $[t_1, t_2]$, where t_1 refers to the cycle when the variable was written for the first time and t_2 equals the cycle when the variable was read for the last time. An algorithm to compute those lifetimes and to optimize the register allocations has been developed and implemented. A detailed description can be found in [Goo87].

The application of this algorithm to the AmplitudeSpectrum processor

```
PROGRAM proc1;
BEGIN
i[0]:reg_1_acu_1 <- #0:reg_1_acu_1 | acu_1=PASS1, bus_2_acu_1 = i[0];
thresh:reg_1_ram_2 <- max[64]:reg_2_comp_1 | comp_1=PASS2,bus_1_comp_1 = thresh;
'*[fimag,0]':reg_2_acu_1 <- '*[fimag,0]':reg_1_rom_ctrl | bus_1_rom_ctrl = '*[fimag,0]', mux2_acu_1=bus_1_rom_ctrl;
'*[freal,0]':reg_2_acu_1 <- '*[freal,0]':reg_1_rom_ctrl | bus_1_rom_ctrl = '*[freal,0]', mux2_acu_1=bus_1_rom_ctrl;
max[0]:reg_2_comp_1 <- #0:reg_2_comp_1 | comp_1=PASS2, bus_2_comp_1 = max[0];
'*thresh':reg_2_ram_2 <- '*thresh':reg_1_rom_ctrl | bus_1_rom_ctrl = '*thresh';
thresh:reg_3_ram_2 <- thresh:reg_1_ram_2,'*thresh':reg_2_ram_2 | ram_2 = WRITE;

FOR i := 1..64 HOLDS BEGIN
    IF 'ampl[i]>=max[i-1]'[i] AND  NOT 'ampl[i]==max[i-1]'[i] THEN BEGIN
        max[i]:reg_2_comp_1 <- ampl[i]:reg_1_comp_1 | comp_1=PASS1, bus_2_comp_1 = max[i];
    END
    ELSE BEGIN
        max[i]:reg_2_comp_1 = max[i-1];
    END;
    ci[i]:creg_1_acu_1 <- i[i-1]:reg_1_acu_1 | acu_1 = INC;
    'ampl[i]>=max[i-1]'[i]:creg_1_comp_1 <- ampl[i]:reg_1_comp_1, max[i-1]:reg_2_comp_1 | comp_1 = MIN;
    'ampl[i]==max[i-1]'[i]:creg_2_comp_1 <- ampl[i]:reg_1_comp_1, max[i-1]:reg_2_comp_1 | comp_1 = MIN;
    i[i]:reg_1_acu_1 <- i[i-1]:reg_1_acu_1 | acu_1 = INC, bus_2_acu_1=i[i];
    'freal[i]*freal[i]'[i-1]:preg_3_macc_1 <- freal[i-1]:reg_1_macc_1, freal[i-1]:reg_2_macc_1 | macc_1 = MPY;
    ampl[i]:reg_1_comp_1 <- fimag[i-1]:reg_1_macc_1, fimag[i-1]:reg_2_macc_1,  'freal[i]*freal[i]'[i-1]:preg_3_macc_1 |
        macc_1 = MACC, bus_1_macc_1 = ampl[i];
    fimag[i-1]:reg_1_macc_1 <- fimag[i-1]:reg_3_ram_1, '*[fimag,i-1]':reg_2_ram_1 | ram_1 = READ, bus_1_ram_1 = fimag[i-1];
    freal[i-1]:reg_1_macc_1 <- freal[i-1]:reg_3_ram_1,'*[freal,i-1]':reg_2_ram_1 | ram_1 = READ, bus_1_ram_1 = freal[i-1];
    fimag[i-1]:reg_2_macc_1 <- fimag[i-1]:reg_3_ram_1,'*[fimag,i-1]':reg_2_ram_1 | ram_1 = READ, bus_1_ram_1 = fimag[i-1];
    freal[i-1]:reg_2_macc_1 <- freal[i-1]:reg_3_ram_1,'*[freal,i-1]':reg_2_ram_1 | ram_1 = READ, bus_1_ram_1 = freal[i-1];
    '*[fimag,i-1]':reg_2_ram_1 <- i[i-1]:reg_1_acu_1, '*[fimag,0]':reg_2_acu_1 | acu_1 = ADD, bus_1_acu_1='*[fimag,i-1]';
    '*[freal,i-1]':reg_2_ram_1 <- i[i-1]:reg_1_acu_1, '*[freal,0]':reg_2_acu_1 | acu_1 = ADD, bus_1_acu_1='*[freal,i-1]';
    END;
END.
```

Figure 8.22 ATOMICS RT description of AmplitudeSpectrum function after data path synthesis. Note: * variable denotes the RAM address of that particular variable.

resulted in a total of nine registers: three on the comparator, two on the multiply/accumulator, and four on the ACU.

The Bus Assignment
As stated previously, dedicated buses are assigned to all variables for the generation of the initial register transfer description of the controller. With this description, the register transfer code is scheduled in time (see Section 8.4.5).

After the scheduling, the minimal number of cycles needed for the execution of the algorithm on the particular architecture is known. Another output of the scheduler is the occupation of the different buses in time (as shown in Figure 8.23a). In order to minimize the number of buses inside a processor (and thus also the area), a bus-merging algorithm is called. This procedure tries to minimize the number of buses while staying within

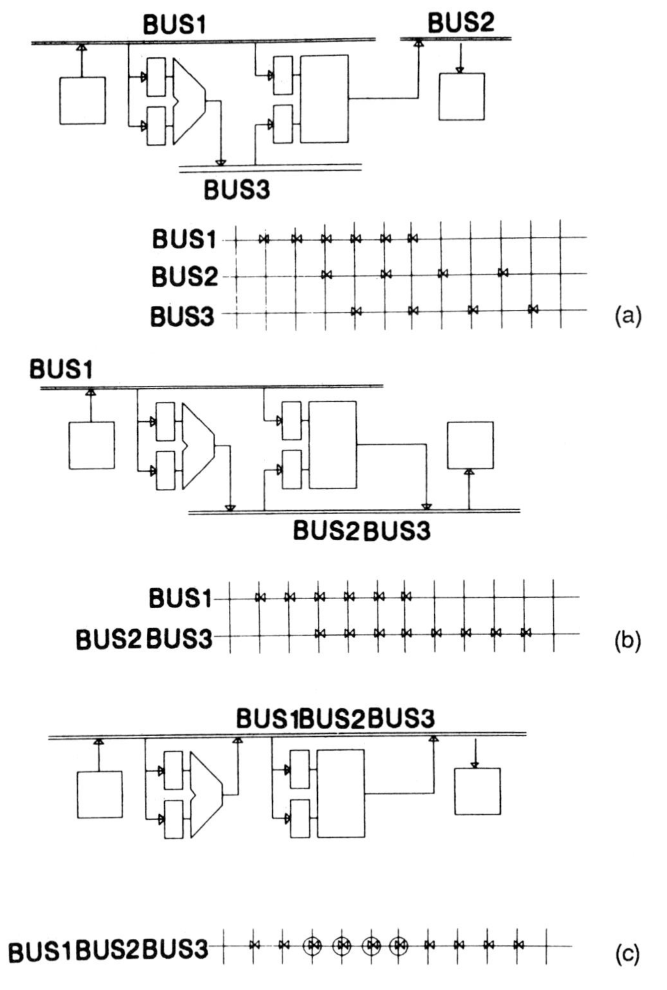

Figure 8.23 Bus Assignment: occupation and conflict diagrams.

the timing specifications. In the example of Figure 8.23, bus 2 and bus 3 can be merged without any conflicts as shown in Figure 8.23b. When, however, an attempt is made to merge all three buses into one (Figure 8.23c), four bus contention problems occur. These collisions will probably result in an increase of four in the number of cycles. A rescheduling of the register transfer operations is needed to compute the exact increase and to check if the total number of cycles is still within the allowed limits.

The bus-merging operation tries to determine which buses should be merged in order to minimize the total number of cycles without exceeding the allowed number of cycles. The total number of buses is limited to a maximum equal to the number of free metal tracks running over an EXU. An iterative optimization algorithm has been implemented:

For (i = 1 to NrOfBuses), perform the following steps:

- Order all buses, the one with the largest occupation on top.
- Take the i most-occupied buses.
- Try to merge all other buses (rest set) with the selected buses. Add the largest of the rest set to the smallest of the selected set (unless another merger causes less conflicts). This is repeated until the rest set is empty.
- If the obtained solution is in agreement with the timing constraint, exit; otherwise continue with the next iteration.

After the merging, a rescheduling of the register transfer operations has to be performed. This defines the actual number of cycles needed. Based on these results, the designer can decide on extra iteration steps if needed.

It is important to notice that precise determination of the allowed number of buses is only possible after the floor planning phase (when the precise placement of the different EXUs is decided upon). An extra iteration step might be needed.

For the AmplitudeSpectrum example, the bus merger reduced the number of buses from the initial number of seven to four without an increase in the total number of cycles.

Present Results

The techniques presented above have been implemented and have demonstrated the feasibility to realize design synthesis in an acceptable and yet flexible way. The rule-based data path synthesizer is flexible enough to accommodate architectural changes or additions in a simple and straightforward way. The introduction and debugging of a new function in the system just takes a couple of hours. The concept of the pragma-based user interference has proven to be very useful and efficient.

It was found, however, that the present synthesis process (which can be called an architecture knowledge base) should be preceded by a more global synthesis step. In this step, an initial solution is proposed in the form of data path allocation (types and number of EXUs based on complexity and timing considerations). This initial step would allow for a faster convergence towards the optimal solution. Therefore, we are presently studying extensions of the preprocessor to a more global optimization, based on critical path analysis and estimation of the mathematical complexity of the algorithm. This will result in a set of preallocations for the data path mapper that will fill in the architectural details (register transfers, macroexpansion, bus assignments, distribution of noncritical operations, etc.).

8.4.5 The Microcode Scheduling and EXU Assignment Operation

As a result of the translation step, the SILAGE description has been transformed into a register transfer (RT) description of the algorithm in the ATOMICS syntax (see Figure 8.22). The RTs are fundamental operations from a control point of view, since each RT is to be realized in a single machine cycle. For each transfer, the RT statement contains also a number of hardware requirements: source and destination register elements, the required operation, and the bus utilization.

In the RT language, no timing of the operations has been imposed. A number of assignments (e.g., the binding of an operation to a particular EXU) are also left open. The tasks of the scheduling operation are thus:

- Given the structure of the synthesized data path, order the RT operations on the time axis in such a way that the execution of the algorithm takes a minimal number of cycles. The derived schedule also has to comply with a number of external I/O timing constraints.
- At the same time, fill in the undefined assignments using the allocated hardware in an optimal way.

It must be mentioned that the scheduling process is influenced by the architecture of the controller, especially by the amount of pipelining. The multibranch controller, presented in Section 8.3, counts three pipeline stages.

The result of the scheduling operation is a set of RTs from which a symbolic microcode can be derived. This mainly involves:

- Transformation of all RTs scheduled at the same cycle into a set of instructions (control actions) for that cycle.
- Computation of conditional jump addresses for every instruction. This microcode serves as the input to the controller generation environment described in Section 8.4.6.

In this section, we will discuss the optimization constraints to be handled by the scheduler, the scheduling techniques used when no repetitive constructs are present, the extension toward loop constructs, and some perspectives. A more detailed and extended treatment of the scheduling problem can be found in [Goo87].

Optimization Constraints
Several timing constraints have to be taken into account during the computation of the optimal schedule. The following types of constraints can be recognized (as illustrated in Figure 8.24):

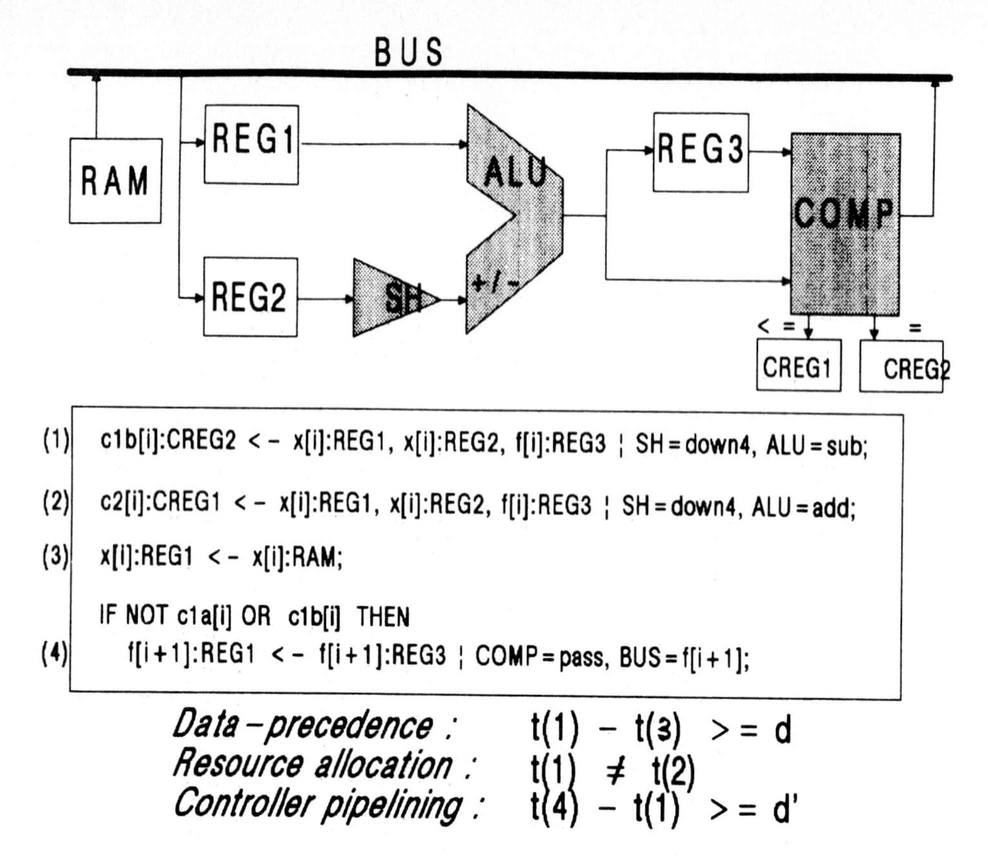

Figure 8.24 Constraints in the microcode scheduling.

- *Data precedence constraints:* An operation writing a variable in a register file is to be scheduled before any operation reading this variable from the register file. For example, in Figure 8.24 the RT operation (1) must be executed after RT operation (3), since (1) uses the result of (3).
- *Resource allocation constraints:* Different RTs that require the same resources (e.g., source register files, destination register files, buses, operators) may not be scheduled in the same machine cycle. Writing in and reading from the same register file is allowed in the same cycle. The number of available hardware units is constrained by the allocation process. In the example shown in Figure 8.24, the RTs (1) and (2) cannot be scheduled at the same time, since they use the same hardware units.
- *Controller pipelining constraints:* Conditional operations are executed depending on an evaluation of status signals produced by operators in the data path. A minimal time difference between the instructions generating the status signals and the conditional instruction based on these status signals is required (e.g., three cycles for the currently used multibranch controller). This figure is dependent on the data path and controller pipelining scheme. This type of constraint can be modeled as a data precedence constraint, in which the depending variable corresponds

to the status signal. For example, the RT (4) of the example in Figure 8.24 must be executed at least three cycles after RT (1), since the result of RT (1) is used in the condition of RT (4).

- *Looping data precedence constraints:* These constraints express a precedence relation between a writing RT in iteration i of a loop and a reading RT in iteration $i + 1$.

These constraints can be derived from the RT description of the algorithm, using a set of data precedence and resource allocation conflict rules. An efficient data structure allowing for fast constraint detection is required.

In the remaining discussion, we will limit the resource allocation constraints to the resource conflict type (assuming complete assignment). Extension of the technique to cover also the assignment problem (maximum number of units available) is trivial.

Scheduling Techniques for Nonrepetitive Programs

In a first step, the scheduling technique used for nonrepetitive programs will be explained (where nonrepetitive means that no loop constructs are allowed). A mathematical model for the scheduling of nonrepetitive programs will be described. In this model, an optimization function is provided and the previously mentioned constraints are described. We will denote an operation by o_i, and the cycle number in which this operation is scheduled by t_i.

Data precedence constraints can be modeled as weighted edges in a directed loop free graph (data precedence graph), in which the vertices represent the operations. The weights are the minimal time differences between the operations. The data precedence constraints are labeled m, with $m = 1..D$. For any m, the data precedence constraint between operation $o_k(m)$ (to be executed at cycle $t_k(m)$) and operation $o_q(m)$ (at $t_q(m)$ where $o_k(m)$ precedes $o_q(m)$ with at least $d(m)$ cycles, can be expressed by the following set of inequalities:

$$t_k(m) - t_q(m) \leq -d(m) \text{ for all } m := 1..D \qquad (8.1)$$

As indicated before, controller pipeline constraints can be modeled as data precedence constraints with an appropriate choice for the $d(m)$ value.

Constraints to prevent resource allocation conflicts can be modeled as difference equations. The pairs of potentially conflicting operations are labeled n, with $n = 1..C$. A minimal number of conflict pairs can be found by excluding those pairs for which there is a directed path in the data precedence graph. The nth pair is denoted $(o_i(n), o_j(n))$, the corresponding cycle numbers are $t_i(n)$ and $t_j(n)$. For every pair of possibly conflicting operations $o_i(n)$ and $o_j(n)$, only one node pair is considered. The resource allocation restrictions can then be formalized as:

$$t_i(n) \neq t_j(n) \text{ for all } n := 1..C \qquad (8.2)$$

The difference constraints (e.g., 8.2) are hard to handle by optimization programs, but can be replaced by a set of linear inequalities as suggested in [Pap82].

The goal of the scheduling operation now is to minimize, within the search space set by the constraints (8.1) and (8.2), the function Pout-Pin, where Pout and Pin are the cycle numbers of dummy RTs that respectively succeed and precede any other RT with a minimal delay of 0. Note that the formulated optimization problem can be easily extended to cope with I/O timing constraints. These constraints can be added as an extra set of inequalities.

It is interesting to see that the problem previously defined can be transformed into an integer linear program (ILP) [Pap82], for which the global optimum can be computed using standard mathematical techniques. Experiments have shown that this is a valuable approach for RT programs of moderate complexity, such as programs of small size with a poor amount of inherent parallelism. In other cases, excessive CPU times have been observed that cannot be tolerated in the CATHEDRAL-II environment, in which the scheduler is part of an iterative design cycle.

Therefore, a heuristic scheduling algorithm, based on graph models, has been developed. The technique is based on Zeman's scheduling algorithm [Zem85], which was originally developed for project management.

The scheduling proceeds on a step-by-step basis: at each point in time, a list of candidate transfers is generated, based on the precedence and pipelining constraints. Due to resource constraints (versus allocation), conflicts can occur between candidate transfers. If this is the case, we select the transfer that is part of the longest critical path and postpone the conflicting transfers until later. Numerous tests have shown that this technique delivers optimal or near optimal results in run times that are far better compared to the ILP approach.

Scheduling Techniques for Repetitive Programs

Due to their applicative nature, program loops do not have any control meaning in the RT description and can only be considered as compact notational formats (equivalent to the SILAGE loop construct). However, the area efficiency of the controller can be improved drastically (especially the ROM area), when loop constructs are allowed in the microcode. Therefore, a FOR-loop construct has been introduced in the RT input language that can be considered as a user pragma, that is, as a hint to the compiler to exploit the repetitivity of the algorithm in order to reduce the controller area. In order to maintain the uniqueness of the variable names (as needed in the applicative context), indexed notations are used. In a first phase, the looping mechanism of the RT language will be copied directly into the microcode without an attempt towards optimization.

Besides the area gain for the controller blocks, the introduction of FOR-loops also has the advantage that it results in a reduction of the scheduling problem. The loop construct imposes a limited ordering of the operations and reduces the number of RT instructions to be scheduled.

The techniques to schedule repetitive programs can be derived by extending the mathematical model for nonrepetitive programs. We only indicate the most important extensions:

- Schedule times of operations are replaced by so-called "potentials" of operations. Operations in successive loop iterations, represented by the same RT statement, have the same potential.
- Scheduling is now performed in a hierarchical fashion: first the loop at the highest level of nesting is scheduled. The results of this operation are frozen and used as a fixed unit of the next hierarchy level.
- The scheduling problem is now redefined as the computation of an optimal set of potentials for the RT operations, so that the overall cycle count is minimized. An optimization function can be derived, as a linear function of the potentials of a set of dummy operations, representing initial or final operations in the bodies of the respective loops.
- The set of data precedence and resource allocation constraints has to be extended with the looping data precedence constraints.

The presence of the looping constraints (that can be considered as data precedence constraints with negative weight) makes it impossible to solve the optimization problem in one step. Therefore, an iterative version of the graph-based scheduling algorithm has been developed and implemented [Goo87].

Results
The ATOMICS tool has been used to schedule the AmplitudeSpectrum function, described by the register transfer description of Figure 8.22. Inspection of the scheduled microcode, given in Figure 8.25, shows that the execution of the cycle of the algorithm takes 516 machine cycles, (4 + 64 * 8). The total microcode program needs 12 horizontal microcode words (or equivalent ROM words). This example clearly shows how the scheduling algorithm is able to exploit the architectural parallelism in an optimal way. For example, during potential 4, simultaneous operations are scheduled on the multiplier, the ACU, and the RAM.

The NOPs on potentials 7 and 8 are caused by the controller pipelining constraint between the generation of the status bits in potential 6 and the conditional operations in potential 9 (delay of 3).

It is interesting to notice that the optimal scheduling for this example counts only 264 cycles in total. The difference is caused by the hierarchical loop scheduling technique, which prevents overlaps between the execution of the consecutive loops. It is possible to manipulate the RT description in such a way that loop overlaps are introduced and that the scheduler also generates the minimal number of cycles. Techniques to perform this manipulation automatically are currently under investigation.

Perspectives
Topics that are currently under investigation include the automatic handling of loop overlaps. We are also considering a redefinition of the optimization function, such that the number of registers instead of the cycle count is minimized and the number of cycles is limited to an upper boundary as defined by the specifications.

Finally, we are also studying how the techniques presented can be used

```
Block       | Operation | Operation
hierarchy   | number    | description
------------+-----------+-----------------------------------------------------

Potential    0 :
----------------
   0.0      |          5 | MAX[0]:REG_2_COMP_1 <- #0:REG_2_COMP_1 | BUS_2_COMP_1=MAX[0], COMP_1=PASS2 <0>;
   0.0      |          3 | '*[FIMAG,0]':REG_2_ACU_1 <- '*[FIMAG,0]':REG_1_ROM_CTRL | MUX2_ACU_1=BUS_1_ROM_CTRL, BUS_1_ROM_CTRL=
            |            | '*[FIMAG,0]' <0>;
   0.0      |          1 | I[0]:REG_1_ACU_1 <- #0:REG_1_ACU_1 | BUS_2_ACU_1=I[0], ACU_1=PASS1 <0>;

Potential    1 :
----------------
   0.0      |          4 | '*[FREAL,0]':REG_2_ACU_1 <- '*[FREAL,0]':REG_1_ROM_CTRL | MUX2_ACU_1=BUS_1_ROM_CTRL, BUS_1_ROM_CTRL=
            |            | '*[FREAL,0]' <0>;

Potential    2 :
----------------
   1.0      |         21 | '*[FREAL,I-1]':REG_2_RAM_1 <- '*[FREAL,0]':REG_2_ACU_1, I[-1+I]:REG_1_ACU_1 | BUS_1_ACU_1='*[FREAL,I-1]',
            |            | ACU_1=ADD <0>;

Potential    3 :
----------------
   1.0      |         20 | '*[FIMAG,I-1]':REG_2_RAM_1 <- '*[FIMAG,0]':REG_2_ACU_1, I[-1+I]:REG_1_ACU_1 | BUS_1_ACU_1='*[FIMAG,I-1]',
            |            | ACU_1=ADD <0>;
   1.0      |         19 | FREAL[-1+I]:REG_2_MACC_1 <- '*[FREAL,I-1]':REG_2_RAM_1, FREAL[-1+I]:REG_3_RAM_1 | BUS_1_RAM_1=FREAL[-1+I],
            |            | RAM_1=READ <0>;
   1.0      |         17 | FREAL[-1+I]:REG_1_MACC_1 <- '*[FREAL,I-1]':REG_2_RAM_1, FREAL[-1+I]:REG_3_RAM_1 | BUS_1_RAM_1=FREAL[-1+I],
            |            | RAM_1=READ <0>;

Potential    4 :
----------------
   1.0      |         18 | FIMAG[-1+I]:REG_2_MACC_1 <- '*[FIMAG,I-1]':REG_2_RAM_1, FIMAG[-1+I]:REG_3_RAM_1 | BUS_1_RAM_1=FIMAG[-1+I],
            |            | RAM_1=READ <0>;
   1.0      |         16 | FIMAG[-1+I]:REG_1_MACC_1 <- '*[FIMAG,I-1]':REG_2_RAM_1, FIMAG[-1+I]:REG_3_RAM_1 | BUS_1_RAM_1=FIMAG[-1+I],
            |            | RAM_1=READ <0>;
   1.0      |         14 | 'FREAL[I]*FREAL[I]'[-1+I]:PREG_3_MACC_1 <- FREAL[-1+I]:REG_2_MACC_1, FREAL[-1+I]:REG_1_MACC_1 | MACC_1=MPY <
            |            | 0>;
   1.0      |         13 | I[I]:REG_1_ACU_1 <- I[-1+I]:REG_1_ACU_1 | BUS_2_ACU_1=I[I], ACU_1=INC <0>;
   1.0      |         10 | CI[I]:CREG_1_ACU_1 <- I[-1+I]:REG_1_ACU_1 | ACU_1=INC <0>;

Potential    5 :
----------------
   1.0      |         15 | AMPL[I]:REG_1_COMP_1 <- 'FREAL[I]*FREAL[I]'[-1+I]:PREG_3_MACC_1, FIMAG[-1+I]:REG_2_MACC_1, FIMAG[-1+I]:
            |            | REG_1_MACC_1 | BUS_1_MACC_1=AMPL[I], MACC_1=MACC <0>;

Potential    6 :
----------------
   1.0      |         12 | 'AMPL[I]==MAX[I-1]'[I]:CREG_2_COMP_1 <- MAX[-1+I]:REG_2_COMP_1, AMPL[I]:REG_1_COMP_1 | COMP_1=MIN <0>;
   1.0      |         11 | 'AMPL[I]>=MAX[I-1]'[I]:CREG_1_COMP_1 <- MAX[-1+I]:REG_2_COMP_1, AMPL[I]:REG_1_COMP_1 | COMP_1=MIN <0>;

Potential    7 :
----------------
   1.0      | NOP

Potential    8 :
----------------
   1.0      | NOP

Potential    9 :
----------------
   1.0      |          9 | IF (NOT (NOT 'AMPL[I]==MAX[I-1]'[I] AND 'AMPL[I]>=MAX[I-1]'[I])) THEN
            |            |   MAX[I]:REG_2_COMP_1=MAX[-1+I];
   1.0      |          8 | IF ((NOT 'AMPL[I]==MAX[I-1]'[I] AND 'AMPL[I]>=MAX[I-1]'[I])) THEN
            |            |   MAX[I]:REG_2_COMP_1 <- AMPL[I]:REG_1_COMP_1 | BUS_2_COMP_1=MAX[I], COMP_1=PASS1 <0>;

Potential   10 :
----------------
   0.0      |          6 | '*THRESH':REG_2_RAM_2 <- '*THRESH':REG_1_ROM_CTRL | BUS_1_ROM_CTRL='*THRESH' <0>;
   0.0      |          2 | THRESH:REG_1_RAM_2 <- MAX[64]:REG_2_COMP_1 | BUS_1_COMP_1=THRESH, COMP_1=PASS2 <0>;

Potential   11 :
----------------
   0.0      |          7 | THRESH:REG_3_RAM_2 <- '*THRESH':REG_2_RAM_2, THRESH:REG_1_RAM_2 | RAM_2=WRITE <0>;

Total of 12 potentials

Total of 516 machine cycles
```

Figure 8.25 Scheduled register transfer description of the AmplitudeSpectrum function.

for the scheduling of operations on a general-purpose signal processor such as a TMS320.

8.4.6 The Controller Generation Environment

The result of the scheduling process is a symbolic microcode description of the processor controller. This description is functionally equivalent to a finite state machine (FSM) and can be implemented on a variety of controller architectures (as long as these architectures have the same amount of internal pipelining, since this influences the scheduling process and thus also the obtained microcode). The same microcode can be implemented on a single FSM or on a ROM-and-PC-based controller. The efficiency of a certain controller depends upon the type of the algorithm under consideration. It is therefore important to be able to compare the implementations of a controller description on a number of available architectures.

This requires a unified environment, as pictured in Figure 8.26. A unique input language is used to describe the requested behavior of the controller. This description tries to capture the FSM nature of the controller, in terms

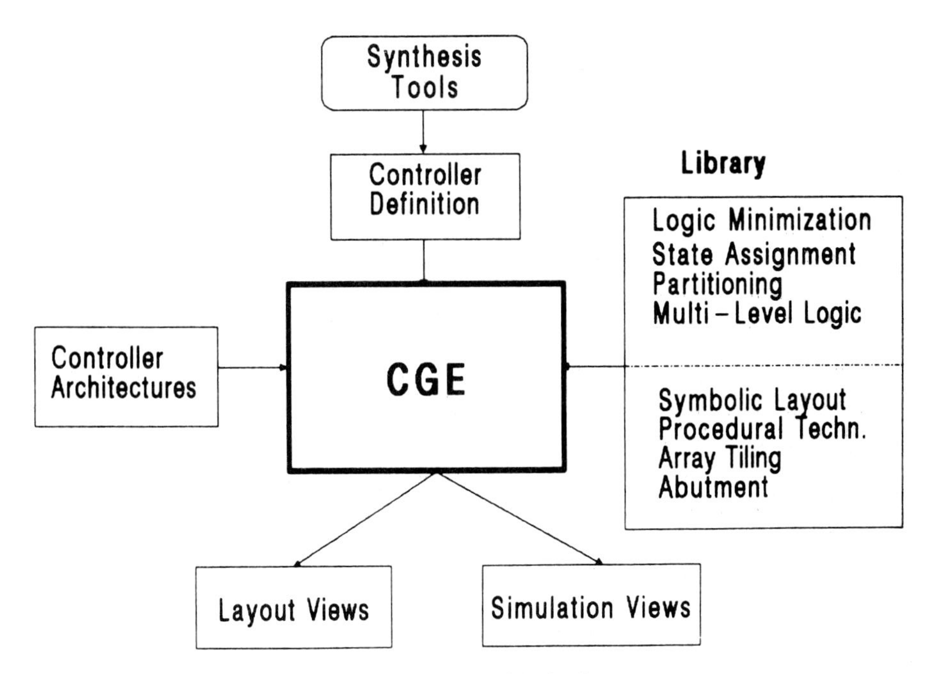

Figure 8.26 Controller Generation Environment: block diagram.

of inputs, outputs, states, and next states. It supports symbolic covers for states and symbolic descriptions of the output behavior. An example of such an input description is given in Figure 8.27. Alpha, beta, and so on, are symbolic notations for the different states of the controller (note that no coding of the states is implied). Constructs such as "alu : inc" are symbolic output covers (macros) used to compact the size of the description and to

```
# CONTROL 'pitch_extractor Processor 1';

# LIBRARY 'cathedral.lib';

/* The library contains the macro-models of the control for the
   different EXU's */

# FSM
    <Alpha>
           UPD cstart := start_flag;
           JMP if (cstart) then Alpha
                                 else Beta;
    <Beta>
           UPD state[1] := alu_sign & acu_overflow;
           CTL alu:inc, acu:modulo,
               acu_reg1:r[1]w[2];
           JMP Gamma;
    <Gamma>
           ...
```

Figure 8.27 Input description for CGE.

avoid the repetition of identical output patterns. A table-driven preprocessor is used to expand these covers into the appropriate output signals (e.g., alu_add = 1, alu_carry = 1, alu_inpa = 1, alu_inpb = 0 for the alu : inc operation).

The description of Figure 8.27 has to be understood as follows: in state beta, perform the output actions described under the CTL label (defined with the aid of the macro mentioned previously). At the same time, the internal state of the FSM is updated under influence of a number of input variables (UPD). Finally, the FSM goes to a next state as defined under the JMP label.

The core of the controller generation environment (CGE) consists of a number of procedures that describe how an input description can be mapped into a particular controller architecture. For each architecture such a procedure must be provided. Such an assembly procedure normally consists of a number of optimization and minimization steps (state assignment, logic minimization, etc.) and a layout generation phase. The goal of the CGE is to ease the introduction of new architectures and to avoid the implementation of the same functions over and over again. Therefore, the CGE provides a library of common functions such as state assignment, logic minimization, multilevel logic synthesis, and partitioning. It also supports constructs to assemble regular arrays using symbolic layout techniques.

It is important to notice the similarity between the controller generation environment and the module generation environment. In both cases, the environment is used for the simple introduction of new regular constructs, which will be used multiple times. The underlying library of support tools is, however, of a different nature.

Presently, we are working toward a prototype version of the CGE, using the multibranch controller of Section 8.3 as a target. The tools used for the generation of this controller are basically the PLA minimization and generation tools, described in [Bar85].

8.5 Conclusions

The present results, obtained from a number of test designs, indicate that our synthesis approach as outlined in Figure 8.17 is a viable strategy that is capable of producing acceptable designs. The system has been conceived as an open architecture that allows the experienced designer to interfere at the various levels and to perform further optimizations if necessary. Until now, the major attention of our synthesis research has been oriented toward the translation and assignment phases that map a higher-level description onto a roughly defined data path architecture (type of EXUs, amount of parallelism). We feel however that further research is needed in following areas:

- Extension of the synthesis techniques to include more global optimization techniques with more elaborated cost functions.

- Extended flexibility with respect to architectural changes or additions.
- Algorithm partitioning and load balancing.

In this way, an initial architecture proposal can be generated automatically that can then be refined and worked out using the already implemented techniques.

These subjects will be the main target of our synthesis research and development efforts in the near future.

REFERENCES

1. [All85] J. Allen, "Computer Architecture for Digital Signal Processing," *Proc. IEEE,* vol. 73, no. 5, May 1985, pp. 854–873.
2. [Bar85] M. Bartholomeus, L. Reynders, M. Pauwels, and H. De Man, "PLASCO: A Procedural Silicon Compiler for PLA Based Systems," *Proc. IEEE CICC Conf.,* Portland, May 1985, pp. 226–229.
3. [Cam85] R. Camposano, "Synthesis Techniques for Digital Systems Design," *22nd Design Automation Conf.,* Las Vegas, 1985, pp. 475–481.
4. [Goo87] G. Goossens, J. Rabaey, J. Vandewalle and H. De Man, "An Efficient Microcode-Compiler for Custom DSP-Processors," IEEE IACCAD Conf., Santa Clara, Nov. 1987.
5. [Hil85] P. Hilfinger, "A High-Level Language and Silicon Compiler for Digital Signal Processing," *Proc. IEEE CICC Conf.*, Portland, May 1985, pp. 213–216.
6. [Jai86] R. Jain, F. Catthoor, J Vanhoof, B. Deloore, G. Goossens, N. Goncalves, L. Claesen, J. Van Ginderdeurer, J. Vandewalle, and H. De Man, "Custom Design of a VLSI PLM-FDM Transmultiplexer from System Specifications to Layout Using a CAD System," *IEEE Journal of Solid State Circuits,* vol. SC21, no. 1, February 1986, pp. 73–85.
7. [Lip86] R. Lipsett, E. Marschner, and M. Shahdad, "VHDL—The Language," *IEEE Design & Test,* April 1986.
8. [Mag84] S. Magar, E. Caudel, and A. Leigh, "A Microcomputer with Digital Signal Processing Capabilities," *Proc. ISSCC 1982,* February 1982, pp. 32–33.
9. [Mari86] E. Marien and H. De Man, "Manual of LOGMOS V4.2: A Simulator Covering Register Transfer, Functional Gate and Switched Level," available from IMEC, Kapeldreef 75, B-3030 Heverlee, Belgium.
10. [Marw84] P. Marwedel, "The MIMOLA Design System: Tools for the Design of Digital Processors," *21st Design Automation Conference,* 1984, pp. 587–593.
11. [Pap82] C. Papadimitriou, K. Steiglitz, *Combinatorial Optimization, Algorithms and Complexity,* Englewood Cliffs, N. J., Prentice-Hall, 1982, pp. 421-424.
12. [Pat81] D. Patterson and C. Sequin, "RISC I: A Reduced Instruction Set VLSI Computer," *Proc. 8th Int. Symposium on Archtiecture,* Minneapolis, May 1981, pp. 443-457.
13. [Pop84] S. Pope, J. Rabaey, and R. Brodersen, "Automated Design

of Signal Processors Using Macrocells," in *VLSI Signal Processing,* IEEE Press, 1984, pp. 239–251.

14. [Rab85] J. Rabaey, S. Pope, and R. Brodersen, "An Integrated Automated Layout Generation System for DSP Circuits," *IEEE Trans. on CAD,* vol. CAD-4, July 1985, pp. 285–296.
15. [Sis82] J. Siskind, J. Southard, and K. Crouch, "Generating Custom High Performance VLSI Designs from Succint Algorithmic Descriptions," *Proc. Conf. on Advanced Research in VLSI,* Cambridge, January 1982, pp. 28–40.
16. [Six86a] P. Six, L. Claesen, J. Rabaey, and H. De Man, "An Intelligent Module Generator Environment," *Proc. 23rd Design Automation Conference,* Las Vegas, July 1986, pp. 730–735.
17. [Six86b] P. Six, I. Vandeweerd, and H. De Man, "An Interactive Environment for Creating Module Generators," *Proc. ESSCIRC 1986,* Delft, September 1986.
18. [Slu80] R. Sluyter, H. Kotmans, and A. Van Leeuwaarden, "A Novel Method for Pitch Extraction from Speech and a Hardware Model Applicable to Vocoder Systems," *Proc. IEEE ICASSP Conf.,* April 1980, pp. 45–48.
19. [Tho83] D. Thomas, C. Hitchcock, T. Kowalski, T. Rajan, and R. Walker, "Automatic Data Path Synthesis," *Computer,* vol. 16, no. 12, December 1983, pp. 59–70.
20. [Tuc85] B. W. Tucker, "Electronic CAD-CAM—Is it Revolution or Evolution," *Proc. 22nd ACM/IEEE DA Conf.,* Las Vegas, June 23–26, 1985, pp. 830–834.
21. [Tze85] C. Tzeng, "Timing Recovery in Digital Subscriber Loops," Ph. D. diss., Mem. No. UCB/ERL M85/29, April 1985.
22. [Van86] E. Vanden Meersch, L. Claesen, and H. De Man, "SLOCOP: A Timing Verification Tool for Synchronous CMOS Logic," *Proc. ESSCIRC 1986,* Delft, September 1986.
23. [Zem85] Jan V. Zeman, "Synthese und Praktische Realisation von Systeme und Algorithmen fur Digitale Signalverarbeitung," Ph. D. diss., Zurich, 1985.

CHAPTER 9

The Genesil Silicon Compiler

Edmund K. Cheng
Stanley Mazor

9.1 Introduction

As integrated circuit (IC) technology entered the VLSI era, the cost, size, performance, and power advantages become so enormous that implementation of major portions of most digital system designs in application specific integrated circuits (ASIC) became imperative. Silicon Compiler Systems (SCS) was formed at a time when digital system designers had only two choices in using ASICs: (1) full-custom IC, which requires the specialized expertise and tools of IC circuit and layout design and (2) semi-custom IC, such as gate arrays, which does not require significant IC design skill and has shorter fabrication turn-around time, but does not use silicon area very efficiently.

Clearly, designing VLSI chips that contain anywhere from 20,000 to 1,000,000 devices represented a major new challenge in complexity management, with some similarities to the software development field. New tools and methodologies were needed to raise the level of design abstraction and maintain a set of design disciplines. These issues were not adequately addressed by the full-custom or semi-custom approaches to ASIC.

In early 1985, SCS introduced a silicon compiler product called the Genesil System [John84]. Genesil was the commercial realization of earlier work done by Dave Johannsen and Carver Mead at the California Institute of Technology. Although Johannsen and Mead coined the term *silicon compiler* while working on the Bristle Blocks [Joha79] research project at Caltech, there is considerable diversity of opinion within the CAD community on the definition of what a silicon compiler is [Gajs85], [VLSI84a], [VLSI84b].

The Genesil system is significant in that it was the first to achieve substantial commercial acceptance, making it a landmark in the evolution of the silicon compilation technology.

Some of the design objectives and assumptions of the system are briefly stated as follows:

- Integrated design system for specification, verification, and tapeout.
- Competitive in density and performance to gate arrays and standard cells.
- Multiple foundries, processes, and technologies.
- Reduced design effort.
- Complexity management for very large designs.
- Cell-based tiling with design rule mapping for multiple foundries.
- Conservative design practices for high success rate.
- User-defined pads at chip periphery.
- Menus and forms on low cost terminals for user input.
- UNIX operating system and C language for platform portability.

Clearly, these are not the final words on what a silicon compiler should be.

An overview of the Genesil system is shown in Figure 9.1 that illustrates the system as having three major software tools: (1) the Genesil silicon compiler, (2) the compiler development tools, and (3) the Genecal process calibration tools.

The rest of this chapter is divided into two parts. Sections 9.2 to 9.4 describe the overall external view of the Genesil system. They contain some typical inputs and outputs and an application example. Sections 9.5 to 9.7

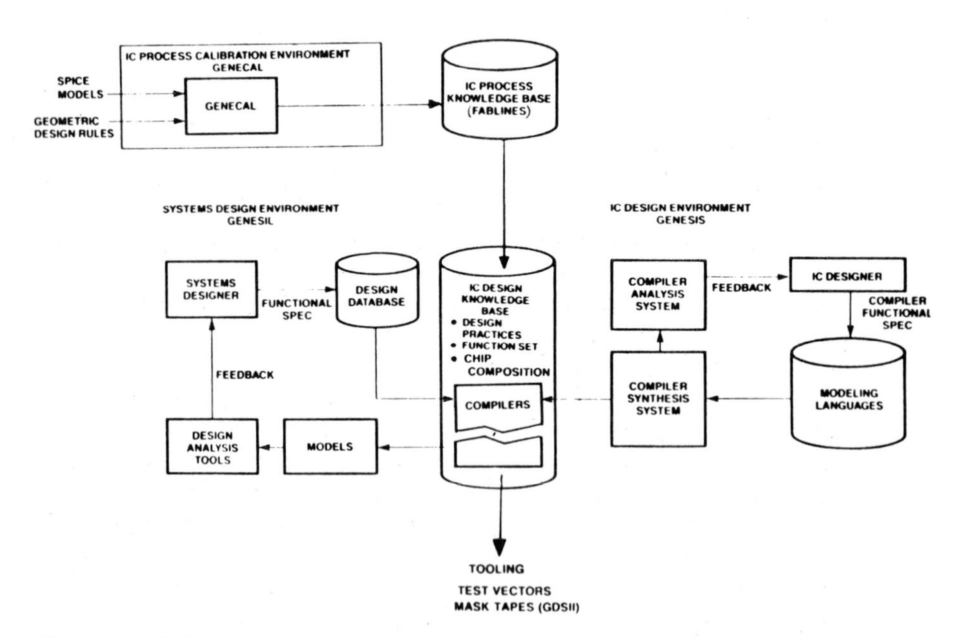

Figure 9.1 Silicon compiler design tools.

describe more detailed aspects of using the system, some of the internal views and underlying principles, and the way in which new compilers can be created.

The VLSI CAD customer base can be divided into two segments: the systems designer and the VLSI chip designer. Systems designers are generally designing logic with conventional TTL circuitry and gate arrays. Typically, they are unfamiliar with VLSI design and layout. The Genesil system described in Section 9.2 is targeted for this kind of engineer.

On the other hand, the system can be used and enhanced by VLSI chip design engineers, and the compiler development system is aimed at these engineers, who can apply their IC design expertise to enhance the compiler by creating new block compilers. These tools for IC designers are described in Section 9.7. The Genecal tools are not described in this chapter.

9.2 Genesil System

The Genesil silicon compiler is a complete system for ASIC chip design using high-level structural elements. The system consists of a collection of block compilers for RAM, ROM, PLA, data paths, and random logic. It also contains tools for net listing, placement, routing, functional simulation, and timing analysis.

The system addresses VLSI design and verification simultaneously, and supports a design style best characterized as "design by successive refinement," or incremental design.

The system provides design leverage by focusing the process of designing VLSI chips toward the architectural level and away from the classical IC design considerations. The term *design leverage* refers to the improvements in design engineering productivity. The improvements are gained because the user does not need to synthesize the design down to transistor and layout levels, at which the design and verification are very error-prone due to high complexities.

Since the physical circuitry and layouts are algorithmically generated from circuit primitives that have been previously verified, this design methodology is often characterized as "correct by construction." Aside from saving the manual labor involved in synthesizing the circuit and layout designs, this method also makes it viable for the systems designer to synthesize a working VLSI chip, even though he or she has neither the tools nor the expertise to design and debug the chip using the conventional method. Of course, the user may not be able to achieve the absolutely optimum design that might be possible by custom tailoring the circuit design at the geometric level.

With Genesil, architectures may be specified hierarchically, and the layout may be synthesized either hierarchically (connecting blocks together to form modules and modules to form chips) or with the flat approach (working at the block level, wiring directly from one block to another).

The user both specifies and debugs the design at the architectural level. The functional and circuit models that are used for design verification are automatically synthesized from the design specifications and are targeted for a user-specified foundry. Design functionality is verified by running functional simulation, while performance is verified by running timing analysis. Typically, these two verification activities are the most time-consuming portion of the chip development, because the specification and implementation of the design is considerably automated. Furthermore, the system provides functional simulation at any level of the design hierarchy. It also produces critical delay path analysis and timing performance reports, using circuit parameters from the completed layout.

The system also uses an integrated hierarachical data base. This custom made data base system is built using UNIX directories and files. A special system file contains tags and pointers to allow both shared files, automatic currency checking (similar to the UNIX "makefile" facility), and automatic bidirectional tracing.

9.2.1 User's Input Specification

The system displays dynamic menus and forms for specifying design details interactively, although other current CAD/CAE workstations primarily accept graphic input (drawing pictures) for the user's specifications. One of the reasons for choosing forms is that high-level functions typically require many parameters. Figure 9.2 is an example of a RAM specification form which exemplifies the number of parameters required for specifying a RAM block.

The system displays both textual forms and complex graphics to document the input specification for a logic block within a chip. Although visual icons are a good way of displaying information, block diagram symbols in the form of nondescript rectangles do not convey enough information for high-level structures. For easy comprehension, schematics for complex objects must both be rich in graphics and contain textual information. Therefore, the system draws diagrams for engineering documentation as a result of analyzing specification forms for objects such as latches and ALUs, as shown in the example (Figure 9.3) in this chapter.

The user's input specification form is checked interactively for syntactic and semantic errors. Many types of logic design and bookkeeping errors can be detected by the system as they are entered. These general types of specification checks are called logic design rule checks (LDRC). For example, dangling outputs or undefined inputs are noted. To ensure correctness of systems that use multiple clocks and clock phases, the compiler identifies which clock phase is used in a register and remembers data bus transfer times; output block diagrams have color coded signals to help the user track clock phases, just as color coded wires aid a technician in tracing connections, to help eliminate timing errors when using signals based on different clock phases. LDRC checks are also run during net listing.

RAM FUNCTIONAL SPECIFICATION

Width:	> 8		
Depth:	> 64		
Timing:	FULL CYCLE	HALF_CYCLE	
Phase A	>PHASE A		(PHASE_A)
Phase B	>PHASE B		(PHASE_B)
		Slicing:	
Slicing Mode:	NONE	COMMON	
Number of Slices:	> 0		
		PORT A:	
Data Flow:	READ	WRITE	READ_WRITE
Drive Type:	DIRECT	PRECHARGED	
Address:	>A ADDR		(A_ADDR)
Data Out:	>A OUT		(A_OUT)
Drive Enable:	>TRUE		(A_DRIVE)
		PORT B:	
Data Flow:	READ	WRITE	READ_WRITE
Address:	>B ADDR		(B_ADDR)
Data in:	>B IN		(B_IN)
Write Enable:	>B WE		(B_WE)
		Address Encoding:	
Address Width:	> 6		
Number of Ranges:	> 1		
Range 0: From:	> 0 To: > 63		

Figure 9.2 RAM functional specification (64 by 8).

9.2.2 Block Compilers

The Genesil function set contains a variety of block compilers, as detailed in Section 9.3. Blocks are instantiated by the block compilers according to the specifications and parameters supplied by the user. The block compilers produce: (l) a foundry specific layout, (2) a timing model for performance evaluation, (3) a functional model for simulating the behavior of the circuit, and (4) a power model for computing the power consumption and transient currents that is used by the router. The internal views of block compilers are presented in Section 9.7, under the discussion of the Genesis system.

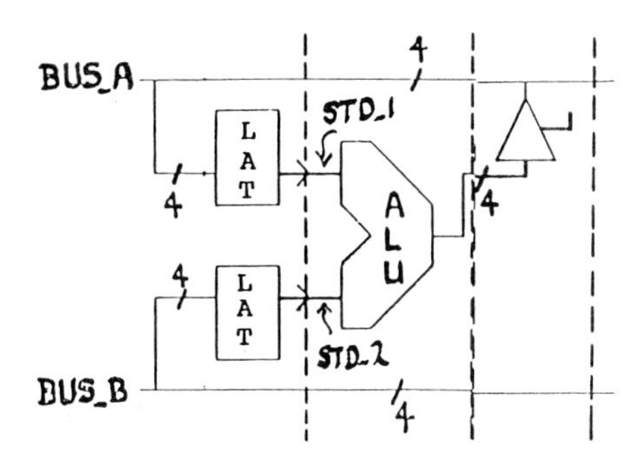

Figure 9.3 Four-bit wide data path block diagram.

Circuit Layout

When the user is satisfied with the specification and block diagram, the compiler generates the layout for the specified foundry process. Typical blocks compile in less than one minute and chip area is reported along with the block's dimensions. The layout model is represented as a proportionately-sized rectangular symbol, that the user can later manipulate during the chip floor planning and placement activities. In addition, the total size of all blocks is accumulated as the design proceeds so that the user can monitor the chip cost and consider tradeoffs in function for die area.

The user can request a color display of the transistor layout on the graphics terminal (as shown in Figure 9.4) or a hard copy color plot on a plotter. In general, detailed layout information is needed only after the chip design is complete, when a tape is written to carry the IC mask tooling data to the foundry; systems designers do not usually need detailed layout plots when designing chips on a silicon compiler because the design is done at the higher functional level. However, those familiar with IC layouts can gain additional insights to make better use of the compiler.

This high-level system does not permit the user to directly edit the physical layout, such as moving a wire or changing a transistor. Since the final tape output is in a standard format, the user has the prerogative to move the design to a physical design system and edit the layout in that environment. However, to do so would obviously jeopardize the integrity of the compiled output.

Timing Model

The speed performance is a vital design parameter in any VLSI development project, and the block compilers produce timing models for the block instances, using the foundry process selected by the user. The timing analyzer (TA) uses this model to analyze circuit delays and calculate the maximum operating frequency (see Section 9.6.4). The TA performs an exhaustive trace of all paths to find the longest, and no stimulating vectors

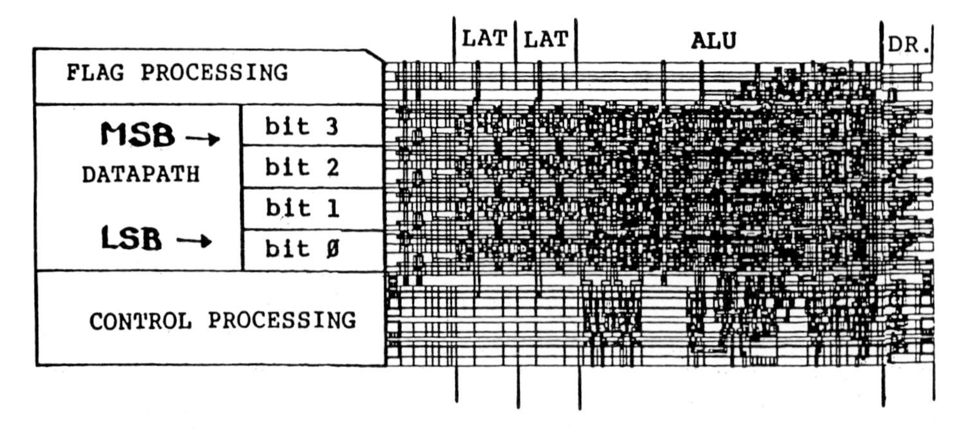

Figure 9.4 Example four-bit wide datapath layout.

or test cases are needed. The timing can be computed over the range of temperature and supply voltage variations. If the estimated performance of a chip does not meet the user's needs, then the microarchitecture or the foundry process can be changed. Figure 9.5 shows the output report from a sample timing run, with the calculated cycle times at the top.

Functional Simulation Model

The user can simulate the behavior of the logic design using the functional simulator (see Section 9.6.2). Functional simulation is much like logic simulation except that the circuit models generated by the block compilers are independent of circuit delays and use high-level logic primitives. This is considerably more efficient and faster than modeling strictly with simple Boolean gates.

The simulator permits a wide variety of output formats on the screen. In preparation for simulation, the user can "paint" the screen to identify where and in what format the simulated data outputs are to be displayed. The display can be created with timing waveforms for those who prefer to see the data in logic analyzer format. Figure 9.6 shows a sample output in waveform format from a simulation run. Input test vectors may be entered interactively at the terminal or read from a file in batch mode.

9.2.3 Geometric Layout Tools

After the functional blocks have been compiled from the user's specifications, they are then compiled into a VLSI chip using the physical layout tools: net listing, placement, and routing.

Net Listing

Block compilers produce instances of rigid rectangular blocks; the net list specifies the desired interconnections between the blocks. The net list editor permits the user to form arbitrary connections, and its internal IDRC program reports questionable and erroneous electrical connections,

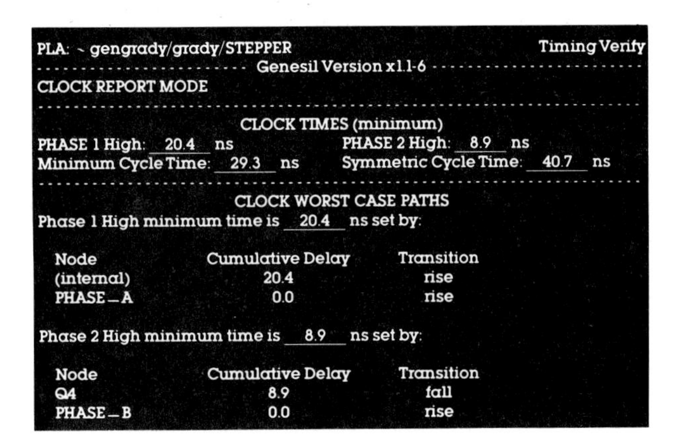

Figure 9.5 Typical TA output report.

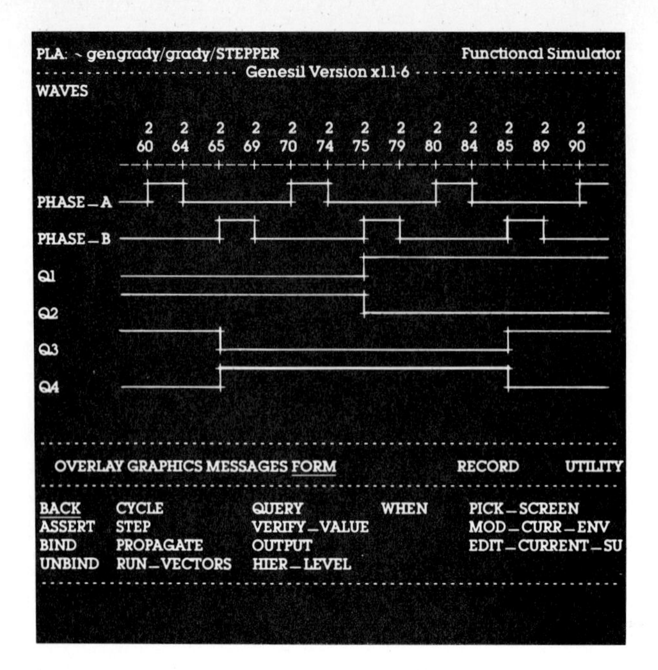

Figure 9.6 Sample simulator output.

for example, wiring two output drivers together or not having a source for multiple sinks.

The system supports hierarchical net listing and signal naming. Blocks have signal connectors, and they are logically grouped into "modules"; a module can contain arbitrary numbers of blocks and other modules. Modularity facilitates sharing designs and building chips from elements that the users have designed previously. Signal names that are completely internal to a module are hidden from other modules and will not cause name conflicts. Signal names that are used as connectors at the module level are visible to other blocks and modules at the same level.

Floor Planning and Placement

The blocks and modules are composed into a chip by placing the compiled blocks and modules relative to one another in the floor planning process, as shown in Figures 9.7a and b. Several macrocell placement algorithms have been reported in the literature ([Sech 85], [Sech 87], [Souk 81]).

To allow the user to interactively influence the floor planning and placement process, the system provides graphic placement aids by generating a recommended best placement for a block, displaying connectivity of an edge to other blocks, and showing the center of gravity for interconnections.

Routing

The router uses the relative placement as its starting position, and moves the blocks apart as needed to complete the wiring channels. It uses binary fusion as the methodology for identifying the order of channel assignment

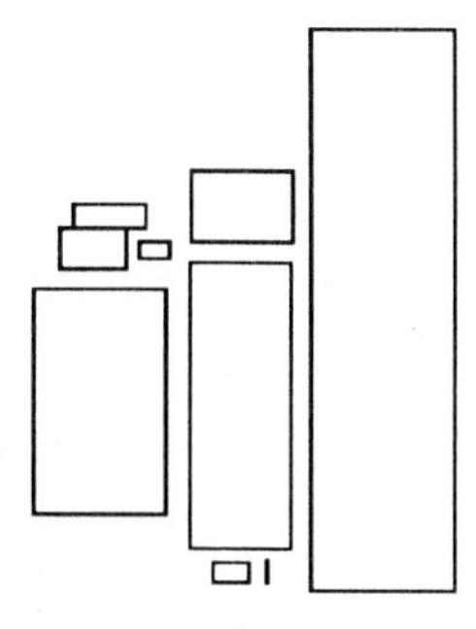

Figure 9.7a Floor plan with 64 by 8 RAM.

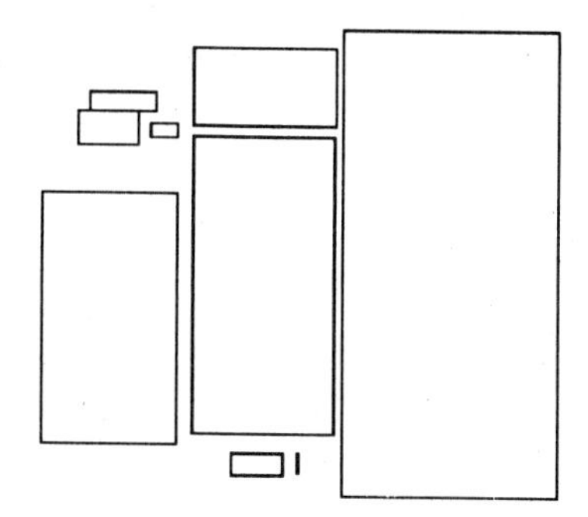

Figure 9.7b Floor plan using 32 by 16 RAM.

and sequencing the global routing strategy; the concept of binary fusion is analogous to slicing structures ([LaPo86], [Laut 80], [Otte 82], [Supo 83]) and bipartitioning placement algorithms [Souk 81].

The system uses a nongridded channel router that works in multiple phases. During channel assignment, the relative block placement can be altered by the router to allow for more space for routing wires. The router can use several layers of interconnection material, for example, polysilicon and metallic layers. The relative priority for the interconnection layers (for example, metal 1 versus metal 2) is process specific. There is a priority system for signal wires so that users can optionally specify their relative priorities.

Because the router can move blocks apart, it can always route 100 percent of the chip; however, the chip size is only known after the routing is complete. The router calculates the parasitic loads and makes the capacitance value available to the timing analyzer so that accurate delays can be calculated. The routing layouts are procedurally generated, so the precise design rules for the interconnecting layers can be observed, such as metal-to-metal spacing and minimum metal width. There are conventions for the location of power and ground ports. The router uses the power consumption and transient current calculations for the block instances, and generates the proper line widths for VCC and ground supply lines using conservative assumptions on resistive voltage drop and metal migration.

9.2.4 Design Activity

Typical compilation speeds are one thousand transistors per minute, running on a workstation. This speed allows the user to explore several architectures in just a few days, which compresses the design cycle and lowers the design cost, making it economically feasible to develop chips that are to be used in small quantities.

The design activity falls into two major phases: exploratory design and final design. During the exploratory phase, the user experiments with different microarchitectures, and the system estimates the chip area, power consumption, and chip speed. During this phase, the compiler provides a large number of default values for details that the user need not be bothered with yet, such as default signal names and approximations for routing area. The timing analyzer estimates the chip's performance and critical paths, so that the design can be tuned to meet the speed requirements. Some examples of tuning involve the comparison of choices such as: choosing a counter instead of an adder, using one adder twice instead of two separate adders, using a shift register instead of a barrel shifter, adding stronger driver-buffers to reduce loading related delays, breaking up large arrays into smaller ones. Using the outputs of the compiler as a basis, the user makes design tradeoffs with choices such as these that are customary to systems designers and do not involve modifying the designs at the transistor or layout level. Hence, global optimization can be done during the exploratory phase and detailed tradeoffs can be made near the end of the design process.

After the user is satisfied with the architecture, the logic details are fully specified. Now the user assigns signal names and spends more time in net listing pieces together. To verify the correctness of the logic, the user runs functional simulation on test cases. Floor planning and routing are used to generate the final layout. The router uses the power model created by the block compiler to determine the width of power lines, which are routed first. In addition to ordering signal wires by assigning a routing priority, other manual controls allow the user to influence the channel assignment of signal runs. A postroute timing model is generated according to the actual layout and parasitic loading parameters, so the timing analyzer can gather all necessary timing information, including delays due to connecting wires. Finally an output tooling tape for the process is generated in the GDS-II or CIF format.

9.2.5 Adding New Compilers

The Genesil compiler provides the systems designer with a rich function set that can produce millions of different block instances according to the parameter values that are specified by the user. Although SCS will enhance the function set from time to time, it is unlikely that any single company can develop all the possible block compilers.

The compiler development system permits those users who are prepared to tackle silicon designs to extend the system by adding new block compilers (see Section 9.7).

9.2.6 Foundry Interface

To service the systems designers who are interested in developing ASIC designs and having them produced commercially, the system interfaces to a number of different silicon foundries that can fabricate state of the art CMOS and NMOS technologies. In addition, SCS has established business relationships with many commercial foundries so that the users of the Genesil system have multiple sources for producing their ASIC designs. These relationships include getting information about their masking, processing, and design rules to ensure the accuracy of the compiled chips. The system also interfaces with the proprietary processes of a number of companies with internal IC fabrication facilities.

The user selects the foundry process on the Process Menu, which includes dozens of foundry and process combinations. The range of selections allows the user to compare die size, speeds, and costs of fabricating the same chip in different processes and foundries, thereby ensuring that the most advanced and competitive processes are available. Furthermore, the user can tradeoff functional features for cost and/or performance based on accurate evaluations and compiler output.

The fabrication and process specific information is entered into the system data base using the Genecal software tool. This information includes layout and process rules.

9.3 Function Set

Just as software compilers synthesize machine language code from programs written in high-level languages, the Genesil silicon compiler synthesizes relevant views of an IC, such as layout, functional description, performance, and power requirements, from high-level architectural descriptions. These descriptions are entered in terms of structures known to the compiler, such as PLA, ROM, RAM, ALU, register, or complete data paths. Structures can be described functionally, in quantity, or in terms of mathematical, logical, or state equations.

The system produces chip layout that is represented as placement and connection of predefined layout primitives (called tiles or cells), without necessarily producing geometric layouts directly. This process is analogous to software compilers generating code for the instruction set of a target computer system.

This section provides more detailed information on the Genesil function sets that are implemented with a number of block compilers, such as the RAM compiler, the ROM compiler, and the PLA compiler, each of which implements one high-level function in efficient circuitry. The random logic compiler contains a large number of general purpose gates and flip-flops which allow the user to implement other functions at the logic-gate level.

Note that an instance of a function is called a block and the capability to create different instances of a block is called a function.

The two following sections describe some examples of control logic blocks and data-related blocks.

9.3.1 Control Logic

Control logic is typically described in terms of flowcharts, state diagrams, and/or equations. The function set includes random logic gates, programmed logic arrays (PLAs), and read only memories (ROMs) for implementing control logic. The user can obtain estimates of the area, power, and speed of the control logic blocks without actually specifying all the details (such as the PLA equations) by specifying the approximate number of ROM words, or PLA min-terms, or estimates of the number of gates. After the final design is complete, then the exact numbers can be calculated by the compilers.

Random Logic

The random logic compiler is used to generate the "glue" logic that is so often needed in completing a design. Like a standard cell system, random logic includes small to medium scale complexity blocks (SSI and MSI) that are parameterized for function and bit width. The compiler puts these blocks in linear strips and the router interconnects them. Most of these blocks are dual-ported, with wiring connectors on both sides to reduce the amount of wire to connect them. Pass-through elements provide for passing signals across random logic "strips." The component family includes most of the popular types of gates, flip-flops, multiplexers, decoders, and drivers. The use of optimizing compilers can automate the process of trading off among speed, power, size, aspect ratio, and so forth.

ROM

ROM blocks are useful for storing constant data and programs. Programmed controllers can be implemented using both horizontal and vertical microprogramming. The ROM compiler includes a flexible macroassembler in which the user can describe a tailored source language for specifying the contents of a ROM array. The ROM itself can be up to 64 bits wide and thousands of words deep. Other physical options can be specified, such as address decoding range, output buffer type, and clock phases. In some cases, these options and sizes are dependent on the technology.

PLA

PLA blocks offer a regular structure for implementing control logic. The system provides a number of physical forms of PLAs and six different symbolic languages for specifying PLA equations. One of the useful linguistic forms is the finite state machine (FSM). The high-level description is input to the compiler and serves as excellent documentation for other designers to read. Furthermore, the compiler's verification tools permit simulating the FSM description before committing to silicon.

Finite State Machine Example

Digital system designers often start with a state diagram, as shown in Figure 9.8b, that describes the various (finite) states for a simplified change-making

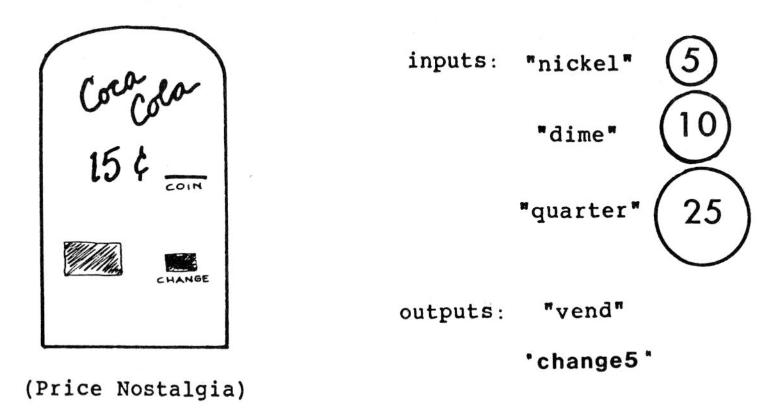

Figure 9.8a State machine example.

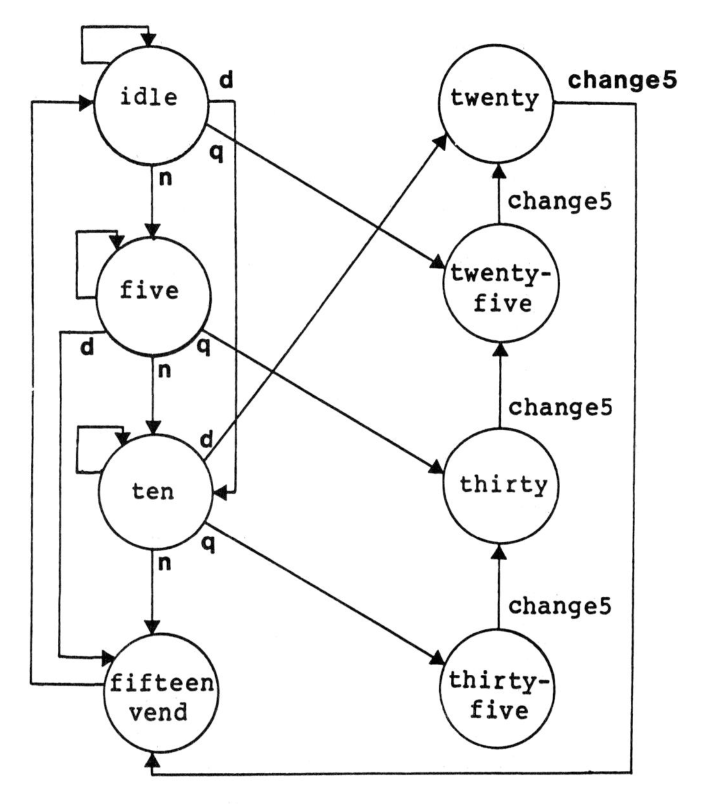

Figure 9.8b Example state diagram for vending machine.

soft drink vending machine (Figure 9.8a). Each state is usually assigned a symbolic state name and transitions between states occur depending on input signals. Arrows from one state to another are shown on the diagram with accompanying input conditions. Outputs can occur while in a state or upon a state transition. In this nostalgic example, 15 cents is required for "vending" a drink. The state diagram illustrates that the "vend" output occurs only in state "fifteen." As coins are deposited (nickels, dimes, and quarters), the machine state changes. In other words, the amount of money entered determines the state of the machine. When two dimes are successively input, the machine goes from the "idle" state to state "ten" and then to state "twenty." In the latter state, the "change5" output is activated, giving the user a nickel back. To simplify the overall design of this system, only nickels are returned. Notice that from state "twenty-five," two state transitions occur, each returning a nickel change ("change5").

High-Level Description Language

The high-level description language is input to the PLA compiler, using a conventional text editor file. An example source file is shown in Figure 9.9; the syntax can be found in the user manual [Sili86]. The description is broken into several sections including the declarations of the inputs, outputs, state variables, state assignments, and the state transition equations (see Figure 9.9). To represent the eight states, the designer used three state variables: $t2$, $t1$, and $t0$. Each state has been manually assigned a unique state value; for example, state fifteen is represented by state value $t2 = 0$, $t1 = 1$, $t0 = 1$. The three specified input signals are each activated for a particular coin: nickel, dime, and quarter. The two output signals signal to vend a soft drink and to return a nickel—change5.

The equation section (see Figure 9.9) indicates the actions for each of the various states. From the idle state, on detecting a dime input, state ten is entered. A second dime input causes a change to state twenty, at which a state transition "always" occurs to state fifteen with five cents change. In state fifteen, the vend output is driven and a transition to the idle state occurs.

Much richer conditions and equations are possible using Boolean expressions and/or *if-then* type statements. These logical specifications can be optimized to reduce the amount of chip area used, using the built-in logic minimizer. Other enhancements would allow mapping between random logic and PLAs and other forms of logic such as domino logic.

9.3.2 Data-Related Blocks

Data paths are typically described in terms of block diagram and buses. The function set includes RAM, data path, multiplier, FIFO, and so forth.

RAM

RAM blocks provide read/write data storage and register arrays. The RAM block compiler supports single, double, and triple port static memory arrays. These arrays can be sliced to permit partial word writing such as loading an 8-bit byte into a 32-bit word. The address decoders are programmable to respond to arbitrary but fixed addresses that are specified as compile time parameters.

Other options allow selection of the output driver type and the clock timing. A block diagram for the RAM is output as documentation. Figure 9.2 shows a sample RAM block specification form. The RAM compilers

```
PLA_SOURCE
        INPUTS nickel, dime, quarter ;
        OUTPUTS vend, change5 ;
        FEEDBACK t0, t1, t2 ;
        STATE
                NAME = coke;
                SIGNALS = t2, t1, t0;
                VALUE = idle, 000;
                VALUE = five, 001;
                VALUE = ten, 010;
                VALUE = fifteen, 011;
                VALUE = twenty, 100;
                VALUE = twenty_five, 101;
                VALUE = thirty, 110;
                VALUE = thirty_five, 111;
        ENDSTATE
        EQUATIONS
        FSM coke:
                STATE idle
                        ON nickel GOTO five
                        ON dime GOTO ten
                        ON quarter GOTO twenty_five
                STATE five
                        ON nickel GOTO ten
                        ON dime GOTO fifteen
                        ON quarter GOTO thirty
                STATE ten
                        ON nickel GOTO fifteen
                        ON dime GOTO twenty
                        ON quarter GOTO thirty_five
                STATE fifteen
                        DRIVE vend
                        ALWAYS GOTO idle
                STATE twenty
                        ALWAYS GOTO fifteen DRIVING change5
                STATE twenty_five
                        ALWAYS GOTO twenty DRIVING change5
                STATE thirty
                        ALWAYS GOTO twenty_five DRIVING change5
                STATE thirty_five
                        ALWAYS GOTO thirty DRIVING change5
        ENDFSM
        END
```

Figure 9.9 Example PLA source file.

have been optimized for register arrays and are not as dense as specialized RAM designs that are optimized for density in a fixed process; of course, these could also be added to the system.

Data Paths

The data flow paths of a design are often described by block diagrams to indicate the types and connectivity of logic blocks. Typical functions are registers, latches, multiplexers, and drivers. Previously, register transfer design languages have been used for specifying data paths; the data path compiler can work from such text language specifications or from specification forms.

The data path compiler implements the registers and bus-wide paths using abutting parallel data path structures and memory arrays. The user enters the specifications into the data path compiler through a dynamic menu and form-driven interface to customize for the number of bits in the data path, the number of registers, and specific data path functions. For example, many designers use bit-slice parts such as the AMD2901. A typical data path of the complexity of the 2901 can be entered in about five minutes. While the 2901 comes only in groups of 4 bits, the system can generate custom VLSI chips that contain paths of any bit width (up to 64 bits); 18 bits or 19 bits is completely acceptable. Frequently, the precision of the calculations depends on the algorithm or the application, and since the designer is working at a high level, it is easy to respecify the number of bits needed as the design progresses and the requirements change.

The data path compiler supports a number of functional blocks such as: ALU, adder/subtractor, incrementer, barrel shifter, logic gates, and priority encoder. It achieves high transistor and wiring density because the various blocks in a data path abut one another; that is, the local outputs of a block connect directly to the local inputs of the adjacent block. One can imagine these as pieces of an interlocking puzzle, or as toy Lego blocks, snapping together.

The internal tiles of the data path are designed around a global two bus system that runs horizontally. Control signals run vertically with their inputs on the bottom and output flags coming out of the top. Each block in the data path has two global buses running across, and in addition each has two local bus inputs and outputs for connecting with a neighbor as shown in Figure 9.3. Actual data flow can be in either direction (left/right) within the data path.

The data path compiler includes an optimizer that reduces the size of the data path instance whenever possible, by removing redundant logic such as duplicated inverters, or removing strong buffers when it is only required to drive a small load, such as a short connection to its immediate neighbor. It uses a "peephole" optimizing algorithm that only looks at an object and its immediate neighbors and executes a set of pattern matching rules. The pattern matching rules can be easily changed or extended by the compiler writer; hence, the optimizer will become increasingly sophisticated in future versions of the data path compiler software.

Figure 9.4 shows an example layout produced by the data path compiler. The global buses labeled BUS_A and BUS_B in this example are four bits wide; they allow for passing data beyond just the adjacent blocks and are constructed of metal lines on the chip. The two local buses are labeled STD_1 and STD_2, and provide for communicating data with the adjacent blocks in the data path.

This arithmetic data path with latches is a typical user application of the parallel data path. It consists of three abutting blocks: two latches in the first block; an ALU in the middle block; and a driver in the third block; latches are on the left, followed by the ALU and finally the driver circuit.

The global BUS_A is used to connect data from the last block to the first block bypassing the middle block.

Four parallel bit slices, each containing four metal lines, run horizontally through the data path. The control logic lies beneath the bit slices and generates signals to operate various portions of the bit-sliced logic.

Multiplier

This compiler produces an M by N first quadrant parallel multiplier circuit. In addition to the parameters for the multiplier and multiplicand size (*M* and *N*), the user can specify pipeline latches and recirculate logic.

The multiplier block has been optimized for small size rather than for maximum speed. In typical designs, a 16 by 16 multiplier will occupy less than 10 percent of the chip area. Hence, system designers can build useful functions containing one or more multipliers in a single chip. Of course the speed of the multiplier will also depend on the technology used.

9.4 Example Design: Audio Generator

This section describes a design example to illustrate the use of the various block compilers and other analysis tools as applied to the design of a complete chip. It is a simple application of a custom chip in a computer-controlled audio signal generator. Some applications for this low-cost audio generator include toys, telephones, music instruments, and test instruments.

Some of the design goals include: (1) use low cost standard parts whenever available, and these include the single-chip 8-bit microprocessor, the monolithic D/A converter, and the R/C reconstruction filter; (2) the capability for programmable wave shapes that include triangular, sinusoidal, square, and other arbitrary wave shapes; (3) programmable frequency and duty cycle, with frequency range of DC to 50 kHz; (5) output voltage range of −5 to +5 volts in 0.1 volt steps.

In Figure 9.10, the function generator contains a commercially available single-chip microprocessor (similar to that in a typical personal computer)

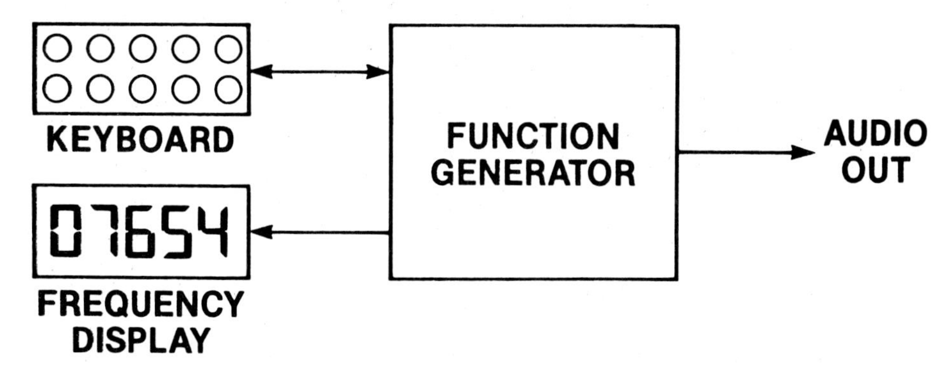

Figure 9.10 Function generator I/O diagram.

that runs the software to control the sound or music generation. It also includes a keyboard for the user to enter a selected frequency, type of waveform, and duty cycle; a visual display for the selected frequency; and an audio output. Although the design could be simplified by using the microprocessor to perform signal generation, adding a custom chip to the system greatly enhances the speed performance and frees the microprocessor for other software tasks.

Figure 9.11 shows the first pass system diagram, which contains a microprocessor chip and the custom function generator (FG) chip driving the D/A converter and the filter. The microprocessor chip accepts keyed commands and loads the RAM table for waveform patterns on the FG chip.

9.4.1 Programmable Waveform

Figure 9.12 shows how the programmable waveform capability is implemented. The FG chip provides the quantized voltage steps to synthesize the waveforms. For example, to generate a sine wave, the microprocessor would first store all the sampled and quantized amplitudes of one complete cycle of the sinusoidal waveform in the RAM on the FG chip, and these data would come from the ROM tables in the microprocessor system; it would then read out the amplitude values sequentially over time. To change to a square wave, triangle wave, or other arbitrary wave shapes, the microprocessor would reload the waveform RAM table with the appropriate cycle pattern from its ROM storage.

9.4.2 Programmable Frequency

As shown in Figure 9.13, the output frequency is programmed by varying the rate at which the waveform RAM table in the FG chip is read out (or

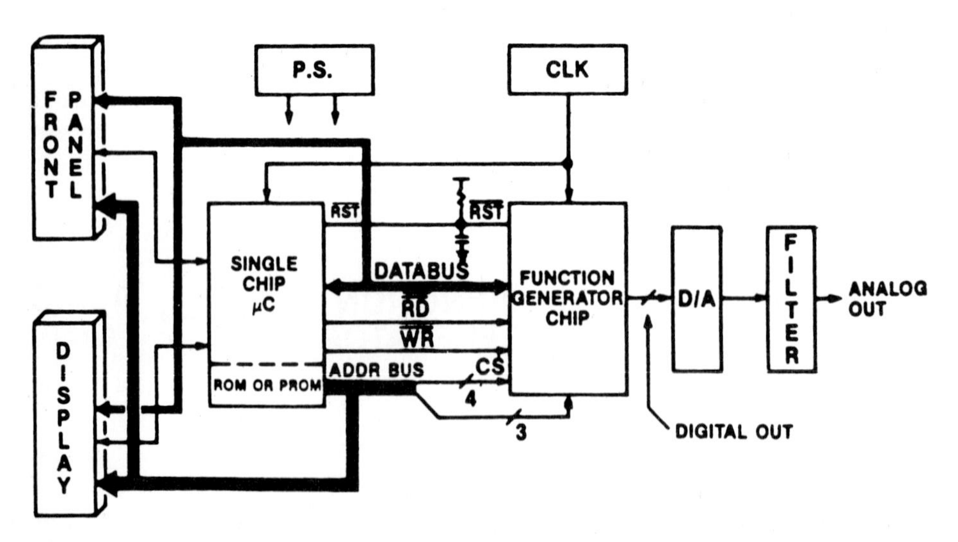

Figure 9.11 Block diagram of function generator system.

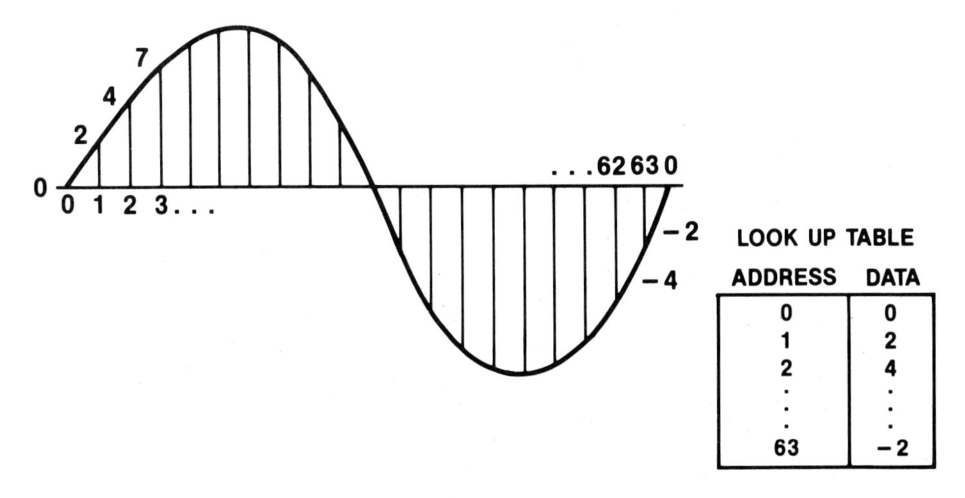

Figure 9.12 One wave cycle stored in table.

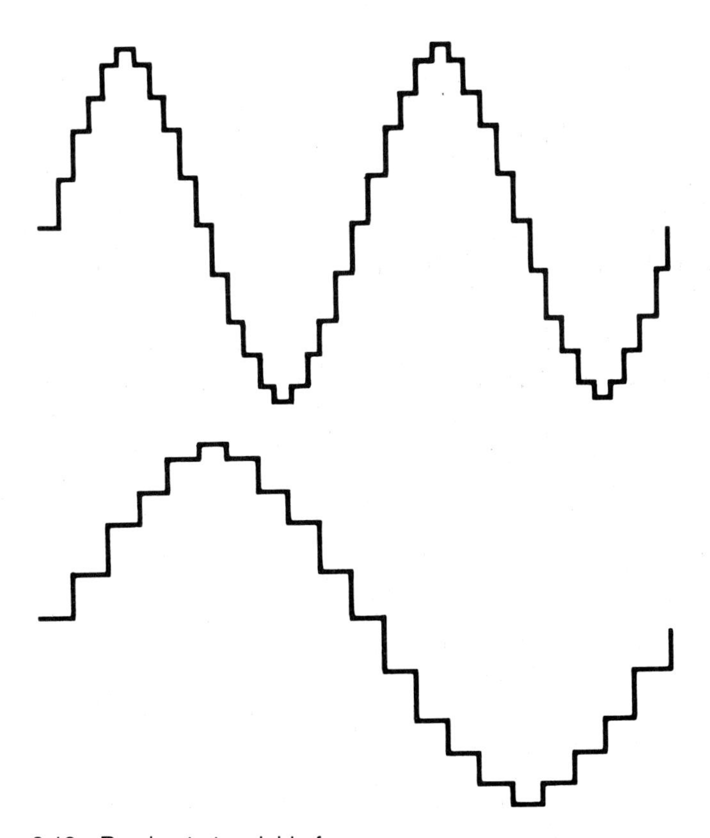

Figure 9.13 Read-out at variable frequency.

scanned). The microprocessor specifies the FG RAM scan rate by writing into the frequency register in the FG chip. To achieve the maximum frequency of 50 kHz, a new sample needs to be produced approximately every 310 nsec. This data rate far exceeds the capability of the typical commercial 8-bit microprocessor chip, and for this reason the RAM on the microprocessor system cannot be used for this purpose, and a separate waveform RAM has to be implemented on the FG chip. For slower frequencies, the scan rate of the waveform RAM is reduced.

Figure 9.14 shows the digital integrator circuit used for the RAM scan-rate generation, which permits 256 discrete frequencies to be programmable by the user via the microprocessor. The content of the frequency register is added to the phase accumulator once per clock cycle. Every time a carry (or overflow) is produced, the address counter (incrementer) steps to the next RAM memory location to access the next amplitude value. For example, if the frequency register is set to 1, then the RAM addresses are incremented every 256 clock cycles; if, on the other hand, the constant is set to 255 (0ffh), then the address counter is incremented every clock cycle.

Figure 9.15 shows how the integrator and incrementer functions are implemented together as a 14-bit arithmetic data path, with the upper six bits used as the RAM address and the lower eight bits used as the frequency accumulator.

9.4.3 Function Generator Chip Design

Figure 9.16 is the block diagram of the FG chip. It includes: the phase accumulator, microprocessor interface circuitry, the waveform RAM, and another adder that can be used to scale the output voltage.

Figure 9.2 shows the fields and functions in the RAM specification form. In this example, Figure 9.16, the RAM contains 64 bytes.

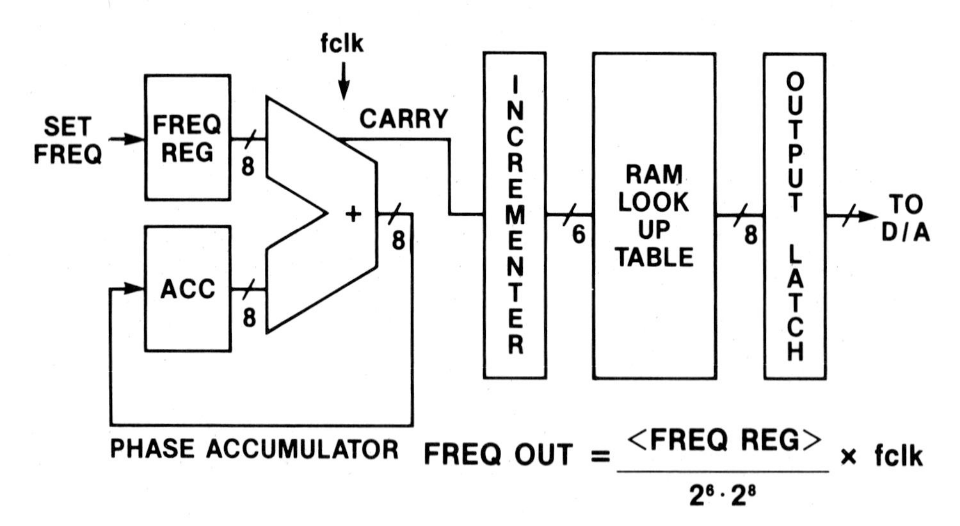

Figure 9.14 Digital integrator implementation.

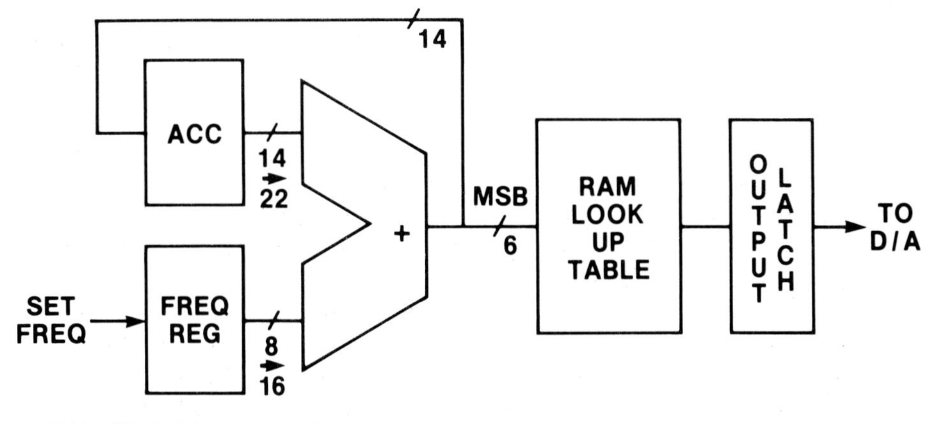

Figure 9.15 Phase accumulator reimplementation.

Figure 9.7a (see page 369) shows a chip floor plan, in which all the functional blocks are shown in their relative placement, proportional sizes, and aspect ratios. If desired, the user can rearrange the relative placement of the blocks; for example, if the time delay in a signal net is required to be short, then the blocks on either end of the net can be placed adjacent to each other to minimize wiring capacitance.

At this point in the design example, suppose the user decides that the aspect ratio of the RAM is too tall and cannot yield a balanced layout. So, the user edits the form to change the RAM size to be 32 by 16 bits and recompiles the block. Figure 9.7b shows the new floor plan, with the revised RAM shown wider and shorter.

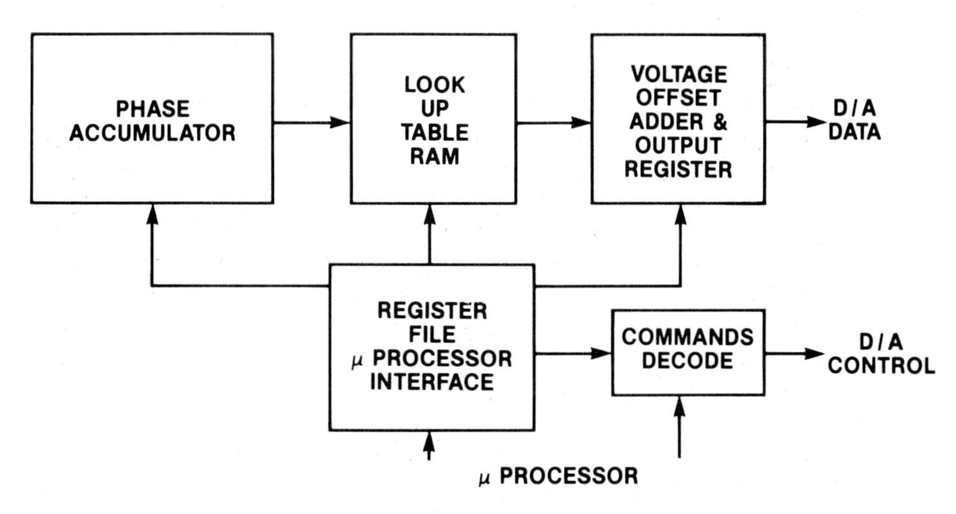

Figure 9.16 Custom function generator circuit.

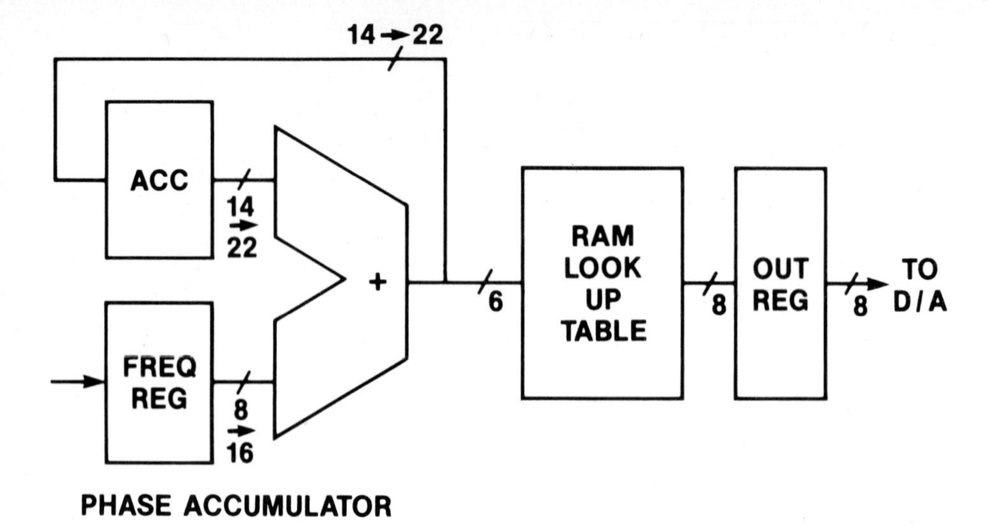

Figure 9.17 Phase accumulator 22 bits.

9.4.4 Final Design

As in many typical design scenarios, changes are made to the specification just when the engineer is satisfied with the design choices. In this case, it is decided that the frequency resolution is inadequate and more bits of accuracy are required for the frequency register and phase accumulator logic.

Figure 9.17 shows the block diagram of the FG chip after the user doubles the bit resolution. The FG chip now uses a 16-bit frequency register and a 16-bit phase accumulator, which are implemented in a 22-bit data path (16 + 6 = 22).

Figure 9.18 shows the block diagram of the microprocessor interface circuitry that was changed to include a second 8-bit frequency register, since it is more convenient for the microprocessor to deliver each byte separately, as it uses an 8-bit data bus. The new arithmetic data path has a different layout size, and Figure 9.19 shows the new chip floor plan after the placement and routing are completed.

The compilation of the chip (Figure 9.19) requires the routing of all the power, ground, clock, and signal wires, which for typical chips of this

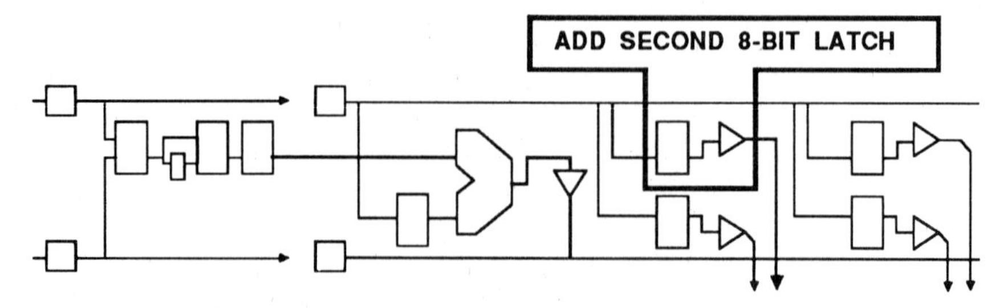

Figure 9.18 Microprocessor interface with added second block.

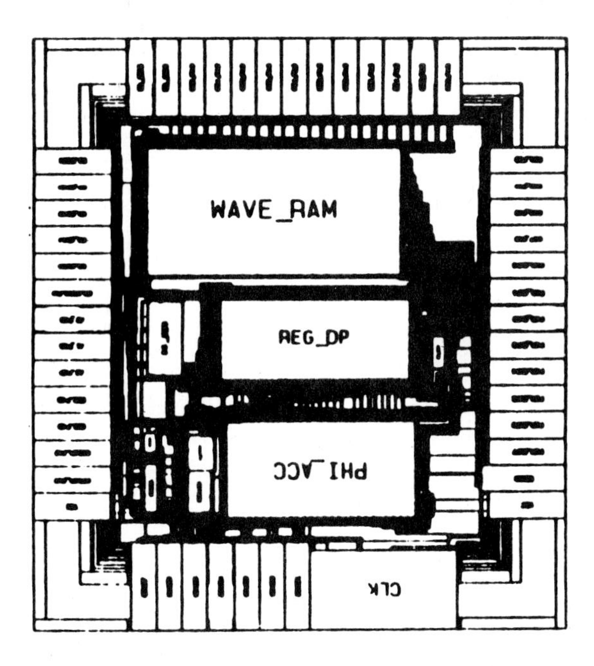

Figure 9.19 FG chip after floor plan and route.

complexity takes about 20 minutes on a workstation. Complete functional simulation and timing analysis can be done now that all of the interconnection loadings are known. The test vectors can be saved for screening the completed wafers or packaged units.

This simple example has illustrated the steps that a digital systems designer takes to design an ASIC chip with a silicon compiler.

9.5 Design Methodology

Certain safe design practices are embedded in the compiler system to (1) ensure that the resultant chips will operate correctly and (2) provide a framework or methodology for the new block compiler that is added to the system.

Designers of complex systems typically employ practices or methodologies that govern the implementation of the system. These methodologies provide a mechanism by which individual subsystems may be designed independently and then integrated into the final working system. By adhering to a unified design methodology (a common set of rules and design practices), a team of designers can develop and manage the implementation of a complex system that a single designer would not be able to accomplish in a reasonable period of time.

The challenge facing the VLSI designer is very similar to the complexity management problem faced by the digital system designer or the software development engineer. VLSI chips contain anywhere from 20,000 to

1,000,000 devices. In order for the completed design to function in the desired manner, the design team must develop and adhere to a unified design methodology that governs the implementation.

To help designers in implementing complex VLSI designs, the system uses a unified methodology for power distribution, physical layout, and signal interconnect. The embedded design practices provide functions that not only automate much of the physical layout and design of a VLSI circuit, but result in circuits that are correct by construction. These embedded functions incorporate the practices of expert IC designers.

Consider the problem of power distribution within a VLSI circuit. The standard practice adopted by the system is to route all VCC and ground interconnects in metal and to provide separate power lines for the core circuitry and the I/O pads. In addition, the system automatically calculates both the current requirement (DC and AC) of each functional block in the design and the IR voltage drop that accumulates in the power interconnects, and uses the information to develop the routing and sizing of the VCC and ground lines.

For interconnect, prudent engineering practices ensure that all clocks and signals are properly buffered and that output drive requirements are met. Standard practices for setting device ratios and for controlling clock skew are also employed.

The other major component of the methodology is comprised of rigorous checking and verification of the functionality and performance (timing) of the design. The system provides interactive checking, verification, and implementation strategies that flag potential errors and hazards or prevent them from being entered into the design.

9.5.1 Signal Types

In order to facilitate the analysis, verification, and checking of the many signals within a VLSI design, the system assigns a "type" to each signal. Typing allows the compiler to automatically classify and validate signals across all levels of a design.

There are three primary signal types: CLOCK, STROBE, and DATA. Two of these signal types, CLOCK and STROBE, convey timing information; the third signal type, DATA, does not. The CLOCK signal type defines the master timing for a circuit or subcircuit. CLOCK signals convey timing information only and provide the point of reference for all timing analysis within the system.

STROBE signals convey both timing information and data and are used to synchronize particular events. STROBES are typically generated by sampling an input signal during a specified CLOCK phase; if the input is true, a STROBE pulse is generated during the following phase. The system provides means for generating STROBES from external inputs, from internal signals, and from other STROBES. This provides a wide variety of synchronization capabilities.

DATA signals are assigned timing attributes based upon the times (referenced by a clock) during which they will be valid. For example, if a given DATA signal is valid during the latter half of Clock Phase A, then that signal is given the attribute of "Valid Phase A." Conversely, if the signal is valid during the latter half of Clock Phase B, the signal is given the attribute of "Valid Phase B." There are five standard DATA attributes and an additional four attributes that deal with precharged timing and multiple clocking schemes.

By applying a set of rules based on these attributes, the system is able to enforce a consistent timing methodology across all levels of a design. For example, it will prevent a signal that is Valid Phase A from being connected to an input that is classified as Valid Phase B. Color coded signals in schematic views let the user rapidly identify the various signals and their types.

In addition to signal typing, the unified methodology must also provide standard checks on signals. For example, the user is alerted to situations in which outputs are tied together, inputs are unconnected, or outputs are left unconnected. The system provides immediate feedback to the designer if such potential errors are discovered in the design.

9.5.2 System Timing and Synchronization Problems

The unified design methodology incorporates a detailed clocking strategy for the implementation of synchronous designs. By applying and enforcing the clocking strategy across an entire design, it prevents system timing errors and hazards from being introduced into the design. Before examining the clocking strategy it is useful to look at some of the potential problems that implementation of such a strategy can overcome.

Asynchronous feedback loops in digital circuits can be a source of unpredictable behavior. Consider a simple latch comprised of two cross-coupled NAND gates. The behavior of this latch is referred to as unclocked since any change in its input values will have an immediate effect on its output. This behavior results in two potential problems. Firsts, glitches present on either input may cause the latch to be set to an erroneous or unanticipated state. This means that the designer must ensure that the inputs to this circuit are glitch free.

There are other well understood problems with clocked state circuits. Consider a situation in which the gating signal to a clocked state circuit is terminating a sampling period while the input signal is transitioning through the illegal region. The output of the state circuit may assume an illegal value. This not only affects the circuit in question, but will likely affect other logic connected to the output. In fact, depending on the value the output assumes and the logic thresholds of the circuits connected to it, some circuits may interpret the output as high, while others will interpret it as low. Additionally, some circuits may themselves assume an illegal output value.

This potential problem makes synchronization of large systems difficult unless a consistent clocking methodology is followed. From this discussion, it is clear that inputs to logic state circuits must be constrained such that they are unchanging at the end of the sampling period.

9.5.3 Clocking Strategy

A well-defined clocking strategy overcomes the aforementioned timing and synchronization problems by establishing timing references for all signals within the system. Such a global clocking strategy is necessary to manage and synchronize communications between the elements of a large and complex VLSI system.

The methodology's embedded design practices take into account logic delays and clocking to implement a clocking strategy that avoids these timing problems. In addition, the methodology inhibits the designer from implementing constructs such as gated clocks or asynchronous feedback loops that introduce undesirable effects such as clock skew and unpredictable timing behavior into the system.

The unified methodology employs an internal two-phase clocking strategy that implements synchronous logic via transparent latches. By utilizing a synchronous design approach, the system can perform rigorous design checking and timing analysis and prevent timing hazards from arising. Although the methodology encourages synchronous design, it does provide support for communications between asynchronous systems.

The system implements synchronous logic as shown in Figure 9.20. PH_A and PH_B represent the two nonoverlapping clocks. In a typical application, the designer supplies the chip with an input clock. The internal two-phase clocks are usually generated from the externally supplied clock via a two-phase clock generator block supplied with the system.

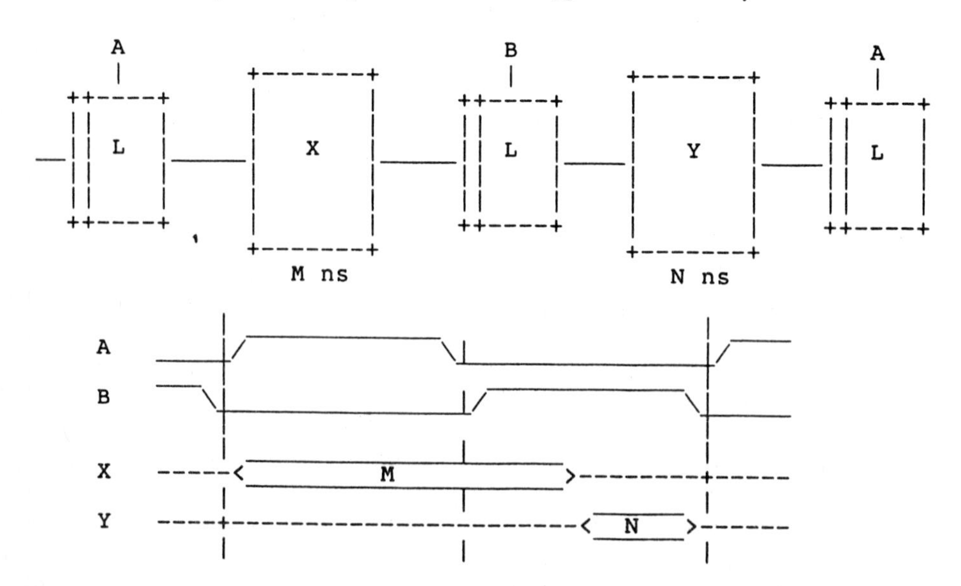

Figure 9.20 Cycle sharing for a two-phase methodology.

During PH_A, all PH_A latches are transparent and sample the outputs of PH_A logic blocks (logic blocks are classified by the clock phase during which their outputs are sampled). The PH_A logic blocks' inputs come from stable PH_B latches. At the end of PH_A, the PH_A latches hold their outputs stable. A similar action occurs during PH_B.

If the clock phases overlap, circuit malfunctions may result since the timing analyzer assumes that there is a guaranteed nonoverlap. The two-phase clock generator guarantees that the phases are nonoverlapping.

Since transparent latches are utilized rather than master-slave latches, it is possible for cycle sharing to be utilized. For example, if PH_A logic computes a final output value before the end of PH_A, the PH_B logic can begin computing its new value since the PH_A latch is transparent. In essence, this gives the PH_B logic more time to compute its final result. A similar argument can be applied to PH_A logic during PH_B.

9.6. Design Verification

The large number of components in a VLSI chip and their complex interrelationships make it almost impossible for the designer to foresee all the consequences of each design decision. In addition, the cost of making a mistake has also skyrocketed, even in the prototyping stages of the design. To minimize the likelihood of costly errors, the designers are increasingly dependent on computer tools to synthesize and verify the designs before committing to the costly silicon fabrication.

In the conventional IC design methodology, the design flow progresses from a set of design wishes to (1) an architectural specification, then to (2) a logic design, and then to (3) a circuit design that is in turn converted into (4) a layout topology design (see Table 9.1). To verify the equivalence of the lower-level designs with the architectural specification, the same set of test vectors would be used to run a gate-level simulation from the logic design or a switch-level simulation from the circuit design. To verify the

Table 9.1 Levels of Design and Simulation

Conventional design steps	*Conventional IC design verifications*	*Silicon compilation verifications*
Wish		
	RTL simulation	
Architecture		Functional simulation
	Logic simulation	
Logic		
	Switch simulation	
Circuit	Layout extraction	Timing analysis
	DRC, circuit	
Layout	simulation	

equivalence between the layout topology and the circuit schematics, a layout extraction program would be used to convert the layout into a transistor net list format that is either compared to the circuit design or used to run a switch-level simulation. These traditional steps require much cross checking and rechecking at various levels.

Most of the verification efforts in conventional IC design are concentrated on checking the equivalence between the various design levels; however, since silicon compilation replaces all the manual steps between architecture design and layout design, the associated equivalence checking becomes obsolete. Hence, only functional simulation at the architectural level (to verify the architectural design) and timing analysis at the layout level (to verify the timing performance) are needed [Chen84].

9.6.1 Separation of Functional Simulation and Timing Analysis

It is a common design practice that functional and timing designs are handled separately, even though there is some intermingling of the two activities in the iteration loop that consists of: do functional design, then do timing analysis, and iterate. When working with synchronous circuits, engineers like to treat functional and timing designs as separable processes, because it is much easier to deal with only one variable at a time.

Also, computing timing delays concurrently with evaluating functional behavior may not yield absolutely accurate timing results because of circuit interactions and can further burden the computational requirements for each simulation cycle.

However, several popular logic simulators do compute timing delays concurrently with functional evaluations by working with estimated gate propagation times; this is also not a very reliable method for uncovering all the critical delay paths.

For a large digital network, a number of test vectors on the order of 100,000 is not uncommon. If the circuit is modeled at the gate or switch level, the CPU run time and memory requirement for each cycle of simulation is very high. Table 9.2 quantifies this statement for one instruction cycle in a VLSI chip [Klec82].

Thus, there is a good match of capabilities because the majority of the design is modeled at the high level that can be most efficiently simulated.

Table 9.2 Comparison of Simulation Run Time and Memory Requirement for One Instruction Cycle in VLSI Chip

Level of simulation	*CPU time*	*Memory*
RTL	<<1 min	<<0.5 MB
Logic	10 min	4 MB
Timing	8 hours	30 MB
Circuit	6 months	250 MB

On the other hand, timing analysis must be done at the lowest physical level, including parasitic capacitance and resistances.

While the compute-time resource is a problem when many test vectors must be used to verify the functionality, the engineering resource in generating those vectors is an even bigger problem. When a design engineer is running a functional or logic simulator, he or she is focused on searching out the problems in the functionality of the design. Since the main concern is to design the test vectors for function, the test vectors may not exercise all of the critical timing delay paths in the circuit. Hence, some critical short paths (for race conditions) and long paths (for delay times) may go unnoticed.

While the test vectors are driving the functional simulation to exercise all the functionalities of the circuitry, timing delay paths could be computed repetitiously or redundantly if they were tagged along in the calculations. On the other hand, a dedicated timing analysis program that evaluates each timing delay path only once would be much more efficient in terms of compute time.

The timing analyzer uses an algorithm that is compatible with the clocking strategy and exhaustively analyzes all possible timing paths in the circuit. With pruning of irrelevant paths by the design engineer, this approach has been found to be both feasible and desirable. On the other hand, it is not feasible to automatically generate test vectors for functional simulation because it is not possible for a machine to know the intent of the design.

Another difference is that functional simulation can usually be computed independently of the process technology, but the timing analysis calculations are very dependent on the physical parameters of the process technology that will build the circuitry.

9.6.2 Functional Simulator

Compared to conventional logic simulators, the functional simulator is more efficient in terms of run time and computer resources because, instead of modeling strictly with simple Boolean gates, its circuit models are generated by the block compilers, and they use high-level logic primitives such as gates, flip-flops, adders, and decoders, and they are coded without any embedded circuit delay information. It is also easier to deal with since the number of structures being simulated in a typical compiled chip will be fewer than one hundred, rather than tens of thousands.

The user can simulate the compiled circuit at any level of the design hierarchy, since the system supports hierarchical signal naming. Signal names that are completely internal to a module are invisible to other modules and will not cause name conflicts. Signal names that are used as connectors at the module level are visible to other blocks and modules at the same level.

The simulator is controlled using a menu and form-driven interface. Input test vectors may be entered interactively at the terminal or read from a file in the batch mode. It has a very flexible screen format for displaying the simulation outputs, much like a modern logic analyzer, so the user can choose where and in what format the data are to be shown.

The input and output test vectors can also be collected for screening and testing the chips after they have been fabricated and packaged.

The user does not have to use the simulator provided by the compiler system. The system has interfaces to several commercial logic simulators, hardware simulation accelerators, and fault graders by providing gate-level models of the compiled circuits.

9.6.3 Timing Analysis

Timing performance in conventional VLSI design has frequently been determined by extracting all the parametric details of the layout of a circuit, and then feeding these parameters into a circuit simulation model of the entire circuit, or the portion under study. Table 9.2 also indicates an exorbitant amount of compute time needed to simulate VLSI chips at the circuit level. This makes it impractical to use circuit-level simulation to obtain timing data, except for small, isolated pieces of circuitry.

Usually, the design engineer manually identifies portions of the circuit whose delay times are deemed to be the worst cases. Due to the complexity of VLSI designs, such manual efforts are very prone to missing some critical paths, either because of oversight in path selection or mistakes in the circuit design. Furthermore, circuit simulators are data dependent; the timing information that they produce depends on the input stimuli data.

To eliminate these problems, analysis must be performed on the entire chip in a value-independent fashion. Several value-independent timing tools have been reported in the literature ([Hite82], [Joup83], [McWi80], [Murp85]).

However, in order for automatic path enumeration to be practical, the timing analysis must run much faster than circuit simulation. The analysis program runs up to 10,000 times faster than a circuit simulation program such as SPICE, while yielding an accuracy that is very close to what a circuit simulator would estimate for most delay paths.

Because of its speed, it is very practical to use a timing analyzer for checking all possible timing paths in a VLSI chip. By operating on computer-extracted data from the actual physical layout, its network model is very accurate. Unlike a circuit simulator, which is driven by external stimuli that are provided by the design engineer, a timing analyzer automatically enumerates all possible timing paths, analyzes their timing delays, and reports on the worst of them; hence, it is not vulnerable to the design engineer's fallibility in picking out all the critical paths.

9.6.4 Timing Analyzer

The timing analyzer (TA) has been used on a variety of circuits generated by the Genesil silicon compiler. Run-time statistics for various designs are presented in Table 9.3 [Lin86]. It can be seen in the bottom two lines of

Table 9.3 Experiences with the TA

	CPU	*Graphics*	*Buffer*
Transistors	50,000	27,000	33,000
Primary clocks	3	1	2
Processing time (CPU minutes VAX 11/785)	6.3	2.1	2.5
Memory required (kbytes)	9,700	5,300	6,700
TA cycle time (ns)	151	94	57
SPICE cycle time (ns)	157	83	53

the table that the three different applications reported by the TA come quite close to an analysis done by SPICE. The largest error shown in the table is about 10 percent. The processing time for the SPICE run is not shown but is measured in days. Note that the times indicated do not include the time to compile and extract the network information for each block. However, the extracted block network information is updated only when the block definition is modified, resulting in the need to compile only the modified blocks between design iterations.

The TA provides the timing information necessary to allow intelligent design tradeoffs. Using an efficient algorithm to traverse a circuit network composed of extracted component values, the TA quickly produces an accurate timing report in a familiar data sheet format, and the menu driven interface simplifies interaction with the timing analyzer.

The combination of the silicon compiler's ability to generate layout information based on high-level specifications and the TA's ability to rapidly generate accurate timing values for the generated design with a minimum of user direction allows the designer to explore design alternatives efficiently.

Timing Analyzer versus Timing Verifier

TA is a timing analyzer while most other tools are timing verifiers. Timing verifiers check for circuit timing violations, given the timing characteristics and constraints of the components. A timing analyzer derives the timing characteristics and constraints, and checks possible violations of a circuit, given its layout information. The automatic derivation of timing characteristics is very important in the silicon compilation environment because the circuit configuration varies according to the user specification; there is potentially an infinite number of configurations, which makes precalibration extremely difficult. The TA copes with this problem by using an RC-based switch-level timing model [Lin84], [Rubi83], which is capable of handling all possible circuit configurations, and arbitrary resistance and capacitance distributions. The calibration is performed on a transistor-type basis, in which the composition of transistor characteristics into circuit characteristics is determined algorithmically. With careful classification of transistor types, this model can be made very accurate, typically within 5 percent to 10 percent of circuit simulation results (see Table 9.3). Twenty different types are adequate to cover all the transistors in the NMOS and CMOS libraries. Calibration based on transistor types, rather than components, makes inclusion of new foundry processes into the silicon compiler's data base straightforward.

Timing Report

The reports generated by other timing tools are generally not aimed at systems designers, often causing important timing information to be overlooked. On the other hand, the report generated by the TA is similar to a TTL data sheet and easy to comprehend. Given a phase pair, the TA reports the minimum phase times and the cycle time of the pair, the setup and hold time of the input connectors, and the longest and shortest delays of the output connectors, and so forth. For combinational circuits, the propagation delay between input and output pairs can be determined using query commands. This set of timing numbers allows a complete external timing characterization of the design. Additionally, these timing numbers can be compared against user-specified requirements to generate a list of violations.

The user may specify environmental variables such as temperature, supply voltage, and capacitive loading of outputs. If any of the timing numbers needs to be further investigated, the details of the limiting path associated with the timing number can be generated, and the circuit simulation input deck can also be produced in order to check the accuracy of the timing analyzer program itself.

Cycle and Phase Timing Calculations

One of the primary concerns of chip and system designers is the maximum frequency at which their design can operate, as frequency often determines the performance of the system being designed. The TA rapidly delivers performance information to the design engineer, allowing a short design iteration cycle. Use of a two-phase clocking methodology results in four primary clock parameters of interest: the Phase_A time, the Phase_B time, the symmetric cycle time, and the asymmetric cycle time, as shown in Figure 9.5.

In many timing analysis tools, the duration of each phase is usually calculated by computing the maximum delay from an element clocked by the phase of interest through combinational logic to the input of an element clocked by the opposite phase plus the clock skew. The asymmetric cycle time is simply the sum of each of the phase times. The symmetric cycle time would be stated as twice the maximum phase time. In the TA, however, phase times are determined by paths that, beginning from a latch that is transparent during a given phase, pass through intervening combinational logic to a circuit element that requires a valid input during the same phase or a stable input for the following phase. These phase times set bounds on the duty cycle of the primary input clock. Note that delay between an element clocked by one phase and an element clocked in the opposite phase does not determine the phase time.

A phenomenon known as cycle sharing is taken into account during cycle time computations. *Cycle sharing* refers to the ability to share time between temporally sequential phase segments. This ability is very applicable in MOS designs, because transparent latches are a natural fit for MOS technologies. Transparent latches allow logic signals to propagate through the latch when it is enabled, thus permitting signal propagation through a long phase segment to continue into the next phase if the next phase requires a shorter propagation time.

Cycle sharing is illustrated in Figure 9.20, in which blocks labeled L are transparent latches clocked by the indicated phase clocks and blocks X and Y are combinational circuits. If block X has a long propagation delay and block Y has a short delay, Phase_A may end and Phase_B may begin before the output of logic block X has settled to its final value. During Phase_B the output of block X reaches its final value, propagates through the Phase_B latch and logic block Y and sets up to the second Phase_A latch. The use of cycle sharing during the design process allows the designer to take advantage of the clocking methodology by allowing long paths started in one phase to be completed in the next phase. Allowing cycle sharing across an arbitrary number of phase segments greatly increases the analysis complexity since combinations of all phase segments across the entire circuit need to be considered. Based on the observation that long segments are usually followed by short segments in practical circuits, the approach taken by the TA is to fully handle cycle sharing between any two adjacent phase segments using the technique of slack propagation. Cycle sharing between more than two segments is only partially supported. Use of cycle sharing techniques can easily reduce reported cycle time requirements, resulting in a reduction of up to 30 percent over cycle times reported without considering cycle sharing.

Both symmetric and asymmetric cycle time values are reported, allowing the designer to choose a clocking structure to most closely match the system environment. In addition to calculating the phase times, TA also determines the longest two-phase path. Denoting the phase time of one phase as A, the phase time of the other phase as B, and the longest path through two temporally sequential phase segments as C, the symmetric and asymmetric cycle times are given by:

$$\text{Symmetric cycle time} = \max(2A, 2B, C)$$

$$\text{Asymmetric cycle time} = \max(A + B, C)$$

Asymmetric clocking permits each phase to be at a minimum value independent of the duration of the opposite phase, allowing the chip to operate at its maximum frequency but setting strict requirements on the input clock duty cycle. Symmetric clocking typically uses a clock divider to generate equal duration phases, freeing the design from the duty cycle of the input clock at the expense of a slight performance degradation.

Human Interface

The TA is controlled using a menu driven interface to select the timing parameters to be generated or checked. Results are presented in a tabular format in Figures 9.21 and 9.22. Timing requirements, input and output system parameters, temperature, and supply voltage level may be specified using menu selections or from setup files. The ability to include these system variables allows the timing analysis to be conducted on realistic worst case operating conditions. Any grouping of setup files may be included when calculating timing parameters, allowing some setup files to contain essentially constant information (e.g., pin loading), while others contain more frequently varying parameters.

Capable of supporting an arbitrary number of clocks, the TA generates the phase times and the symmetric and asymmetric cycle times for each clocking regime. In addition to generation of the longest paths that determine the phase and cycle times, an ordered list of the paths that determine these values is maintained. This list of paths is easily traversed by the user to investigate successively longer or shorter paths or may be indexed

CLOCK TIMES (minimum)

Phase 1 High: 79.2 ns | **Phase 2 High: 66.2 ns**
Minimum Cycle Time: 145.4 ns | **Symmetric Cycle Time: 158.4 ns**

- -

CLOCK WORST CASE PATHS

Phase 1 High minimum time is 79.2 ns set by:

Node	Cumulative Delay	Transition
PHI_ACC_RN/MSLT_11B/(internal)	79.2	rise
PHI_ACC_RN/MSLT_11B/O10	69.5	fall
PHI_ACC_RN/ADD_11B/O10	69.3	fall
PHI_ACC_RN/ADD_11B/CIN	50.9	rise
PHI_ACC_RN/ADD_11B/COUT	49.9	rise
PHI_ACC_RN/ADD_11B/I0	19.9	fall
REG_DP/INTER11_VAL2[0]	18.6	fall
REG_DP/PHASE_A	10.2	rise
CLK/PHASE_A	8.3	rise
CLK/CLK_IN	0.0	fall

Phase 2 High minimum time is 66.2 ns set by:

Node	Cumulative Delay	Transition
PHI_ACC_RN/ADD_11B/COUT	66.2	rise
PHI_ACC_RN/ADD_11B/CIN	49.0	rise
PHI_ACC_RN/ADD_11B/COUT	48.0	rise
PHI_ACC_RN/ADD_11B/J0	18.1	fall
PHI_ACC_RN/MSLT_11B/P0	17.5	fall
PHI_ACC_RN/MSLT_11B/PHASE_B	9.2	rise
CLK/PHASE_B	7.6	rise
CLK/PHASE_A	6.1	fall
CLK/CLK_IN	0.0	fall

- -

Figure 9.21 Clock report mode.

Output	OUTPUT DELAYS (ns)			
	Ph1(r) Delay		Ph2(r) Delay	
	Min	Max	Min	Max
CC0_PAD/CC0	26.5	34.6	26.5	34.6
CC1_PAD/CC1	26.8	34.8	26.8	34.8
CC2_PAD/CC2	26.8	34.9	26.8	34.9
CC3_PAD/CC3	26.9	35.0	26.9	35.0
CC4_PAD/CC4	26.9	35.1	26.9	35.1
CC5_PAD/CC5	27.0	35.1	27.0	35.1
CC6_PAD/CC6	27.1	35.4	27.1	35.4
DBUS0/DB0	16.2	32.0	29.9	29.9
DBUS1/DB1	16.2	32.0	29.9	29.9
DBUS2/DB2	16.1	31.9	29.8	29.8
DBUS3/DB3	16.1	31.9	29.8	29.8
DBUS4/DB4	16.0	31.8	29.7	29.7
DBUS5/DB5	16.1	31.9	29.8	29.8
DBUS6/DB6	16.1	31.9	29.8	29.8
DBUS7/DB7	16.1	31.9	29.8	29.8
DS0_PAD/DS0	27.3	35.9	27.3	35.9
DS1_PAD/DS1	27.6	36.0	27.6	36.0
DS2_PAD/DS2	27.5	36.2	27.5	36.2
DS3_PAD/DS3	27.5	36.3	27.5	36.3
DS4_PAD/DS4	27.5	36.0	27.5	36.0
DS5_PAD/DS5	27.3	36.0	27.3	36.0
DS6_PAD/DS6	28.7	37.1	28.7	37.1
DS7_PAD/DS7	28.9	37.6	28.9	37.6
DS8_PAD/DS8	28.9	37.6	28.9	37.6
DS9_PAD/DS9	28.9	37.6	28.9	37.6
PAR_OUT0/DAT_OUT0	18.1	32.0	29.0	43.6
PAR_OUT1/DAT_OUT1	18.1	32.0	29.0	43.7
PAR_OUT2/DAT_OUT2	18.0	31.9	29.0	43.7
PAR_OUT3/DAT_OUT3	18.0	31.9	29.0	43.7
PAR_OUT4/DAT_OUT4	17.9	31.8	29.0	43.7
PAR_OUT5/DAT_OUT5	18.0	31.9	29.0	43.7
PAR_OUT6/DAT_OUT6	18.0	31.9	29.0	43.7
PAR_OUT7/DAT_OUT7	18.0	31.9	29.0	43.7

Figure 9.22 Output delay mode.

directly. In a single timing analysis session, the user can investigate several paths in addition to the maximum length path that determines any of the clocking periods. The ability to easily observe all paths resulting in out-of-specification timing values allows the designer to make several circuit improvements in a single design iteration, greatly reducing the overall design time.

Input setup times, hold times (see Figure 9.23), and output delay times for all chip primary inputs and outputs relative to a specified input clock are automatically generated. These features allow complete external timing characterization of a chip design. In addition, if the user specifies input and output timing requirements for the chip, these values can be compared against the user-specified requirements to generate a list of violations that clearly indicates which signals exceed specification by displaying the actual and required values.

The user has the ability to query the path determining any timing result value. Cumulative or incremental delay values may be selectively displayed for each node in a chosen path, quickly identifying the portions of a path needing the closest scrutiny. Setup, hold, and delay values may

Input	INPUT SETUP AND HOLD TIMES (ns)			
	Setup Time		Hold Time	
	Ph1(f)	Ph2(f)	Ph1(f)	Ph2(f)
A0_PAD/A0	9.0	19.6	−4.4	−0.1
A1_PAD/A1	9.1	19.7	−4.5	0.1
A2_PAD/A2	9.2	20.0	−4.6	0.1
ADEC0_PAD/AD0	32.0	35.6	−24.1	−18.1
ADEC1_PAD/AD1	32.0	35.6	−24.1	−18.0
ADEC2_PAD/AD2	31.8	35.4	−23.9	−17.9
CSB_PAD/CSB	27.5	29.3	−11.5	−5.4
DD_DSC0/DD0	—	—	—	—
DD_DSC1/DD1	—	—	—	—
DD_DSC2/DD2	—	—	—	—
DD_DSC3/DD3	—	—	—	—
DD_DSC4/DD4	—	—	—	—
DD_DSC5/DD5	—	—	—	—
DD_DSC6/DD6	—	—	—	—
RDBAR_PAD/RDB	—	5.2	—	1.0
WRBAR_PAD/WRB	—	−1.0	—	—

Figure 9.23 Setup and hold timing output.

be generated for any arbitrary internal node. Also, path delays between any two arbitrary nodes may be generated by simply specifying the end points of the path. These capabilities provide quick feedback about the design, allow the designer to judge the effects of proposed design modifications before implementation, and provide a basis for weighing different design alternatives. TA can also generate circuit simulation paths and stimuli for timing critical paths.

9.6.5 Key Parameters

All of the vital statistics from the block compilers and the router are collected in a set of key parameters that gets updated as the design progresses. Results from timing analysis and simulations are also recorded. They are available to the user upon a simple interrogation.

Logic design is a process of incremental refinement. It is an exploratory process in which design specifications and design goals coevolve. The use of the TA and the resulting key parameters allow the system designer to explore and compare architectures.

The user need only perform a functional simulation at the architectural level of abstraction, and a subsequent static timing analysis based upon a thorough analysis of a model of the final layout, while dispensing with everything in between.

9.7 Compiler Writing

If the user needs to implement a certain logic function that is not directly available in the function set, he can use the random logic compiler to generate a wide variety of general-purpose gates and flip-flops, and use the standard-cell methodology to implement any arbitrary logic function. However, this method frequently results in circuitry that is not as dense as the higher-level compiled blocks because of routing channel areas and a large number of interconnect wires.

In some cases it may be desirable for the user to design circuitry at the transistor layout level using an external physical design system. The system supports the importing of the layout of blocks that have been designed in this manner; it needs the block signal names at the block periphery and the connector locations, and the layout data can be read in such industry-standard formats as GDSII, CIF, and EDIF. Hence, a designer can combine both compiled blocks and hand-crafted blocks in the same chip design.

If the user plans to use these special functions over and over again, or they are needed in many different chips, it may be better to extend the function set rather than importing the special block each time. This is especially useful if various configurations of these special blocks are required. The user can extend the function set or add a new compiler in a way similar to the way that the Genesil function set was originally created, using the compiler development system.

Some of the reasons for doing so are: to extend the function set; to replace a given block compiler with the user's own personalized version; and to generate stand-alone block compilers.

The system allows users who are prepared to tackle silicon designs to add compilers to the compiler system and extend the Genesil system. These new compilers can be tailored for a specific type of circuit and, if desired, for a particular foundry's process. As the popularity of silicon compilation grows, the compiler system will grow in power as users add new capability. In turn, this will add new usages and capabilities.

9.7.1 Block Compiler

A block compiler accepts parameters for a function and produces an instance of a logic block. As an example, a RAM compiler accepts the user's specification for the number of words and size of each word and produces the circuitry for a particular RAM for a given chip use. The compiler designer decides what the parameters are, their ranges, and what circuitry to generate; the chip designer chooses particular parameter values and compiles a particular instance of the block.

In the RAM example, the block compiler might allow just three parameters: width, depth, and output type. Often, circuit design choices place limits on the actual parameter values that can be used, such as a word size between 2 and 100 bits, and the compiler checks that parameters are in range. The compiler could generate different-sized transistors in the row buffers, depending on the word size, or could be designed to use a fixed transistor size, depending on the user's design strategy. The compiler might allow the user to select the output drive strength (e.g., fast, average, or low power), and these choices would affect transistor size.

A block compiler can be divided into five distinct component pieces: (1) the editor, (2) the block parameter structure, (3) the functional simulation model compiler, (4) the layout compiler, and (5) the tile elements. Each piece implements a different aspect of the block compiler, and they are described in the following sections.

Block Editor

The block editor program interacts with the user via menus and forms on the CRT. To create a particular instance of a block, the user selects and enters parameter values in the menus and forms. The parameters can be entered via fill-in-the-blank, multiple-choice, or yes-no fields in the forms that are parsed by the block editor and checked for numerical ranges, correctness, consistency, and completeness. The system provides a programming language (called ICX) for writing the block editor programs.

After the user enters parameters in the form, the editor stores the data in the data structure called the block parameter structure that is used by other parts of the compiler.

Figure 9.24 shows a typical screen menu for a simple RAM, with parameters for width and depth, and Figure 9.25 shows the block editor program that generates that screen menu. The default values for the form and the limit checks are also specified in this program. The numerical range or limit checks are part of the user input checking that helps to detect clerical and logical errors early in design entry. Other types of checks are built into the net listing and the functional simulator.

Block Parameter Structure

The block parameter structure (BPS) is an internal binary data structure used by a block compiler to store the various parameters that characterize a particular block instance, and it tells the block compiler how to instantiate it. This set of block-specific data is shared among the various parts of the compiler.

For example, the BPS for a RAM block may contain the number of words, the word size, and the output drive type.

```
EXAMPLE RAM

WIDTH  ___4___
DEPTH  ___4___

DRIVE TYPE:   DIRECT  TRISTATE
DATA IN:  ____________
DATE OUT: ____________
```

Figure 9.24 Example RAM specification form.

```
ICX Example of RAM editor

func defaults {   /* set defaults */
        set bps_width 4
        set bps_depth 4
        set bps_drtype DIRECT
        set bps_address ADDRESS   }

func buildform {
        /* create form on screen */
        clean_screen s_only

        title "EXAMPLE RAM"

        blank_line
        int_field "Width" bps_width 4 40
        int_field "Depth" bps_depth 2 32
        button_field "Drive type" bps_drtype '(DIRECT TRISTATE)
        bus_connector "ADDRESS  vx(t)"   bps_address
        bus_connector "DATAIN   vx(t)"   bps_din
        bus_connector "DATAOUT  sx(t+1)" bps_dout   }
```

Figure 9.25 Example block editor program.

Functional Simulation Model Compiler

It is not mandatory for a block compiler to contain a functional simulation model compiler; it is only necessary if the user wants to generate block instances that can be verified using the Genesil functional simulator.

To program the simulation model compiler, the system provides a modeling language (called FCX) that provides many useful modeling primitives such as gates, flip-flops, and latches, as well as high-level primitives such as adders and decoders. The simulation model compiler reads the BPS and instantiates a simulation model that corresponds to the block instance.

Figure 9.26 shows an example simulation model compiler program for a simple circuit consisting of two latch primitives to hold the write enable and select enable signals. The simulation compiler reads in the actual parameter values from the BPS and configures a block model with a net list connecting the logic block with the other circuit models.

Layout Compiler

The layout compiler program generates the circuit layout of block instances; the circuit layout will later be merged with all the other blocks and the

```
FCX EXAMPLE for RAM

func model {
        bport (bps datain)  datain<@Width>
        bport (bps dataout) dataout<@Width>
        sig  ramout<@Width>
        if(!= @Drtype @DRDIRECT) {   * add a drive connector *
                port (bps drive) drive
                sig  drive_y     /* sampled drive signal */
        }

        LATCH welatch we_y we phy /* sample write enable */
        LATCH selatch select_y select phy /* sample select bus */
```

Figure 9.26 Example simulation model of RAM.

routing to produce the chip layout that is used to produce the tooling masks for fabricating the chip. The block layout is made up of predefined layout tiles or cells (called *tile elements*) that are assembled into arrays according to a tiling algorithm or program (called *build rules*). The layout compiler uses the tile element library and the BPS information to configure the tile elements. To program the layout compiler, the system provides a tiling language (called GCX).

Many of the function blocks are made with array structures that are implemented with iterative loops in the layout compiler program to build the rows and columns of the tiled arrays according to the BPS parameters. For multiple choice parameters the compiler selects a particular set of tile elements based upon the BPS parameter value. The tiling language also has provisions for "rotates" and "flips" and built-in references for directions such as north, south, east, and west.

In addition, the layout compiler specifies the physical connector information and composes the timing model of the block instance for use by the timing analyzer. For this feature, the type and usage of each transistor and a net list need to be coded in each tile element. For the RAM compiler, the tile elements consist of memory cells, row drivers, and column sense amplifier cells. The template and the tiling scheme control how these elements are placed. Figure 9.27 shows a layout compiler example for a 10-words by 5-bits RAM, in which the template describes what a 1-word by

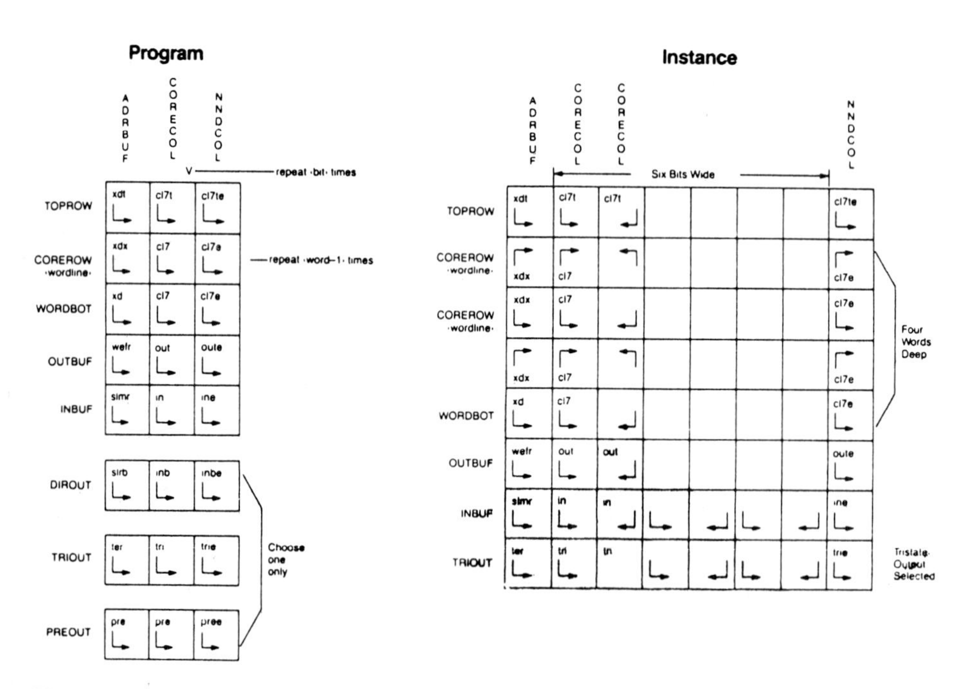

Figure 9.27 RAM compiler template and example.

1-bit RAM looks like, and the tiling scheme indicates that this pattern is to be reproduced ten times in one direction and five times in the other direction in order to produce the layout with the 10 row drivers, 5 column sense cells, and 50 memory cells.

The system does not contain a layout editor. The user needs to enter the graphical layout designs using a conventional layout system. The user then brings the tile elements into the system and writes the tiling program that puts these tile elements together. Hence, circuit design and layout proceed in the traditional manner, and it is easy to take advantage of existing designs. Clearly a block compiler can only generate circuitry that the user has entered and only for configurations that the compiler expects. Therefore it is incumbent on the user to consider all of the combinations of parameters and error conditions if the block compiler is to operate correctly.

One of the major objectives of developing the silicon compiler is that it can be verified in advance to be correct. This influences the choice primarily to use larger tiles and simpler tiling programs, because it is easier to verify the correctness of tile elements than software programs. To illustrate the reasoning used, let's take things to the extreme; if the tile elements are very small (only one or two transistors, for example), the layout has to be controlled by a very complex tiling program. Obviously, the compiler cannot be correct unless the tiling program and all the tile elements are correct. Since it is in general not possible to verify in advance the correctness of a complex program, the compiler's validity also cannot be verified. On the other hand, by using larger tile elements, the tiling program becomes more trivial and can be verified to be correct.

To verify the correctness of the compiler, the user needs to devise many different test instances to exercise the software and tile elements. Since the building of a block compiler is considerably more complex than just doing the circuit design for a specific block, and there may be thousands of potential combinations to consider, the system also provides an efficient verification tool for checking all possible adjacent pairs of tile elements for connector consistency and design rule errors.

Tile Elements

In general, a block compiler's function is parametric and produces different circuitry depending on user-specified parameter values. One way of implementing this is to divide the circuitry into tile elements that are manipulated by a tiling program according to the input specifications.

These tile elements are layout primitives that contain the layout geometries, connector, and transistor usage type information. The circuit design of a logic block is done in the conventional way using CAD and circuit analysis tools. The user creates tile elements using his or her own physical design system. Tile elements are generally created with rectangular outlines and are abutted by the layout compiler; they can also be overlaid on one another. Tile elements can contain structures such as drivers, amplifiers, and memory cells. They can also contain very simple primitives such as wires, transistors, and inverters. These tile elements are designed using either a generic set of design rules that will allow them to be modified and transformed for different target processes (process portable) or a particular set of process specific design rules.

The transistor usage type information is used by the net list, timing model, and power calculations. The system uses about twenty different transistor usage types (e.g., pull-up, sense-amp, push-pull, pull-down, bootstrap, etc.), as well as other codes indicating transistor types such as n channel and p channel.

The system maintains a tile element library that is used by the block compilers to generate circuit layout. The tile elements are stored in file format called CDL, which can be converted into a number of standard formats such as GDS-II and CIF.

9.7.2 Process Portability

One of the advantages of silicon compilation is that VLSI chips can be designed for fabrication by more than one foundry process. The system generates process-portable circuitry because all the tile elements are designed according to a set of generic layout rules that allows them to be transformed or retargeted for different processes. The set of generic design rules consists of about 50 physical parameters dealing with spacing, widths, and overlaps of the various diffusion, polysilicon, and metal layers.

Alternatively, a compiler and tile set may be designed to use only one specific set of process rules, in which case the layout could not be automatically transformed for another process.

The transformation from the generic design rules to a particular foundry process occurs during the tapeout activity. To target the layout of the function blocks to a particular set of layout rules, a combination of linear scaling and "CD-sizing" (shrinking or bloating the geometries) operations are performed on the individual mask layers in the data base. In addition, logical operations can be performed on the different mask layers; for example, one mask layer can be "anded" with another and the resulting layer can be CD sized.

These transformation steps are very commonly used in polygon-based design rule checking (DRC) programs and are therefore familiar to semiconductor masking engineers. They are programmed when a new process is entered on the system. The creation of a process file is done with the Genecal calibration tool, which is beyond the scope of this chapter.

This method of targeting foundry processes does not always achieve the smallest layout size when compared with what an expert layout designer can accomplish. Typically, this type of size penalty is small (less than 15 percent) and it does allow the creation of a silicon compiler without having to manually lay out different tile elements for every supported process. Also, if the block compilers were to procedurally generate the circuit layout, the current state of the art in such software technology would impose an even bigger size penalty.

On the other hand, the interconnection wires placed by the router are procedurally generated in a process specific way according to the exact layout rules. Since most ASIC chips are about 50 percent wiring, this means that this portion of the chip takes full advantage of the design rule capabilities.

9.8 Future Directions of Silicon Compilation

The following is a list of issues and directions for the future development of silicon compilation. It is not shown in any priority ordering.

- Application specific block compilers (e.g., analog, radiation hardened)
- More foundry processes and different technologies (GaAs)
- Better and faster hardware platforms and accelerators
- Automatic block placement
- High-level input description and logic synthesis
- Improved minimization and optimization in synthesis
- Increased use of AI/expert system for handling complex heuristics
- Parametric layout generation
- Test support and ATG
- Interactive router

9.9 Summary

Success with gate arrays accelerated the use of ASICs at systems companies for space, power, and cost reduction, as well as performance enhancements. Silicon compilation claimed improvements in density, performance, and design time by allowing all masks custom chips to be implemented by logic designers who are not familiar with IC circuit and layout design.

A year after the Genesil system was introduced, there were about 100 stations installed at customer sites and ten chips were in production. These chips contained between 30,000 and 80,000 transistors and were generally done by engineers without prior experience with VLSI design.

The system contains more than 1,000,000 lines of C source code and occupies more than 60 megabytes of UNIX file space. It took about 30 programmers and IC designers more than three years to develop.

Some users' experiences are summarized as follows. The average number of chips being designed per user station is two, by two design engineers. The fastest completion time has been about two months, the average around six months, and the longest nearly ten months. The most complex chip design is over 100,000 transistors, and the average is around 35,000. Typical applications include: floating point arithmetic, CPU, peripheral controller, graphic terminal circuit, and signal processing. Most of the new designs are targeted for 2 micron and 1.2 micron CMOS processes.

[Chen84] E. K. Cheng, "Verifying Compiled Silicon," *VLSI Design*, vol. 5, no. 10, October 1984, pp. 70–74.

[Gajs85] D. D. Gajski, "Silicon Compilation," *VLSI Systems Design*, November 1985, pp. 48–64.

[Hitc82] R. B. Hitchcock, "Timing Verification and Timing Analysis Program," *19th Design Automation Conf.*, June 1982, pp. 594–604.

[Joha79] D. Johannsen, "Bristle Blocks: A Silicon Compiler," *16th Design Automation Conf.*, June 1979.

[John84] S. C. Johnson and S. Mazor, "Silicon Compiler Lets System Makers Design Their Own VLSI Chips," *Electronic Design*, vol.32, no.20, October 1984, pp. 167–181.

[Joup83] N. P. Jouppi, "TV: An nMOS Timing Analysis," *Third Caltech VLSI Conf.*, March, 1983, pp. 71–85.

[Klec82] J. E. Kleckner, et al., "Electrical Consistency in Schematic Simulation," *Proc. ICCC–82*, 1982, pp. 30–33.

[LaPo86] D. P. La Potin and S. Director, "Mason: A Global Floorplanning Approach for VLSI Design," *IEEE Trans. on CAD*, vol. CAD–5, no.4, October 1986, pp. 477–489.

[Laut80] U. Lauther, "A Min-Cut Placement Algorithm for General Cell Assemblies Based on a Graph Representation," *Journal of Digital Systems*, vol.4, no.1, 1980, pp. 21–34.

[Lin84] T. M. Lin and C. A. Mead, "Signal Delay in General RC Networks," *IEEE Trans. on CAD*, vol. CAD–3, no.4, October 1984, pp. 331–349.

[Lin86] T. M. Lin and L. Fiance, "Timing Analysis in a Silicon Compilation Environment," *Applications Memo #26, Silicon Compilers Inc.*, January 1986.

[McWi80] T. M. McWilliam, "Verification of Timing Constraints on Large Digital Systems," *Proc. 17th Design Automation Conf.*, June 1980, pp. 139–147.

[Murp85] B. J. Murphy, J. E. Kleckner, and K. K. Tam, "STA: A Mixed-Level Timing Analyzer," *ICCAD Digest*, 1985, pp. 176–178.

[Otte82] R. Otten, "Automatic Floorplan Design," *Proc. Design Automation Conf.*, June 1982, pp. 261–267.

[Oust83] J. K. Ousterhout, "Crystal: A Timing Analyzer for nMOS VLSI Circuits," *Proc. Third Caltech VLSI Conf.*, March 1983, pp. 57–69.

[Rubi83] J. Rubinstein, P. Penfield, and M. A. Horowitz, "Signal Delay in RC Tree Networks," *IEEE Trans. on CAD*, vol. CAD–2, no.3, July 1983, pp. 202–211.

[Sech85] C. Sechen, *The Timberwolf Macro/Custom Cell Placement and Chip-Planning Program*, Technical Report, University of California, Berkeley, December 1985.

[Sech87] C. Sechen, "TimberWolfMC: Chip-Planning, Placement, and Global Routing of Macro/Custom Cell Integrated Circuits Using Simulated Annealing," *Proc. Design Automation Conf.*, June 1987.

[Sili86] Silicon Compilers, Inc., "Genesil System Function Set Specification," *Users Manual*, 1986.

[Souk81] J. Soukup, "Circuit Layout," *Proc. IEEE*, vol. 69, October 1981, pp. 1281–1304.

[Supo83] K. J. Supowit and E. A. Slutz, "Automatic Floorplan Design," *Proc. Design Automation Conf.*, June 1983, pp. 164–170.

[VLSI84a] VLSI Design Staff, "Part 1: Drawing a Blank: Silicon Compilers," *VLSI Design*, September 1984, pp. 54–58.

[VLSI84b] VLSI Design Staff, "Part 2: Casting an Image: Silicon Compilers," *VLSI Design*, October 1984, pp. 65–68.

CHAPTER 10

Design Methodology of the Concorde Silicon Compiler*

Vince Corbin
Warren Snapp

10.1 Introduction

This chapter discusses the approach toward compilation adopted by Seattle Silicon Corporation as demonstrated by the Concorde ASIC compiler. It explains what differentiates this approach from previous design automation-techniques and how this approach differs from other silicon compilers. A sample circuit design illustrates the method. Finally, topics of function block synthesis, irregular data paths, and available function blocks are discussed in more detail.

10.1.1 What Is New About Concorde?

The Concorde compiler was developed to automate application-specific IC design for the system-level logic designer and to produce higher-quality designs with higher levels of productivity and production economics than standard cell and gate array design automation software. The goal in chip quality has been to produce designs of clearly superior density and performance to standard cells and gate arrays. The goal in productivity has been to increase productivity of the overall design process (concept through mask tape) from the 30 to 40 transistors per workday of standard cells to 200 to 300 transistors per workday. The goals in economics have been to guarantee first-pass silicon success, to lower the production cost, to make true second

* The research discussed in this chapter was done under the auspices of the Seattle Silicon Corporation.

sourcing an everyday reality, and to create designs that migrate to new technologies with little additional cost and risk.

These goals have been accomplished by a silicon compiler approach using as input a high-level structural description, synthesizing high-density functional blocks, and using sophisticated general block autoplacement and routing software with manual guidance and optimization. All parts of the design process are controlled to maximize correctness and minimize errors; additional analysis and verification tools are provided to increase design confidence to nearly 100 percent.

Previous design automation techniques have assumed a homogeneous set of primitives, such as fixed height standard cells or rows of identical uncommitted transistors. This makes the algorithms easier, but it produces designs far inferior in density and speed to handcrafted layout. In contrast, human full-custom designers do not work with a homogeneous set of primitives. They divide the process into two distinct phases: (1) the definition of functional blocks that take advantage of as much architectural regularity as possible and (2) the planning of global and local floor plan so that blocks will fit and route together with minimal wasted space and wire length. The individual function blocks are then designed with optimal transistor sizing and layout topology to achieve maximum speed and density while adhering to constraints imposed by adjacent blocks. These blocks are then placed according to the floor plan and interconnected by abutment and with high-density routing, using techniques such as contour routing and over-the-cell routing.

The essence of the Concorde approach is two-level physical design synthesis, analogous to handcrafted design.

Function Block Synthesis

The first level is synthesis of function blocks. Maximum benefit is obtained from blocks of high enough level to capture significant architectural regularity, such as generalized data path structures, memory, and multiplier arrays. These blocks can be synthesized using abutment and over-the-cell routing techniques, achieving density and performance close to optimized hand layout.

Function blocks in Concorde are synthesized by function block generators (FBGs), written in a C language–based development environment called the compiler development system (CDS). Each FBG is fully customized to generate various design views of that specific function. The primary views generated include layout geometry, transistor net list, gate-level model, and timing model.

Because of the high level of these function blocks, it is important that the FBGs be parameterized in functional options (e.g., bit width), electrical characteristics (e.g., output drive), and floor plan configuration (e.g., aspect ratio and pin locations) in order to be appicable to as many designs as possible and to fit flexibly into chip floor plans. Without such parameterization, an extremely large library of function blocks would be required, numbering in the hundreds of thousands; this is why extending standard cells to include libraries of fixed higher-level function blocks is impractical. The Concorde design system has less than 30 highly parameterized digital FBGs and 20 parameterized analog FBGs.

Like handcrafted circuits, blocks synthesized by FBGs are optimized for density and performance. Building an FBG begins with an expert IC designer designing a circuit and a corresponding handcrafted polygon layout. The circuit and layout views are then encoded within CDS with parameterization for function, electrical characteristics, floor plan, and one additional feature not yet mentioned—that is, process variation. The encoding techniques were developed to produce the absolute minimum possible compromise in layout density. Consequently, the synthesized geometries maintain the density of the original handcrafted layouts.

This density is also maintained to a remarkable degree as the target process is changed. For instance, in changing the target CMOS process from *n*-well to *p*-well, or from 3 micron to 2 micron, or even to 1.25 micron, the density remains close to handpacked density for the new technology. As an example, Figure 10.1 shows the area difference between identical flip-flop, FIFO, adder, and small RAM functions synthesized for five dif-

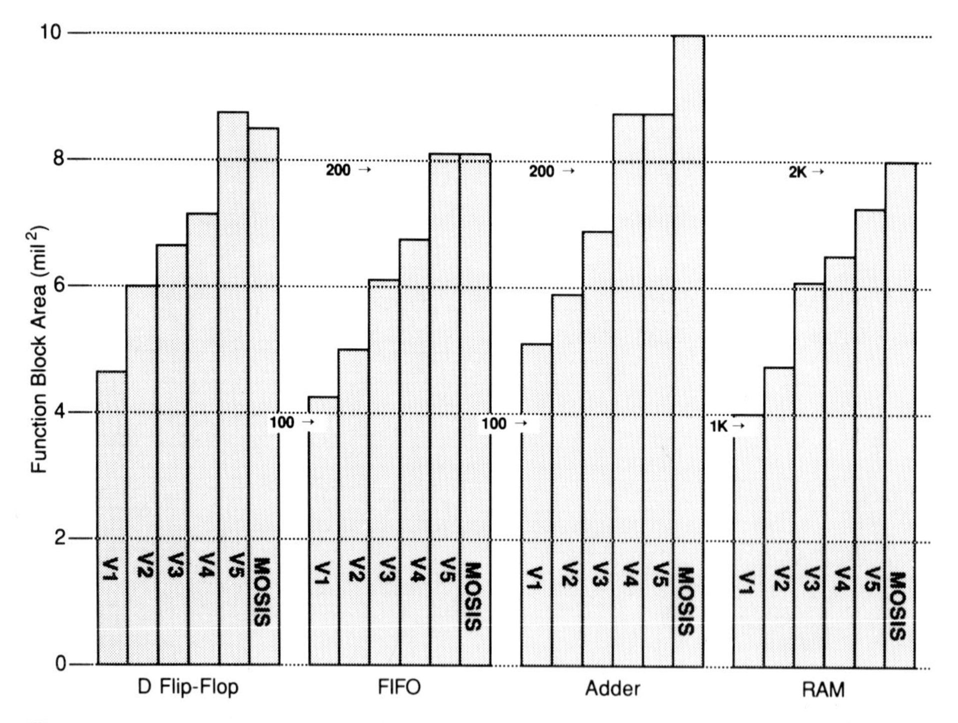

Figure 10.1 Area of example Concorde function blocks in five different 1.25-micron VHSIC or VHSIC-like processes and the MOSIS 1.25-micron "shrinkable" process. By synthesizing layouts from over 80 individual process rules, density improvements of up to 50 percent are obtained over linear scaling.

ferent 1.25-micron processes and the MOSIS 1.25-micron process. Process V1 results in nearly twice the density of V5 and MOSIS, although they are all 1.25-micron processes. This density difference is not and cannot be accomplished by linear scaling and biasing of fixed cells. Rather, the layout synthesis is driven by tables of geometry design rules and process parameters. More than 80 geometry design rules are individually considered, such as metal 1 spacing, metal 2 width, cut size, and poly overlap of cut in the direction of metal. Consequently, Concorde blocks can take advantage of density improvements in advanced processes. Details of this synthesis and spacing adjustment technique are discussed later.

General Block Placement and Routing

The second level in the Concorde physical design approach is the placement and routing of general blocks.

Blocks vary in size from large data path assemblies and memory to individual gates and clock buffers or flip-flops and counters. Fixed height cells are used for gates and other smaller blocks. They are placed between large blocks in groups of rows of abutted cells, resembling standard cells. Two different autoplacement algorithms are used, one for general blocks and another for optimizing placement within groups of smaller cells. An important part of the system is that the placement can be modified manually through interactive graphical manipulation with the aid of force vectors and "rats nest" visualization of interconnects.

A channel definition and routing scheme that guarantees 100 percent completion is used. Channels are defined as a slicing sequence, and are ordered in such a way that all core routing can be completed using channel routers; switchbox routers are not required. Detailed channel routing is accomplished using a rectilinear channel router. Gridded, gridless, and contour versions are available. Placement is adjusted to minimize area and guarantee completion. Global routing is accomplished using a heuristic algorithm to generate a near-optimal Steiner tree on a region adjacency graph for each multiterminal net. Power and ground are routed as planar tapered trees in one level of metal, with the degree of tapering determined by power consumption of individual blocks.

Routing can be done either hierarchically or flat. If a design is built up hierarchically, the routing can later be optimized automatically, eliminating the redundancies and circuitous routing that naturally arise in hierarchical routing.

After automatic routing has been completed on a section or on the entire circuit core, critical nets can be optimized by an interactive symbolic wiring editor if desired. This editor guarantees that nets will not be disconnected or shorted. The wiring editor can also be used for fully manual routing, which may be desirable in specific applications.

After completion of the core, the packager synthesizes the pad ring routes and the pad ring to the core using a moat router. The moat router is capable of keeping power and ground, and specified other signals, on a single layer of metal, if that is possible.

Efficient placement and routing may require aspect ratio changes in certain function blocks. After designers decide where such changes are advantageous, they can recompile function blocks, specifying new aspect ratios.

Data Path

Another more specialized placement and routing tool for bus-oriented architectures is available. The data path compiler is a combination of FBGs for individual functions plus an automatic composition tool that arranges and interconnects the generated cells in rows and columns much as a human layout designer would. The data path composition tool uses over-the-cell routing for two-layer metal processes and achieves density that is 30 to 50 percent higher than that attainable by the best channel routing techniques on fixed height cells, even with optimal placement.

Design Entry and Analysis Tools

Above, we explained that the essence of the Concorde approach is two-level physical design, analogous to handcrafted design. This physical design approach accomplishes the chip quality and economic goals but is not sufficient in itself to achieve the 5 to 10 times productivity improvements desired over standard cells, nor the 10 to 50 times productivity improvements desired over full custom. Increasing the productivity has required some new approaches to design entry and analysis. These approaches include (1) the user interface, (2) aids to rapid design space exploration, and (3) a static timing analyzer.

USER INTERFACE The user interface was designed to minimize human error, minimize the amount of information the user needs to remember, and automatically invoke data conversion software when moving from one part of the design process to another. This is done by making the user interface a tree of menus and forms. As much as possible, the user is pushing buttons and selecting options rather than invoking programs through operating system textual command lines. The user follows a menu tree to select the FBG desired and then fills out a form selecting functional, electrical, and floor plan options. Menus are selected for specifying the technology and for selecting which FBG views should be generated (geometry, simulation, symbol, etc.). Menus also specify which views should be automatically converted to the host workstation format (Mentor Graphics or Valid Logic Systems schematic and simulation formats). Pushing the "create" button in an FBG form invokes creation of all activated views and conversions. Pushing the "status" button displays when various views of the FBG were last created or converted.

The user can select menus for placement and routing. Pushing one button converts the interblock schematic from the workstation format to internal format and executes automatic placement. Pushing another button invokes the interactive placement and wiring editor, PRIDE. Actions with PRIDE are menu driven with optional short and long text commands. Within PRIDE, the automatic routing software can be invoked by pushing menu buttons. The circuit appears in the wiring editor in symbolic form after the routing is complete. Pushing another button generates final layout geometry and a schematic symbol. It also back annotates the workstation simulation model automatically. Similar pushbutton operations are invoked to bring up text editors to modify textual inputs to compilers, to specify and generate the I/O pad ring, and to invoke the static timing analyzer. In general, the entire design process can be executed almost entirely by pushbutton operations.

A second element of the human interface is provided by the host workstation. Concorde relies on either Mentor Graphics or Valid Logic Systems software for schematically specifying interconnections between function blocks and for gate-level logic simulation. Both of these systems have quality user interfaces, providing easy-to-use menu capabilities and minimization of human error. Transmission of data between Concorde and the Mentor or Valid data base is transparent. For instance, when the "create" button is pushed in an FBG form, the simulation model and a schematic symbol for that function block are automatically generated and placed in a Mentor or Valid data base library. They can be activated within the workstation schematic entry simply by making a menu selection in the workstation editor. No memory of library search paths or invocation command sequence is required. As a result, errors, frustration, and overall design time are reduced.

RAPID DESIGN SPACE EXPLORATION The second contributor to increased productivity is a series of aids to rapid design space exploration. The first aid is the fact that function blocks correspond to high-level structural objects that are familiar conceptual primitives to the logic or system designer. Designers do not have to decompose ALUs, register files, or FIFOs to gate-level primitives. Secondly, the designer is able to estimate size, performance, and power consumption quickly for each proposed function block in order to evaluate design alternatives. Size and power estimates can be obtained from FBGs before synthesis, or actual size can be found from executing the geometry creation, which requires a few seconds to several minutes depending on the size of the block and the views being created. Performance estimates can be obtained from feedback fields on the FBG input forms and by creating the simulation model and invoking the static timing analyzer. A chip analyzer tool is in development; it will tabulate and summarize estimates for all function blocks in the proposed design and add estimated interconnect area as well, providing accurate estimates for the complete design.

STATIC TIMING ANALYZER The third contributor to productivity is the static timing analyzer. With the static timing analyzer, critical paths and their delays can be found quickly, and set-up and hold as well as other timing constraints can be checked without executing logic simulation. The algorithms used are described later.

10.1.2 How Does Concorde Differ From Other Silicon Compilers?

Figure 10.2 presents a Y diagram [Gaj83] of the Concorde compilation approach. From entry point one, a function block specification is compiled to mask geometries and to gate and circuit level structural views as well as a behavioral model. From entry point two, interblock netlists drive autoplacement and routing of functional blocks, with manual intervention through the placement and wiring editors. The multiple views allow simulation to be performed at levels from behavioral down to circuit.

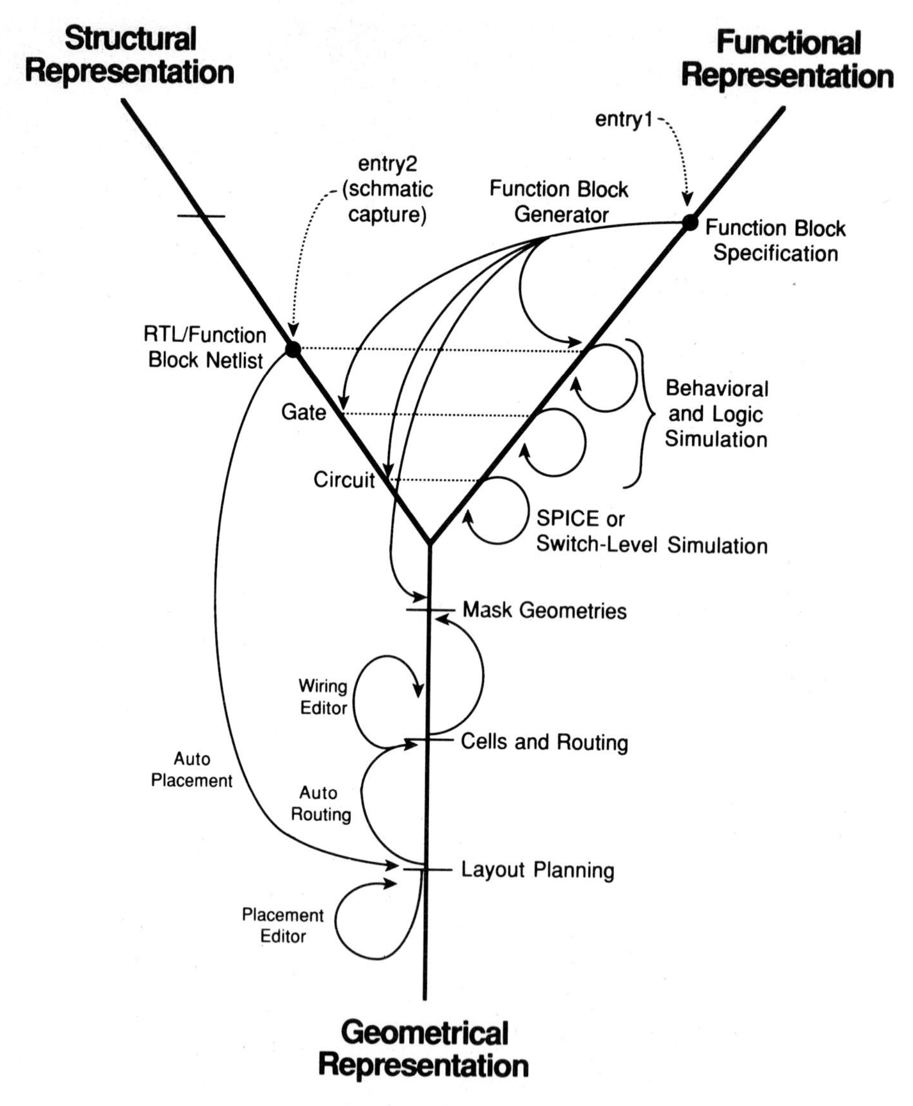

Figure 10.2 Y diagram of Concorde Synthesis.

Various other silicon compiler approaches have been discussed elsewhere ([Bla85], [Bra85], [Gold85], [Jam85], [Jerr86], [Krek85], [Mar86], [Sou83]) including this book, and a further identification is not needed here. However, a useful distinction may be to point out differences between commercial and academic compilation approaches.

Commercial approaches that have survived have generally had the following characteristics: they are applicable to a wide variety of design applications: they use structural inputs (little behavioral synthesis so far); they are applicable to a range of target processes; they consist of a complete and reliable design automation system; and they have some mechanism to guarantee quality results on a large scale. Within this arena, Concorde stands out as providing (1) function block synthesis that is truly process design rule independent, (2) powerful automatic placement and routing capability, (3) interactive optimization features in placement and routing, (4) the ability to handle irregular data paths, (5) no artificial restrictions on design style or clocking scheme, and (6) analog as well as digital CMOS FBGs and analysis mechanisms. The system is transparently integrated to commercial CAE workstation systems (Mentor Graphics and Valid Logic systems), which allows full access to the workstation design and analysis tools. In addition, tools are provided to enable the user to write his or her own function block generators and incorporate previously designed fixed cells or blocks. The result is a tool that is applicable to a wide range of real-world demands and that creates high-quality designs with short development times and excellent production economics.

10.2 The Design Process

This section outlines the general design methodology supported by the Concorde compiler (see Figure 10.3). Each step of the design process is discussed briefly and illustrated with an example of a chip designed using the system. It is important to stress that the design process described, and the example given, assume a top-down design approach, which generally gives the greatest leverage in a compiled design. The Seattle Silicon automated layout tools are specifically designed to generate IC geometry from high-level input rather than from gate-level descriptions of a circuit. However, where required, gate-level input can be used, but generally with no density advantage over standard cell approaches. In the example, although the circuit is built chiefly from higher-level modules, several function areas are specified at the gate level.

The Concorde design process can generally be divided into eight steps, including fabrication and testing steps:

1. *Block diagram.* A block diagram representing the functionality of the circuit is constructed, with the kinds of modules available kept under consideration by the designer. At this stage, the compiler design process differs little from the top level of most VLSI circuit design procedures.

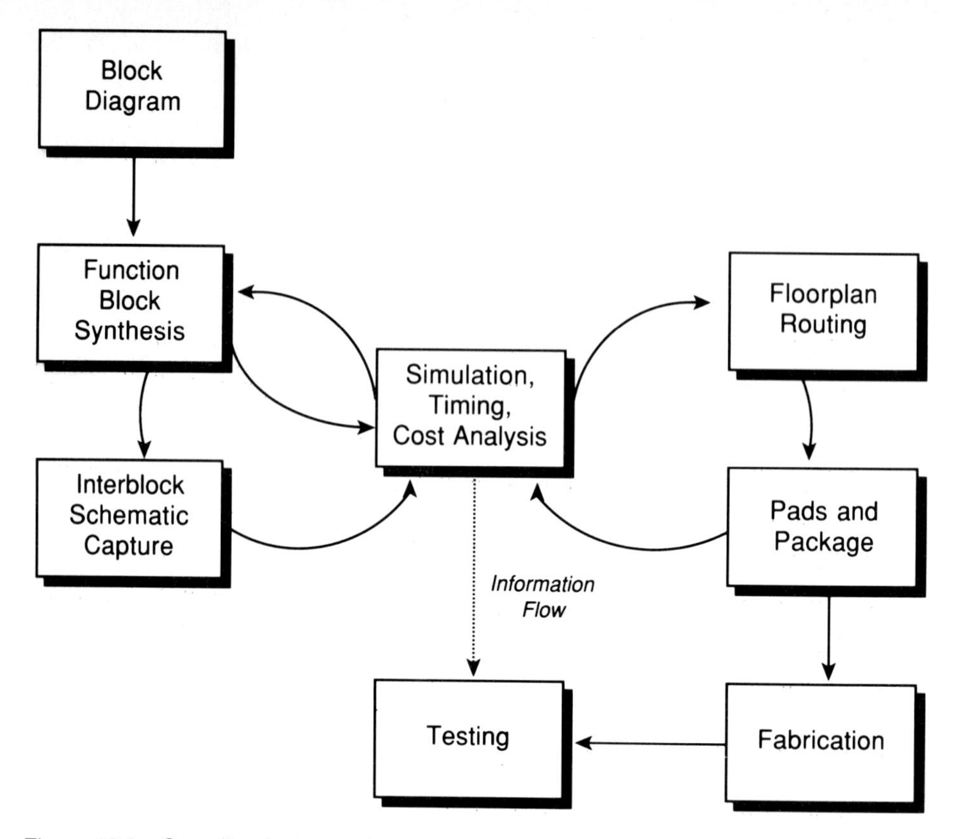

Figure 10.3 Compiler design cycle.

2. *Function block synthesis.* In the second stage, integrated circuit modules are compiled for each block or group of blocks in the diagram. Depending on the kind of function and the approach of the individual designer, each block may be defined using a high-level language, menu input, or a selection of individual logic components. As a first step, estimates of size and performance for a particular process may be gathered from menu feedback in order to evaluate design alternatives. As the design concept solidifies, simulation models and schematic bodies may be compiled, to be used in timing analysis and simulation. Finally, geometry can be synthesized when the functional design is complete. Simulation models and schematic bodies are automatically and transparently translated into host workstation format (Mentor Graphics, Valid Logic, etc.).
3. *Interblock schematic capture.* A detailed logic schematic is constructed using the workstation tools with each symbol on the schematic corresponding to a function block compiled in the second design stage. The connectivity defined in the schematic is available immediately for logic simulation and will later be transferred directly to the automatic placement and routing process.

4. *Simulation and timing analysis.* A functional simulation and dynamic timing analysis of the design based on the schematic is performed using the host workstation simulation tools. This enables the designer to refine and analyze detailed functionality and data-dependent timing. In the process, the designer develops a sequence of simulation vectors that verify the functionality of the design and a static timing analysis can be performed on individual blocks or groups of interconnected blocks using the Concorde Timing Analyzer. Path descriptions and delays along the path will be listed for the slowest paths, and setup and hold violations will be reported. Clocking schemes can be specified as single-phase or any multiple-phase sequence. Delays are checked from inputs to outputs, inputs to latches, and latches to outputs. Based on analysis results, alternative design approaches may be selected that will be used later as a basis for developing the prototype test program.
5. *Floor planning and routing.* The physical layout of the circuit core is specified in a graphics-based placement and routing environment (PRIDE), with module placement based on connectivity defined and simulated in the earlier stages. Initial placement is done automatically during the net list conversion process using algorithims that will be discussed later. The user may graphically move modules into alternate positions. Feedback is provided to guide the designer toward the optimum placement. Routing itself may be carried out by a general block autorouter, by a manual routing editor, or by a combination of the two approaches, including hand optimization of autorouted circuits. A 50,000 transistor circuit with 92 blocks and 498 nets can be fully autorouted on an Apollo DN3000 in less than 11 minute (643 seconds). Once the placement and routing are completed, simulation models are updated to reflect parasitic capacitance loading resulting from the routing. The circuit is then resimulated or reanalyzed for timing. As with the earlier design phase, redefinition, recompilation, rerouting, and resimulation cycles optimize circuit performance.
6. *Packaging.* In the final stage, the package for the IC is specified, the connections between circuit core and the pads are defined, and the routing from the core to the pads is performed. The interconnection to pads is automatically translated from the schematic input, and pads are automatically placed and routed. Autoplacement takes into account user-defined pin-out as well as optimal routing and pad placement relative to optimal bonding wire configuration. A manual override for circuit optimization is available in a graphic editor. After this final stage is complete, a chip-level simulation model is created that incorporates full pin-to-pin loading and interconnect delays.
7. *Fabrication.* After the IC design is complete, its design data base is available in standard GDSII or CIF output format for generation of the mask tape and subsequent manufacture of masks, processing of wafers and packaging of dice. By experience, we have found that the probability of first-pass working circuits can be very high if a final verification step

is performed at this time. This step checks for design configuration consistency, errors in manual editing of wiring and specification files, shortcomings of higher-level simulation models and correct integration of external cells. It can be performed by the user or provided as a service, and consists of (1) a full-chip detailed geometric Design Rule Check (DRC) and Electrical Rule Check (ERC), (2) extraction of the transistor circuit, with parasitics, from the geometry, and (3) execution of a (switch-level) transistor simulation of the entire chip. The simulation is performed against test vectors generated during step 4 above.

8. *Testing.* Engineering prototypes are usually tested after packaging of visually good dice, whereas production parts will use both wafer probe sorting and final test of packaged parts. Prototype testing is performed by translating simulation vectors generated in step 4 above into an automatic test program. A variety of test equipment has been used for this purpose. Production testing can be optimised at the user's discretion.

10.2.1 Block Diagram

The first step of the design process is to construct a block diagram of the functionality of the circuit, making efficient use where possible of available function blocks (see Appendix). In the case of the 2910, a standard top-down design process was appropriate. At the highest level, the functionality of the entire chip maps well into available functions (see Figure 10.4), except that a LIFO stack was not available at that time. The stack was implemented with five D-type registers on a bus, controlled by a stack control logic block, which will be defined later. The 12-bit wide functions in the large dashed rectangle naturally map into a data path compiler block.

The AMD 2910 is a microprogram controller (for use in high performance digital processor applications) that allows addressing of up to 4,096 words of microprogram. The controller contains a four-input multiplexer that selects either the register/counter, direct input, microprogram counter, or stack as the source of the next microinstruction address. Blocks in the diagram are the following:

- *Regcnt (register counter).* This block functions as a latch and decrementer. The load and enable signals are supplied by the reg control block. The register counter is used to implement loop instructions by counting down from the loaded value.
- *Reg control.* This block provides the load and enable signals to regcnt. In essence, this block determines whether the register counter is active.
- *Zero detector.* This block detects when the register counter has reached zero and outputs that as the Zbar signal to the Instruction PLA.
- *Instruction PLA (Uopt).* This control circuit provides signals to various parts of the circuit, based on the values of Zbar, PassFail, and the I< 3..0> input.
- *Multiplexer (MUX).* This block selects between one of four 12-bit inputs, two from off-chip, one from the stack, and one from the Microprogram

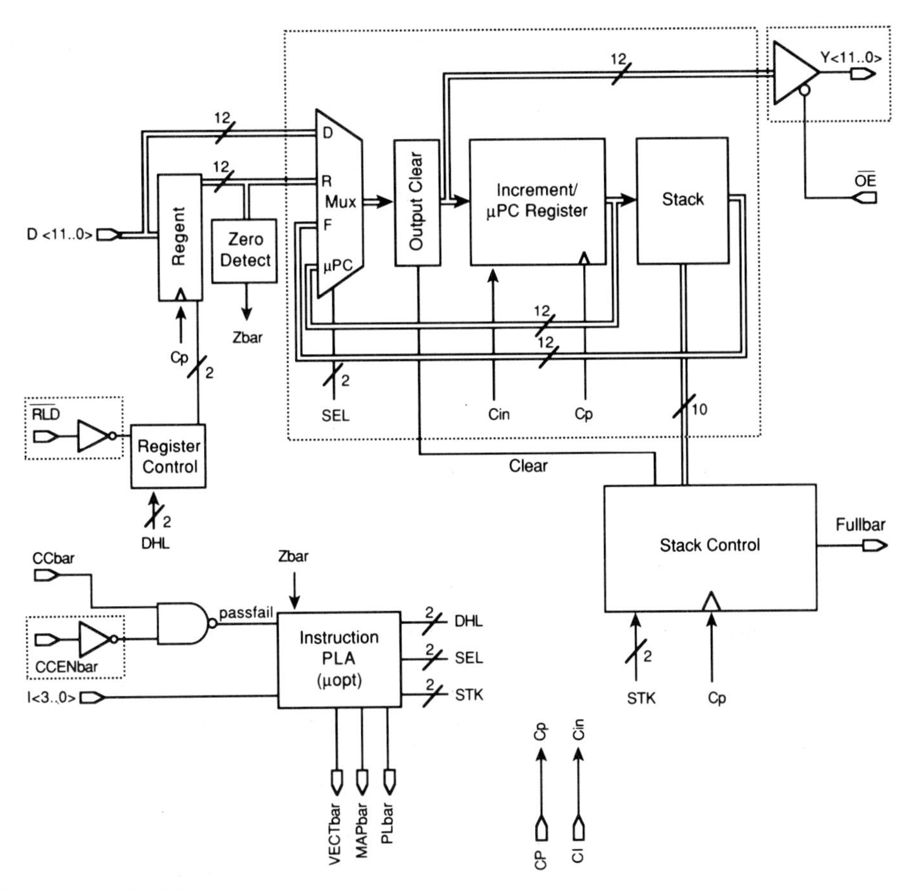

Figure 10.4 BLOCK diagram of 2910.

counter/register (MPC reg). The control signal for the multiplexer comes from Uopt, and the multiplexer output goes to the output/clear generator.

- *Output/Clear generator.* This generator receives the multiplexer outputs and provides the output of the 2910. If Clear is enabled, the output is set to zero, regardless of the input. Otherwise, the signal is passed through.
- *Incrementer.* This block combines the incrementer and MPC reg functions.
- *Stack.* This block contains a five-level, 12-bit stack in a LIFO fashion, with the output sent to the multiplexer.
- *Stack control.* This block contains state machine for controlling the LIFO stack. Operations are controlled by two signals from Uopt.

10.2.2 Function Block Synthesis

The first step in compiling function blocks is to select a process design rule set. Once a set of process design rules is selected, all compiled blocks and routing will be synthesized optimally for those design rules, that is, the compiler ensures that mask geometry is optimally dense and that violations of the design rules do not occur as blocks are compiled and routed. In addition, all simulation and timing analyses are based on the geometric definitions and process parameters associated with the selected process. Upon changing design rules for a particular design, the modules are recompiled for the new rules and an alternative version of the design is stored in the data base. With a detailed system block diagram constructed and a design rule set chosen, the designer specifies and compiles blocks to correspond to each of the functions in the block diagram. A listing of currently available function blocks is in the appendix.

Specifying Modules

Modules can be specified in several ways, as discussed in the following sections.

DIRECTLY VIA FORMS. Modules are usually specified from one or more forms or menus. When a module is built from menu selections, three different types of input options can be supplied. *Functional options* determine the function performed by the module (e.g., the number of bits for an adder or the center frequency for a switched capacitor filter). *Electrical options* determine circuit characteristics and are used to optimize performance (e.g., the drive of output buffers). *Physical options* control the physical characteristics of the layout geometry of the modules. Among the physical options the designer can specify are the aspect ratio of the module, port locations, and folding options. Electrical options are usually changed as a result of simulation to adjust circuit performance, while physical options are typically altered during floor planning to fit blocks together efficiently. Menu input can also be used to specify multiple levels of hierarchically nested modules and the connectivity between them. The data path compiler, for example, uses menu input to define constituent elements and the bus and control signal connections between them.

FROM GATES. The compiler allows designers to create custom modules from gate-level cells using one of several place and route functions and then to include those modules in higher-level designs. One cell of a custom adder, for example, can be built from individual gates, instantiated schematically, and interconnected using automatic or manual routing. The module can then be included within a data path structure along with modules available through the menu.

C5 MODULES. Designers have the option of designing custom modules using the C5 Compiler Development System. Modules created through C5 are compatible with modules provided through the standard digital and analog compilation tools. For more information on C5, see Section 10.3.1.

FOREIGN MODULE INTEGRATION. A custom module in the selected design rule set may also be incorporated within a compiled design using Concorde Foreign Module Integration (FMI) techniques. Any cell designed with a specific rule set using host workstation tools and compatible standard cells, macrocells, or other predefined function blocks can be used in conjunction with compiled designs.

Architectural Exploration During Function Block Synthesis

At the initial compilation stage, the designer often builds several modules that implement the same function in different ways. These modules are then substituted into the circuit to explore which of the architectures is most efficient in terms of performance and silicon utilization. The 2910 design used a variety of modules for implementation of the block diagram, with each block requiring different numbers and types of modules. The zero detector, for example, was simply implemented with a series of five gates, while the stack control finite state machine used a PLA structure with the logic specified through a text file.

For the 2910 model, the designer explored alternative architectural methods of performing different functions in the circuit. The Uopt block, for example, could have been implemented with either a PLA structure or a ROM. While the ROM module would have provided more control over the output of the 2910, it also would have required an extra control signal to enable output, resulting in a more complicated design. The PLA, on the other hand, has fewer connections and is faster; relative size depends on sparsity of the addressing scheme.

Module Synthesis

There were several different components in the circuit involving 12-bit data paths, including a multiplexer, incrementer, and the stack. The designer chose to implement this portion of the circuit as a single data path module (see dashed rectangle in Figure 10.4) consisting of those elements requiring efficient transfer of the full 12 bits at one time. The data path compiler automatically repeats a logic structure for the number of bits desired, rather than requiring the designer to repeat the same circuit logic twelve times. Figure 10.5 shows two different implementations of the 2910 data path—the larger using a channel routing approach for single-metal processes and the smaller using over-the-cell buses when dual metal is available. For the 2910, over-the-cell routing decreases the area by 32 percent, which might favor the use of a dual-layer metal process.

As part of the process of constructing modules, the designer can choose from a variety of clocking schemes depending on the desired behavior of the circuit, and the modules to be incorporated into the circuit. Storage element clocking schemes available include level-sensitive and edge-triggered. System clocking schemes may be (1) single-phase, (2) dual-phase nonoverlapping, or (3) multi-phase. For the 2910, a single-phase, rising edge–triggered clocking scheme was chosen.

Compilation time for modules is workstation dependent. In the internal benchmarks, the time to compile all modules did not exceed two hours of real-time on any workstation.

Once the specification for each module is complete, the module is com-

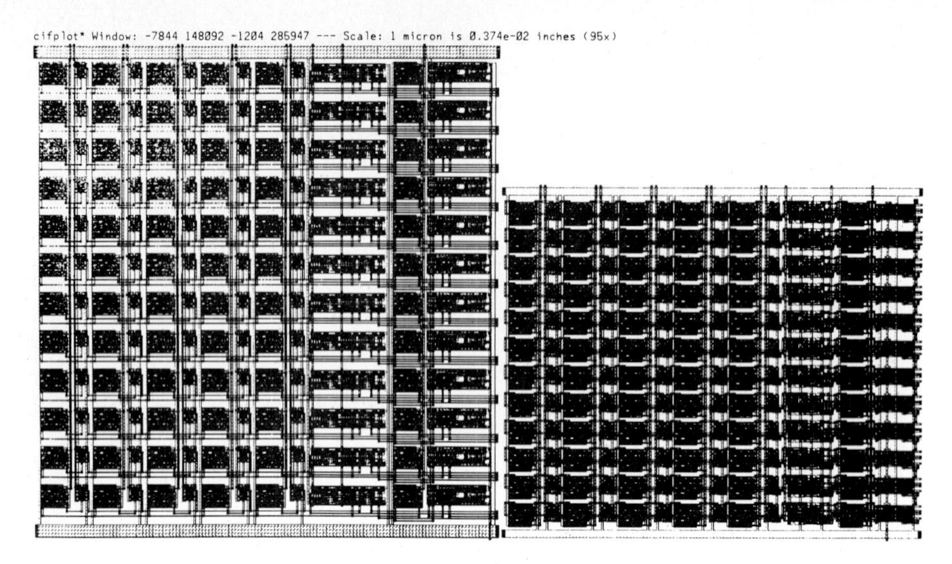

Figure 10.5 Data path function block for 2910, showing comparison between single-metal and dual-metal routing approaches.

piled—a process that generates four different representations of the module: (1) *simulation model,* which contains a net list of behavorial and gate primitives with design rule and process parameter specific information to conduct functional and timing simulation with the host workstation simulation tools and static timing analysis with the timing analyzer; (2) *symbolic model,* a symbolic description of the module created for use by the host workstation schematic editor; (3) *geometry model,* the actual geometry of the module for the specified design rules produced by the compilation process; and (4) *footprint model,* a boundary polygon description of the module, which also includes port information for the terminals on the module, used to represent the module in the graphics-based automatic placement and routing environment.

10.2.3 Schematic Capture, Functional Simulation, and Timing Analysis

Although the 2910 example used in this discussion involves only a single IC, one of the essential features of open system compilation may be understood only by reference to the larger design within which any single ASIC is intended to operate. The compiler creates schematic symbols and simulation models for each compiled module in the same format used by the host workstation. This shared data base structure allows the designer to construct and simulate a system-level design, which may include library components and standard parts along with the compiled modules, even before the ASIC design is completed. In this way, a board-level schematic may be captured and several different integration strategies explored early in the design cycle. System circuit performance may be estimated in an early functional simulation and then resimulated after different elements are integrated and the results compared. Because creating compiled circuitry is rapid, several different combinations of application-specific circuits and standard parts may be evaluated before any commitment to an implementation strategy is made. In this sense, an open system workstation-based compiler provides an implementation-independent design environment.

In the case of the 2910 benchmark, the schematic consisted of symbols representing each of the compiled modules (see Figure 10.6). The initial functional simulation was performed using a standard logic simulator (see

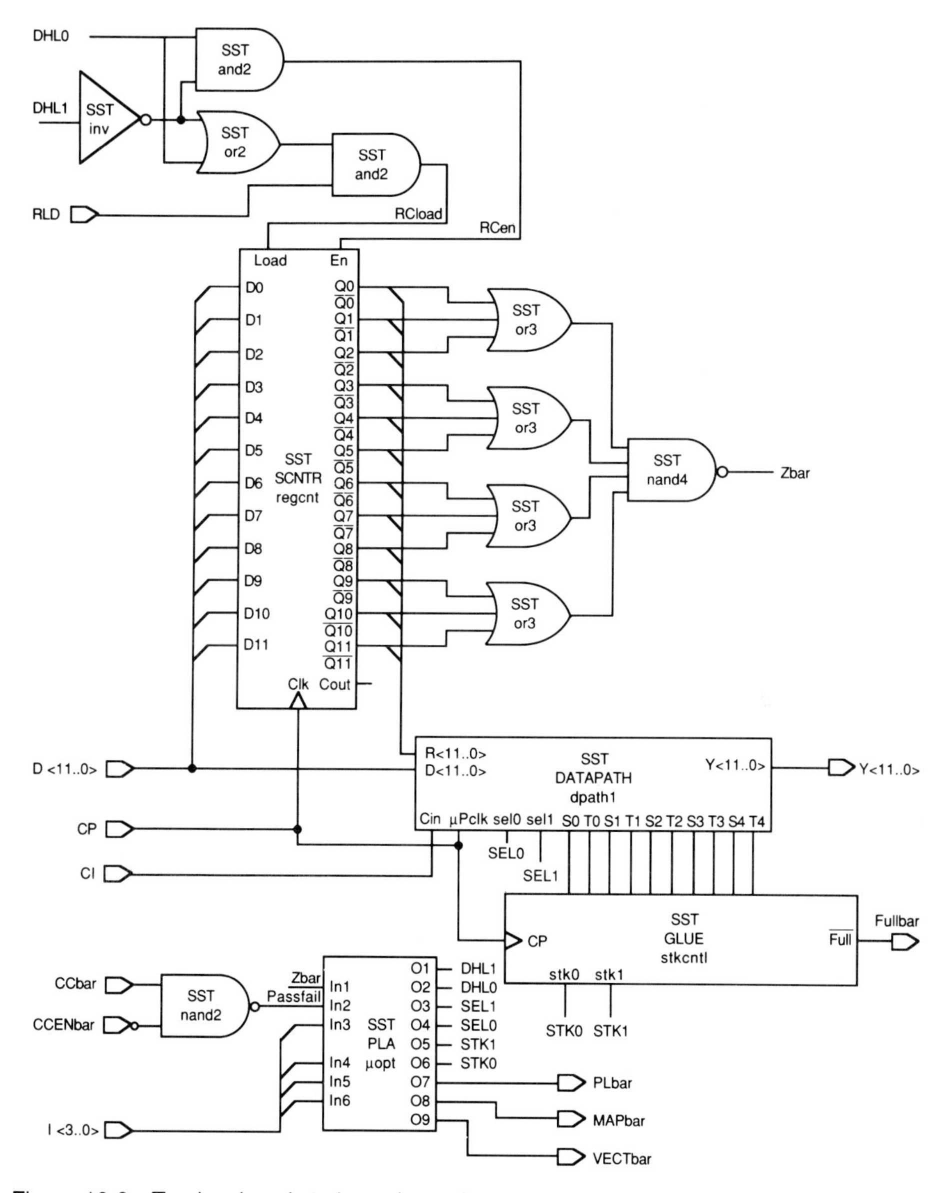

Figure 10.6 Top-level workstation schematic.

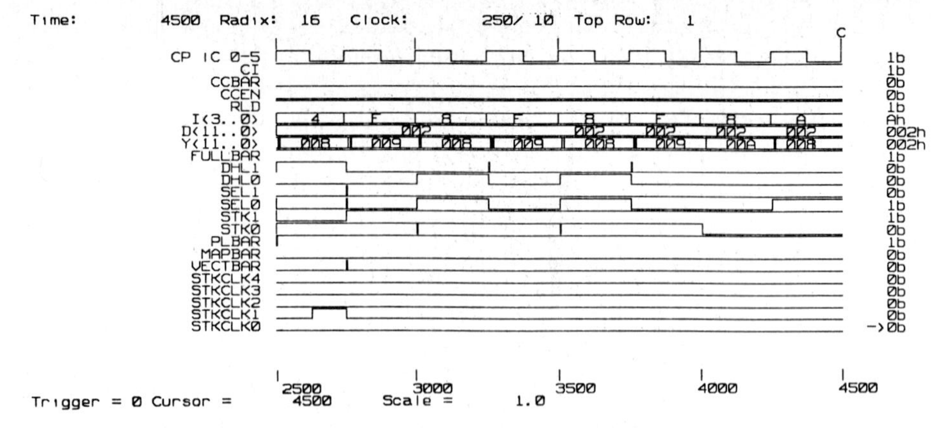

Figure 10.7 Functional simulation using workstation simulator.

Figure 10.7). The actual design cycle involved several iterations at this stage as the functionality of the chip was refined and checked. Additional gates and logic were added to the circuit to achieve the desired behavior, and the new circuit was simulated after each modification. After routing interconnections were made in the next stage, a tabular simulation output was used to analyze the timing based on interconnect parasitics.

A static timing analysis tool is also available to more efficiently provide information on timing delays and possible timing problems for a complete or partial design. In the "path trace" mode [Ben82], the critical path to every node in the network is calculated, and timing errors (e.g., setup and hold time errors) are detected. In the "path query" mode, the critical path between any two nodes, or all paths from a given node, are determined. Path tracing consists of two major phases: (1) clock path tracing (tracing of clock signals to storage element inputs) and (2) data path tracing (tracing of external inputs and storage element outputs to storage element inputs or external outputs). Following the data trace, timing constraint checking is performed. This includes checking for setup and hold time violations, glitches on clock lines, minimum or maximum pulse width constraint errors, and clock skew violations. For setup and hold time checks, common skew in the clock fan-out paths for two devices is considered, in order to avoid generating misleading timing violation reports. The only restriction on multiple system clocks is that the frequency of one must be an integer multiple of the other. Arbitrary phasing and duty cycles are allowed. In the path query mode, the following information is available:

1. All paths from a given input.
2. Longest path from a given input.
3. Critical path between two designated nodes.

10.2.4 Use PRIDE to Route

After the design has been functionally simulated and adjusted for desired performance, the IC layout for the circuit core is defined and routed using

Place and Route Interactive Design Environment (PRIDE). PRIDE provides a full-screen color graphics placement and routing editor that includes multiple autorouters with manual routing override. As the first step in routing, the module footprints based on compiled geometries are brought into the PRIDE editor with the connectivity (which was defined in the schematic design step) shown by stretchable, point-to-point lines. Conversion from the workstation data base and automatic initial placement are invoked by specifying the name of the schematic and selecting the "convert-from-schematic" button.

The general block automatic placement algorithm guarantees 100 percent routing completion by assuring that the final placement if "sliceable." The algorithm begins building a placement tree by recursively partitioning cells into two subsets using the min-cut approach. [Kern70] [Schw72]. Heuristics are used to keep the area of the subsets close, for a better fit. The algorithm then rotates each cell or group of cells and assigns directions to each channel to minimize the bounding box of the layout [LaPo86]. It then collects nearby standard cells into clusters and calls a standard cell optimization algorithm, which does combinatorial optimization with pairwise interchange of cells to minimize net perimeters [Nah86]. It can handle variable-height and variable-width cells.

After automatic placement, if desired, the designer may adjust the modules graphically, using automatic interconnection force vectoring as an aid to optimal placement. The force vectors graphically represent the relative strength and direction of electrical interconnections based on physical module placement both for individual modules and for the design as a whole.

Routing itself can be carried out in several ways using the PRIDE routing toolbox. A full general block autorouter (SuperGlue) performs automatic routing and placement adjustment, using a sliceable floor plan approach, guaranteeing 100 percent completion with very high execution speed. The three major phases of the automatic routing algorithm are channel definition, global routing, and detail routing [Kim83]. Routing channels are automatically defined and ordered by decomposing the layout into a hierarchy of sliceable sublayouts [Dai85]. These channels are highlighted on the screen by green rectangles, allowing the user to optimize channel definition, if desired, by adjusting block positions. The global router uses a heuristic algorithm to generate a near-optimal steiner tree on a region-adjacency graph for each multiterminal net [Chen83] [Tar83]. Detail routing is accomplished by one of several channel routers [Deu76] [Heyn82] [Yosh84]. Nonrectangular channel routing regions are used, greatly improving the density achievable with the sliceable floor plan approach. Techniques of channel compaction such as nongridded and contour routing improve overall density an additional 5 to 10 percent. Once routing for each portion of the chip is complete, the entire chip can be routed to minimize redundancy and allow passthroughs. Although the entire chip can be routed at once, routing portions of it allow the designer to work with manageable portions of the circuit to optimize module placement. Figure 10.8 shows final block routing of the 2910.

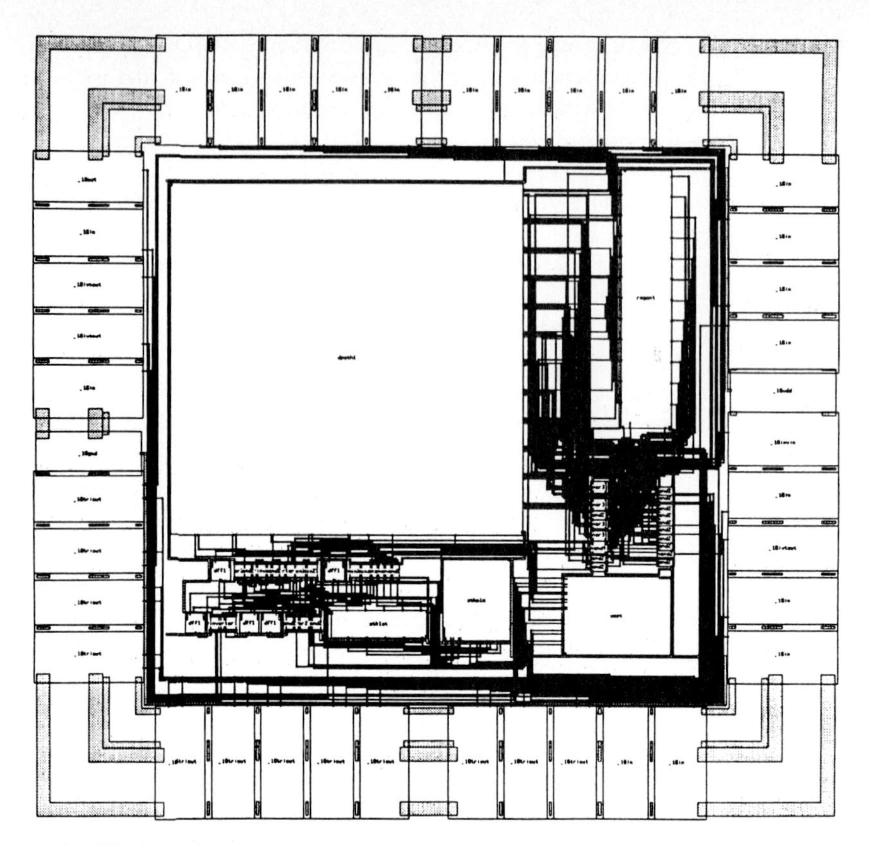

Figure 10.8 Block routing.

An interactive symbolic routing utility allows hand optimization of circuit interconnections and detailed routing of critical portions of a design. Nets are shown as symbolic center lines and are protected against shorts or disconnections. The symbolic routers can also be used to create custom blocks from lower-level cells, with full simulation modeling automatically synthesized. Modules built in this way by the routers can be incorporated into higher-level compilers (e.g., the data path compiler) to create custom-designed modules using the automatic design tools.

After the circuit is routed, a modified simulation model is created that incorporates both loading and delays associated with the routing interconnections. At the conclusion of the circuit routing, circuit timing analysis is performed with the host workstation simulator or the static timing analyzer to verify detailed circuit performance.

10.2.5 Use Packager to Design I/O Circuitry

The packager portion of the compiler allows the designer to specify how the internal circuitry is connected to the external environment. The pad characteristics and the IC package are chosen, and the compiler generates updated simulation information that enables the entire chip, including the effects of routing between the circuit core and the pads, to be simulated and analyzed. Pad characteristics can be selected and compiled along with the rest of the modules for inclusion in the initial schematic diagram and simulation. Pad choices include user-configurable input, ouput, and bidirectional pads in either a standard or high-density pitch. Pad circuits include latch-up and elecrostatic discharge protection. The user-configurable I/O pad features include variable buffering, three-state option pull-up/pull-down impedance, and level shifting functions. Circuit-to-pad connections are translated from the host workstation's schematic file. A package type is selected from a library of packages, which can be extended by the user. Constraints on bonding wire length and angle may also be specified. The compiler verifies that the die size will fit within the package cavity. Package connections are specified through menu input with a graphics-based manual override. The user selects pin-out, with automatic placement and routing of pads; or pad locations can be specified, with automatic pin assignment and routing. Routing from core to the pads is done by a special-purpose "moat router" adapted from channel routing algorithms. Once pads are placed automatically, they conform to the user-definable bonding rules for the specified package. Pads may be placed on only two sides of the die, if desired.

When pad placement and routing is complete, updated simulation information is created for the circuit including the effects of all routing. The designer returns to the simulation stage at this point to once again verify that the circuit meets or exceeds tolerances. This final simulation data also generate the test vectors against which prototype chip functionality and performance is measured. A bonding diagram can be plotted to aid in manual wirebonding of prototype parts.

For the 2910, standard width, standard drive I/O pads were chosen, with three-state pads selected for output functions and placed in a SST4OP250 (four-pin, side brazed ceramic, 250 by 250 mil cavity) package. Pad-to-pin assignments were made to correspond to the pin-out of the merchant part, and the packager was invoked to place the pads and route connections to the IC core. The automatic place and route process was accomplished in less than a minute. Once the pads and pin connections were complete, the entire chip was resimulated for functionality and timing using the workstation's logic simulator. A plot of the completed 2910, including the pads, is shown in Figure 10.9.

10.2.6 Produce Tape for Fabrication

Once the circuit has been designed, simulated, routed, packaged, and resimulated, a GDSII format tape is created. As a final step to guarantee near–100 percent first time working silicon, conversion and verification of

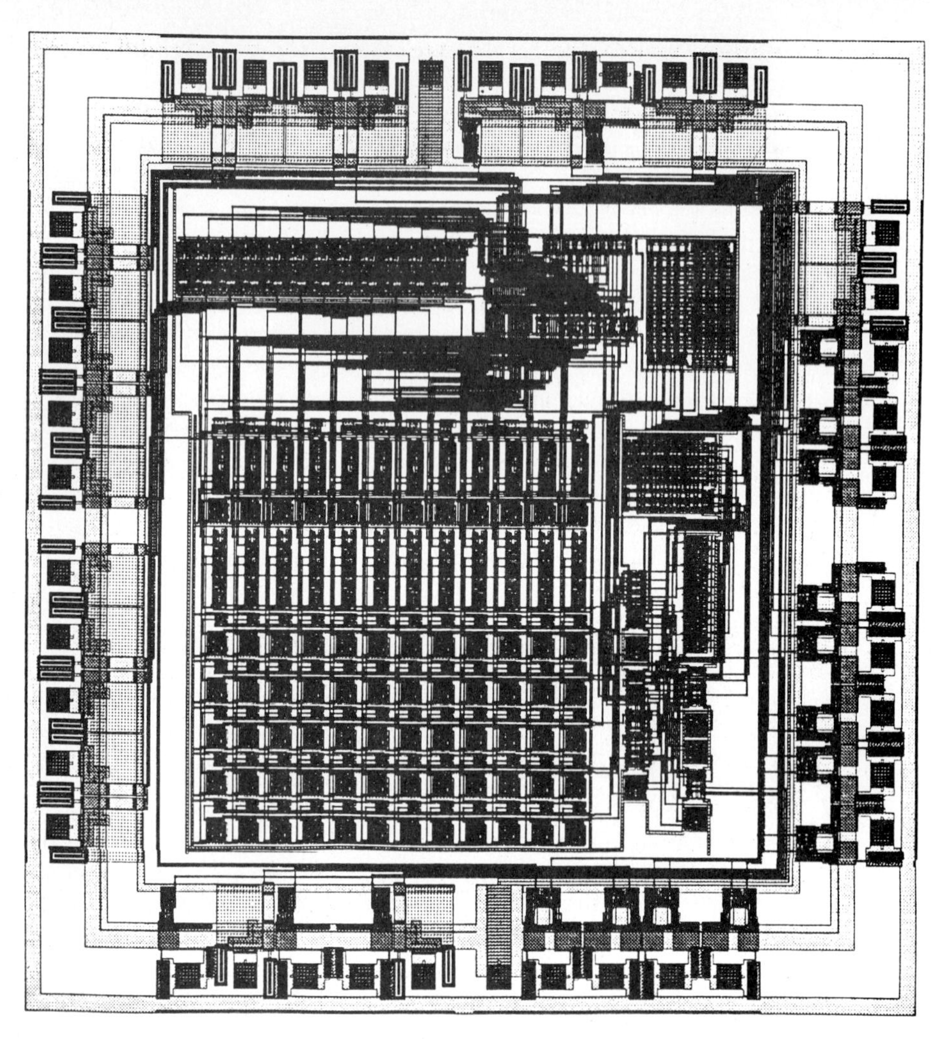

Figure 10.9 Final plot of the 2910.

the mask tape is performed. This verifies design rule correctness, circuit connectivity, and switch-level circuit timing.

10.2.7 A Complex Placement and Routing Example

The example of the 2910 shown throughout this section was discussed because it is a simple design that serves to illustrate many of the features of the compiler. In actual practice, VLSI designs are rarely as simple and straightforward as the 2910, and yet as the complexity of the circuit grows, so does the ability of the compiler to facilitate design and significantly reduce total design time relative to conventional methods.

One example of a chip designed by a user of the compiler, for a confidential commercial application, is shown in Figure 10.10. This chip contains 40,000 transistors, was designed with a productivity of 350 transistors

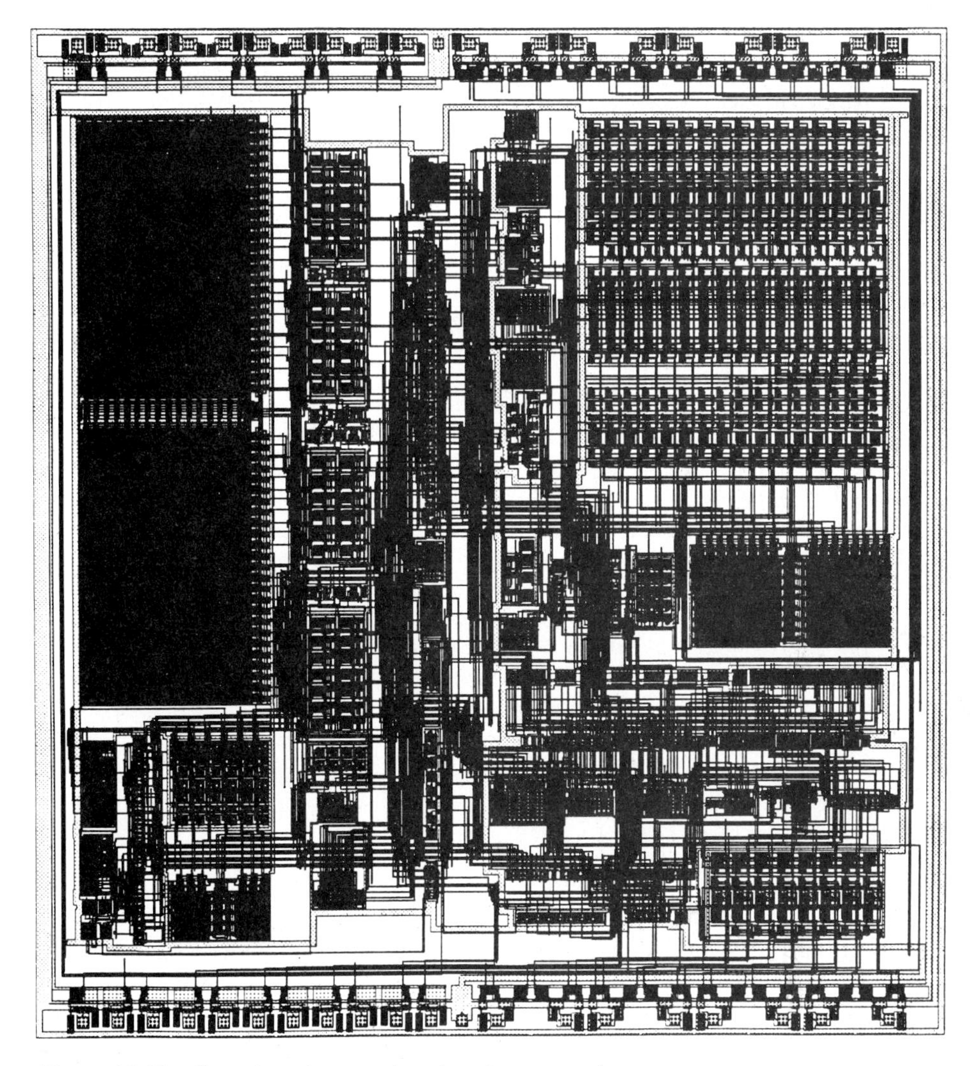

Figure 10.10 Complex placement and routing example.

per worker day (from concept to mask tape), and resulted in working prototypes in less than four months from design start. Density of the entire chip (including I/Os) is 1.75 square mils per transistor in two-micron technology. Figure 10.10 illustrates mixtures of large blocks and standard cells, a sliceable floor plan, nonrectangular channel routing, and I/Os on only two sides of the die.

10.3 Adding New Modules

The function block module compiler set provides a large variety of digital and analog modules that can be customized for specific applications.

However, there are times when a designer needs a particular function not available from the selection of modules or when the function may be inefficient when constructed using existing modules.

In these situations, Concorde makes available two additional methods of constructing modules: the C5 Custom Compiler Development System for custom module creation, and a Foreign Module Integration (FMI) utility that gives designers the ability to bring modules from other systems into a compiled design.

10.3.1 Compiler Development System

The Compiler Development System (CDS) facilitates the development of parameterized module compilers compatible with standard Concorde function blocks and tools. CDS consists of a set of software tools that allows the user to create any desired custom module and make it available to the compiler system.

CDS consists of four distinct pseudolanguages, which are actually sets of function calls in the C programming language:

- SLIC generates layout information that consists of the actual geometry of the cell. It is a distant derivative of the Caltech LAP language.
- gsSLIC generates gate-level simulation information available to the host workstation.
- ssSLIC generates schematic views used by the schematic editor of the host workstation.
- tsSLIC generates a transistor-level SPICE deck for circuit-level simulation.

In addition to these pseudolanguages, a set of utilities is provided. These include a footprint generator, various debugging utilities, and a program for incorporating new design rule sets. Using these languages and utilities, a designer can create at a low level the code necessary to construct parameterized modules that can be stored in the system data base. CDS is intended to be used by experienced IC designers familiar with the C programming language who want to create modules that are not available from Concorde. Designers may use CDS to construct and encode a new custom module or to convert an existing custom module to make it compatible with the compiler system. CDS cells can then be parameterized, as well as made process independent. CDS also provides a system for user entry of new design rules and process parameters based on foundry specifications.

Design Rule Sets

Each semiconductor manufacturer has a set of specifications (process design rules) that determine minimum sizes and relative placement of subtransistor polygons for ICs fabricated in a particular process. Rather than using a linear scaling and biasing technique to change geometries for different processes, Concorde uses individual design rule variables. This allows independent variation of the dimensions to achieve optimum performance and circuit density. By expressing dimensions of the circuit in terms of expressions involving design rule variables rather than absolute values, each dimension changes independently when values for the design rules are established. For each design rule set, approximately 80 independent design rules have been independently formulated. As an example of this nonlinear scaling, compare the two geometries shown in Figure 10.11, which shows a D flip-flop generated using two different rule sets. Notice that the internal proportions of the circuit have been altered, not simply scaled down.

This method of achieving design rule independence can also be used to good advantage with CDS by varying only some of the parameters of the design rule set. A design rule set may be optimized for performance—that is, where high speed is critical even at the expense of yield—by simply altering a few of the design rules. Similarly, a design intended for high-volume production can be compacted by using CDS to optimize the design rules for yield.

As an example of how process independence works, consider the two input NAND gate shown in Figure 10.12. This gate was constructed entirely using commands in SLIC, an extension language of C used to implement all the Concorde function block generators.

In order to provide a virtual grid to describe wire paths and device locations, a series of x and y "landmarks," or reference points, is established as follows:

```
landmark("x1",x1,xorigin)
landmark("x2",x2,x1+MAX(nac2pwc,pac2sbc)) ;
landmark("x3",x3,x2-dac+a2py+ac2gt);
landmark("x4",x4,MAX(x2+ac2py*2+py2py,x3+a 2py+ac2gt-dac));
landmark("x5",x5,x4+ac2pycm);
landmark("y1",y1,yorigin);
landmark("y2",y2,y1+bm2bm);
landmark("y3",y3,y2+MAX(MAX(pwonc+poac,2* ac2py),MAX(nac2pac,bm2b-
m)));
landmark("y4",y4,y3+MAX(bm2bm,2*ac2py));
```

The landmark statement is a function call with three arguments. The first is a string used to label the landmark in documentation plots, the second is the landmark variable to be assigned a value, and the third argument is an expression that will be evaluated at synthesis time and assigned to the landmark variable. The landmarks represent coordinates of an irregular virtual grid. The expressions in the third argument are the spacing adjustment equations, developed by the CDS user, that compute the minimum spacing between adjacent grids for any arbitrary but so-called well-behaved process design rule set. Frequently used compound rule relationships are defined as macros and are used to simplify the development of correct spacing equations. This method is referred to as *dynamic virtual grid compaction*, by which "dynamic" refers to the fact that space compaction is embedded into run-time code and does not require an automatic compaction

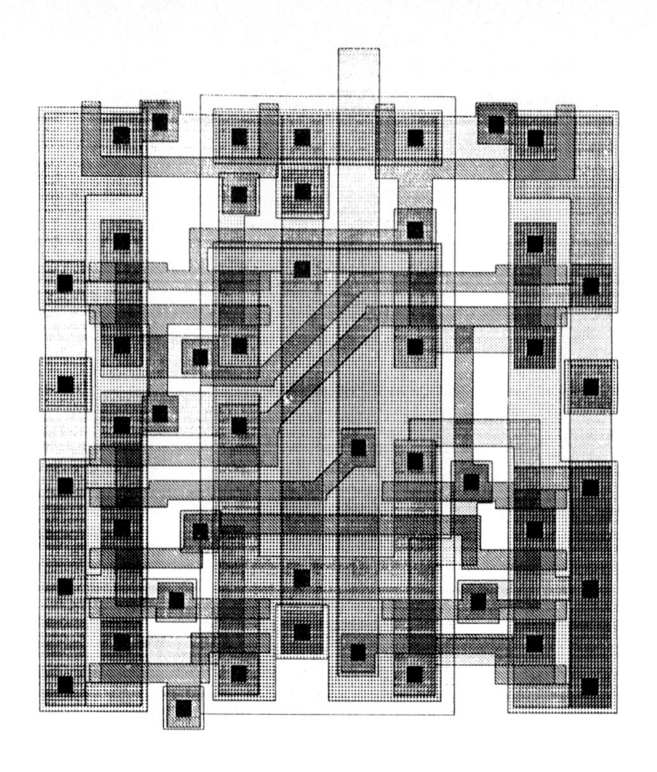

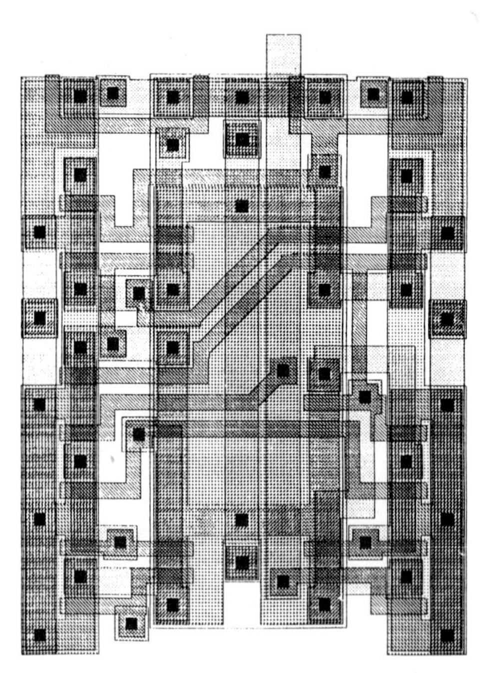

Figure 10.11 D flip-flops generated with different design rule sets.

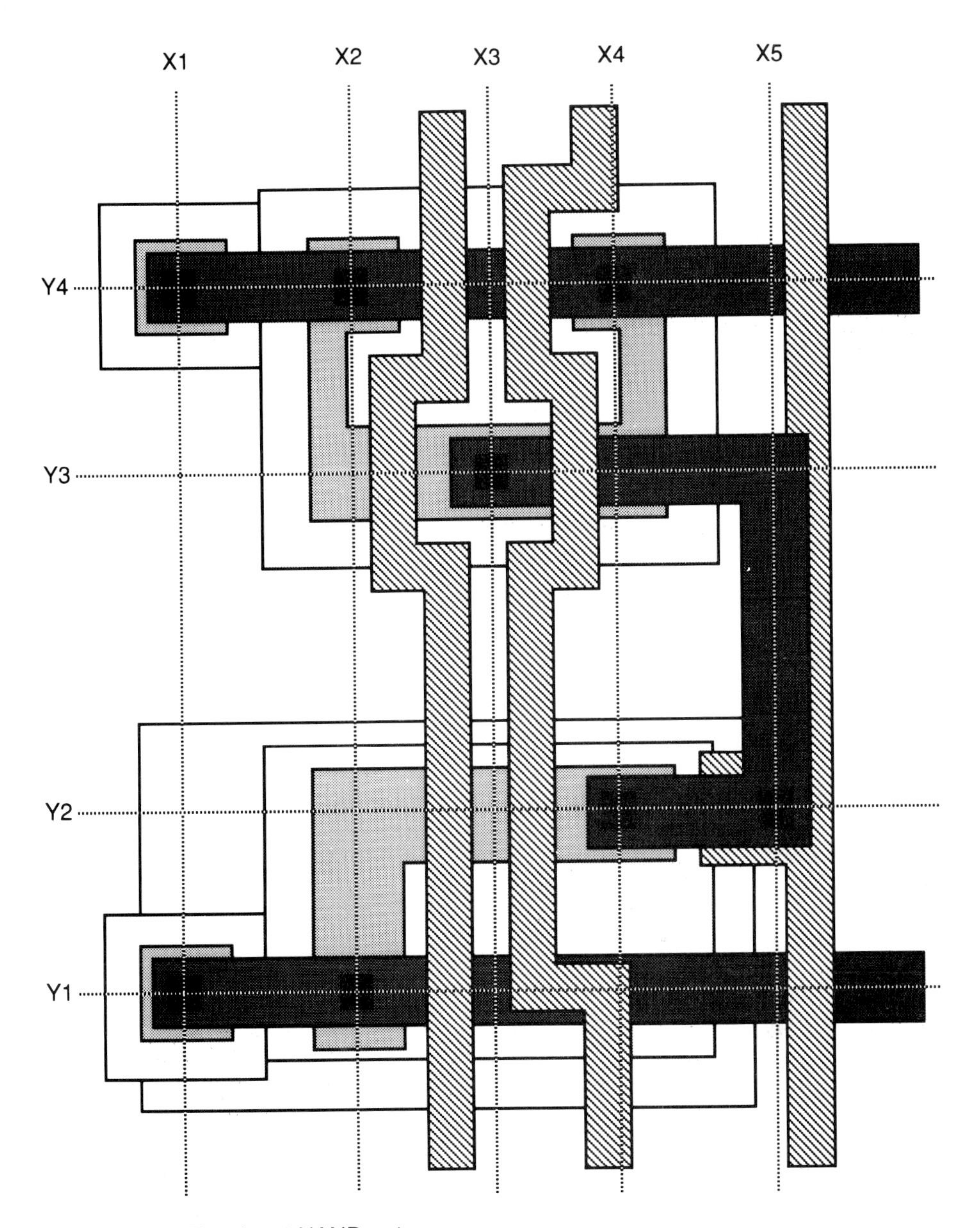

Figure 10.12 Two-input NAND gate.

algorithm. Automatic compaction algorithms may produce poor and even incorrect results in certain cases. *Dynamic virtual grid compaction* gives the compiler writer complete control over the final results.

Note that each landmark, except the first, is defined in terms of a previous landmark offset by design rule variables or macros. For example, the macro "ac2gt" represents the minimum allowable distance from the center of an active contact (ac) to the center of a poly gate (gt), assuming a minimum width poly wire. When the design is actually compiled for a specific process, each of these variables has a value substituted to yield a process-dependent location. In some cases, such as for x2 and x4, the position is determined by which of the two values is the largest (or MAX) in the design rule set.

Three types of shapes can be generated in the SLIC language: wires, boxes, and polygons. Each of these geometric shapes is constructed with a series of function calls. For example, the wire function together with extend functions x and y define a path along which a wire of a specified width is created. The first parameter is the wire width, and the next two are the starting coordinates. The x and y function calls continue the wire in the x or y direction respectively. Wire statements for the metal and active regions of the NAND gate are shown in the following excerpt:

```
layer("metal");
   wire(bm,x1,y1); x(x5+MAX(nac2pwc,pac2sbc));
   wire(bm,x1,y4); x(x5+MAX(nac2pwc,pac2sbc));
   wire(bm,x3,y3); x(x5); y(y2); x(x4);
layer("active");
   wire(aw,x2-dac,y4); y(y3);
   wire(aw,x4+dac,y4); y(y3);
   wire(2*hac,x2,y3); x(x4);
   wire(2*hac,x2,y1); y(y2); x(x4);
```

The macro "nac2pwc" specifies the compound rule for n-active contact to p-well contact spacing. Other macros have related meanings. The contacts for the NAND gate are expressed simply with calls to functions that are passed the desired contact locations; they create all layers of the contact structure in a process-independent manner. Since the landmarks were established as the locations of the contacts, the placement arguments are straightforward. The MA function places a metal-to-active contact cell at a specified coordinate and provides the contact cut and all layers associated with the cut with the minimum overlaps and extensions specified by the design rule set.

```
PWCON(x1,y1);
SUBCON(x1,y4);
MA(x2,y1);
MA(x2,y4);
MA(x3,y3);
MA(x4,y2);
MA(x4,y4);
```

As can be seen, through the judicious use of process variables, modules can be constructed so that design rule independence, as well as correctness, can be easily achieved. The modules used in Concorde are all constructed using a similar methodology.

Additional calls are made to create simulation, symbolic, and transistor-level views using the other pseudolanguages mentioned previously. Because all views are created by C programs, they can be, and typically are, highly parameterized for functional, electrical, and physical characteristics in addition to the parameterization for geometric rules described above.

10.3.2 Foreign Module Integration

Foreign Module Integration (FMI) allows modules developed on or imported to the host workstation and not generated with CDS to be used in a compiled design. A designer might want to bring in a particularly fast design of a circuit element, one for which there is no compiler equivalent, or a standard cell or megacell from a semiconductor company's library.

In order to integrate the foreign module into Concorde, three views of the module must be entered into the data base:

- A schematic symbol for the module
- A simulation model for the module
- A geometry and footprint description for the module

Because Concorde is integrated into the unified design environment of the host workstation, interfacing a foreign module is not difficult. The workstation symbol editor can be used to create the schematic symbol. Since Concorde translates its schematic symbol information to the workstation format, the foreign module can be added to the circuit as though it were a Concorde module. Similarly, Concorde uses the workstation simulation software to verify and test the circuit based on Concorde parameters written in the workstation format. Thus the simulation model created by the designer in constructing the foreign module is inherently available to the workstation. The compiler system merely needs to know the name and search path to the simulation model, which is listed in a simple text file format.

The first two parts of FMI, then, are relatively transparent to the user. However, the last part, placement and routing, does require that information from the workstation be translated into the compiler data base. In order to properly connect and route the module, the compiler system must have information about the footprint of the module and its connection ports. There are two ways in which this information can be communicated through a textual specification file or by direct translation of cell geometry. In the first method, a text that defines the outline geometry and connections of the module is created by the designer. Concorde reads this file and creates a representation of the module used by PRIDE. With the second method, a compiler utility takes cell geometry in GDSII format and converts it directly to internal format. The user then assigns attributes to the foreign module's terminal ports in the Concorde *VIEW* layout display program.

Once this data base integration is accomplished, the foreign module is treated by all the compiler and workstation tools like any of the compiled modules, with one exception. Since custom layout does not support multiple design rule sets as Concorde does, the foreign module must be completely redesigned if a different process is chosen. Concorde modules, including those created with CDS simply, need to be recompiled using the new process information to be available with different rule sets.

10.4 Multiple User Interface Levels

10.4.1 Menu

When specifying modules, the primary user interface to the compiler is a series of interlocking menus from which the user selects options, either directly from the keyboard or by manipulating a mouse or puck. The higher-level menus are constructed in a tree fashion, in which each menu choice leads to further selections. At any menu level, the user can return to a higher-level menu, temporarily leave the compiler to perform file maintenance activities, or quit altogether. The menus are not merely passive acceptors of specifications, they also provide immediate feedback on key performance parameters as a module is being constructed. For example, the designer has four different possible inputs in the construction of an analog OpAmp, and based on the values selected for these inputs, the menu displays 18 key performance characteristics of the OpAmp (see Figure 10.13). The menus also dynamically reconfigure in response to some option selections.

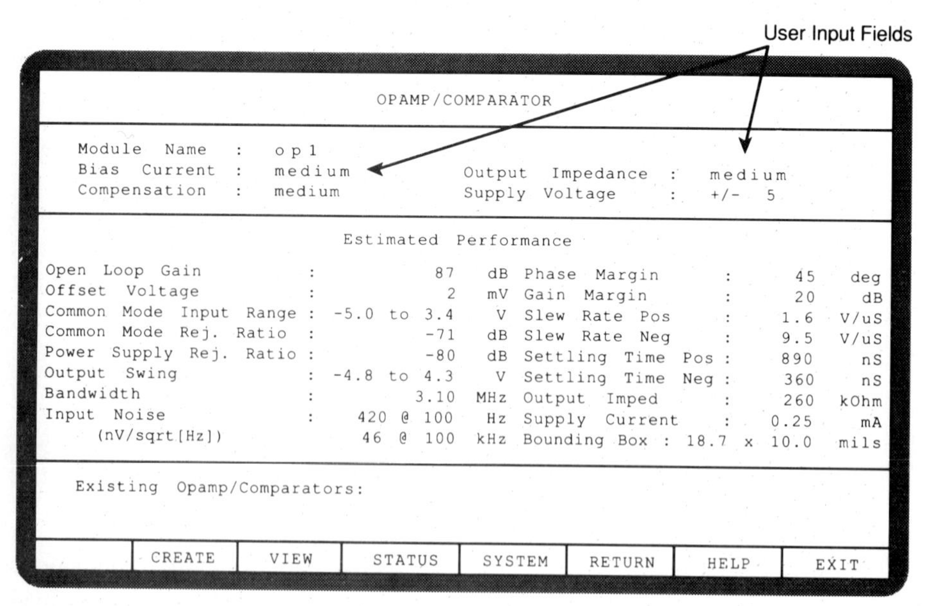

Figure 10.13 Sample Op Amp menu.

The designer can explore the relationships between the different parameters and the performance of the device and can optimize the module for the particular application using these interactive menus. In addition to providing performance feedback, the module customization menus help to prevent specification errors. If data that are inconsistent with other options are entered, two things can happen. Either the entry is refused and the allowable range displayed, or the other characteristics of the device are adjusted to accommodate the option chosen. Changing the center frequency in a notch filter, for example, changes the clock frequency to be consistent with the new value chosen.

10.4.2 Menus that Generate Specification Files

In the case of some high-level compilers (e.g., the datapath compiler) menu input is used to specify bit width, constituent elements, connectivity of bus and control signals, and location of ports to create highly flexible circuit structures. The menu input accommodates data path structures that are regular—that is, those in which standard elements are used, the number of bits is constant for each element, and bus and control signals have similar connections for all bits within an element.

This regular structure accommodates the majority of data path designs, but for irregular applications, a second level of user interface is provided. When a data path module is created, the compiler automatically creates a text format specification file that may be edited by the user by opening an edit window from the menu. By altering the specification file, the user may insert custom elements (e.g., modules constructed from low-level compiled cells combined in the routing environment), alter signal paths within the module, or use different bit widths for various elements.

10.4.3 Direct Specification Using Text Files

The user with specialized demands can bypass menu construction of modules altogether in some cases. For example, in the analog modules, the biquad filters are built up through calls to filter procedures. In a specification file, the experienced analog user can provide transfer functions for the filter directly to these filter procedures to construct custom filters. These filter procedures allow for a wider range of selections and construction of particular types of filters. Other compilers, such as PLA, ROM, and SSTAR use text files for programmable logic input. Random logic can also be specified by text file rather than by schematic, if desired.

10.4.4 Menu Input with Graphic Override

In the specific case of the packager, which specifies I/O pads and circuit-to-pad routing, all required specifications can be entered through a system of menus and all placement and routing is performed automatically. But at that point, as with the routers used to lay out the circuit core, the user has the option of altering placement and routing interactively in a graphics-based editor. This approach gives the designer the simplicity of menu specification coupled with the power of selectively altering options using a visually oriented procedure.

10.4.5 An Example: The Data Path Compiler

To illustrate the flexibility possible with the data path compiler, an example is presented of a data path module in which two four-bit buses are latched and multiplexed, and the resulting signals are provided as output to a D-type register (see Figure 10.14).

This circuit is built by first choosing the data path option from the digital modules menu and specifying a four-bit data path. The register menu is then used to specify two transparent level-sensitive latches, and the connections for each cell are specified. Note that although a four-bit bus is being constructed, the latches need be specified only once, as the three latches for the remaining three bits are identical. The edge-triggered D-type register is also created with the register option menu as shown in Figure 10.15.

The multiplexer portion of the circuit is created from a multiplexer element menu by specifying the cell type (a two-to-one multiplexer for this application) and the buffer size. Finally, in a separate menu, the order of the elements in the data path is given, with the two transparent latches first, then the multiplexer and finally the D-type register. At this point, if no special features of the data path were required, the circuit would be created.

However, assume in this case that the designer wants to tap the signal from Latch2 to the multiplexer for the 0 bit, and provide an external connection for this, as shown in Figure 10.16. This irregular logic can be accomplished by editing the specification file for the data path, which defines all the connections and modules for the data path.

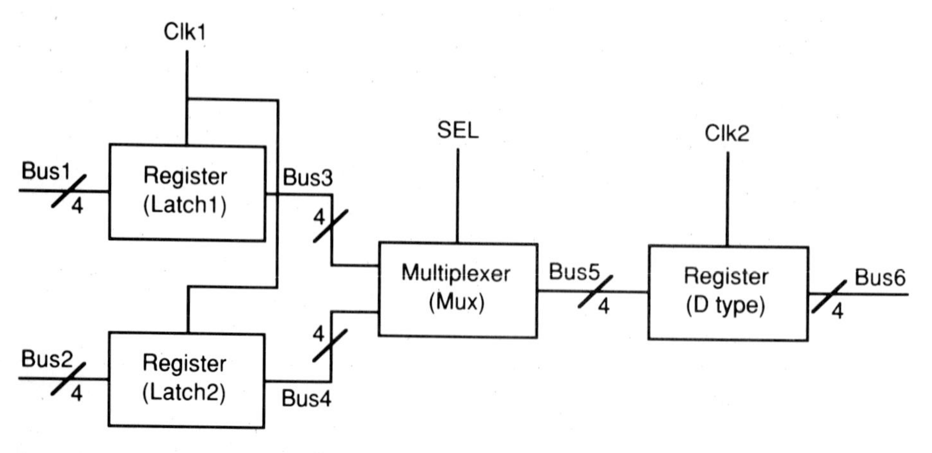

Figure 10.14 Sample data path block diagram.

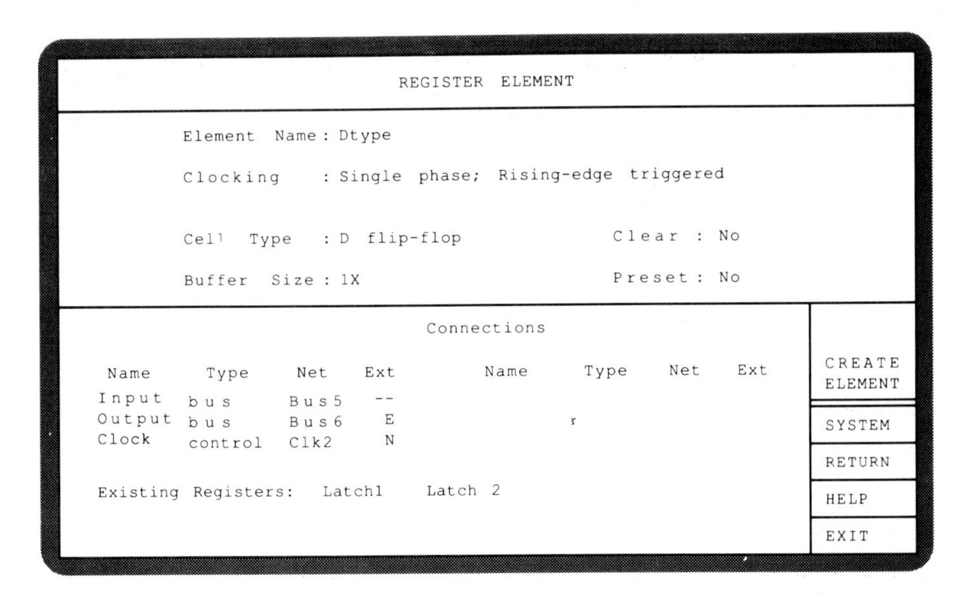

Figure 10.15 Element menu for D-type register.

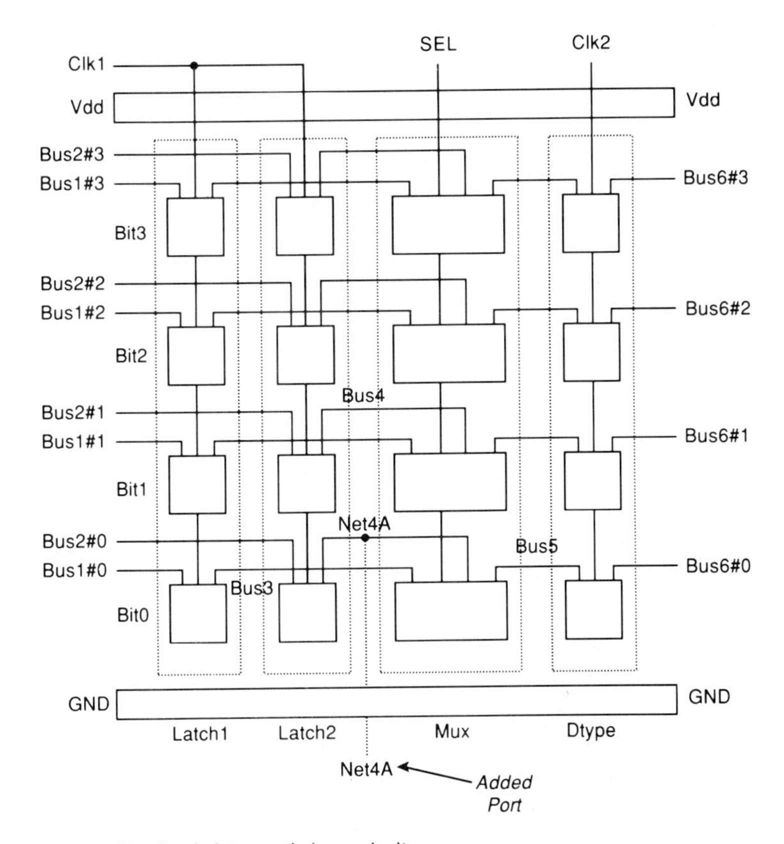

Figure 10.16 Desired data path irregularity.

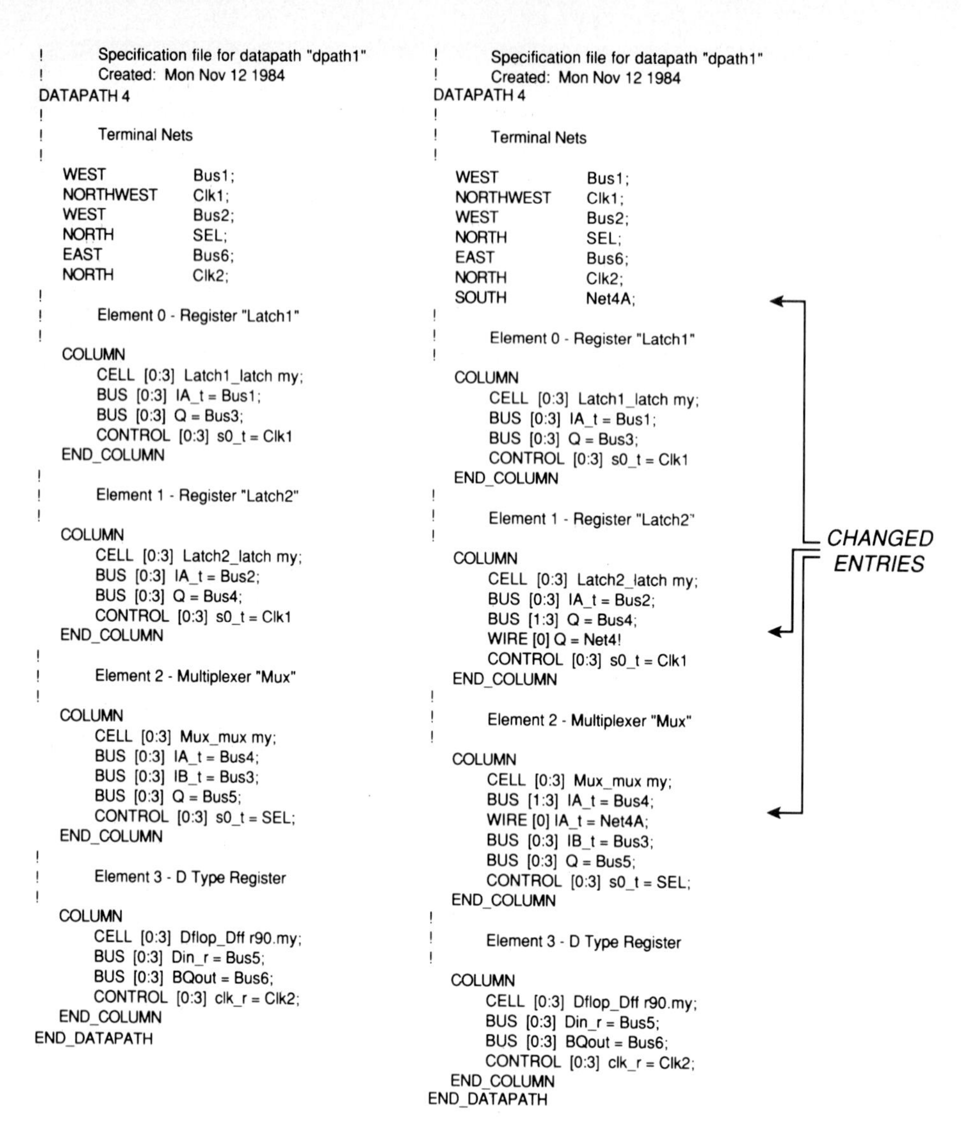

```
!       Specification file for datapath "dpath1"
!       Created: Mon Nov 12 1984
DATAPATH 4
!
!       Terminal Nets
!
   WEST            Bus1;
   NORTHWEST       Clk1;
   WEST            Bus2;
   NORTH           SEL;
   EAST            Bus6;
   NORTH           Clk2;
!
!       Element 0 - Register "Latch1"
!
   COLUMN
        CELL [0:3] Latch1_latch my;
        BUS [0:3] IA_t = Bus1;
        BUS [0:3] Q = Bus3;
        CONTROL [0:3] s0_t = Clk1
   END_COLUMN
!
!       Element 1 - Register "Latch2"
!
   COLUMN
        CELL [0:3] Latch2_latch my;
        BUS [0:3] IA_t = Bus2;
        BUS [0:3] Q = Bus4;
        CONTROL [0:3] s0_t = Clk1
   END_COLUMN
!
!       Element 2 - Multiplexer "Mux"
!
   COLUMN
        CELL [0:3] Mux_mux my;
        BUS [0:3] IA_t = Bus4;
        BUS [0:3] IB_t = Bus3;
        BUS [0:3] Q = Bus5;
        CONTROL [0:3] s0_t = SEL;
   END_COLUMN
!
!       Element 3 - D Type Register
!
   COLUMN
        CELL [0:3] Dflop_Dff r90.my;
        BUS [0:3] Din_r = Bus5;
        BUS [0:3] BQout = Bus6;
        CONTROL [0:3] clk_r = Clk2;
   END_COLUMN
END_DATAPATH
```

```
!       Specification file for datapath "dpath1"
!       Created: Mon Nov 12 1984
DATAPATH 4
!
!       Terminal Nets
!
   WEST            Bus1;
   NORTHWEST       Clk1;
   WEST            Bus2;
   NORTH           SEL;
   EAST            Bus6;
   NORTH           Clk2;
   SOUTH           Net4A;
!
!       Element 0 - Register "Latch1"
!
   COLUMN
        CELL [0:3] Latch1_latch my;
        BUS [0:3] IA_t = Bus1;
        BUS [0:3] Q = Bus3;
        CONTROL [0:3] s0_t = Clk1
   END_COLUMN
!
!       Element 1 - Register "Latch2"
!
   COLUMN
        CELL [0:3] Latch2_latch my;
        BUS [0:3] IA_t = Bus2;
        BUS [1:3] Q = Bus4;
        WIRE [0] Q = Net4!
        CONTROL [0:3] s0_t = Clk1
   END_COLUMN
!
!       Element 2 - Multiplexer "Mux"
!
   COLUMN
        CELL [0:3] Mux_mux my;
        BUS [1:3] IA_t = Bus4;
        WIRE [0] IA_t = Net4A;
        BUS [0:3] IB_t = Bus3;
        BUS [0:3] Q = Bus5;
        CONTROL [0:3] s0_t = SEL;
   END_COLUMN
!
!       Element 3 - D Type Register
!
   COLUMN
        CELL [0:3] Dflop_Dff r90.my;
        BUS [0:3] Din_r = Bus5;
        BUS [0:3] BQout = Bus6;
        CONTROL [0:3] clk_r = Clk2;
   END_COLUMN
END_DATAPATH
```


Figure 10.17 Data path specification file with modifications to incorporate irregularity.

The specification file begins with the external connections, called *terminal nets,* and continues by specifying the cell and connections for each column, as shown on the left side of Figure 10.17. In order to make the desired modification to the circuit, three additional statements must be added to the file, and two statements must be changed, as indicated on the right side of Figure 10.17.

In the terminal net section, the new terminal must be associated with a net and a direction (to indicate the side of the module where the connection will be placed). If we call the net in which the new port will appear "Net4A" and plan to export it to the south side of the module, this statement appears as:

```
SOUTH Net4A;
```

The location of the desired output is between Latch2 and the multiplexer, so the only portion of the specifications to be changed are those specifying connections for this part of the circuit. Latch2 is specified in the second group of statements beginning with the word "COLUMN", and the multiplexer and its connections are delineated in the following set of COLUMN statements. Following the second BUS statement for Latch2, the following statement is inserted:

```
WIRE [0] Q = Net4A;
```

This statement indicates that the output from Latch2 for the first bit is connected to Net4A. Since it is no longer true that all four lines pass from Latch2 directly to the multiplexer, the BUS statement must also be modified to connect only bits 1 through 3 to the multiplexer:

```
BUS [1:3] Q = Bus4;
```

In the third COLUMN statement, which specifies the connections to the multiplexer, a similar set of changes needs to be made, indicating that the input for bit 0 connects to the multiplexer through Net4A and that the other three bits connect through Bus4. Once these changes are made, the specification of the modification to the data path is complete. As is seen in this example, adding custom connections in a data path is a simple process that can often be accomplished with few changes. The designer would now compile the data path and generate geometry, footprint, and simulation information, all of which would incorporate the new port.

State-of-the-art block placement and routing algorithms minimize wasted area between the blocks. But it is also important to allow user guidance and control over placement and routing to keep the user from being constrained by suboptimal automation algorithms. No matter how sophisticated the automation becomes, humans will always be more creative and see some way to improve the implementation.

Finally, capturing the design intent at a higher level of abstraction allows the system designer without IC expertise to use the design system and allows the design system to maintain consistency through the implementation phases of the design. This also provides increased design tool efficiency, because many tools then need only operate on block abstractions and interblock design data. High-level capture of the design intent also gives the capability for automatic and guided optimizations of logical, electrical, and physical levels, and it provides a more effective interface to behavioral-level design and synthesis tools increasingly required for VLSI and ULSI design.

10.5 Future Directions

As ASICS become very complex and are being designed by system designers it becomes imperative that all physical design be done automatically. In order to get the most out of advanced processes the algorithms of automatic physical design will need improvement.

Research in physical design needs to continue in topics of general block placement and routing for extremely large and high-speed circuits ,(hundreds of thousands of transistors) in submicron technologies. This will involve effectively using multiple (three or more) layers of interconnect, developing efficient, over-the-cell routing schemes, and addressing cross-coupling capacitance problems. Automatic changing of aspect ratios to optimize block packing and automatic port location will be topics of interest as will be improvements in floor planning tools. Coupled with that will be tools to more easily develop FBGs with variations in floor plan options and providing a wider degree of technology independence without sacrificing density, performance, or analysis model accuracy.

Another area of research is the extension of compiler concepts to high-speed bipolar and gallium arsenide technologies, including the simulation and timing analysis tools required.

More research needs to be done in front-end tools for top-down design with early design cost estimators, and research work in design synthesis needs to be tied to efficient structural compilers.

In order to open silicon design to the greatest possible number of designers and to reduce design time to the minimum possible, the means of specifying a design will be raised to higher levels of abstraction. In the future designers will specify ASICs at the architectural and behavioral level, possibly by use of a procedural hardware description language.

Another capability needed to allow efficient synthesis from high-level input is automatic optimization. This refers to a number of functions, including logic optimization, automatic clock deskew, timing-driven physical design, and others.

Finally, a research area that will have high payoff is that of high fault coverage automatic test generation for compiled circuits, without the need for full scan path logic. Functions block generators could conceivably generate test views as well as physical and structural views. These views could describe how to test a block, as well as how to pass patterns through it to and from other blocks. The tests could be generated using more realistic fault models than stuck-at-one and stuck-at-zero models and, therefore could detect more subtle reliability problems. Sophisticated software would have to be developed to synthesize a full-chip test program from the individual block tests. It is not yet known under what conditions this is even theoretically possible.

With design productivity via compilers now exceeding 300 transistors per worker day, and increasing, the productivity bottleneck in commercical ASIC design is becoming production test program development for high-reliability circuit requirements. Compilers also offer a solution for higher productivity and superior results in test development if these problems can be solved.

10.6 Summary

Within this chapter a design technique using the Concorde compiler has been described, including the consideration of architectural and design alternatives, capturing of the design intent, design analysis, and floor planning and routing operations. Digital and analog function blocks available have also been described, and techniques for creating additional new function blocks using the Compiler Development System have been shown. Finally, some of the ways that blocks can be customized by the user were illustrated, and examples of the entire technique applied to realistic problems were shown.

As discussed in the introduction, the essence of the methodology is the concept of a high-level structural input with two levels of physical design. The function block generators allow complex design blocks to be reused in many design applications and floor plan configurations. They also improve density and performance by allowing transistor sizes within each block to be small and eliminating use of large transistors except for critical signals and at outputs of blocks. Also, routing within blocks can be done optimally, taking advantage of regularity, abutment, and over-the-cell routing.

By building FBGs in a process-independent fashion, this captured knowledge can be reused even as manufacturing processes continue to change.

Appendix

The typical digital modules available through the Concorde C3 and C4 compilers include, but are not limited to, the following:

1. *Data path.* The data path compiler accommodates *n*-bit register transfer and pipelined data processing architecture, which can be composed of register, adders, 16 function ALUs with a full range of arithmetic add/subtract functions, and 2-, 4-, and 8-input multiplexers, and barrel shifters. Incrementor, decrementor, twos-complementor, zero detect, parity generator, arithmetic comparator, and additional gates and multiplexers are also available. Through the use of the data path compiler, the designer can translate block-diagram circuit descriptions into efficient bus-oriented data path layouts. No limits are placed on the number of internal or external buses or on the clocking scheme. When the data path is compiled from menu options, a specification file that can be modified by the user, is automatically generated. Thus, menu entry can be used for standard *n*-bit-wide bus operations, and the specification file can be edited to handle irregular bus widths or to create

custom functions such as special carry-in and carry-out operations. The data path itself can be composed of up to 64 bits and is completely autorouted using specialized over-the-cell routing algorithms.

2. *Multiplier.* A Carry-Save Adder (CSA) array implementation is provided for separately parameterized bit widths of multipliers and multiplicands (nxm). A carry-look-ahead adder is used for the final sum.
3. *MSI (medium-scale integration).* MSI includes commonly used components optimized for performance and silicon utilization. The MSI set includes user-configurable synchronous counters, ripple counters, adders, shift registers, multiplexers, decoders, and comparators.
4. *PLA (programmed logic array).* A tabular code file is supplied to generate the PLA logic. Outputs can also be directed back into the array as inputs providing dynamic state machine storage elements.
5. *Memory.* A choice of RAM, ROM, and multiport RAM memories is provided with user-specified words and bits per word. The RAM module compiler generates high performance static random access memory. The ROM compiler creates a low power read only memory equipped with dynamic built-in address decoders. Programming of ROM is accomplished through tabular text file input. Both RAM and ROM have a variety of input/output options (e.g. tristate, etc.).
6. *Register file.* An nxm three-port register file is provided with up to two simultaneous read and write operations. Edge-triggered or level-sensitive clocking and options for separate or combined buses and three-state outputs are available.
7. *FIFO buffer.* A first-in-first-out, asynchronous, fall-through buffer is provided, with user-specified number of words and bits per word. Options are available to bring out word lines to detect the degree of fullness.
8. *LIFO stack.* A synchronous nxm last-in-first-out stack is provided, using a high-speed shift register stack pointer implementation. Overflow and underflow are detected as are advanced warnings for each.
9. *SSI (small scale integration).* SSI includes Buffer, Gate, D flip-flop, J-K flip-flop, and Latch components. Edge-triggered and level-sensitive clocking options are provided. Gates include up to 4-input NAND, NOR, AND, OR, as well as XOR, XNOR, and inverters, which are placed as rows of cells at a fixed height. The modules available from this option allow the designer to work at the cell level to generate patterns of random logic not easily grouped into larger function units.
10. *SSTAR (Seattle Silicon Array).* SSTAR is a folded logic array with internal registers well suited to the construction of arbitrary state machines and irregular counters. The SSTAR performs similar functions to a PLA except that the OR plane is folded on top of the AND plane. This allows sections of rows to be isolated, which permits several independent logic operations to occur at the same time, thus increasing area efficiency. Fully static, edge-triggered registers and complementary logic planes provide extremely low static power drain. The SSTAR

is specified through the use of a tabular text file, similar to a highly condensed truth table. The code file describes the relationships between the inputs and the outputs and also gives instructions for creating the circuitry. The basic elements available for SSTAR construction include D flip-flops, J-K master-slave flip-flops, buffered inputs, D flip-flop shift registers, and OR columns. A debugging tool is provided to quickly exercise the array logic behavior before synthesis is executed.

Analog Modules

Concorde also provides a family of analog module compilers that are used for both analog and mixed digital/analog applications.

1. *OpAmps/comparators.* The options the designer can choose for an operational amplifier include compensation, output impedance, and bias current, with dynamic feedback provided for performance parameters.
2. *Filters.* There are four types of switched capacitor filters available:
   a. *Biquad filters.* The biquad filter provides any second order filter function that can be implemented as switched capacitor filters (high, low, band pass, and band elimination). Biquad filters can be cascaded together to provide higher-order filters, thus providing a flexible architecture in a variety of filter types.
   b. *Low pass filter.* The low pass filter is implemented as a doubly terminated ladder filter. Menu options include corner frequency, clock frequency, and filter style (e.g., Butterworth, Chebyshev).
   c. *Band pass filter.* The band pass filter compiler implements a first- or second-order Butterworth filter using a process-insensitive ladder filter configuration.
   d. *Custom filter.* A low-level specification menu is provided for construction of custom filters by the knowledgable designer through direct specification of circuit elements.
3. *Analog Switch.* These compilers create SPST through DPDT analog switches and analog multiplexers. For the regular analog switch, the designer can select the type of switch, regular or inverted control signal, and the switch impedence at a user-specified supply voltage. The multiplexer provides control of up to 16 inputs, and up to 4 multiplexers can be ganged together.
4. *Oscillators.* Designed as a Pierce oscillator, the user provides values for the crystal parameters and passive components. Feedback is provided; and, if the on-line analysis indicates that the circuit does not oscillate, suggestions are provided to achieve oscillation.
5. *Converters (D/A and A/D).* There are three types of D/A and A/D converters available:
   a. *Serial A/D.* The serial A/D converter is slow, has high resolution and linearity, is insensitive to noise and offset voltages, and does not

require precision-machined components. Up to 15 bits of linearity have been achieved with standard CMOS processing.

b. *Successive approximation A/D.* A fast signal converter, capable of 8-bit conversions at 100 kHz, the design is implemented as a charge redistribution A/D using binary weighted arrays of capacitors to perform the successive approximation of the incoming signal.

c. *Binary weighted D/A.* The binary weighted D/A features a bipolar output with single reference and a constant conversion time of 2.5 microseconds.

6. *Discrete components.* Such components available from the analog module menu include FETs with a variety of implementations, resistors, and capacitors.

References

1. [Ben82] L. C. Bening, T. A. Lane, C. R. Alexander, and J. E. Smith, "Developments in Logic Network Path Delay Analysis," *22nd Design Automation Conf.*, 1982, pp. 605–615.
2. [Bla85] T. Blackman, J. Fox, and C. Rosebrugh, "The SILC Silicon Compiler: Language and Features," *22nd Design Automation Conf.*, 1985, pp. 232–237.
3. [Bra85] R. K. Brayton, N. L. Brenner, C. L. Chen, G. DeMicheli, C. T. McMullen, and R. H. J. M. Otten, "The Yorktown Silicon Compiler," *ISCAS '85*, Kyoto, Japan, June 1985.
4. [Chen83] Nang-Ping Chen, "Routing System for Building Block Layout," Ph.D. diss., University of California, Berkeley, May 1983.
5. [Dai85] Wei-Ming Dai, T. Asano, and E. S. Kuh, "Routing Region Definition and Odering Scheme for Building-Block Layout," *IEEE Trans. CAD,* vol. CAD-4, no. 3, July 1985, pp. 189–197.
6. [Deu76] D. N. Deutsch, "A "Dogleg" Channel Router," *13th Design Automation Conf.*, 1976, pp. 425–433.
7. [Gaj83] D. D. Gajski and R. H. Kuhn, "Guest Editor's Introduction, New VLSI Tools," *Computer,* vol. 16, December 1983, p. 12.
8. [Gold85] A. V. Goldberg, S. S. Hirschhorn, and K. J. Lieberherr, "Approaches Toward Silicon Compilation," *IEEE Circuits and Devices Magazine,* May 1985, pp. 29–39.
9. [Heyn82] W. Heyns, "The 1-2-3 Routing Algorithm," *19th Design Automation Conf.*, 1982, pp. 113–120.
10. [Jam85] R. Jamier and A. A. Jerraya, "APOLLON, A Data-Path Silicon Compiler," *IEEE Circuits and Devices Magazine,* May 1985.
11. [Jerr86] A. Jerraya, P. Varinot, R. Jamier, and B. Courtois, "Principles of the SYCO Compiler," *23rd Design Automation Conf.*, 1986, pp. 715–721.
12. [Kern70] B. W. Kernighan and S. Lin, "An Efficient Heuristic for Partitioning Graphs," *Bell System Tech. Journal,* Feb. 1970.
13. [Kim83] S. Kimura, N. Kubo, T. Chiba, and I. Nishioka, "An Automatic Routing Scheme for General Cell LSI," *IEEE Trans. CAD,* vol. CAD-2, no. 4, October 1983, pp. 285–292.

14. [Krek85] D. E. Krekelberg, G. E. Sobelman, and C. S. Jhon, "Yet Another Silicon Compiler," *22nd Design Automation Conference*, Las Vegas, NV, 1985.
15. [LaPo86] D. LaPotin and S. W. Director, "Mason: A Global Floorplanning Approach for VLSI Design," *IEEE Trans. on CAD,* October 1986.
16. [Mar86] T. Marshburn, I. Lui, R. Brown, D. Cheung, G. Lum, and P. Cheng, "Datapath: A CMOS Data Path Silicon Assembler," *23rd Design Automation Conf.,* 1986, pp. 722–729.
17. [Nah86] S. Nahar, S. Sahni, and E. Shragowitz, "Simulated Annealing Using Combinational Optimization," *23rd Design Automation Conf.,* 1986.
18. [Schw72] D. G. Schweikert and B. W. Kernighan, "A Proper Model for Partitioning of Electrical Circuits," *9th Design Automation Conf.,* 1972.
19. [Sou83] J. R. Southard, "MacPitts: An Approach to Silicon Compilation," *Computer,* vol. 16, December, 1983, pp. 74–87
20. [Tar83] R. E. Tarjan, *Data Structures and Network Algorithms*, Society for Industrial and Applied Mathematics, Philadelphia, 1983.
21. [Yosh84] T. Yoshimura, "An Efficient Channel Router," *21st Design Automation Conf.,* 1984.

Index

S

T

V

W

Y